Cees W. Passchier

Rudolph A. J. Trouw

Microtectonics

Cees W. Passchier
Rudolph A. J. Trouw

Microtectonics

2nd, Revised and Enlarged Edition

With 322 Images

Springer

Authors

Professor Dr. Cees W. Passchier

Tektonophysik
Institut für Geowissenschaften
Becherweg 21
Johannes Gutenberg University
55099 Mainz
Germany

Professor Dr. R. A. J. Trouw

Departamento de Geologia
Universidade Federal do Rio de Janeiro
CEP 21949-900 Rio de Janeiro
Brazil

Additional material to this book can be downloaded from http://extras.springer.com.

| ISBN-10 | 3-642-44111-4 | Springer Berlin Heidelberg New York |
| ISBN-13 | 978-3-642-44111-0 | Springer Berlin Heidelberg New York |

Springer is a part of Springer Science+Business Media
springeronline.com
© Springer-Verlag Berlin Heidelberg 2005
Softcover re-print of the Hardcover 2nd edition 2005

The use of general descriptive names, registered names, trademarks, etc. in this publication does not imply, even in the absence of a specific statement, that such names are exempt from the relevant protective laws and regulations and therefore free for general use.

Cover design: Erich Kirchner, Heidelberg
Typesetting: Büro Stasch · Bayreuth (stasch@stasch.com)
Production: Almas Schimmel
Printing: Stürtz AG, Würzburg
Binding: Stürtz AG, Würzburg

Printed on acid-free paper   30/3141/as – 5 4 3 2 1 0

Preface

The origin of this book lies in a practical course in microtectonics started by Prof. Henk Zwart at Leiden University, the Netherlands, in the early 1960s. Both of us were students of Henk Zwart at the University of Leiden and later, as his assistants, charged with the organisation of this course. As such, we became enchanted by the many interesting thin sections of his collection which expanded over the years, as Henk extended his work from the Pyrenees to the Alps, the Scandinavian Caledonides and to many other places in the world. An explanatory text was elaborated and regularly updated by a number of assistants, including us, under Henk's supervision. This text, together with many thin sections of the collection, served as a core for the present book. In the early 1980s, the Geology Department of Leiden University was transferred to the University of Utrecht. The collection was transferred as well, and one of us (C. W. P.) became responsible for its organisation and maintenance. A visit of R. A. J. T. to Utrecht in 1991 with a number of didactic microstructures collected in South America triggered the final effort to build a manual for the study of microtectonics. Because of his contributions to science and his enthusiasm for microtectonics, we dedicate this book to Henk Zwart, who inspired us, taught us the principals of microtectonic analysis, and also furnished many crucial examples of microstructures.

The first edition of "Microtectonics" from 1996 is now outdated, and we felt that the large amount of new work in microtectonics warranted the setup of this second edition. This edition contains nearly all the old material, but adds material on new research of the last ten years and some material that was left out of the first edition for several reasons. As a result, the number of figures has increased from 254 to 322 and the number of references from 659 to 1451.

Few geologists will be able to remember what their first impression was when they were confronted with a deformed rock under the microscope. That is unfortunate, because it inhibits experienced geologists from looking at geometries in thin section in an unbiased way. We commonly think that we 'see' processes such as dynamic recrystallisation, refolding and grain growth, while all we actually see are geometric patterns that may have formed in a number of different ways. In this book, we try to preserve some of the 'first encounter approach' with deformed rocks and follow some guidelines that result from many years of teaching structural geology. First of all, structural geology and microtectonics are visual sciences, and need good and abundant illustrations to be properly understood. The result will be obvious for anyone who pages through this book. Secondly, there is no harm in explaining features at a basic level rather than in full detail for specialists; specialists can skip simple explanations, but all those who are new to the subject have better access to the content of the book if things are explained in a simple way. Finally, there is nothing more frustrating than to know that a certain subject has been covered in the literature, but not to know who wrote about it. Therefore, we have included a large number of references throughout the text.

The following critical readers of the original manuscript for the first edition helped considerably to improve the quality with useful suggestions: Hans de Bresser, Bas den Brok, Paul Dirks, David Gray, Monica Heilbron, Renée Heilbronner, Ralph Hetzel, Kyuichi Kanagawa, Win Means, Uwe Ring, Herman van Roermund, Luiz Sergio Simões, Carol Simpson, Ron Vernon and Janos Urai. Their help is gratefully acknowledged. Photographs, samples, thin sections and material for the text were kindly provided by Ralph Hetzel, Paul Dirks, Domingo Aerden, Michel Arthaud, Coen ten Brink, Hanna Jordt-Evangelista, Reinhardt Fuck, Leo Kriegsman, Gordon Lister, Leo Minnigh, Jin-Han Ree, André Ribeiro, Chris Schoneveld, Janos Urai, Simon Wallis, Klaus Weber and Dirk Wiersma. For the second edition similar help was given by Scott Johnson, Daniel Koehn, Hans de Bresser, Michael Stipp, Holger Stünitz, Manuel Sintubin, Erich Draganitz, Sara Coelho, Renée Heilbronner, Geoffrey Lloyd, Chris Wilson, Anne-Marie Boullier, Lutz Nasdala, Steve Foley, Michael Bestmann, Giorgio Pennacchioni, Chris Ryan, Martyn Drury, David Ferrill, Vincent Heesakkers, Jens Becker, Nico Walte, Claudio Valeriano, Rodrigo Peternel, Camilo Trouw, Felipe Medeiros, Mauro Torres Ribeiro and Margareth Guimarães. Tarcisio Abreu elaborated high quality thin sections. Many other persons helped in one way or the other, either by providing samples with interesting microstructures that were integrated in our collections but not shown, or by discussing the meaning of microstructures. Their help is also gratefully acknowledged.

The Volkswagen Stiftung, the German Science Foundation (DFG), the Schürmann Foundation, the Dutch Royal Academy of Sciences, the Deutscher Akademischer Austauschdienst e. V. (DAAD) and the Brasilian Research Council (CNPq) provided funding for our research, the results of which have been used in this book; this support is gratefully acknowledged. R. A. J. T. thanks the Brasilian Research Council (CNPq) also for financing his stay at Utrecht University.

About This Book

This book deals with the description and interpretation of small scale structures in deformed rocks as seen in thin section through the optical microscope. The book is meant for advanced undergraduate and graduate students, and is best used in combination with a practical course where thin sections can be studied and discussed. In our experience, a collection of 100–200 thin sections with examples from structures treated in our Chapters 3–9 are sufficient for such a course.

In Chapter 1 the 'philosophy' of how we think that microstructures can be understood is discussed, including their usefulness in tectonic studies. Chapter 2 gives a simplified, non-mathematical background in kinematics and rheology, meant to explain the terminology used in the interpretation of microstructures. Deformation on the grain scale and deformation mechanisms are treated in Chapter 3. Chapters 4 to 7 form the core of the book and deal with the most commonly observed microstructures. In Chapter 8 some primary microstructures from igneous and sedimentary rocks are discussed, and in Chapter 9 a brief outline is given of a new development in microtectonics which we called microgauges: structures that can be used to obtain quantitative data from deformed rocks. Chapter 10 describes a number of additional techniques other than optical microscopy. These techniques either use thin sections, or can be used in combination with optical microscopy to obtain additional data. The descriptions are short but should allow the reader to decide if it is advantageous to use an unfamiliar technique, available outside the home department. Chapter 11 gives an overview of the current state of the art of experimental studies. Chapter 12 describes problems involved with sampling and preparation of thin sections, including the problem of the interpretation of three dimensional structures using two-dimensional sections. A glossary and index are given at the end of the book; the definitions in this glossary reflect our opinion on the meaning of the terminology as used at present. A number of boxes explaining subjects in more detail are present in several chapters. Figures in these boxes are numbered separately, starting with "B." and the chapter number.

Throughout the text, words written in italics indicate items that are introduced for the first time, most of which are explained in the glossary and can be found in the index. In the figure captions, PPL and CPL mean plane polarised and crossed polarised light, respectively. Provenance of the photographed thin sections is given where known. For some thin sections in the old collection from Leiden, the provenance is unknown and these were marked as 'Leiden Collection'. The scale of the photographs is given as a width of view.

The *accompanying CD* contains full colour versions of all drawings and of many photographs in the book, animations of figures in the book and videos of deformation experiments, a number of additional colour photographs, figures from the boxes, an explanation on the use of the U-stage, the glossary and 35 photographs of examples of common problems in microtectonics meant to be used as interpretation exercises. The CD also contains a search option for figure captions and the glossary. Since all figures in the book are also on the CD, there is no separate reference for them in the text of the book. Additional colour photographs of items on the CD which do not appear in the book are marked in the text with '⊘Photo'. Where these show similar items as figures in the book, they have been given the same number. Animations and videos are both marked as '⊘Video'. If they animate figures from the book or show similar features, they have been given the same number as relevant figures in the book.

C. W. Passchier
R. A. J. Trouw

To see a world in a grain of sand
and a heaven in a wild flower
hold infinity in the palm of your hand
and eternity in an hour

William Blake

Salve, parens rerum omnium Natura,
teque nobis Quiritium solis celebratam
esse numeris omnibus tuis fave

Gaius Plinius Secundus

To Henk Zwart

for his contribution to structural and metamorphic geology

Contents

A Framework of Microtectonic Studies

Basic concepts of microtectonics are introduced and discussed in this chapter. The study of thin sections of rocks, which was originally mainly petrographic, has evolved during the last century to include a number of structural characteristics, that constitute the microstructure or fabric of a rock. The study of this fabric can be used to reconstruct the structural and metamorphic history, but also to improve the understanding of deformation and metamorphic processes.

Successive stages in the deformational and metamorphic evolution of a rock are commonly preserved as part of a fabric and the recognition and correct interpretation of these fabrics is essential for the understanding of this evolution. Two concepts are important: deformation phases and metamorphic events.

The first relates to specific periods during which a rock is deformed under the influence of a deviatoric stress field, leaving visible records such as folds, cleavages or lineations. Successive deformation phases may be superposed on each other leaving overprinting structures such as folded foliations, refolded folds or folded lineations.

Metamorphic events correspond to the formation of a specific mineral assemblage, thought to reflect particular P-T conditions or the crossing of a reaction equilibrium curve in P-T space. These may also be superposed on each other, leaving only a recognizable record if the equilibrium during the last event(s) was incomplete.

The main objective of a microstructural analysis is therefore to unravel the sequence of deformation phases in a specific area and to link this sequence to metamorphic events, in order to establish a P-T-t path.

1.1
Introduction

From their first use in the 19th century, thin sections of rocks have been an important source of information for geologists. Many of the older textbooks on structural geology, however, did not treat microscopic aspects of structures, while petrologists would describe microscopic structures as, for example, lepidoblastic or nematoblastic without paying much attention to kinematic and dynamic implications. During the last decades, however, structural geologists learned to profit from the wealth of data that can be obtained from the geometry of structures studied in thin section, and metamorphic petrologists have appreciated the relation of structural evolution on the thin section scale and metamorphic processes.

Deformed rocks are one of the few direct sources of information available for the reconstruction of tectonic evolution. Nevertheless, observations on the geometry of structures in deformed rocks should be used with care; they are the end product of an often complex evolution and we can only hope to reconstruct this evolution if we correctly interpret the end stage. Simple geometries such as folds can be formed in many ways and it may seem hopeless to try and reconstruct a complex evolutionary sequence from geometrical information only. However, despite the simple geometry of our face, we can individually recognise most of the six billion people on our planet. It is likely that structures in rocks also contain a large amount of detail, which we cannot (yet) recognise and interpret because we are not trained to do so, and partly because we do not know what to look for. It is interesting to page through old publications on microstructures, e.g. on inclusions in garnet or on porphyroclasts, to see how drawings evolved from simple to complex while understanding of the processes related to the development of these structures increased. At any time, some degree of misinterpretation of structural evolution is unavoidable and part of the normal process of increasing our understanding of the subject. This book is therefore a-state-of-the-art description of microstructures and their interpretation.

Observations on the microstructure or *fabric* of a rock (Box 1.1), specifically in thin section, can be used in two major fields. They can be applied to thematic studies, to understand mechanisms of rock deformation and metamorphism; or they can be used to reconstruct the structural and metamorphic history of a volume of rock. Thin section studies are mostly in the latter field. Because such thin section studies can serve to reconstruct tectonic evolution, we use the term *microtectonics*.

This chapter is not only meant as a general introduction to the subject of microtectonics, but also serves as a definition of the framework within which we see studies in microtectonics. As such, it contains terminology that is explained only later in the text and in the glossary. This

> **Box 1.1   Fabric, texture, microstructure**
>
> In this book we mainly deal with *fabrics*. A fabric "includes the complete spatial and geometrical configuration of all those components that make up a rock. It covers concepts such as *texture, structure* and crystallographic preferred orientation" (Hobbs et al. 1976). The parts that make up a fabric, also known as *fabric elements*, should be penetratively and repeatedly developed throughout a volume of rock; a single fault in a volume of rock is not considered to be part of the fabric, but a large number of parallel foliation planes are. Fabric elements are therefore dependent on scale (cf. Fig. 2.4). A volume of rock may have a *random fabric*, i.e. a random distribution and orientation of its elements or, more commonly, a non-random fabric, including foliations and lineations.
>
> In this book, we mainly deal with fabrics on microscopic scale, or *microfabrics*. Microfabric elements may include grain shape, grain boundaries, deformation lamellae, aggregates of grains with similar shape, and lattice preferred orientation.
>
> In the non-geological literature about metals and ceramics, the term *texture* is generally used for lattice-preferred orientation. On the other hand, most of the older textbooks on metamorphic petrology (e.g. Turner 1968; Miyashiro 1973; Best 1982; Williams et al. 1982; Bucher and Frey 1994) make a distinction between the *texture* and the *structure* of a metamorphic rock. In these texts, *texture* refers to the geometrical aspects of the component particles of a rock including size, shape and arrangement, whereas *structure* usually refers to the presence of compositional layering, folds, foliation, lineation, etc. In fact there is no clear difference between the two concepts and the subcommission on the systematics of metamorphic rocks of the IUGS recommends substituting the term texture by *microstructure*. In this book, we use the terms microstructure and microfabric (see also Sect. 1.1) as synonyms.

chapter can be read before, but also in conjunction with the other chapters.

In theory, one could expect that a sedimentary rock, which is buried, deformed, metamorphosed and brought back to the surface, should have the same mineral composition as the original sediment if perfect equilibrium conditions were to be attained at each stage. A simple fabric should be developed in such a case in response to gradual changes in the stress field and in metamorphic conditions. Fortunately for the geologist, who relies on structures and mineral assemblages in deformed rocks as a source of information, this is almost never the case. In most deformed rocks, structures with different style and orientation and minerals, which represent different metamorphic grades, *overprint* each other. This means that equilibrium is generally not attained at each stage: mineral assemblages representative of different metamorphic conditions may be 'frozen in' at different stages during burial and uplift. With overprint we mean that structures or mineral assemblages are superposed on each other and must therefore differ in age; this may be visible through crosscutting relations, overgrowth, or even differences in deformation intensity. In practice, however, *overprinting relations* can be difficult to establish. This book mainly serves to illustrate the possibilities of recognising overprinting relations in thin section and to determine the conditions at which they formed. The aim is then to translate overprinting relations in terms of *deformation phases* and *metamorphic events*.

Deformation phases are thought to be distinct periods of active deformation of rocks on a scale exceeding that of a single outcrop, possibly separated by time intervals with little or no deformation during which metamorphic conditions and orientation of the stress field may have changed (Sects. 1.2, 2.11). The concept was originally created in relation to groups of structures that can be separated in the field by overprinting criteria (Sect. 1.2). Metamorphic events are episodes of metamorphism characterised by changes in mineral assemblage in a volume of rock. Such changes are thought to reflect changes in metamorphic conditions.

Once deformation phases and metamorphic events are defined, it is necessary to determine to what extent they correspond to *tectonic events* or *metamorphic cycles*, i.e. events on a larger scale such as those associated with plate motion or collision. Finally, *orogenies* (e.g. the Alpine orogeny) may encompass several tectonic events with associated metamorphic cycles. The following example illustrates this concept. In thin sections from several outcrops, a horizontal biotite foliation is overprinted by a steeply dipping chlorite foliation, and both are cut by brittle faults (Fig. 1.1). Based on these overprinting relations we could argue that a first 'deformation phase' with a component of vertical shortening formed a foliation under conditions suitable for growth of biotite; later, a second 'deformation phase' of oblique shortening was accompanied by chlorite growth under lower-grade metamorphic conditions. A third deformation phase affected both earlier structures at very low-grade or non-metamorphic conditions or at high strain rate, to cause brittle faulting.

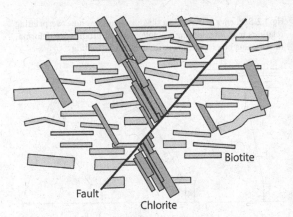

Fig. 1.1. Schematic diagram of a biotite foliation (*horizontal*), a chlorite foliation (*inclined*) and a brittle fault. The sequence of overprinting relations is: biotite foliation-chlorite foliation-fault. The three structures may represent different deformation phases since they overprint each other, have different orientation and represent probably different metamorphic conditions

Time intervals of no-deformation activity are postulated between the deformation phases during which metamorphic conditions changed significantly while the volume of rock under consideration was 'passively' transported to another position in the crust (e.g. by erosion and uplift). The deformation phases are accompanied by metamorphic events, which may lie on the retrograde leg of a single metamorphic cycle (Sect. 1.3). The size of the area over which these deformation phases can be recognised should now be investigated and gradients in style and orientation monitored. Finally, the synchronous or diachronous nature of a deformation phase can in some cases be established by absolute dating of minerals associated with structures visible in thin section, or by dating crosscutting intrusions. Comparison with similar data on a larger scale, either from the literature or by carrying out further field and thin section research, can establish the regional significance of deformation phases with relation to tectonic events. Because such large-scale analysis is not part of the subjects covered in this book, we restrict ourselves to the establishment of overprinting relations, deformation phases and metamorphic events from data obtained in thin section. The following section gives an outline of some of the problems involved in establishing overprinting relations and deformation phases.

1.2
Establishing and Interpreting Deformation Phases

The concept of deformation phases has been used extensively in the geological literature in reconstruction of the structural evolution of rock units with complex deformation patterns (e.g. Ramsay 1967; Hobbs et al. 1976; Ramsay and Huber 1987; Marshak and Mitra 1988). The underlying idea is that permanent deformation in a volume of rock occurs when differential stresses (Sect. 2.11) are relatively high and that the orientation of the stress field may change between such periods of permanent deformation without visible effects on the rock fabric. The older fabric is not always smoothly erased or modified to a new fabric, since deformation in rocks is commonly partitioned (that is: concentrated in certain domains and less concentrated or absent in others); relicts of older fabric elements may be locally preserved. A foliation that is shortened parallel to the foliation plane may develop folds, commonly with a new crenulation cleavage developing along the axial surface. The older foliation will be completely erased only at high strain or by recrystallisation and grain growth under favourable metamorphic circumstances (Box 4.9). Boudins and tight or isoclinal folds may be refolded but remain recognisable up to very high strain. Lattice-preferred orientation may be preserved in less deformed lenses up to high strain and porphyroblasts may preserve relicts of older structures as long as the porphyroblast phase remains intact (Sects. 7.3–7.7).

Although the concept of deformation phases seems fairly simple and straightforward, there are some problems with its general application, as outlined below.

1. Overprinting relations may be produced by a single deformation phase

Non-coaxial progressive deformation (Sect. 2.5.2) may produce overprinting relations between structures without a major change in the large-scale orientation and magnitude of the stress field. Especially in mylonitic rocks developed in shear zones, it is common to find folds (often sheath folds) that deform the mylonitic foliation and which are clearly the result of the same deformation phase that produced the mylonitic foliation in the first place (Sect. 5.3.2); such folds can be formed at any time during progressive deformation (Fig. 1.2, ⊘Video 1.2).

2. Subsequent deformation phases do not necessarily produce overprinting relations

Two subsequent deformation phases with a similar orientation of the stress field and a similar metamorphic grade may be indistinguishable in the final fabric (Fig. 1.3). For example, a 2 400 Ma-old hornblende foliation, formed under amphibolite facies conditions, could be overprinted by a 1 600 Ma phase of amphibolite facies deformation in a stress field with approximately the same orientation. The result would be strengthening of the earlier foliation. In such cases, only detailed microprobe work or mineral dating may reveal the correct sequence of events.

3. Only the relative age of deformation phases can be established

Identical overprinting relations may develop where a Paleo-proterozoic foliation is overprinted by a crenulation cleavage of Phanerozoic age, or in a thrust nappe within an interval of several hundred thousand years only (Fig. 1.4). If metamorphic conditions are significantly different for two deformation phases, a minimum time separation can be established but otherwise absolute age dating is required.

4. The significance of deformation phases depends on the scale of observation

During development of a fold the axial planar foliation may be rotated to such an extent that a crenulation cleavage is locally formed, overprinting the earlier formed foliation (Williams 1972a). Such overprinting relations form during a single phase of deformation. The same effect may occur on a larger scale; consider a volume of rock in a thrust sheet that is transported over a ramp; the sudden changes in orientation of the rock volume when it moves up and over the ramp can induce overprinting relations. In this case, the final fabric will show separable deformation phases on thin section and outcrop scale, but these will be part of a single deformation phase on a regional scale.

5. Deformation phases may be diachronous

Deformation may affect volumes of rock in a progressive way, starting from one side and reaching the other end much later (Hobbs et al. 1976). As a consequence, an overprinting structure labelled D_2 may be older in a certain area than a D_1 fabric in another. A common setting for such an evolution may be in accretionary wedges, where undeformed rocks arrive at a subduction zone, and subsequently become incorporated in the wedge (Figs. 1.5, B.7.1, ⊘Video B.7.1).

Because of the problems mentioned above and also because of the subjective nature of subdivision in sets of structures (e.g. Hobbs et al. 1976), some geologists have become reluctant to use the deformation phase concept

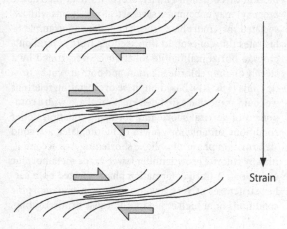

Fig. 1.2. Sequence of events in a shear zone to show how overprinting relations may form during a single phase of progressive deformation if some heterogeneity is present to cause folding

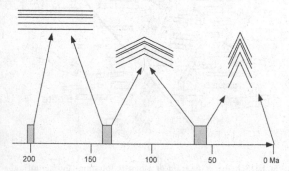

Fig. 1.3. Subsequent deformation phases, represented as *grey blocks* on the *time bar*, do not necessarily produce overprinting relations. If metamorphic conditions and stress orientation are similar, structures like folds may just be further tightened

any longer. However, we feel that the concept continues to be useful to classify structures in a sequential order, if used with care. Deformation phases refer only to the relative age of structures in a limited volume of rock, (commonly in the order of a few hundred km^3) and are generally not equivalent to tectono-metamorphic events of regional significance. It is therefore necessary to determine the tectonic significance of local deformation phases. To establish deformation phases it is important to define sets of structures based on reliable overprinting criteria, such as a foliation (S_n) that has been folded (D_{n+1} folds), and not just on style, orientation, tightness of folds etc., which are criteria that may change from one outcrop to the next in structures of the same age. It is also important to take metamorphic conditions during deformation into consideration, since these are not subject to rapid change (Fig. 1.1). A final warning must be given for the extrapolation of phases from one area to another, or even from one outcrop to the next. The criteria for subdivision remain subjective in the sense that different workers may define a sequence of deformation phases in a different way, resulting in a variable number of phases for the same area. This, however, does not necessarily mean that one of these workers is right, and the others in error; it may just be a matter of different criteria for definition.

For all overprinting relations it is necessary to determine whether they could have formed during a single phase of deformation under similar metamorphic conditions. The following criteria may help to determine whether overprinting relations correspond to separate deformation phases:

a Two overprinting structures composed of different mineral assemblages that represent a gap in metamorphic grade must belong to different deformation phases.
b Foliations that overprint each other commonly represent deformation phases on thin section scale (Sect. 4.2.10.2), but exceptions such as oblique fabrics (Sect. 5.6.2) and shear band cleavages (Sect. 5.6.3) exist.
c Overprinting folds with oblique axial surfaces represent different deformation phases. Care should be taken with refolded folds with parallel axes (Type III of Ramsay 1967), especially in the case of isoclinal folds since these may form during a single deformation phase (Fig. 1.2).
d Shortened boudins are commonly formed by overprinting of two deformation phases (Passchier 1990a; Sect. 5.6.13).
e Some structures preserved in porphyroblasts represent separate deformation phases (Sects. 7.3–7.5).
f Intrusive veins or dykes can be important to separate phases of deformation and their associated foliations.

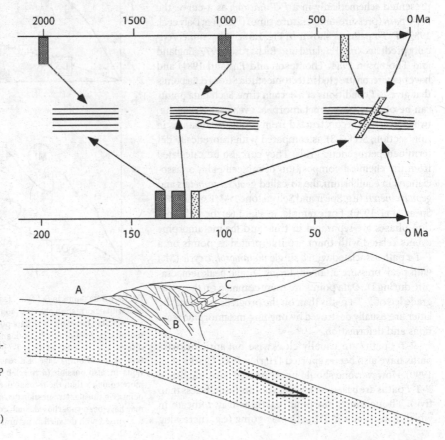

Fig. 1.4.
Only relative ages of deformation can be established by overprinting criteria; overprinting structures shown at *centre right* could form over any time interval, e.g. over 2 000 Ma (*upper bar*) or over 30 Ma (*lower bar*)

Fig. 1.5.
Schematic representation of an active accretionary prism. At *A* no deformation is occurring whereas at *B* a first deformation phase D_1 is responsible for oceanward thrusting, probably accompanied by the development of foliations and folds in deep levels. While such D_1 structures develop at *B*, a second phase of deformation D_2 related to back thrusting is already overprinting D_1 structures at *C* (see also ⊘Video B.7.1)

1.3
Deformation Phases and Metamorphic Events

A metamorphic evolution can be subdivided into *metamorphic events* defined by the growth of particular metamorphic minerals, in a way similar to the concept of deformation phases. Certain fabrics are indicative of growth sequences in metamorphic rocks (e.g. inclusions in porphyroblasts and reaction rims; Sects. 7.6.5, 7.8) and relations between porphyroblasts and foliations commonly reveal the relative time sequence of their generation (Sects. 7.3–7.5). However, one must keep in mind that a metamorphic event is of an essentially different nature than a deformation phase. Whereas the latter is thought to reflect a period of deformation in between intervals of little or no deformation, the former normally reflects only the passing of critical P-T values necessary for a chemical reaction to start and to produce one or more new minerals in the rock. Since deformation often has a catalysing effect on mineral reactions, many such metamorphic events are found to coincide approximately with deformation phases. In other words, many metamorphic minerals are found to have grown during specific deformation phases.

The metamorphic history of a volume of rock can be presented schematically in a P-T diagram as a curve, the P-T-t path (pressure-temperature-time) (Fig. 1.6a; Daly et al. 1989). P-T-t paths as shown in Fig. 1.6 have been theoretically predicted (e.g. England and Richardson 1977; England and Thompson 1984; Thompson and England 1984) and have been reconstructed in tectonic studies from data points that give P-T conditions at a certain time. Such data points can be obtained from metamorphic events (reactions between minerals reconstructed from geometric relations in thin section; Sect. 7.8) as compared with theoretically determined petrogenetic grids. They can also be calculated from the chemical composition of mineral pairs or associations in equilibrium, the so-called *geothermometers* and *geobarometers* (e.g. Spear and Selverstone 1983; Essene 1989; Spear et al. 1990). For example, in Fig. 1.6a, the deformation phases are separated in time and the metamorphic events related with them are interpreted as points on a P-T-t path associated with a single *metamorphic cycle* (M_1) with peak pressure attained during D_1 and peak temperature during D_2. Data points are more common on the retrograde leg of P-T-t paths than on the prograde leg, since the latter are usually destroyed by ongoing metamorphic reactions and deformation.

P-T-t paths are usually clockwise but anticlockwise paths have also been reported (Harley 1989; Clarke et al. 1990). However, one should be aware that most published P-T-t paths are based on few data points (usually less than five). Though metamorphic reactions may indicate in which direction a P-T-t path was going (e.g. increasing

temperature or decreasing pressure), many are based on the passage of a single reaction line and therefore the direction may vary by 180° (Fig. 1.6a). Although P-T-t paths are usually presented as a single smooth curve representing a single metamorphic cycle, possible complex details in the shape of the P-T-t path can rarely be resolved. Real P-T-t paths may have complex shapes with several minor metamorphic cycles and subcycles which can only be reconstructed in rare cases, and then only through detailed combined structural and petrological studies (Fig. 1.6b; Kriegsman 1993; Zhang et al. 1994).

The P-T-t path will generally be valid for only a relatively small volume of rock (at most a few km³), and different paths can often be reconstructed for different crustal units. The way in which these differ gives important information on the regional tectonic evolution. Where paths merge and continue together, rock volumes have been fixed with respect to each other (Figs. B.7.1 and B.7.2,

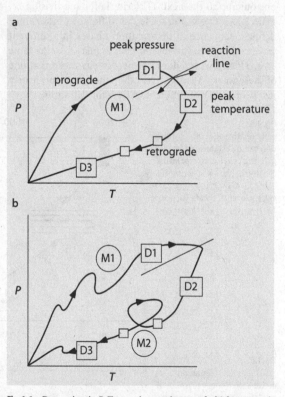

Fig. 1.6. a Data points in P-T space (*squares*) some of which are associated with deformation phases D_1 to D_3. A simple clockwise P-T-t path is postulated based on these data points, representing a single metamorphic cycle of prograde, peak and retrograde metamorphic conditions. The direction of the P-T-t path at peak metamorphic conditions is based on passage of a reaction line; however, other directions are also possible (*arrows*). **b** In reality, the P-T-t path may be more complex than the reconstruction based on the available data points; e.g. during retrogression two metamorphic cycles (*M1* and *M2*) may have been superposed. Such complex paths can sometimes be recognised with detailed structural and petrological work

⊘Video B.7.1). This can be the case if major sections of crustal material such as nappes or terranes are juxtaposed along shear zones. Even within major nappes and terranes the P-T-t evolution may reveal considerable differences when analysed by detailed thermobarometric and microstructural analyses. Especially in complex tectonic domains like the Pennine nappe stack in the Alps, this technique showed to be capable to identify distinct tectono-metamorphic units within an apparently homogeneous basement nappe (Spalla et al. 1999, 2000; Paola and Spalla 2000).

Flow and Deformation

In this chapter, basic principles of continuum mechanics are explained in a non-mathematical way, assuming no previous knowledge of the subject and using simple concepts and illustrations. Continuum mechanics is a subject that is considered to be difficult by many students, and seen as too theoretical to be of practical use in the interpretation of geological structures. It is true that it is rarely possible to make detailed reconstructions of flow and flow history for a rock sample, but it is crucial to have a basic understanding of the mathematical tools to describe the motion of particles in a continuum, and the interaction of forces and motion in a volume of rock. In this book, there is no space to give a detailed treatment of the subject, but we aim to treat at least the basic terminology so that the reader can work through the literature on microstructures unaided.

In the first part of the chapter, reference frames are explained as a necessary tool to describe flow and deformation, and it is important to realize how a choice of reference frame can influence the description of deformation patterns. Then, flow, and deformation are treated including the important concepts of instantaneous stretching axes, vorticity, and the kinematic vorticity number. The central part of this chapter explains kinematics and how to understand the motion of particles in a rock in two and three dimensions. Finally, the concepts of stress and rheology are briefly explained, and basic terminology of these subjects given.

2.1

2.1
Introduction

A hunter who investigates tracks in muddy ground near a waterhole may be able to reconstruct which animals arrived last, but older tracks will be partly erased or modified. A geologist faces similar problems to reconstruct the changes in shape that a volume of rock underwent in the course of geological time, since the end products, the rocks that are visible in outcrop, are the only direct data source. In many cases it is nevertheless possible to reconstruct at least part of the tectonic history of a rock from this final fabric. This chapter treats the change in shape of rocks and the methods that can be used to investigate and describe this change in shape. This is the field of *kinematics*, the study of the motion of particles in a material without regard to forces causing the motion. This approach is useful in geology, where usually very little information can be obtained concerning forces responsible for deformation. In order to keep the discussion simple, the treatment is centred on flow and deformation in two dimensions.

2.2

2.2
Terminology

Consider an experiment to simulate folding using viscous fluids in a shear rig. A layer of dark-coloured material is inserted in a matrix of light-coloured material with another viscosity and both are deformed together (Fig. 2.1). The experiment runs from 10.00 to 11.00 h, after which the dark layer has developed a folded shape. During the experiment, a particle P in one of the fluids is displaced with respect to the shear rig bottom and with respect to other particles. At any time, e.g. at 10.10 h, we can attribute to P a velocity and movement direction, visualised by an arrow or *velocity vector* (Fig. 2.1). If we follow P for a short time, e.g. for 5 s from 10.10 h, it traces a straight (albeit very short) line parallel to the velocity vector. This line is the *incremental displacement vector*. At another time, e.g. 10.40 h, the velocity vector and associated incremental displacement vector of P can be entirely different (e.g. related to the folding of the dark layer). This means that the *displacement path* followed by P to its final position at 11.00 h is traced by a large number of incremental displacement vectors, each corresponding to a particular

Fig. 2.1. Schematic presentation of the velocity, incremental displacement and finite displacement of a particle *P* in a deformation experiment in a shear box. Velocity of P at 10.10 h and 10.40 h can be illustrated as a velocity vector. If deformation proceeds over 5 seconds, the incremental displacement vector will be parallel to the velocity vector. The sequence of incremental displacement vectors gives the finite displacement path. The finite displacement vector is different and connects initial (10.00 h) and final (11.00 h) positions of P

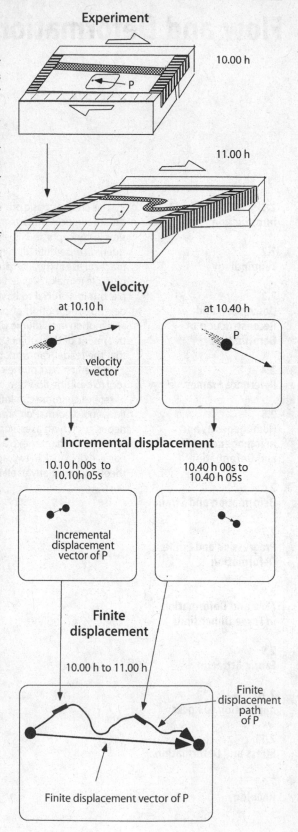

Box 2.1   Terminology of deformation and flow; a traffic example

The difference between flow and deformation can be visualised by the example of cars in a town. If we compare the positions of all red cars in a town on aerial photographs at 8.30 and at 9.00 h, they will be vastly different; the difference in their initial and final positions can be described by *finite displacement vectors* (Fig. B.2.1a). These describe the *finite deformation pattern* of the distribution of cars in the town. The finite deformation pattern carries no information on the *finite displacement paths*, the way by which the cars reached their 9.00-h position (Fig. B.2.1b). The finite displacement paths depend on the velocity and movement direction of each individual car and its change with time. The velocity and movement direction of each car at 8.33 h, for example, can be described by a velocity vector (Fig. B.2.1c). The combined field of all the velocity vectors of all cars is known as the *flow*

pattern at 8.33 h (Fig. B.2.1c). *Flow* of the car population therefore describes the pattern of their velocity vectors. At 8.52 h (Fig. B.2.1d) the flow pattern will be entirely different from that at 8.33 h and the flow pattern is therefore described only for a specific moment, except if the cars always have the same direction and velocity. If we register the displacement of cars over 2 seconds, as a vector field starting at 8.33 h, this will be very similar to the velocity vector field at 8.33 h, but the vectors now illustrate displacement, not velocity. These vectors are *incremental displacement vectors* that describe the *incremental deformation pattern* of the distribution of cars in the town. The incremental deformation pattern is usually different from the finite deformation pattern. If we add all incremental displacement vectors from 10.00 to 11.00 h, the sum will be the finite displacement paths (Fig. B.2.1b).

Fig. B.2.1a–d.
Illustration of the concepts of flow and displacement or deformation using cars in a town

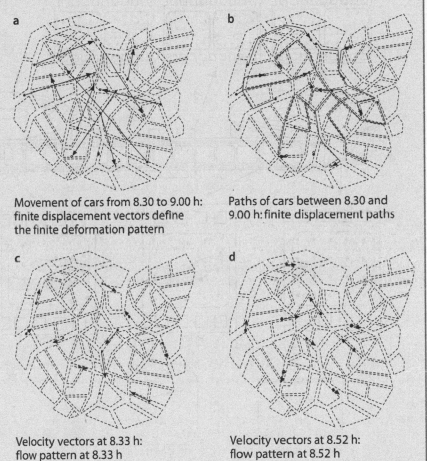

a Movement of cars from 8.30 to 9.00 h: finite displacement vectors define the finite deformation pattern

b Paths of cars between 8.30 and 9.00 h: finite displacement paths

c Velocity vectors at 8.33 h: flow pattern at 8.33 h

d Velocity vectors at 8.52 h: flow pattern at 8.52 h

velocity vector. The displacement path is also referred to as the *particle path*. We can also compare the positions of the particle P at 10.00 and 11.00 h, and join them by a vector, the finite *displacement vector* (Fig. 2.1). This vector carries no information on the displacement path of P.

If the behaviour of more than one particle is considered, the pattern of velocity vectors at a particular time is known as the *flow pattern* (Fig. 2.2). The pattern of in-

cremental displacement vectors is known as the *incremental deformation pattern*. The pattern of displacement paths is loosely referred to as the *deformation path* and the pattern of finite displacement vectors is the *finite deformation pattern*. The process of accumulation of deformation with time is known as *progressive deformation*, while *finite deformation* is the difference in geometry of the initial and final stages of a deformed aggregate.

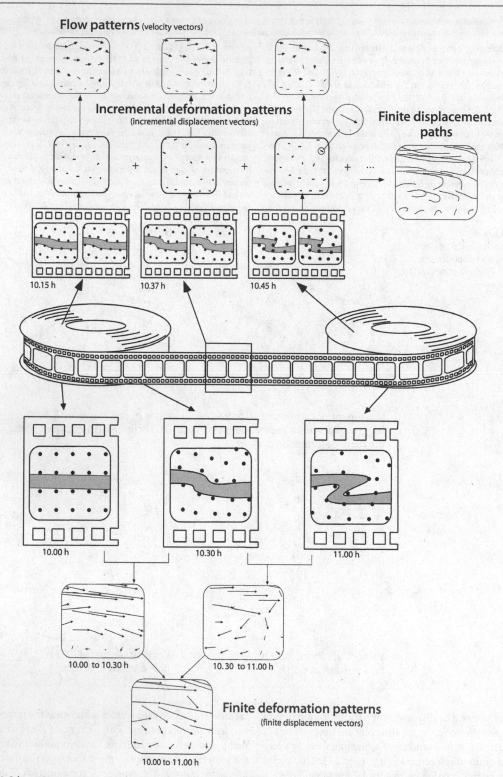

Fig. 2.2. Schematic presentation of the reconstruction of patterns of flow, incremental deformation and finite deformation based on a film of the experiment in Fig. 2.1. At the *top* is shown how incremental deformation patterns can be determined from adjacent images on the film: flow patterns and finite displacement paths can be constructed from these incremental deformation patterns. At the *bottom* is shown how finite deformation patterns can be constructed from images that are further separated in time. *Black dots* are marker particles in the material

2.3
Description and Reconstruction of Deformation

It is interesting to consider how we could accurately describe velocities and displacement of particles in the experiment of Fig. 2.1 using a film (Fig. 2.2). Intuitively, one would assume that the film gives a complete and accurate picture of the experiment, and that no further problems occur in reconstruction of flow and deformation. However, such a reconstruction is more difficult than it would seem. If we compare stages of the experiment that are far apart in time, e.g. at 10.00, 10.30 and 11.00 h, we can connect positions of particles by vectors which define the finite deformation pattern (Fig. 2.2 bottom). However, these finite deformation patterns carry no information on the history of the deformation, i.e. on the displacement paths of individual particles. Finite displacement paths have to be reconstructed from incremental deformation patterns; if we take two stages of the experiment that are close together in time, e.g. two subsequent images of the film (Fig. 2.2 top), these can be used to find the incremental deformation pattern. Finite displacement paths can be accurately reconstructed by adding *all* incremental deformation patterns. This is obviously impossible in practice. An approximation can be obtained by adding a selection of incremental deformation patterns, or a number of finite deformation patterns which represent short time periods. The flow pattern at particular stages of the deformation can be reconstructed from the incremental deformation patterns since these have the same shape.

2.4
Reference Frames

The flow, incremental and finite deformation patterns in Fig. 2.2 were produced with a camera fixed to an immobile part of the shear rig. The shear rig acts as a *reference frame*. However, the patterns would have a different shape if another reference frame were chosen. Figure 2.3 shows three possible arrangements for reconstruction of finite deformation patterns from two photographs taken at 10.10 and 10.50 h (Fig. 2.3a). For most studies of flow and deformation it is advantageous to choose a reference frame fixed to a particle in the centre of the domain to be studied, since this produces symmetric patterns around the central particle. An example is shown in Fig. 2.3d, where one particle P is chosen to overlap in both photographs, and the edges of the photographs are parallel; we have now defined a reference frame with orthogonal axes parallel to the sides of the photographs (and therefore to the side of the shear box), and with an origin on particle P. The patterns in Fig. 2.3b and c are not wrong, but less useful; they have additional translation and rotation components that have no significance

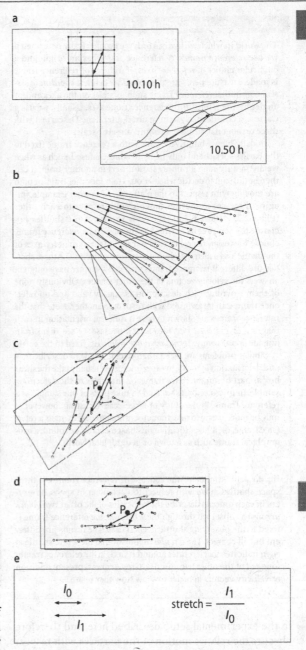

Fig. 2.3. Illustration of the influence of different reference frames on the finite deformation pattern for two stages in the experiment of Fig. 2.1. **a** Two photographic enlargements of the same segment of material at 10.10 h and 10.50 h. *Arrows* indicate the distance between two particles in the two deformation stages. **b, c** and **d** show three different ways of constructing finite deformation patterns from the two images. In **b** no particle in both photographs is overlapping and the finite deformation pattern has a large component of translation. In **c** and **d** one particle P is chosen to overlap in both photographs. In **d**, the sides of the photographs are chosen parallel as well. Since the photographs were taken with sides parallel to sides of the shear box, **d** is selected as the most useful presentation of the finite deformation pattern in this case. **e** Illustration of the concept of stretch. l_0 is original length; l_1 is final length

---

Box 2.2   How to use reference frames

The world in which we live can only be geometrically described if we use *reference frames*. A reference frame has an origin and a particular choice of *reference axes*. If a choice of reference frame is made, measurements are possible if we define a *coordinate system* within that reference frame such as scales on the axes and/or angles between lines and reference frame axes. Usually, we use a Cartesian coordinate system (named after René Descartes) with three orthogonal, straight axes and a metric scale.

In daily life we intuitively work with a reference frame fixed to the Earth's surface and only rarely become confused, such as when we are in a train on a railway station next to another train; it can then be difficult to decide whether our train, the other train, or both are moving with respect to the platform. As another example, imagine three space shuttles moving with respect to each other (Fig. B.2.2, ⊘Video B.2.2). The crew in each of the shuttles can choose the centre of its machine as the origin of a reference frame, choose Cartesian reference axes parallel to the symmetry axes of the shuttle and a metric scale as a coordinate system. The three shuttles use different reference frames and will therefore have different answers for velocity vectors of the other shuttles. Obviously none of them is wrong; each description is equally valid and no reference frame can be favoured with respect to another. Note that the reference frames are shown to have a different orientation in each diagram of Fig. B.2.2 (⊘Video B.2.2), because we see them from outside in our own, *external reference frame*, e.g. fixed to the earth.

Similar problems are faced when deciding how to describe flow and deformation in rocks. In experiments, we usually take the shear box as part of our reference frame, or the centre of the deforming sample. In microtectonics we tend to take parts of our sample as a reference frame. In the study of large-scale thrusting, however, it may be more useful to take the autochthonous basement as a reference frame, or, if no autochthonous basement can be found, a geographical frame such as a town or geographical North.

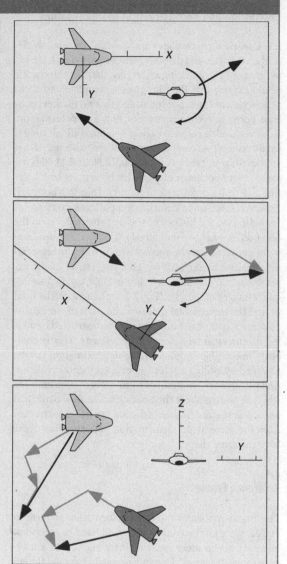

Fig. B.2.2. Illustration of the concept of reference frames. If three space shuttles move with respect to each other in space, observers in each one can describe the velocities of the other two (*black arrows*) as observed through the windows; the reference frame is fixed to the observing shuttle in each case. The results are different but all correct. The *circular arrow* around the white shuttle at *right* indicates that it rotates around its axis in the reference frames for each of the other two shuttles. *Grey arrows* represent addition of velocity vectors in order to show how they relate

---

2.5

in the experimental setup described here, and therefore obscure the relative motion of the particles with respect to each other.

Flow and deformation patterns have certain factors that are independent of the reference frame in which they are described. For example, the relative finite displacement of two particles in Fig. 2.3 can be found from the distance between pairs of particles in both photographs. The final distance divided by the initial distance is known as the *stretch* of the line connecting the two particles (Fig. 2.3e); this stretch value does not change if another reference frame is chosen (cf. Fig. 2.3b, c and d). In the case of flow, *stretching rate* (stretch per time unit) is equally independent of reference frame.

2.5
Homogeneous and Inhomogeneous Flow and Deformation

2.5.1
Introduction

Usually, flow in a material is *inhomogeneous*, i.e. the flow pattern varies from place to place in the experiment and the result after some time is inhomogeneous deformation (e.g. Fig. 2.2). The development of folds and boudins in straight layering (Figs. 2.1, 2.2) and the displacement pattern of cars in a town (Fig. B.2.1) are expressions of inhomogeneous deformation. However, the situation is not

a

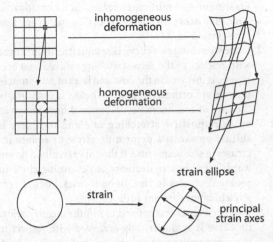

inhomogeneous
deformation

homogeneous
deformation

strain ellipse

strain

principal
strain axes

b

1000 m

10 m

10 cm

1mm

1μm

Fig. 2.4. Illustration of the concepts of homogeneous and inhomogeneous deformation. **a** For homogeneous deformation, straight and parallel lines remain straight and parallel, and a circle deforms into an ellipse, the axes of which are finite strain axes. Inhomogeneous and homogenous deformation occur on different scales. **b** Five scales of observation in a rock. From *top* to *bottom* Layering and foliation on a km scale – approximately homogeneous deformation; layering and foliation on a metre scale – inhomogeneous deformation; foliation on a cm scale – approximately homogeneous deformation; thin section scale – inhomogeneous deformation; crystal scale – approximately homogeneous deformation

as complex as may be supposed from Fig. B.2.1 since, contrary to cars, the velocities of neighbouring particles in an experiment or deforming rock are not independent.

Flow in nature is generally inhomogeneous and difficult to describe in numbers or simple phrases. However, if considered at specific scales (Fig. 2.4), flow may be approximately homogeneous with an identical flow pattern throughout a volume of material, wherever we choose the origin of the reference frame (Fig. 2.4a). The result after some time is homogeneous deformation.

Characteristic for homogeneous deformation is that straight and parallel marker grid lines remain straight and parallel, and that any circle is deformed into an ellipse. Homogeneous flow or deformation can (in two dimensions) be completely defined by just *four* numbers; they are *ten-*

Box 2.3   Tensors

All physical properties can be expressed in numbers, but different classes of such properties can be distinguished. Temperature and viscosity are independent of reference frame and can be described by a single number and a unit, e.g. 25 °C and 10^5 Pas. These are *scalars*. Stress and homogeneous finite strain, incremental strain, finite deformation, incremental deformation and flow at a point need at least four mutually independent numbers to be described completely in two dimensions (nine numbers in three dimensions). These are *tensors*. For example, the curves for the flow type illustrated in Fig. 2.6a need at least four numbers for a complete description, e.g. amplitude (the same in both curves), elevation of the $\dot{S}$-curve, elevation of the ω-curve, and orientation of one of the maxima or minima of one of the curves in space. We might choose another reference frame to describe the flow, but in all cases four numbers will be needed for a full description.

Homogeneous deformation can be expressed by two equations:

$$x' = ax + by$$

$$y' = cx + dy$$

where (x', y') is the position of a particle in the deformed state, (x, y) in the undeformed state and a, b, c, d are four parameters describing the *deformation tensor*. Homogeneous flow can be described by similar equations that give the velocity components v_x an v_y in x and y direction for a particle at point x, y:

$$v_x = px + qy$$

$$v_y = rx + ty$$

p, q, r, t are four parameters describing the *flow tensor*. Both tensors can be abbreviated by describing just their parameters in a *matrix* as follows:

$$\begin{pmatrix} a & b \\ c & d \end{pmatrix} \text{ and } \begin{pmatrix} p & q \\ r & t \end{pmatrix}$$

Multiplication of these matrices with the coordinates of a particle or a point in space gives the complete equations. Matrices are used instead of the full equations because they are easier to use in calculations.

sors (Box 2.3). It is therefore attractive to try and describe natural flow and deformation as tensors. This is possible in many cases, since deviation of flow from homogeneity is scale-dependent (Fig. 2.4b); in any rock there are usually parts and scales that can be considered to approach homogeneous flow behaviour for practical purposes (Fig. 2.4b).

2.5.2
Numerical Description of Homogeneous Flow and Deformation

Imagine a small part of the experiment in Fig. 2.1 that can be considered to deform by homogeneous flow (Fig. 2.5a). At 10.33 h, a regular pattern of velocity vectors defines the flow pattern (Fig. 2.5b). How to describe such a flow pattern numerically? Imagine pairs of material points to be connected by straight lines or *material lines* (Fig. 2.5c), and register the *stretching rate* ($\dot{S}$) and *angular velocity* (ω) of these connecting lines (Fig. 2.5d). The stretching rate can be measured without problems, but in order to measure the angular velocity, a reference frame is needed; the edges of the shear box can be used as such. Stretching rate and angular velocity can be plotted against line orientation, since all parallel lines give identical values in homogeneous flow (Fig. 2.4a). Two regular curves result that have the same shape for any type of flow, but are shifted in a vertical sense for different flow types (Fig. 2.5e). The amplitude of the curves may also vary, but it is always the same for both curves in a single flow type. Maxima and minima always lie 45° apart. If the curves have another shape, flow is not homogeneous. It is now possible to define certain special characteristics of homogeneous flow, as follows (Figs. 2.5e, 2.6):

1. Two lines exist along which stretching rate has its maximum and minimum value, the *instantaneous stretching axes* (ISA). They are orthogonal in any flow type (Figs. 2.5f, 2.6).
2. If the stretching rate curve is symmetrically arranged with respect to the zero stretching rate axis, no area change is involved in the flow, and lines of zero stretching rate are orthogonal (Fig. 2.6); flow is *isochoric*. In the case of area increase, all material lines are given an extra positive stretching rate and the curve is shifted upwards; a deforming circle or square increases in size in this case. If the curve is shifted downwards there is area decrease. A deforming circle or square decreases in size. In both cases, lines of zero stretching rate are not orthogonal.
3. If in a reference frame fixed to ISA the angular velocity curve is symmetrically arranged with respect to the zero angular velocity axis, no 'bulk rotation' is involved in the flow, and lines of zero angular velocity (irrotational lines) are orthogonal. Flow is said to be *coaxial* because a pair of lines that is irrotational is parallel to the ISA (Fig. 2.6). This flow type is also known as *pure shear flow* and has orthorhombic shape symmetry (Fig. 2.6 top). If all material lines are given an identical extra angular velocity, the angular velocity curve is shifted upwards (dextral rotation) or downwards (sinistral rotation). In both cases, flow is said to be *non-coaxial* since irrotational lines are no longer parallel to ISA (Fig. 2.6 centre). The deviation of the angular velocity curve from the axis is a measure of the rotational character of the flow, the *vorticity* (Figs. 2.5e, 2.6; Box 2.4). A special case exists when

Fig. 2.5.
a Sequence of stages in a deformation experiment (small part of the experiment of Fig. 2.1). Deformation is homogeneous. **b** Two subsequent stages are used to determine the velocity field at 10.33 h. **c** Marker points in the flow pattern can be connected by lines. **d** For each line a stretching rate ($\dot{S}$) and angular velocity (ω) are defined, which can (**e**) be plotted in curves against line orientation. In the curves, special directions can be distinguished such as the instantaneous stretching axes (*ISA*) and irrotational lines. The amplitude of the $\dot{S}$-curve is $\dot{S}_k$, a measure of the strain rate, and the elevation of the symmetry line of the ω-curve is a measure of the vorticity. Orientations of ISA and irrotational lines (**f**) can be found from the graphs

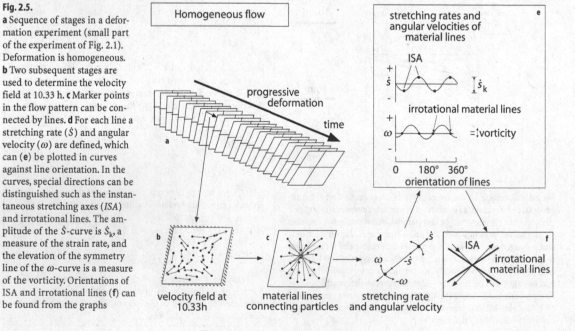

the angular velocity curve is just touching the zero angular velocity axis and only one irrotational line exists; this flow type is known as *simple shear flow* (Fig. 2.6 bottom). All non-coaxial flows have a monoclinic symmetry.

Since flow can be visualised by two simple curves, it must be possible to describe flow using parameters of these curves such as their amplitude, the elevation of each of the curves with respect to the horizontal axis, and the orientation of special directions such as ISA in the chosen reference frame. This orientation can be expressed

by the angle α_k between one of the ISA and the side of the shear box. The first three of these parameters are defined as (Fig. 2.6):

$$\dot{S}_k = \dot{S}_1 - \dot{S}_2 = \omega_1 - \omega_2$$

$$W_k = (\omega_1 + \omega_2)/\dot{S}_k$$

$$A_k = (\dot{S}_1 + \dot{S}_2)/\dot{S}_k$$

$\dot{S}_k$ is a measure of the *strain rate* (the amplitude of the stretching rate curves in Fig. 2.5 and 2.6). W_k is

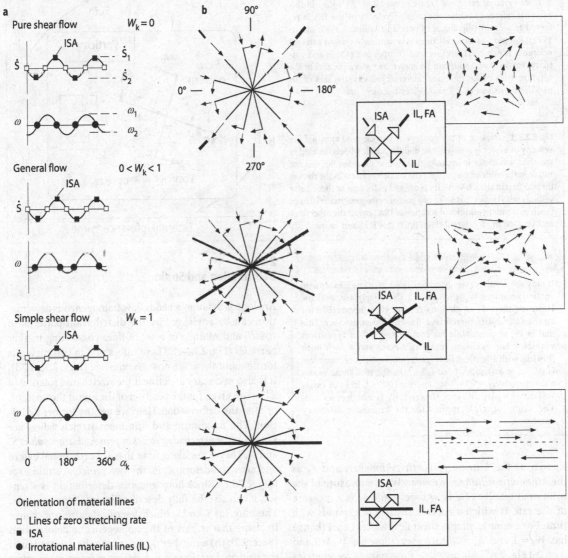

Fig. 2.6. Three types of isochoric flow, represented by: **a** Graphs of stretching rate ($\dot{S}_k$) and angular velocity of material lines (ω) against line orientation α_k. **b** Spatial distribution of material lines with *arrows* indicating sense of stretching rate and angular velocity. **c** Velocity vectors (flow pattern). W_k Kinematic vorticity number; *ISA* instantaneous stretching axes; *IL* orientation of irrotational material lines; *FA* fabric attractor, one of the *IL* (Sect. 2.9)

Box 2.4   Vorticity and spin

Vorticity is the 'amount of rotation' that a flow type possesses (Means et al. 1980). The concept of vorticity can be illustrated with the example of a river (Fig. B.2.3a, ⊘Video B.2.3). In the centre, flow is faster than near the edges. If paddle wheels are inserted in the river along the sides, they will rotate either sinistrally or dextrally; the flow in these domains has a positive or negative vorticity. In the centre, a paddle wheel does not rotate; here the vorticity is zero. Rotation of material lines must be defined with respect to some reference frame (the edges of the river in Fig. B.2.3a, ⊘Video B.2.3) and the same therefore applies to vorticity. In this book, we define vorticity as the summed angular velocity of any two orthogonal material lines in the flow with respect to the ISA (i.e. the ISA act as our reference frame; Fig. B.2.3b).

If an external reference frame is used and ISA rotate in this external reference frame, the angular velocity of the ISA is referred to as *spin* (Fig. B.2.3b; Lister and Williams 1983; Means 1994). Rotation of material lines in a randomly chosen external reference frame can therefore have components of spin and vorticity. Vorticity and spin can be shown as vectors parallel to the rotation axis of the orthogonal material line sets (the axes of the paddle wheels in Fig. B.2.3a, ⊘Video B.2.3).

Fig. B.2.3. Illustration of the concept of vorticity and spin. **a** If the velocity of a river is fastest in the middle, paddle wheels inserted in the river will rotate in opposite direction at the sides, but will not rotate in the middle; they reflect the vorticity of flow in the river at three different sites. **b** Vorticity is defined as the sum of the angular velocity with respect to ISA of any pair of orthogonal material lines (such as *p* and *q*); additional rotation of ISA (and all the other lines and vectors) in an external reference frame is known as spin

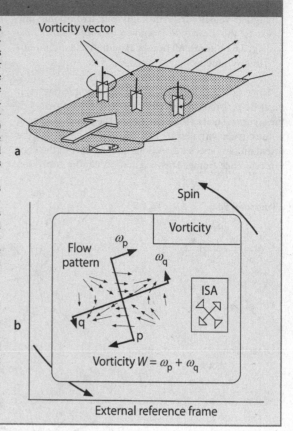

2.6

Box 2.5   Vorticity and kinematic vorticity number

It may seem unnecessarily complicated to define a kinematic vorticity number W_k when we can also simply use vorticity. However, there is an obvious reason. W_k is *normalised* for strain rate and is therefore a dimensionless number. This makes W_k more suitable for comparison of flow types than vorticity. For example, imagine a river and a rock both flowing with identical flow patterns. Vorticity in the river is $0.2 \ s^{-1}$ at a strain rate of $0.3 \ s^{-1}$. In the rock these values are respectively $4 \times 10^{-14} \ s^{-1}$ and $6 \times 10^{-14} \ s^{-1}$. In both cases, vorticity is vastly different. However, W_k is in both cases 0.66. The same principle applies for the kinematic dilatancy number A_k.

known as the *kinematic vorticity number*, and A_k as the *kinematic dilatancy number*. W_k is a measure of the rotational quality of a flow type, while A_k is a measure of the rate at which a surface shrinks or expands with time. For example, simple shear flow without area change has $W_k = 1$ and $A_k = 0$. Pure shear flow has $W_k = 0$ and $A_k = 0$ (Fig. 2.6). All possible flow pattern geometries can be defined by just W_k or A_k, while $\dot{S}_k$ defines how fast deformation is accumulated for a particular flow type and α_k describes its orientation in an external reference frame.

2.6
Deformation and Strain

Analogous to homogeneous flow, homogeneous deformation can be envisaged by the distribution patterns of *stretch* and *rotation* of a set of lines connecting marker particles (Fig. 2.7a–e). These values plot in two curves as for flow, but these are now asymmetric (Figs. 2.5e, 2.7d). It is also necessary to define if the stretch and rotation of a line are given for the position of the line at the onset of, or after the deformation. Here, we use the former definition. The maximum and minimum stretch values are known as the *principal stretches* or *principal strain values* S_1 and S_2. They occur along lines that are orthogonal before and after the deformation, the two *principal strain axes* (Fig. 2.7d,e). Since homogeneous deformation is a tensor, it can also be fully described by just four numbers. These are: (a) S_1 and S_2 which describe the *strain* or change in shape that is part of the homogeneous deformation (Sect. 9.2); (b) a number β_k describing the orientation of the principal strain axes in a reference frame at the onset of deformation, and (c), ρ_k, the rotation of the principal strain axes in the reference frame between the initial and the final state (Fig. 2.7e). Note that *deformation* is normally composed of *strain* (which only describes a change in shape) and a *rotation* component ρ_k. Therefore, deformation and strain should not be used as synonyms.

In homogeneous deformation, a circle is deformed into an ellipse (Figs. 2.4a, 2.7e). The shape of such an ellipse is a measure of the strain; the principal strain axes are the long and short axes of this ellipse. If the original circle has a radius 1, the ellipse is known as the *strain ellipse* and the length of the principal strain axes is S_1 and S_2 respectively.

2.7
Progressive and Finite Deformation

A homogeneous pattern of flow leads to accumulation of homogeneous deformation. Figure 2.8a shows how the total stretch and rotation of material lines with respect to each other can be identical in deformation states accumulated by pure shear flow and simple shear flow. Homogeneous finite deformation carries no information on

Fig. 2.7.
a Two stages of the deformation sequence in Fig. 2.5a that are far apart in time can be used to reconstruct **b** the deformation pattern. **c** Sets of *marker points* can be connected by material lines and the rotation (*r*) and stretch (*S*) of each line monitored. **d** These can be plotted against initial orientation of the lines. In the curves, principal strain axes can be distinguished. **e** Finite deformation as deduced from these curves contains elements of strain and rotation (ρ_k). β_k defines the orientation of a material line in the undeformed state that is to become parallel to the long axis of the strain ellipse in the deformed state

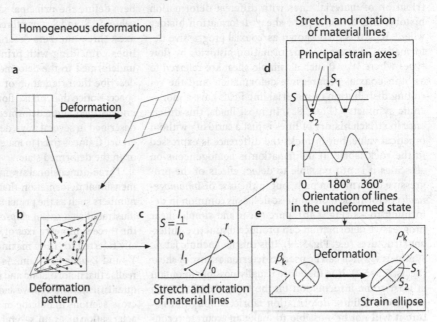

Fig. 2.8.
The effect of deformation history. **a** Two identical squares of material with two marker lines (*black* and *grey lines*) are deformed up to the same finite strain value in simple shear and pure shear progressive deformation respectively. The initial orientation of the squares is chosen such that the shape and orientation of deformed squares is identical. **b** The finite stretch and relative orientation of both marker lines is identical in both cases, but the history of stretch and rotation of each line (illustrated by the *curves*) is different. **c** *Circular diagrams* show the distribution of all material lines in the squares of **a**. *Ornamentation* shows where lines are shortened (*s*), extended (*e*) or first shortened, then extended (*se*) for each step of progressive deformation. The orientation of ISA is indicated

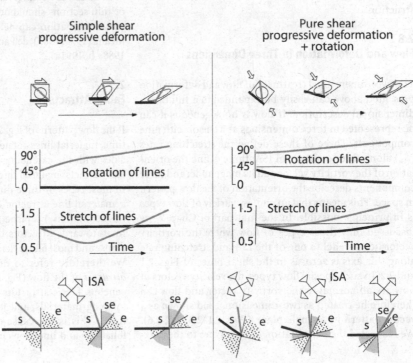

the deformation path or on progressive deformation. However, the stretch and rotation *history* of material lines does depend on the flow type by which it accumulated. This is illustrated in Fig. 2.8b by the stretch and rotation history of two lines in Fig. 2.8a. If the stretch behaviour of all material lines is studied, the difference in pattern is even more obvious (Fig. 2.8c); if deformation accumulates by pure shear flow, the orthorhombic symmetry of the flow pattern is reflected in the symmetry of the distribution of material lines with different deformation history (Fig. 2.8c). A pure shear deformation history where $W_k = 0$ is also known as 'coaxial progressive deformation'. Progressive deformation histories by flow types where $W_k \neq 0$ such as simple shear are referred to as 'non-coaxial progressive deformation' and the resulting distribution of material line fields have a monoclinic symmetry (Fig. 2.8c). In most fluids, this difference in stretch history of lines is just a curiosity without practical value, but in rocks the difference is expressed in the rock fabric. If deformation is homogeneous on all scales it is not possible to detect effects of the progressive deformation path, but in the case of *inhomogeneous* deformation on some scales, as is common in deforming rocks (Fig. 2.4b), pure shear and simple shear progressive deformation can produce distinctive, different structures (e.g. Fig. 5.39). It is this monoclinic fabric symmetry, which can be used to determine sense of shear (Sects. 5.5–5.7). It is therefore usually possible to obtain at least some information on the type of deformation path from a finite deformation fabric, although in nature it will not be possible to make an accurate reconstruction.

2.8
Flow and Deformation in Three Dimensions

The two-dimensional treatment of flow and deformation presented above can easily be expanded to a full three-dimensional description. If flow is homogeneous it can be represented in three dimensions as a tensor with nine components. Three of these define the stretching rates ($\dot{S}_k$) along three orthogonal ISA; three define the orientation of the vorticity vector and its magnitude; and three components describe the orientation of the flow pattern in space. This means that an endless variety of flow types is in principle possible. In the first part of Chap. 2, we discussed only those types of flow where the vorticity vector lies parallel to one of the ISA and stretching rate along this axis is zero, as in the shear boxes of Figs. 2.1 and 2.2. In such special flow types, the velocity vectors of flow are all normal to the vorticity vector, and flow can therefore be treated as two-dimensional and shown as a vector pattern in a single plane (Figs. 2.1–2.3, 2.5, 2.6). We restricted the presentation of flow types to these ex-

amples since they suffice to illustrate the principle, and may indeed represent some flow types that occur in nature, such as simple shear in ductile shear zones between rigid wall rocks. However, it is important to realise that flow is a three-dimensional phenomenon and that two-dimensional simplifications may be unsuitable to describe certain details correctly.

Homogeneous deformation in three dimensions is also expressed as a tensor with nine numbers. Three numbers define the principal stretches or principal strain values S_1, S_2 and S_3 along three orthogonal principal strain axes; three numbers describe the rotation of material lines coinciding with principal strain axes from the undeformed to the deformed state; and three numbers describe the orientation of the principal strain axes in space. Notice that, unlike flow, deformation compares an undeformed and a deformed state, and can therefore be described in several ways, depending on whether the reference frame is fixed to material lines in the undeformed or in the deformed state.

Three-dimensional strain is a component of three-dimensional deformation that can be described by three numbers such as the principal stretches S_1, S_2 and S_3. It is illustrated as a *strain ellipsoid*; principal strain axes are the three symmetry axes of this ellipsoid. They are usually referred to (from maximum to minimum) as the *X-, Y-* and *Z-axes* of strain. As for flow, it is important to realise that deformation and strain are three-dimensional quantities, although we usually see two-dimensional cross-sections in outcrop or thin section; for a full characterisation of strain, several orthogonal outcrop surfaces or thin sections should be studied. More details on flow and deformation can be found in Means et al. (1980), Means (1979, 1983), de Paor (1983) and Passchier (1987a, 1988a,b, 1991a).

2.9
Fabric Attractor

If the flow patterns of Fig. 2.6 work on a material for some time, material lines rotate towards an axis, which coincides with the extending irrotational material line; this axis 'attracts' material lines in progressive deformation. In most types of three-dimensional homogeneous flow, a 'material line attractor' exists in the form of a line or (less commonly) a plane (Fig. 2.9). Since material lines rotate towards attractors, the long axes of the finite strain ellipse and most fabric elements in rocks will do the same. We therefore refer to these directions as the *fabric attractor* of the flow (Fig. 2.9). Even if flow is not homogeneous, fabric attractors may occur as contours in deforming materials, and fabric elements will approach them. This is the cause of the development of many foliations and lineations in deformed rocks.

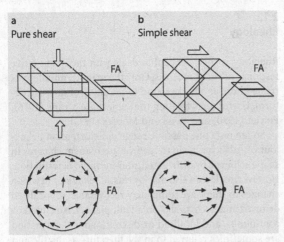

a Pure shear   **b** Simple shear

FA   FA

FA   FA

Fig. 2.9. Concept of the fabric attractor. In both pure shear **a** and simple shear **b** deformation, material lines rotate towards and concentrate near an attractor direction, as shown in the stereograms. This line is the fabric attractor (*FA*). Both foliations and lineations rotate permanently towards this attractor

2.10
Application to Rocks

The observations on flow and deformation presented above for experiments in a shear box are directly applicable to any surface within a deforming rock, although velocities in rocks are obviously very small. The homogeneous flow model allows us to predict what will happen if a rock undergoes progressive deformation by operation of particular flow types provided that it deforms as a continuum without faults on the scale of observation. Unfortunately, in rocks we lack the video camera and we have to make all reconstructions of finite deformation and the deformation path from the end-fabric in outcrop. If the initial configuration of the deformed material is known (e.g. lengths and angles in the case of fossils or minerals), it is possible to determine the magnitude and orientation of finite strain, but without supplementary information we can say little about the deformation path.

2.11
Stress and Deformation

Although in microstructural analysis it is usually only possible to reconstruct aspects of kinematics, it is useful to consider briefly how forces in rocks can lead to flow and deformation. A study of the relationship between forces and changes in shape is the study of *dynamics*.

Deformation of rocks is associated with forces in the earth's crust. It is advantageous to describe such forces independently of the size of the volume of material we are dealing with, i.e. using force per unit area (Nm^{-2}). Although we are used to think of forces in terms of simple

numbers (scalars), they are in fact defined by size *and* direction, and can be drawn as vectors. In a continuum, the force-vector on a surface has a direction and size that are dependent on the orientation of that surface (Fig. 2.10a, ⊘Video 2.10). Therefore, it is not possible to define the forces in a rock at a particular point by a single vector; each surface through the point has a different force vector associated with it. The relation between these values is ex-

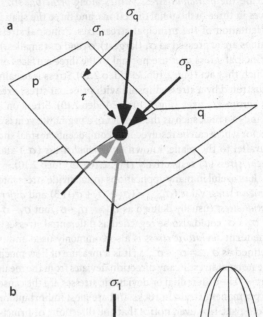

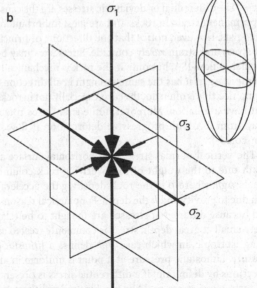

Fig. 2.10. a Illustration of the concept of stress. Surfaces *p* and *q* through a point in a rock under stress each have a different stress vector σ_p and σ_q associated with them. Each stress vector can be decomposed into a normal stress (σ_n) and a shear stress (τ) on the plane. **b** The complete stress state at the point is a tensor that can be represented by three orthogonal principal stress vectors, which operate on three orthogonal surfaces. These principal stress vectors are symmetry axes of a stress ellipsoid as shown at *right*

2.10

2.11

pressed as the *stress* at that point in the material. Notice that stress is defined only for a particular point, since it is usually different from place to place in a material.

Like flow and deformation, stress is a tensor which, in three dimensions, needs nine numbers for its complete characterisation. However, since stress is taken to be symmetric in geological applications, six independent numbers are usually sufficient. Of these, three numbers describe the *principal stress values* along *principal stress axes* in three orthogonal directions, and three the spatial orientation of the principal stress axes. Principal stress values are expressed as σ_1 (largest), σ_2 and σ_3 (smallest). Principal stress axes are normal to the three surfaces on which they act (Fig. 2.10b, ⊘Video 2.10). Stress is usually illustrated by a stress ellipsoid with principal stress axes as symmetry axes (Fig. 2.10b, ⊘Video 2.10). Stress on a plane in a rock such as the contact of a pegmatite vein is a vector which can be resolved into components normal and parallel to the plane, known as *normal stress* (σ_n) and *shear stress* (τ) respectively (Fig. 2.10a, ⊘Video 2.10).

It is useful in many applications to subdivide stress into a *mean stress* value ($\sigma_{mean} = (\sigma_1 + \sigma_2 + \sigma_3) / 3$) and *differential stress* (usually defined as $\sigma_{diff} = \sigma_1 - \sigma_3$, but $\sigma_1 - \sigma_2$ or $\sigma_2 - \sigma_3$ could also be regarded as differential stresses). The term *deviatoric stress* is also commonly used and is defined as $\sigma_{dev} = \sigma_n - \sigma_{mean}$; it is a measure of how much the normal stress in any direction deviates from the mean stress. The differential or deviatoric stresses are the cause of permanent strain in rocks and are most important for geologists. However, notice that the directions of principal stress and strain *rarely* coincide. Stress axes may be parallel to flow-ISA, but only if the rock is mechanically isotropic, e.g. if it has the same strength in all directions; in practice, this is often not the case, especially not in rocks that have a foliation. Moreover, finite strain axes rotate away from ISA with progressive deformation if flow is non-coaxial.

The vertical normal stress on a horizontal surface at depth due to the weight of the overlying rock column equals ρgh, where ρ is the rock density, g the acceleration due to gravity and h the depth. For practical reasons, and because differential stresses are thought to be relatively small at great depth, stress is commonly treated as being isotropic, in which case ρgh defines a *lithostatic pressure*. Lithostatic pressure at a point is uniform in all directions by definition; if a differential stress is present, the term mean stress could be used instead of lithostatic pressure. If pores are open to the surface, a *fluid pressure* may exist in the pores of the rock that is 2.5–3 times smaller than a lithostatic pressure at the same depth. If the pores are partly closed, the fluid pressure may approach the magnitude of the lithostatic pressure or σ_3. In that case rocks may fracture, even at great depth (Etheridge 1983); this is one of the reasons for development of veins (including fibrous veins) in many metamorphic rocks (Sect. 6.2).

2.12
Rheology

Rheology is the science that deals with the quantitative response of rocks to stress. Only the main terminology is treated here as a background to the study of microstructures. Useful texts treating the subject are Means (1976), Poirier (1980) and Twiss and Moores (1992).

So far, only one possible range of deformation behaviour of rocks has been treated, i.e. permanent changes in shape achieved by distributed, non-localised deformation. However, rocks can also display elastic behaviour in which changes in shape are completely recoverable, or localised deformation such as slip on a fault plane. Distributed or continuous, and localised or discontinuous deformation are sometimes referred to in the literature as ductile and brittle deformation (Rutter 1986; Schmid and Handy 1991; Blenkinsop 2000, p 4). However, the terms ductile and brittle are scale-dependent, since flow in a deformation band would be brittle on the grain scale, but ductile on the metre scale. In order to avoid this problem, we prefer another use of the terminology where ductile and brittle refer to deformation mechanisms (Chap. 3). In this book, *brittle deformation* is used for fracturing on the grain scale and frictional slip on discrete faults and microfault surfaces around rock or grain fragments. These processes are not much influenced by temperature, but strongly pressure-dependent. Brittle deformation is commonly associated with volume change. *Ductile deformation*, also known as *viscous flow* is produced by thermally activated deformation mechanisms such as intracrystalline deformation, twinning, kinking, solid-state diffusion creep, recovery and recrystallisation. Depending on scale, it can also be localised.

All minerals and rocks can deform in both a brittle and a ductile way, and in general ductile deformation occurs at higher temperature and lithostatic pressure than brittle deformation, i.e. at deeper levels in the crust (Sect. 3.14). For ductile deformation, the rheology of rocks is usually described in terms of strain rate/stress relations. Stress is usually given as a shear stress (τ) or as a single 'differential stress value' ($\sigma_1 - \sigma_3$) since in experiments on rheology, symmetric stress tensors are imposed on the rock. There are several possible types of ductile rheological behaviour. Any rock will show *elastic behaviour* under mean stress by a small decrease in volume, and under differential stress by a small change in shape (usually less than 1%). Such an elastic strain is completely recoverable if the stress is released (Figs. 2.11a, 3.15). Mean stress increase in rocks will not lead to permanent deformation, even at very high values, unless the rock has a high porosity, or transformation to mineral phases with a higher density can take place. However, if elastic strain in response to differential stress exceeds a limit that the rock can support (the *yield strength*), ductile flow

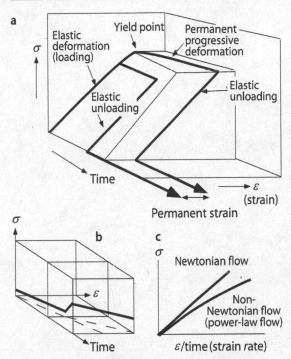

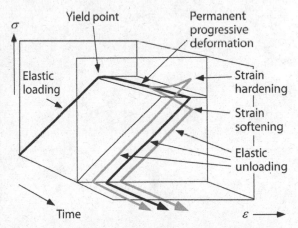

Fig. 2.11. Illustration of some concepts of rheology in space, plotting differential stress (σ) and strain (ε) against time. **a** The *bold curves* illustrate loading and unloading of a sample in an experiment; when differential stress is applied, behaviour is first elastic till a yield point is reached, beyond which permanent deformation begins. When stress is released, the elastic strain is relaxed and permanent strain remains. **b** Graph for permanent deformation in σ-ε-time space. Permanent deformation will proceed at a certain strain rate but if differential stress is increased, the strain rate will increase as well. **c** The way in which strain rate increases with stress can be linear (Newtonian flow) or exponential (non-Newtonian or power-law flow)

and accumulation of strain as described above can occur. Beyond this limit, rocks will deform permanently and if the differential stress is released, only elastic strain will be recovered (Fig. 2.11a). The speed at which the rock changes shape permanently (the strain rate) increases with increasing differential stress, but the relationship between stress (σ) and strain rate ($\dot{\varepsilon}$) can be variable. If strain rate increases in a linear fashion with differential stress ($\dot{\varepsilon} \propto \sigma$) the rock is said to show linear or *Newtonian flow* behaviour. Most of the fluids that we know from daily use such as water, oil and honey are Newtonian. If strain rate increases exponentially with stress ($\dot{\varepsilon} \propto \sigma^n$), flow behaviour is said to be *non-Newtonian* or *power law* (Fig. 2.11b). Both types of flow are probably common in

rocks. The stress exponent n is known as the strain-rate sensitivity of the flow stress and is 1 for Newtonian behaviour and higher than one for power-law behaviour, though not usually exceeding 5 for rocks. Although there are exceptions, Newtonian flow is thought to represent diffusion-accommodated processes (including pressure solution; Sects. 3.3, 3.8), while power-law flow is typical of processes involving dislocation creep (Sect. 3.4). In simple cases, rocks show *steady state flow*, meaning that, if the differential stress is not varied, they will deform at a constant strain rate. The ductile strength of rocks generally decreases with increasing depth in the crust if other factors do not change. Mean stress does not have much influence on the ductile rheology of rocks. However, grain size can under many circumstances be important (Sect. 3.8).

Most rocks do not show steady state flow during the entire deformation history because the fabric of the rock changes with progressive deformation. Both strain hardening and strain softening behaviour occur in rocks (Fig. 2.12). In Fig. 2.12 strain *hardening* and *softening* are indicated for constant strain rate, as may happen in an experiment. In nature, hardening may be a process of decreasing strain rate and increasing differential stress and softening may be associated with strain rate increase and a drop in differential stress. Strain hardening may lead to brittle fracturing of the rock or cessation of deformation; softening may lead to localisation of the deformation in shear zones (Sect. 5.3.4).

Fig. 2.12. Graph for permanent deformation in stress (σ)-strain (ε)-time space. If differential stress increases with time at constant strain rate of permanent deformation, the material is subject to strain hardening; if it decreases, it is subject to strain softening

Deformation Mechanisms

Chapter 3 deals with deformation structures on the scale of individual grains. Grain scale brittle deformation and cataclastic flow occur in the upper crust or at high strain rate. At deeper crustal levels, rocks deform by ductile flow through a range of mechanisms of ductile grain scale deformation such as dissolution-precipitation, intracrystalline deformation by dislocation glide and creep, diffusion creep, twinning and kinking. Ductile deformation in rocks could not lead to high strain if it was not accompanied by mechanisms that reduce the damage imposed during the deformation process. There are two main groups of such mechanisms; recovery, which removes dislocations inside the crystal lattice, and recrystallisation that operates by migration of grain boundaries. Three main types of dynamic recrystallisation are treated; subgrain rotation, bulging and high temperature grain boundary migration. After deformation slows down or stops, grain boundary migration can continue by grain boundary area reduction, and so-called foam textures can develop. In absence of deformation this process is know as static recrystallisation.

The second part of this chapter discusses grain scale deformation processes for a number of rock-forming minerals. This is necessarily a short description of what is presently known, with a large number of references for further reading. Treated are quartz, calcite, dolomite, feldspars, micas, olivine, pyroxenes, garnet and amphibole. Finally, a short outline is given of the deformation of polymineralic rocks with quartz-feldspar aggregates as an example. The final section of this chapter treats flow laws and deformation mechanism maps.

3.1
Introduction

Deformation in rocks is achieved by a large number of processes on the scale of individual grains. The actual processes involved depend on factors such as mineralogy, composition of the intergranular fluid, grain size, lattice-preferred orientation, porosity and permeability; and on external controls such as temperature, lithostatic pressure, differential stress, fluid pressure and externally imposed strain rate. In this chapter, we will briefly introduce the most important rock deformation processes in a sequence from low temperature-high strain rate to high temperature-low stain rate. Grain-scale microstructures that are thought to be formed in response to these processes are highlighted, and it is shown how such microstructures can be used to identify deformation processes that have been operating.

Grains are volumes of crystalline material separated from other grains of the same or different minerals by a grain boundary. If a grain boundary separates grains of the same mineral, they must have a significantly different lattice orientation. Some authors restrict the use of the term grain boundary for surfaces separating grains of like minerals, and use the term *interphase boundary* for surfaces separating different minerals (Fliervoet et al. 1997). In practice, it is difficult to maintain this distinction when describing aggregates composed of many grains, and we therefore use grain boundary for both types of surfaces. Structures visible within grains are known as *intracrystalline deformation structures*.

Although we treat deformation processes and microstructures one by one, this does not mean that they occur isolated in deformed rocks. Most deformed rocks have a long and complicated history of burial, deformation, metamorphism and uplift, and several stages of this process may have contributed to the final fabric. Since peak metamorphic conditions tend to erase earlier structures, most overprinting structures tend to be higher temperature features which are overprinted by lower temperature ones.

3.2
Brittle Fracturing – Cataclasis

At low temperature or high strain rate, rocks change shape by brittle deformation, i.e. by fracture formation and propagation associated with movement along faults. In the terminology of brittle deformation a *fracture* is a planar discontinuity usually with some dilation, including *cracks*, *joints* (large cracks) and *faults*. A crack or joint opens at right angles to the plane of the fracture and has no displacement (Fig. 3.1a); a fault has lateral displacement (Fig. 3.1b). A propagating fault has a *progress zone* at its tip (Fig. 10.9) where isolated *microcracks* form and propagate, microcrack density gradually increases, and finally microcracks link to form a through-going fault

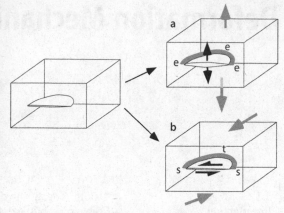

Fig. 3.1. Microcrack propagating in extension **a** and shear **b**. When the crack opens, the tips propagate in extension mode (*e*), sliding mode (*s*) or tearing mode (*t*)

(Fig. 5.1; Hallbauer et al. 1973; Blenkinsop and Rutter 1986; Lloyd and Knipe 1992; Moore and Locker 1995). Motion on the fault then gradually separates grain segments and a volume of *brittle fault rock* is produced along the active fault (Fig. 5.1).

Microcracks are planar discontinuities in rocks on the grain scale or smaller, commonly with some dilation but with negligible displacement. They may nucleate on minor flaws in the crystal lattice, fluid or solid inclusions in crystals, or on grain boundaries (Tapponier and Brace 1976). Microcracks propagate laterally by movement of their tips into intact surrounding material. When the crack opens the walls can be displaced in a tensional regime, in a shear regime or in a combination of both. If a shear component is present the structure is better referred to as a microfracture, and motion can be towards a tip line, or parallel to it (Fig. 3.1b). In all cases, elastic displacement creates a differential stress increase at the tip of the fracture that depends on fracture length, applied bulk stress, elastic properties of the material and resistance to breaking atomic bonds at the crack tip, known as *fracture toughness*. Displacement on a microfracture can lead to fracture propagation if a certain critical differential stress is reached, in extension, sliding or tearing mode (Fig. 3.1). This displacement is usually in the plane of the microfracture if it lies isolated in a homogeneous isotropic material such as glass (Fig. 3.1). However, microfractures may also obtain a curved shape if the stress field at the tip interferes with that of a neighbouring fracture or another inhomogeneity such as an inclusion (Fig. 3.2). In rocks, most minerals are mechanically anisotropic and microfractures commonly form along certain crystallographic directions such as the cleavage direction in micas (Wong and Biegel 1985), feldspars, amphiboles, pyroxenes (Williams et al. 1979; Brown and Macaudiere 1984; Tullis and Yund 1992) and calcite; even quartz is slightly anisotropic for fracturing (Vollbrecht et al. 1991). If there

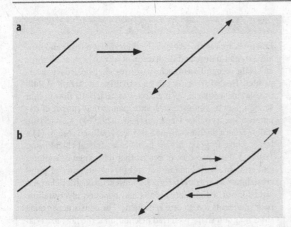

Fig. 3.2. a An isolated crack in an isotropic material propagates radially. **b** If cracks are close together they may obtain a curved shape because the stress fields at the tips of the cracks influence each other

is a shear component along the microfracture but it cannot propagate laterally for some reason, e.g. when the fracture lies along a short grain boundary, horn-shaped *wing cracks* may form (Horii and Nemat-Nasser 1985; Fig. 3.3). Microfractures are called intragranular if they only affect a single grain. Fractures that transect several grains are known as *intergranular* or *transgranular* fractures (Fig. 5.1).

Fracture propagation as described above is valid for continuous media such as single grain interiors of non-porous polycrystalline rocks. In porous rocks, the situation is slightly different. Fractures mostly form and propagate at sites where grains touch (Fig. 3.4). In poorly or unconsolidated porous material, compression normal to the contact of impinging grains leads to fractures which radiate out from the edge of contact sites, known as *impingement microcracks*. These are either straight and *diagonal* or occur in a cone-shaped pattern known as *Hertzian fracture* (Figs. 3.4, 5.1) (Dunn et al. 1973; Gallagher et al. 1974; McEwen 1981; Zhang et al. 1990; Menéndez et al. 1996). Impingement microcracking may induce splitting of grains or shedding of fragments from the sides of grains.

When a critical differential stress is reached in the fracture tip, fractures can grow laterally with a velocity that is a significant fraction of the velocity of elastic waves in solids, as anyone will recognise who has seen glass shatter. Alternatively, stress at the fracture tip can induce slow growth of a microfracture known as *subcritical microcrack growth* (Atkinson 1982; Darot et al. 1985). The speed of subcritical microcrack growth does not only depend on stress, but also on temperature and chemical environment, especially of the fluid in the crack. Subcritical microcrack growth can happen by volume change due to phase change (Blenkinsop and Sibson 1991) but most commonly by *stress corrosion cracking* due to breaking of bonds in the crystal at the crack tip by chemical reaction (Atkinson 1984; Kerrich 1986). Subcritical micro crack growth is probably faster than processes like dissolution-precipitation (Sect. 3.3).

Fig. 3.3.
Wing-cracks form at the tip of a non-propagating microfracture

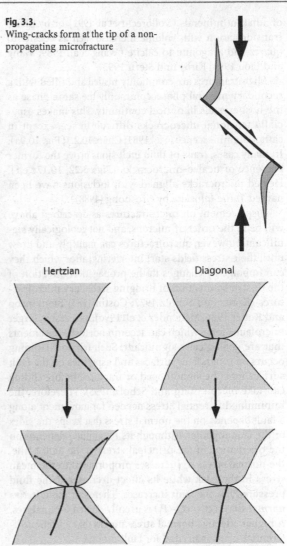

Hertzian   Diagonal

Fig. 3.4. In porous rocks, impingement microcracks can form at contact points. Two examples are given, Hertzian- and diagonal intragranular microcracks

Microfractures described above can form by stress enhancement at their nucleation sites in response to high bulk differential stress or, in the case of porous rocks, due to lithostatic pressure and pore collapse in the absence of bulk differential compressive stress. Other possible causes for microfracture nucleation and propagation are elastic or plastic mismatch, where two mineral phases have different rheological properties and local stress concentration builds up (Tapponier and Brace 1976; Wong and Biegel 1985; Hippertt 1994). Common examples are cracks at corners of mica grains in quartz and fractured feldspar grains in ductile quartz mylonite (Chap. 5). Cracks may also form as accommodation features related to other structures such as twins or kinks (Carter and Kirby 1978; Sect. 3.5), by different thermal expansion or contraction of adjacent minerals (Vollbrecht et al. 1991) or by phase transformation with volume increase such as coesite to quartz and aragonite to calcite (Wang et al. 1989; Wang and Liou 1991; Kirby and Stern 1993).

Microfractures are commonly healed and filled with a secondary mineral phase, commonly the same phase as the host crystals in optical continuity. This makes especially tensional microcracks difficult to see, except in cathodoluminescence (Stel 1981; Chap. 10.2.1; Fig. 10.9a). In many cases, trails of fluid inclusions prove the former presence of healed microcracks (Figs. 3.22, 10.17c,e,f). Healed microcracks aligned with inclusions have been named *Tuttle lamellae* by Groshong (1988).

Displacement on microfractures as described above will be in the order of microns, and not geologically significant. However, microfractures can multiply and grow until their stress fields start interfering, after which they can impinge by changes in the propagation direction of the tips, or by creation of bridging secondary microfractures (Kranz and Scholz 1977; Costin 1983; Blenkinsop and Rutter 1986; Menéndez et al. 1996). As a result, larger microfaults form which can accommodate displacements that are geologically significant. Such frictional sliding occurs on rough fault surfaces and asperities on the fault surface must be smoothened or fractured before sliding can take place (Wang and Scholz 1995). Therefore, the minimum differential stress needed for movement along a fault depends on the normal stress that keeps the sides of the fault together. Although its magnitude depends on the orientation of the principal stress to the fault plane, the normal stress σ_n increases proportional to the mean stress in the rock while its effect decreases if the fluid pressure P_f in the fault increases. Therefore the *effective normal stress* ($\sigma_e = \sigma_n - P_f$) is usually quoted for analyses. A higher effective normal stress means that a higher differential stress is needed for fault motion.

Sliding on faults and fracturing of wall rock forms a volume of *brittle fault rock* such as *gouge*, *cataclasite* and *breccia* (Figs. 3.5, 5.3; Box 3.1; Sect. 5.2) along the fault

plane. Fracturing can operate fast, approaching seismic velocity, or slow by fracturing of individual grains. Fracturing can be transgranular, breaking grains into everfiner fragments in a process called *constrained comminution* (Sammis et al. 1987; Antonellini et al. 1994; Menéndez et al. 1996). In this case the final *particle size distribution* (PSD) can be *fractal* (Sammis et al. 1987; Blenkinsop 1991b). In sediments, however, especially poorly lithified ones deformed at shallow depth, fracturing can also occur by rupture in grain contact cement, or by flaking of grains, in which case grains show conchoidal fracture surfaces and intermediate size particles are underrepresented (Rawling and Goodwin 2003). Commonly, slow transgranular fracturing is aided by processes such as pressure solution, intracrystalline deformation (Lloyd 2000; Hadizadeh and Tullis 1992), chemical reactions and mineral transformation (Atkinson 1982; Blenkinsop and Sibson 1991).

Fig. 3.5.
a Cataclasite fabric – angular fragments of all sizes, some transecting grain boundaries, are embedded in a fine-grained matrix. Many larger fragments are crossed by healed fractures, aligned with fluid- and solid inclusions. **b** Recrystallised fabric of small new grains that grew at the expense of old grains. The new grains show little variety in grain size

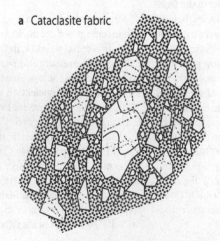

a Cataclasite fabric

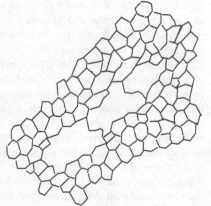

b Recrystallised fabric

Movement on a fault can proceed along distinct sliding planes or slickensides in or at the edge of the produced volume of cataclasite or gouge, but also by distributed *cataclastic flow* within the mass of fractured material. Cataclastic flow operates by sliding and rotation of the fragments past each other, and further fragmentation of these into smaller particles (Sibson 1977b; Evans 1988; Blenkinsop 1991b; Rutter and Hadizadeh 1991; Lin 2001). Rotation of fragments can be suppressed if fracturing is along crystal cleavage planes as in feldspar and amphibole, and in such cases a crystallographic preferred orientation can result in the cataclasite or gouge (Tullis and Yund 1992; Hadizadeh and Tullis 1992; Imon et al. 2004). In the fractured material, cataclastic flow can occur by grain boundary sliding with limited or no further fracturing of grains, or, at the other extreme, fracturing and other grain deformation processes may limit the rate at which cataclastic flow can occur (Borradaile 1981). During cataclastic flow, voids are created that may be filled with vein material precipitated from solution, which is subsequently involved in the cataclasis; as a result, most cataclasite and breccia contains abundant fragments of quartz or carbonate derived from these veins (Sect. 5.2). Fluid migration through cataclasite may also cause lithification by cementation, so that the cataclasite may be inactivated and fault propagation, fracturing and cataclasis migrates into another part of the rock volume. Even if this does not occur, cataclastic flow is usually instable and terminates by localisation of deformation into slip on fault planes after which new breccia, cataclasite or gouge can be produced. Cataclastic flow usually occurs at diagenetic to low-grade metamorphic conditions. The conditions also depend on the type of minerals involved (Sect. 3.12) and on fluid pressure; high fluid pressure promotes cataclastic flow in any metamorphic environment and is responsible for the common occurrence of veins in cataclasite and breccia.

3.3 Dissolution-Precipitation

An important deformation mechanism in rocks that contain an intergranular fluid is *pressure solution*, i.e. dissolution at grain boundaries in a grain boundary fluid phase at high differential stress. Pressure solution is localised where stress in the grain is high, mostly where grains are in contact along surfaces at a high angle to the instantaneous shortening direction (Figs. 3.6–3.8, ⊘Video 3.6). Selective pressure solution at grain contacts occurs because the solubility of a mineral in an aqueous fluid is higher where a crystal lattice is under high stress than at localities where stress is relatively low (Robin 1978; Wheeler 1987a, 1992; Knipe 1989). For example, in a sandstone where grains are in contact (Fig. 3.6, ⊘Video 3.6) the grain lattice near contact points is more strongly compressed than elsewhere; as a result, material will dissolve near these contact points and be redeposited at sites of low differential stress. A locally higher density of crystal defects near contact points may also enhance solubility (Spiers and Brzesowsky 1993). In this way, grains will change shape by local dissolution and redeposition without internal deformation (Fig. 3.6b, ⊘Video 3.6).

Fig. 3.6.
a Oolites surrounded by a pore fluid. At contact points, differential stresses are relatively high, as indicated by *shading*.
b Pressure solution changes the shape of the grains. Material dissolved at the contact points is redeposited in adjacent pore spaces, indicated by *dark shading*

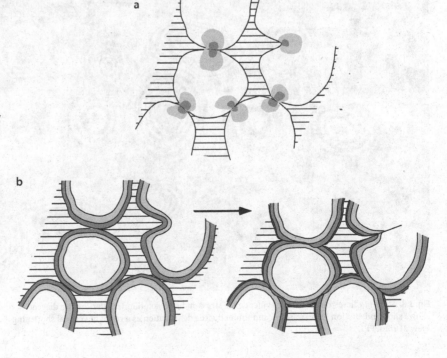

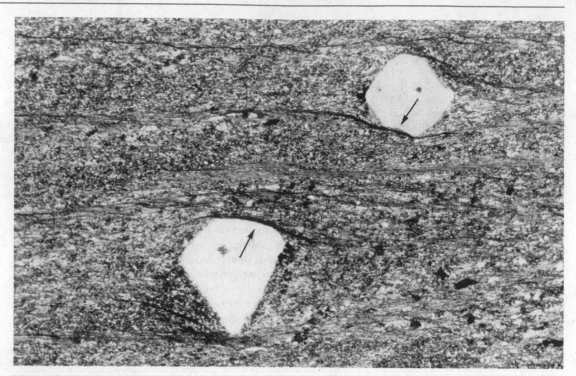

Fig. 3.7. Dissolution of single idiomorphic quartz crystals (*arrows*) in an ignimbrite. *Dark horizontal seams* consist of insoluble material that became concentrated during dissolution. Leonora, Yilgarn Craton, Australia. Width of view 4 mm. PPL

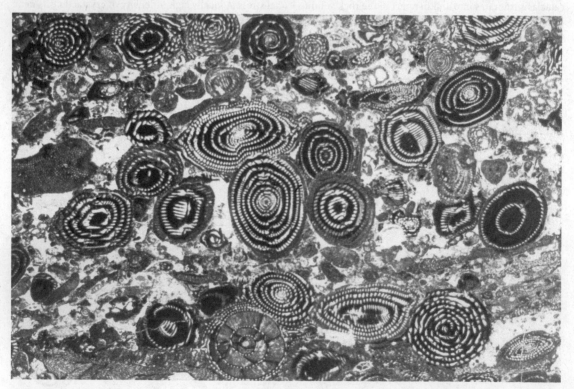

Fig. 3.8. *Alveolina* limestone showing evidence for stress-induced solution transfer during diagenetic compaction. The four fossils in the *centre* show indentation by dissolution and minor ductile deformation as a result of vertical shortening. Eastern Pyrenees, Spain. Width of view 21 mm. PPL

Along grain contacts, pressure solution may occur in a thin fluid film between grains (Rutter 1976), possibly enhanced by an etched network of microcracks in the contact surface (Gratz 1991; den Brok 1998), or it may occur by dissolution undercutting of 'island structures' that are surrounded by fluid-filled channels, forming a stress-supporting network between grains (Ray 1982; Spiers et al. 1990; Lehner 1995). The dissolved material can diffuse away from the sites of high solubility down a stress-induced chemical potential gradient to nearby sites of low solubility by stress-induced solution transfer, usually referred to as *solution transfer*. Redeposition of the dissolved material may occur at free grain boundaries that are in contact with the fluid. Newly precipitated material may be of a different mineral composition or phase as compared to the dissolved material; this is known as *incongruent pressure solution* (Beach 1979; McCaig 1987). Alternatively, the fluid with dissolved material can migrate over a larger distance and deposit material in sites such as veins or strain shadows (Chap. 6), or even migrate out of the deforming rock volume[1]. Pressure solution and solution transfer of material are dominant at diagenetic to low-grade metamorphic conditions where fluids are abundant and deformation mechanisms favoured at higher temperatures, such as intragranular

deformation, are hampered. However, the process may also be important at higher metamorphic grade (Wintsch and Yi 2002). The effect of pressure solution is particularly clear in the development of differentiated crenulation cleavage at low to medium metamorphic grade, as explained in Sect. 4.2.7.3 (see also Bell and Cuff 1989). Pressure solution in quartz or calcite seems to be enhanced by the presence of mica or clay minerals at grain boundaries of these minerals (Houseknecht 1988; Hippertt 1994; Dewers and Ortoleva 1991; Hickman and Evans 1995). Details of pressure solution are described in Durney (1972), Elliott (1973), Gray and Durney (1979a), Rutter (1983), Groshong (1988), Knipe (1989), den Brok (1992, 1998), Wheeler (1992), Shimizu (1995), and den Brok et al. (1998a,b, 2002). Box 3.2 lists evidence for pressure solution in thin section.

3.4
Intracrystalline Deformation

Crystals can deform internally without brittle fracturing by movement of so-called *lattice defects*, a process known as *intracrystalline deformation* (Figs. 3.9, 3.10; Box 3.3). Lattice defects in crystals can be grouped into *point defects* and *line defects* or *dislocations* (Figs. 3.11, 3.13, 3.14). Point defects are missing or extra lattice points (atoms or molecules) known respectively as *vacancies* and *interstitials* (Fig. 3.11a). Line defects may be due to an 'extra' half lattice plane in the crystal. The end of such a plane is known as an *edge dislocation* (Fig. 3.11b). Besides edge dislocations, *screw dislocations* exist where part of a crystal is displaced over one lattice distance and is therefore twisted (Fig. 3.11c). Edge and screw dislocations can be interconnected into *dislocation loops* (Figs. 3.11d, 3.14); they are end members of a range of possible dislocation types. Dislocations can also split into *partial dislocations*, separated by a strip of misfitted crystal lattice known as a *stacking fault*.

Dislocations cannot be directly observed by optical microscopy, only by TEM (McLaren 1991; Sect. 10.2.4; Figs. 3.13, 3.14, 10.11). However, they can be made visible indirectly by etching of pits where they transect a polished surface, or by decoration techniques; in olivine, decorated dislocations can be made visible by heating a sample in an oxidizing environment (Kohlstedt et al. 1976; Karato 1987; Jung and Karato 2001).

A dislocation is characterised by a *Burgers vector* (Figs. 3.11d, 10.11), which indicates the direction and minimum amount of lattice displacement caused by the dislocation. The Burgers vector can be imagined by drawing a square circuit around the dislocation from atom to atom, with an equal number of atoms on each side of the square; in an intact crystal this circuit would be closed, but around a dislocation the loop is not closed – the missing part is the Burgers vector (Fig. 3.11d).

Box 3.2   Evidence for pressure solution

Evidence for the action of pressure solution is the presence of truncated objects such as fossils, detrital grains, pebbles and idiomorphic phenocrysts (McClay 1977; Rutter 1983; Houseknecht 1988; Figs. 3.7, 3.8, ⊘Photo 3.8), truncation of chemical zoning in crystals such as garnet or hornblende (Berger and Stünitz 1996) and the displacement of layering on certain planes (Figs. 4.4, 4.21). In the latter case, however, the possibility of slip along the contact should also be considered; if the contact is indented, the displacement is most probably due to pressure solution (Fig. 4.21). Spherical grains may form indenting contacts. Equally sized grains will be in contact along relatively flat surfaces, while small grains tend to indent into larger grains (Blenkinsop 2000). Planes on which pressure solution occurred are commonly rich in opaque or micaceous material, which is left behind or deposited during the solution process (Figs. 3.7, 4.20). A spectacular example are *stylolites*, highly indented surfaces where material has been dissolved in an irregular way, allowing the wall rocks to interpenetrate (Box 4.3; Fig. B.4.4).

The opposite process, deposition of material from solution, can be visible as new grains, fibrous vein fill or fibrous overgrowth of grains in strain shadows (Chap. 6). New grains grown from solution may be recognised by lack of intracrystalline deformation structures (Sect. 3.4), well defined crystallographically determined crystal faces, and growth twins. New grown rims of material in optic continuity with older parts of a grain are also common but may be difficult to distinguish, except by cathodoluminescence (Sect. 10.2.1). Fluid inclusion trails (Sect. 10.5) can also reveal the presence of overgrowths.

---

[1] Something to remember when drinking mineral water.

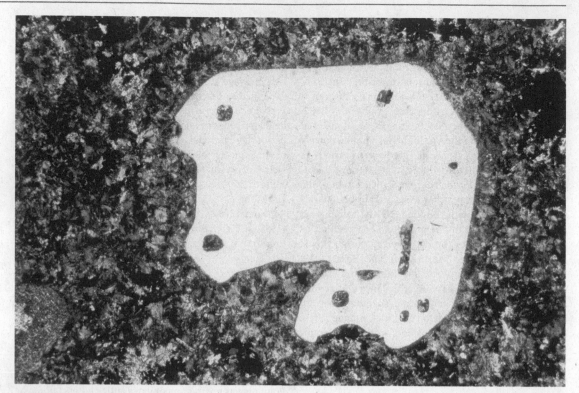

Fig. 3.9. Subhedral quartz crystal in an undeformed ignimbrite. Ornica, Southern Alps, Italy. Width of view 4 mm. CPL

Fig. 3.10. Quartz crystal flattened by intracrystalline deformation in a deformed ignimbrite. The crystal is boudinaged and the fragment on the *right hand side* shows deformation lamellae and undulose extinction. Argylla Formation, Mount Isa, Australia. Width of view 4 mm. CPL

Box 3.3   Evidence for intracrystalline deformation

Individual dislocations cannot be observed with an optical microscope. However, the effect of the presence of dislocations in a crystal lattice may be visible. A crystal lattice which contains a large number of similar dislocations can be slightly bent; as a result, the crystal does not extinguish homogeneously as observed with crossed polars; this effect is known as *undulose extinction* (Figs. 3.10, 3.17, ⊘Video 3.17). Undulose extinction can be 'sweeping' when it occurs as large-scale, regular bending of the crystal due to the presence of dislocations, but can also be patchy and irregular, when it is associated with (microscopically invisible) small fractures and kinks besides dislocation tangles (Hirth and Tullis 1992). *Microkinks* occur as small isolated structures in quartz and feldspars. They are probably associated with cataclastic failure at sites of dislocation tangles (Tullis and Yund 1987) and are therefore indicative of dislocation glide.

Another effect that is commonly observed in crystals deformed at low temperature by intracrystalline deformation are lamellae with a high optical relief which usually have a distinct preferred orientation, known as *deformation lamellae* (Fairbairn 1941; Ingerson

and Tuttle 1945; Carter 1971; Christie and Ardell 1974; Drury 1993; Figs. 3.10, 3.18, ⊘Video 3.18) also known as Fairbairn lamellae (Groshong 1988; Wu and Groshong 1991a). Deformation lamellae consist of dislocation tangles, small elongate subgrains (Blenkinsop and Drury 1988; McLaren 1991; Trepmann and Stöckhert 2003), and arrays of very small solid or fluid inclusions that are only visible by TEM. Deformation lamellae are particularly common in quartz, where they usually have a sub-basal orientation. How deformation lamellae actually develop and how they should be interpreted is only partly understood.

Finally, the presence of a lattice preferred orientation has been suggested as evidence for deformation by dislocation creep, although in some minerals (calcite) it can also form by deformation twinning. At elevated temperature, intracrystalline microstructures such as undulose extinction and deformation lamellae may be absent due to recovery or recrystallisation (see below). In this case, the presence of a strong lattice preferred orientation can be taken as evidence for dislocation creep.

Fig. 3.11.
a Lattice with two types of point defects. **b** Edge dislocation defined by the edge of a half-plane in a distorted crystal lattice. **c** Screw dislocation defined by a twisted lattice. **d** Dislocation with edge and screw dislocation regions in a crystal. A square itinerary of *closed arrows* around the dislocation is used to find the Burgers vector of the dislocation, indicated by *open arrows*

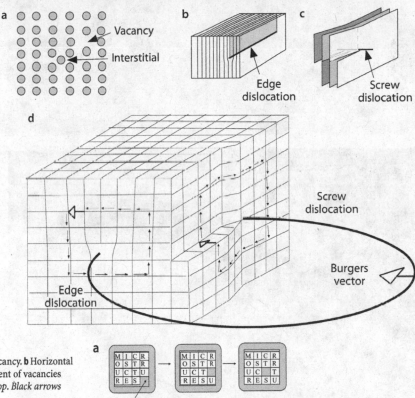

Fig. 3.12.
a The principle of movement of a vacancy. **b** Horizontal shortening of a crystal by displacement of vacancies from *right side* of the crystal to the *top*. *Black arrows* indicate movement of vacancies

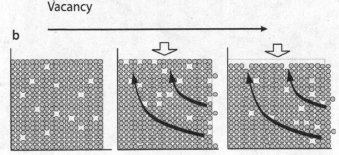

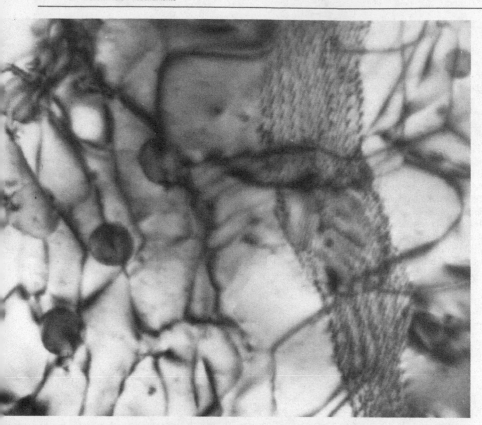

Fig. 3.13.
TEM image of a network of free dislocations decorated by microfluid inclusions (*grey circles*), and a subgrain wall defined by a closely spaced array of dislocations. This type of microstructure is typical for dislocation creep deformation of quartz. Bayas Fault zone, Spain. Width of view 2 µm. (Photograph courtesy Martyn Drury)

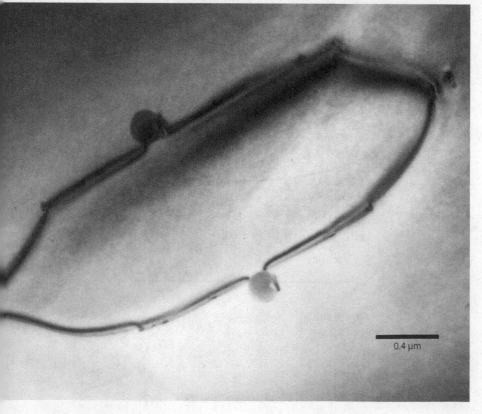

0.4 µm

Fig. 3.14.
TEM image of a dislocation loop in olivine from a kimberlite xenolith. The dislocation loop is pinned by small particles of ilmenite and glass. (Photograph courtesy Martyn Drury)

The shape of a crystal cannot be permanently changed by just squeezing it; the distance between lattice points can only be changed by a very small amount, leading to elastic deformation. If stress is released, the original shape is recovered. A permanent change in shape can only be achieved by a change in the relative positions of molecules or atoms. This happens by movement of lattice defects through a crystal in the process of intracrystalline deformation (Poirier 1985; Hull 1975).

Consider the vacancies in Fig. 3.12. If neighbouring atoms occupy the vacancy sites, vacancies are moving through the crystal and the crystal may change shape permanently (⊘Videos 3.12a, 3.12b). Moving dislocations can also cause relative displacement of parts of a crystal lattice. Figure 3.15a (⊘Videos 3.11, 3.15a) shows how movement of a dislocation displaces parts of crystals without actually separating one part of the crystal from the other. Dislocations can be generated in a crystal at so-called *dislocation*

sources. An example is a *Frank-Read source* (Fig. 3.15b, ⊘Video 3.15b).

Ductile deformation of rocks is to a large extent achieved through the migration of dislocations and vacancies. Lattice defects can cause significant strain in crystals only if new defects are continuously created; this can happen at dislocation sources and vacancy sources within the crystal or at crystal boundaries.

Intracrystalline deformation by glide of dislocations alone is known as *dislocation glide*. Dislocations have a distinct orientation with respect to the crystal lattice and can move only in specific crystallographic planes and directions (Fig. 3.11d). A specific slip plane coupled with a slip direction (the Burgers vector) is known as a *slip system*. Slip systems (Box 3.4) for minerals are normally determined by TEM (Sect. 10.2.5; Fig. 10.11). In most common rock-forming minerals such as quartz, feldspars, calcite and olivine, several slip systems of different orientation can be active (Sect. 3.12). The type of slip system that will be active in a crystal depends

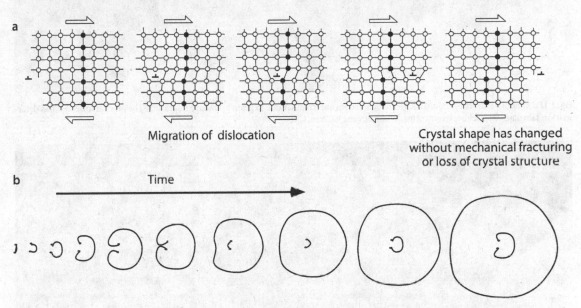

a   Migration of dislocation   Crystal shape has changed without mechanical fracturing or loss of crystal structure

b   Time

Fig. 3.15. a Deformation of a crystal by movement of an edge dislocation; the *top half* of the crystal is translated over one lattice unit to the *right* as a result of the passage of a single dislocation *from left to right*. View normal to the edge dislocation. One lattice plane is marked to show the relative displacement of the upper part of the crystal with respect to the lower part **b** Operation of a Frank-Read dislocation source. A short dislocation segment between two inclusions in a crystal, with Burgers vector in the plane of the paper, is displaced under influence of a differential stress in the crystal. The dislocation propagates into a kidney shape until the ends meet and annihilate; a dislocation loop as in Fig. 3.11d is formed, and the remaining dislocation segment can migrate again to form more dislocation loops

Fig. 3.16.
a Dislocation blocked by an impurity in the crystal. **b** Migration of vacancies to the dislocation plane can cause climb of the dislocation away from the obstruction. **c** After climb, the dislocation is no longer blocked and can pass the obstruction

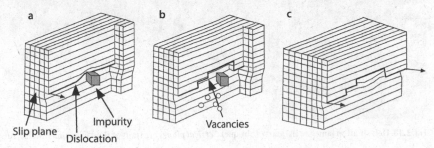

a   Slip plane   Dislocation   Impurity   b   Vacancies   c

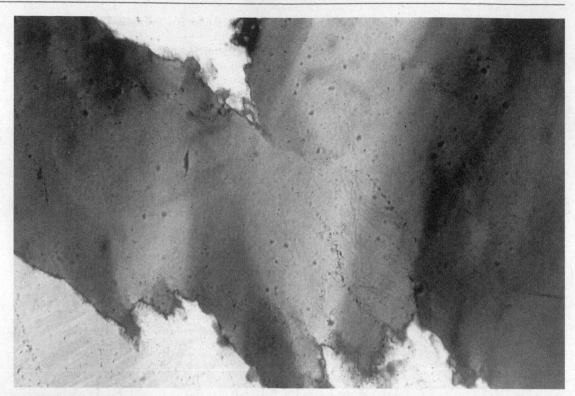

Fig. 3.17. Undulose extinction in quartz. Grain boundaries are irregular due to grain boundary migration. The grain at *lower left* has deformation lamellae. Micaschist, Orobic Alps. Width of view 0.6 mm. CPL

Fig. 3.18. Deformation lamellae in quartz (oblique). *Vertical planes* are trails of fluid inclusions. Mt Isa, Australia. Width of view 1.8 mm. CPL

A slip system in a crystal is defined by a slip plane and a direction of slip (the Burgers vector) within this plane. These elements are usually indicated by Miller indices of the plane, followed by the indices of the slip direction vector, e.g. (001)[010]. Instead of indices, standard abbreviation letter symbols are used in some cases. Notice the shape of the brackets used; if a specific plane and direction are indicated this is done as (plane)[direction]. A set of symmetrically equivalent slip systems is indicated as {planes}<directions>. <f∩r> indicates the intersection line of f- and r-planes. If the Burgers vector does not correspond to a unit cell length, the length can be indicated with the indices, e.g. as {110}1/2<–110>.

3.5
Twinning and Kinking

Some minerals can deform by *deformation twinning* (or mechanical twinning) in addition to dislocation creep and glide (Fig. 3.19; Jensen and Starkey 1985; Smith and Brown 1988; Burkhard 1993; Egydio-Silva and Mainprice 1999). Twinning can accommodate only a limited amount of strain and always operates in specific crystallographic directions, so that additional pressure solution, dislocation creep or recrystallisation (see below) is needed to accommodate large strains. In general, twinning occurs in the lower temperature range of deformation (Sect. 3.12). Twinning is most common in plagioclase and calcite, but also occurs in dolomite, kyanite, microcline (Eggleton and Buseck 1980; White and Barnett 1990), biotite (Goodwin and Wenk 1990), quartz (Dauphiné twinning; Barber and Wenk 1991; Lloyd 2004), diopside (Raleigh 1965; Raleigh and Talbot 1967) and jadeite (Ferrill et al. 2001). Deformation twins are commonly wedge-shaped or tabular and can propagate by movement of the twin tip, or by movement of the twin boundary into the untwinned material, where the twin boundary remains straight. At elevated temperatures, twin boundaries can bulge into the untwinned crystal, except where they are pinned by grain boundaries or other, crosscutting twins (Sect. 9.9; Fig. 9.7). This process of *twin boundary migration* recrystallisation (Vernon 1981; Rutter 1995; Figs. 3.20, 9.7) can completely sweep the untwinned parts of grains. In this sense, it resembles other recrystallisation mechanisms but it only occurs within grains; twin boundary migration does not cause grain growth, since grain boundaries are not affected.

Kinking resembles twinning but is not so strictly limited to specific crystallographic planes and directions. Kinking is common in crystals with a single slip system such as micas but also occurs in quartz, feldspar, amphibole, kyanite and pyroxenes at low temperature (Sect. 3.12; Bell et al. 1986; Nishikawa and Takeshita 1999, 2000; Wu and Groshong 1991a). Box 3.5 lists evidence for deformation twinning in thin section.

on the orientation and magnitude of the stress field in the grain and on the *critical resolved shear stress* (CRSS) τ_c for that slip system; τ_c must be exceeded on the slip system to make the dislocation move. The magnitude of τ_c depends strongly on temperature, and to a minor extent on other factors such as strain rate, differential stress and the chemical activity of certain components such as water that may influence the strength of specific bonds in a crystal. For each slip system this dependence is different. As a result, the types of dominant slip system that are active in a crystal change with metamorphic and stress conditions (Sect. 3.12).

When different slip systems intersect in a crystal, migrating dislocations can become entangled and their further movement is obstructed. Dislocations may also become *pinned* by secondary phases in the crystal lattice (Fig. 3.14). Such dislocation 'tangles' can inhibit movement of other newly formed dislocations, which pile up behind the blocked ones. The crystal becomes difficult to deform and hardens (Fig. 3.24). This process is referred to as *strain hardening*.

If we twist a piece of steel wire, it is difficult to bend it back into its original shape, and the wire becomes harder to deform upon renewed bending. Eventually the wire may snap; by bending the wire, we have caused migration and entanglement of dislocations in the lattice of the metal crystals. Strain hardening occurs also in rocks, and can enhance brittle failure. There are, however, mechanisms that work against strain hardening and allow ductile deformation to continue. One important mechanism that allows dislocations to pass obstruction sites is the migration of vacancies to dislocation lines (Fig. 3.16); this effectively displaces the dislocation, and allows it to 'climb' over a blocked site. The mechanism of dislocation glide with climb of dislocations is known as *dislocation creep*. The term *crystal plastic deformation* is used to describe deformation by dislocation creep.

An important effect of intracrystalline deformation is the development of a *lattice-preferred orientation* (LPO). Since dislocations move only in specific lattice planes, a rock deforming by movement of dislocations may develop a preferred orientation of the grains that make up the rock. The development and interpretation of lattice preferred orientation is discussed in Sect. 4.4.

Deformation twins can commonly be distinguished from growth twins by their shape; deformation twins are commonly tapered, while growth twins are commonly straight and stepped (Figs. 3.19, 9.7). Twins may be restricted to certain parts of a crystal. Growth twins are commonly bounded by zoning, while deformation twins can be concentrated at high strain sites such as the rim of crystals or sites where two crystals touch each other. In plagioclase, growth and deformation twins occur. Deformation twins commonly taper towards the crystal centre (Fig. 3.19; Sect. 3.12.4). In calcite, most twins are deformation twins that tend to taper towards the grain boundary (Sects. 9.5.1, 9.6.3; Fig. 9.7a).

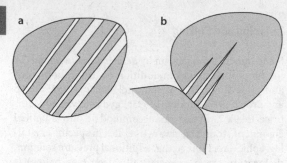

Fig. 3.19. a Growth twins in plagioclase with steps. **b** Deformation twins in plagioclase, with tapering edges nucleated on a high stress site at the edge of the crystal

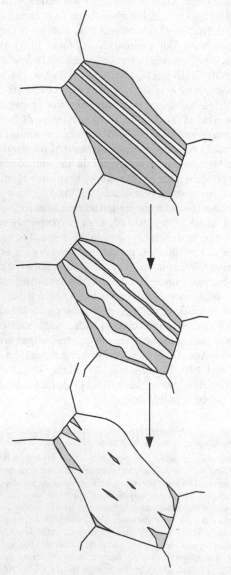

Fig. 3.20. Twin boundary migration recrystallisation in calcite can sweep whole crystals by migration of twin boundaries. Grain boundaries are not affected by this recrystallisation mechanism

3.6
Recovery

Any crystal can be imagined to possess a certain amount of '*internal strain energy*', which is at its minimum when the crystal lattice is free of dislocations. If we deform a crystal and induce dislocations, we increase this internal strain energy by local changes in the distance between atoms; the increase in internal energy is proportional to the increase in total length of dislocations per volume of crystalline material, also known as the *dislocation density*. Dislocations and dislocation tangles are formed in response to imposed differential stress (Figs. 3.13, 3.24). Other processes tend to shorten, rearrange or destroy the dislocations. Vacancies can migrate towards dislocation tangles and straighten the blocked sections, thus annihilating the tangles; bent dislocations can straighten, and dislocations can be arranged into networks. These processes can decrease the total dislocation length and hence the internal strain energy of crystals and will therefore operate following the thermodynamic principle to minimise total free energy in a system. During deformation, dislocation generation and annihilation mechanisms will compete while after deformation stops, dislocation annihilation mechanisms progress towards an equilibrium situation with the shortest possible length of dislocations

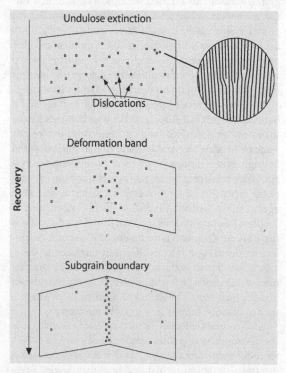

Fig. 3.21a–c. Schematic illustration of the recovery process. **a** Dislocations distributed over the crystal give rise to undulose extinction. **b** Recovery causes concentration of dislocations in deformation bands and eventually **c** in a subgrain boundary (tilt wall)

Fig. 3.22. Subgrains in quartz (*horizontal*), orthogonal to trails of fluid inclusions (*vertical*). Quartzite, Mt Isa, Australia. Width of view 1.8 mm. CPL

Fig. 3.23. Typical chessboard subgrains in quartz. Orthogneiss of the Nagssugtoquidian belt, W-Greenland. Width of view 0.7 mm. CPL. (Photograph courtesy H. Stünitz)

in the crystal lattice. The general term *recovery* is commonly used to cover these mechanisms of reducing dislocation density.

Dislocations in a crystal can be grouped into regular planar networks as a result of recovery (Figs. 3.13, 3.21; ⊘Video 3.21a). These networks are known as *subgrain walls* or *subgrain boundaries* (Fig. 3.13, 3.21). Such boundaries separate crystal fragments known as *subgrains*, which are slightly misoriented with respect to their neighbour subgrains or to the host grain (Fig. 3.22, 3.23). The orientation of a subgrain boundary depends on the orientation of the slip system of the dislocations that accumulate in it (Trepied et al. 1980).

A subgrain boundary can be imagined as a plane separating two crystal fragments that have rotated slightly with respect to each other; such boundaries can therefore be classified according to the orientation of the rotation axis. Subgrain boundaries with rotation axes parallel and normal to the boundary are known as *tiltwalls* and *twistwalls* respectively. A tiltwall is shown in Fig. 3.21 and consists of an array of edge dislocations with the same Burgers vector. A twistwall consists of two intersecting sets of screw dislocations with different Burgers

Fig. 3.24. TEM image of a small dislocation free amphibole grain in a ▶ plastically deformed old amphibole grain with high dislocation density (Cumbest et al. 1989). This microstructure is consistent with static recrystallisation. Dynamically recrystallised grains would be deformed and contain some free dislocations. Senja, Norway. Recrystallisation at 520–540 °C from amphibole plagioclase geothermometry. (Photograph courtesy Randy Cumbest and Martyn Drury)

vectors. *Complex walls* have an oblique rotation axis and consist of networks of dislocations having two or more different Burgers vectors.

Once formed, subgrain boundaries can migrate to some extent (Means and Ree 1988) or evolve into grain boundaries by addition of more dislocations (Sect. 3.7.3). Box 3.6 lists evidence for recovery in thin section.

3.7
Recrystallisation

3.7.1
Grain Boundary Mobility

Besides recovery another process, *grain boundary mobility*, can contribute to the reduction of dislocation density in deformed crystals (Poirier 1985; Gottstein and Mecking 1985; Drury and Urai 1990; Jessell 1987). Imagine two neighbouring deformed crystals, one with high and one with low dislocation density (Figs. 3.24, 3.26a). Atoms along the grain boundary in the crystal with high dislocation density can be displaced slightly so that they fit to the lattice of the crystal with low dislocation density. This results in local displacement of the grain boundary and growth of the less deformed crystal at the cost of its more deformed neighbour (Figs. 3.24, 3.26a, inset). The process may increase the length of grain boundaries and thereby increase the internal free energy of the crystal aggregate involved, but the decrease in internal free energy gained by removal of dislocations is greater. As a result, new small grains may replace old grains. This reorganisation of material with a change in grain size, shape and orientation within the same mineral is known as *recrystallisation* (Poirier and Guillopé 1979; Urai et al. 1986; Hirth and Tullis 1992). In solid solution minerals such as feldspars, recrystallisation may be associated with changes in composition, which may be an additional driving force for the process (Sect. 3.12.4). There are three different mechanisms of recrystallisation that can operate during deformation depending on temperature and/or flow stress. With increasing temperature and decreasing flow stress these are: bulging, subgrain rotation, and high temperature grain boundary migration recrystallisation (Figs. 3.25, 3.26; Urai et al. 1986; Wu and Groshong 1991a; Hirth and Tullis 1992; Dunlap et al. 1997; Stipp et al. 2002).

Box 3.6   Evidence for recovery

In response to recovery, dislocations tend to concentrate in planar zones in the crystal, decreasing dislocation density in other parts. In thin section, this results in the occurrence of zones in the crystal which have approximately uniform extinction, and which grade over a small distance into other similar crystal sectors with a slightly different orientation. These transition zones are known as *deformation bands* (Fig. 3.21). They can be regarded as a transitional stage between undulose extinction and subgrain boundaries (⊘Video 3.21b).

Subgrains (Figs. 3.13, 3.22, 4.26, ⊘Video 3.22) can be recognised as parts of a crystal which are separated from adjacent parts by discrete, sharp, low relief boundaries. The crystal lattice orientation changes slightly from one subgrain to the next, usually less than 5° (Fitz Gerald et al. 1983; White and Mawer 1988). Subgrains can be equant or elongate (Fig. 4.26). In many cases, subgrain walls laterally merge into deformation bands or high-angle grain boundaries (Fig. 3.28).

It is also important to note that recovery in bent crystals as described above is only one of the possible mechanisms to form subgrains; alternative, though possibly less common mechanisms are sideways migration of kink band boundaries, the reduction of misorientation of grain boundaries and impingement of migrating grain boundaries (Means and Ree 1988).

Fracturing, rotation and sealing by growth from solution may also play a role in the development of some subgrains in quartz (den Brok 1992). If crystals are separated into strongly undulose subgrains of slightly different orientation but with fuzzy boundaries, and if such crystals contain fractures, the 'subgrain' structure may be due to submicroscopic cataclasis of the grains (Tullis et al. 1990); such subgrain-like structures and even undulose extinction can form by dense networks of small fractures. Only TEM work can show the true nature of the structure in this case.

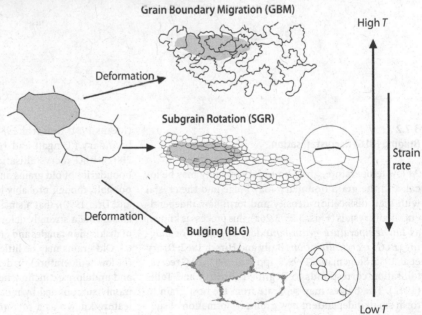

Grain Boundary Migration (GBM)

High *T*

Deformation

Subgrain Rotation (SGR)

Strain rate

Deformation

Fig. 3.25.
The three main types of dynamic recrystallisation in a polycrystal. The substance of one of two large grains that recrystallise is indicated by *shading*, before and during recrystallisation

Bulging (BLG)

Low *T*

Fig. 3.26.
Three mechanisms of dynamic recrystallisation on the grain scale. **a** Bulging (BLG) recrystallisation. If two neighbouring grains have different dislocation density, the grain boundary may start to bulge into the grain with the highest density (*inset; grey straight lines* in crystals indicate crystal lattice planes). On the scale of individual grains, the grain with higher dislocation density (*shaded*) is consumed by bulging of the less deformed grain; the bulge may eventually develop into an independent grain. **b** Subgrain rotation (SGR) recrystallisation. Rotation of a subgrain in response to migration of dislocations into subgrain walls during progressive deformation can cause development of high angle grain boundaries and thus of new grains. *Bars* in the subgrains indicate lattice orientation. **c** High-temperature grain boundary migration (GBM) recrystallisation. At high temperature, grain boundaries become highly mobile and may sweep the material in any direction to remove dislocations and subgrain boundaries. Subgrain rotation also occurs, but where subgrain boundaries (*s*) are transformed into grain boundaries, the latter become also highly mobile

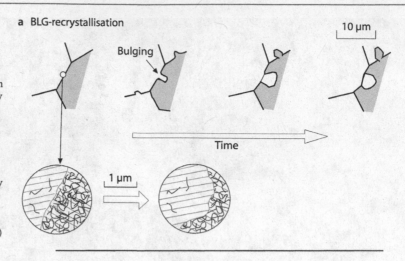

a BLG-recrystallisation

b SGR-recrystallisation

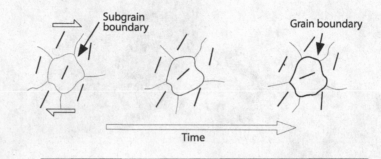

c GBM-recrystallisation

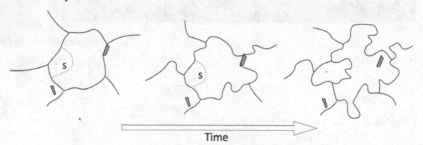

3.7.2
Bulging (BLG) Recrystallisation

At low temperature, grain boundary mobility may be local, and the grain boundary may bulge into the crystal with high dislocation density and form new, independent small crystals (Figs. 3.25, 3.26a); this process is known as *low-temperature grain boundary migration* or *bulging (BLG) recrystallisation* (Baily and Hirsch 1962; Drury et al. 1985; Shigematsu 1999; Stipp et al. 2002). BLG recrystallisation corresponds to Regime 1 of Hirth and Tullis (1992). The bulges may separate from the host grain to form small independent new grains by formation of subgrain boundaries, which evolve into grain boundaries

(Means 1981; Urai et al. 1986), or by migration of a grain boundary (Tungatt and Humphreys 1984; Stipp et al. 2002). BLG recrystallisation occurs mostly along the boundaries of old grains and at triple junctions. It is also possible, though probably less common in rocks (Drury and Urai 1990), that a small dislocation-free core nucleates inside a strongly deformed grain with high density of dislocation tangles and grows at the cost of the old crystal. Old grains may be little deformed or show fractures (at low temperature) or deformation lamellae (Box 3.3) and undulose extinction. Remains of old grains are commonly surrounded by moats of recrystallised grains, a feature known as a *core-and-mantle structure* (Gifkins 1976; White 1976; Shigematsu 1999; Fig. 5.20; Sect. 5.6.5).

3.7.3
Subgrain Rotation (SGR) Recrystallisation

A special recrystallisation process occurs when dislocations are continuously added to subgrain boundaries. This happens only if dislocations are relatively free to climb from one lattice plane to another. The process is known as *climb-accommodated dislocation creep*. In such cases, the angle between the crystal lattice on both sides of the subgrain boundary increases until gradually the subgrain can no longer be classified as part of the same grain (Fig. 3.26b; Box 3.6); a new grain has developed by progressive misorientation of subgrains or *subgrain rotation*. This process is known as *subgrain rotation-recrystallisation* (abbreviated SGR recrystallisation) and generally occurs at higher temperature than BLG recrystallisation. SGR recrystallisation corresponds to Regime 2 of Hirth and Tullis (1992). Old grains tend to be ductilely deformed and elongate or ribbon-shaped, with numerous subgrains. Core-and-mantle structures form at low temperature and low strain, but generally subgrains and new grains occur in "sheets" between old grain relicts, or old grains may be entirely replaced by subgrains and new grain networks. All gradations between subgrains and grains of the same shape and size occur (Nishikawa and Takeshita 2000; Nishikawa et al. 2004). Subgrains and grains are commonly slightly elongate. Characteristic is a gradual transition from subgrain- (low angle-) to grain- (high angle-) boundaries (Fig. 3.25).

3.7.4
High-Temperature Grain Boundary Migration (GBM) Recrystallisation

At relatively high temperature, grain boundary mobility increases to an extent that grain boundaries can sweep through entire crystals to remove dislocations and possibly subgrain boundaries in a process called high-temperature grain boundary migration (GBM) recrystallisation (Figs. 3.25, 3.26; ⊘Video 11.6a,b,e; Guillopé and Poirier 1979; Urai et al. 1986; Stipp et al. 2002). GBM recrystallisation corresponds to Regime 3 of Hirth and Tullis (1992). Subgrain formation and rotation is normally active during this process, but once grain boundaries are formed by this process after a certain amount of rotation of former subgrains (Lloyd and Freeman 1991, 1994), they can become highly mobile. Grain boundaries are lobate and grain size is variable. New grains tend to be larger than coexisting subgrains. It is difficult to distinguish new grains from relicts of old grains, except possibly by the distribution of fluid and solid inclusions. If secondary phases are present in an aggregate, pinning or grain boundary mobility structures (Fig. 3.34) are common. At very high temperature, grains have highly loboid or amoeboid boundaries, but may be nearly "strain free", i.e. devoid of undulose extinction and subgrains.

3.8
Solid-State Diffusion Creep, Granular Flow and Superplasticity

If the temperature in a deforming rock is relatively high with respect to the melting temperature of constituent minerals, crystals deform almost exclusively by migration of vacancies through the lattice. This process is known as *grain-scale diffusive mass transfer*. There are two basic types: *Coble creep* and *Nabarro-Herring creep*. The former operates by diffusion of vacancies in the crystal lattice along grain boundaries; the latter by diffusion of vacancies throughout the crystal lattice (Knipe 1989; Wheeler 1992).

Especially in fine-grained aggregates, crystals can slide past each other in a process known as *grain boundary sliding* while the development of voids between the crystals is prevented by solid state diffusive mass transfer, locally enhanced crystalplastic deformation, or solution and precipitation trough a grain boundary fluid. This deformation process is referred to as *granular flow* (Boullier and Gueguen 1975; Gueguen and Boullier 1975; Stünitz and Fitz Gerald 1993; Kruse and Stünitz 1999; Fliervoet and White 1995; Paterson 1995; Fliervoet et al. 1997). Since grain boundary sliding is rapid, it is the accommodation mechanism that normally determines the strain rate of granular flow (Mukherjee 1971; Padmanabhan and Davies 1980; Langdon 1995). In metallurgy, some fine-grained alloys can be deformed up to very high strain in tension without boudinage, a process known as *superplastic deformation* (Kaibyshev 1998; Zelin et al. 1994). The term *superplasticity* has also been used in geology (Schmid 1982; Poirier 1985; Rutter et al. 1994; Boullier and Gueguen 1998b; Hoshikuma 1996) and refers to very fine-grained aggregates (1–10 μm) of equidimensional grains, which deformed to very high strain without developing a strong shape- or lattice-preferred orientation. Grain boundary sliding is thought to play a major role in such deformation (Boullier and Gueguen 1975; Allison et al. 1979; Schmid 1982; van der Pluijm 1991; Rutter et al. 1994). Grain size seems to be the major parameter in determining whether an aggregate will deform by dislocation creep or by solid state diffusive mass transfer and grain boundary sliding (Schmid et al. 1977; Behrmann 1983). Small grain size favours grain boundary sliding since diffusion paths are relatively short. Presence of a second mineral phase can also enhance the process since it hampers grain growth (Kruse and Stünitz 1999; Newman et al. 1999; Krabbendam et al. 2003).

Many geologists use *diffusion creep* as a collective term for Coble- or Nabarro-Herring creep and superplasticity or granular flow, since rheological flow laws (Box 3.11) for these processes are very similar. Box 3.7 lists evidence for diffusion creep processes in thin section.

Box 3.7   Evidence for solid-state diffusion creep and grain boundary sliding

A process like solid-state diffusion creep is expected to leave few traces. Therefore, few microstructures have been proposed as evidence for diffusion creep. The process may give rise to strongly curved and lobate grain boundaries between two different minerals at high-grade metamorphic conditions (Fig. 3.33; Gower and Simpson 1992). Another possible effect is the erasure or modification of chemical zoning or fluid inclusion density and content in grains. Ozawa (1989) suggested that sector Al-Cr zoning observed in spinel grains in peridotite is formed by unequal diffusivity of these ions when spinel is deformed by solid-state diffusion creep in the Earth's mantle.

Grain boundary sliding has also been suggested as a deformation mechanism in rocks but is equally elusive. Solid-state diffusion creep combined with grain boundary sliding may prevent development or cause destruction of a lattice-preferred orientation. If a fine-grained mineral aggregate has undergone high strain but consist of equant grains and lacks a clear lattice-preferred orientation, or has a lattice-preferred orientation that cannot be explained by dislocation activity, this may be taken as indirect evidence for dominant grain boundary sliding as a deformation mechanism (White 1977, 1979; Boullier and Gueguen 1975; Allison et al. 1979; Padmanabhan and Davies 1980; Behrmann 1983, 1985; Behrmann and Mainprice 1987; Stünitz and Fitz Gerald 1993; Rutter et al. 1994; Fliervoet and White 1995; Fliervoet et al. 1997; Bestmann and Prior 2003). On the other hand, the presence of a preferred orientation cannot be used as proof against the action of grain boundary sliding (Rutter et al. 1994; Berger and Stünitz 1996).

Other, less reliable evidence for grain boundary sliding is:

- Linking up of grain boundaries along several grain widths (White 1977; Stünitz and Fitz Gerald 1993; Zelin et al. 1994).
- Diamond-shaped or rectangular grains formed by straight and parallel grain boundary segments, often in two directions throughout a sample (Lister and Dornsiepen 1982; Drury and Humphreys 1988; Fliervoet and White 1995; Hanmer 2000). This is also known as a *reticular grain aggregate*. Such boundaries are especially conspicuous in monomineralic aggregates of minerals such as quartz or calcite, for which this structure is unusual.
- The presence of diffuse contacts between strongly flattened fine-grained monomineralic aggregates of two different minerals. This may be a mixing effect of grain-boundary sliding (Tullis et al. 1990; Fliervoet et al. 1997; Hanmer 2000; Brodie 1998b).
- The presence of anticlustered distribution of mineral phases in a deformed fine-grained aggregate (Boullier and Gueguen 1975, 1998a; Rubie 1983, 1990; Behrmann and Mainprice 1987; Stünitz and Fitz Gerald 1993; Newman et al. 1999; Brodie 1998b, p 403). This may be due to selective nucleation of one phase in triple junction voids between grains of another phase, formed during grain boundary sliding (Kruse and Stünitz 1999).
- In the TEM, possible indications for grain-boundary sliding are a low dislocation density in grains; a lath shape of grains, and the presence of voids along grain boundaries (Fig. 3.36; Gifkins 1976; White and White 1981; Behrmann 1985; Behrmann and Mainprice 1987; Tullis et al. 1990; Fliervoet and White 1995).

3.9
Competing Processes During Deformation

At low temperature, minerals deform by brittle deformation but there are many indications that pressure solution and brittle processes occur together in low-grade deformation. Pressure solution is slow and may not be able to accommodate faster bulk strain rates, especially if diffusion paths increase in length when solution surfaces become more irregular with time, as in stylolites (Gratier et al. 1999). Fracturing can temporarily relieve stresses and increase possibilities for pressure solution. Such combined slow and fast processes may also cooperate in other combinations. Kinking and twinning commonly are associated with brittle fracturing as well.

At more elevated temperature, crystalplastic deformation is initiated but the conditions at which this happens not only depend on temperature, but also on strain rate and fluid pressure in the rock. During deformation of a crystalline material, continuous competition exists between processes that cause distortion of the crystal lattice and processes such as recovery and recrystallisation that reduce the dislocation density. Recrystallisation during active deformation such as the BLG, SGR and GBM recrystallisation discussed above are known as *dynamic recrystallisation* (Figs. 3.27–3.33; ⊘Video 11.6a,b,e). Box 3.8 lists evidence for dynamic recrystallisation in thin section.

Although the three mechanisms of dynamic recrystallisation are described separately, there are transitions and they can operate simultaneously under certain conditions, (Sect. 9.9; Fig. 9.10; Lloyd and Freeman 1994). In solid solution minerals such as feldspars, amphiboles and pyroxenes, however, BLG and GBM recrystallisation is not only driven by internal strain energy (Sect. 3.12), but also by chemical driving potentials associated with differences in composition of old and new grains (Hay and Evans 1987; Berger and Stünitz 1996; Stünitz 1998). This is specifically important where recrystallisation takes place at other (lower) metamorphic conditions as during the formation of the older, recrystallising minerals. Any link between temperature and changes in recrystallisation mechanism as outlined above depends on these differences in composition.

There are two main types of deformation based on dislocation creep, depending on the accommodating process (Sellars 1978; Zeuch 1982; Tullis and Yund 1985); climb-accommodated dislocation creep (Yund and Tullis 1991) associated with SGR recrystallisation (Guillopé and Poirier 1979), and recrystallisation-accommodated dislocation creep where grain boundary migration is the accommodating mechanism (Tullis and Yund 1985; Tullis et al. 1990). There are indications that, with increasing temperature, the accommodating mechanism in quartz is first BLG recrystallisation when dislocation climb and recovery is difficult, which then switches to SGR recrystallisation at the onset

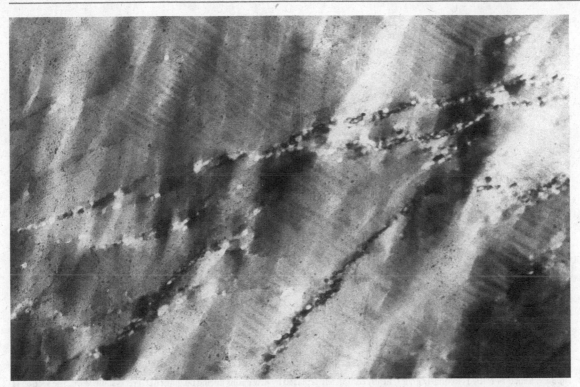

Fig. 3.27. Quartz grain with deformation lamellae (*top left* to *lower right*) and subgrain boundaries (*subvertical*) transected by bands of new grains (*lower left* to *top right*) formed by bulging (BLG)-recrystallisation. Quartz vein in micaschist. Southern Alps, Italy. Width of view 1.8 mm. CPL

Fig. 3.28. Typical fabric of dynamic recrystallisation in quartz. Relicts of large old quartz grains with undulose extinction and elongate subgrains pass laterally into domains of small, new grains formed by bulging (BLG) recrystallisation. St. Barthélemy, Pyrenees, France. Width of view 1.8 mm. CPL

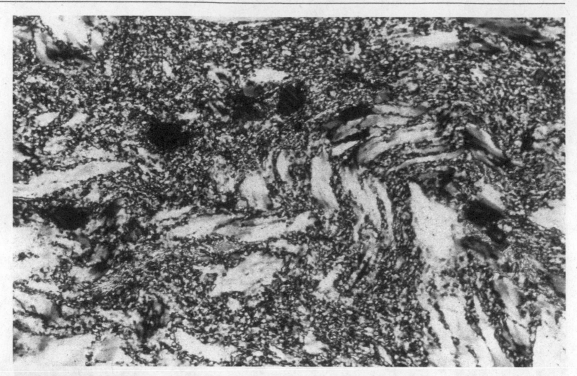

Fig. 3.29. Relicts of folded old quartz grains, nearly completely replaced by new grains during bulging (BLG) recrystallisation. The section is taken normal to the aggregate lineation in a quartzite mylonite with a dominantly linear shape fabric. Aston Massif, Pyrenees, France. Width of view 1.4 mm. CPL

Fig. 3.30. Polycrystalline quartz aggregate, probably developed predominantly by subgrain rotation (SGR) recrystallisation. Transitions exist between grains surrounded by high angle boundaries and subgrains. A relict of a deformed old quartz grain occurs at *upper left*. St. Barthélemy, Pyrenees, France. Width of view 1.8 mm. CPL

Fig. 3.31. Typical fabric of dynamically recrystallised quartz formed by subgrain rotation (SGR) recrystallisation. Grains have a weak shape-preferred orientation that defines a continuous foliation. Granite mylonite. Qin Ling mountains, China. Width of view 1.8 mm. CPL

Fig. 3.32. Polycrystalline quartz with irregular grain boundaries formed in response to grain boundary migration (GBM) recrystallisation. The *light-grey central grain* is bulging into the *dark grain* at *lower right*. Quartzite, Yilgarn Craton, Australia. Width of view 1.8 mm. CPL

Fig. 3.33. Gneiss with lobate grain boundaries, especially between quartz and feldspar. The rock has been deformed at high-grade metamorphic conditions and GBM-recrystallisation affected quartz and feldspar. St. Barthélemy Massif, Pyrenees, France. Width of view 4 mm. CPL

Fig. 3.34.
Four microstructures, which indicate movement direction of a migrating grain boundary during GBM recrystallisation (after Jessell 1987). *Solid arrows* indicate the movement direction of the grain boundary by growth of the light grain into the shaded grain. Further explanation in text

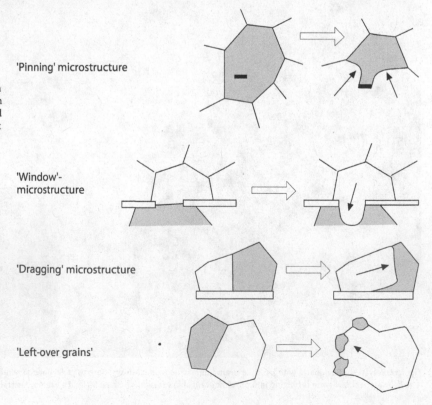

'Pinning' microstructure

'Window'-microstructure

'Dragging' microstructure

'Left-over grains'

Fig. 3.35.
Pinning structure of quartz
grain boundaries on white mica.
Amphibolite facies micaceous
quartzite. Southern Minas Gerais,
Brazil. Width of view 2.5 mm. CPL

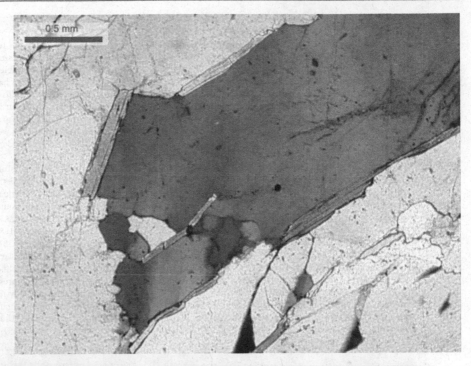

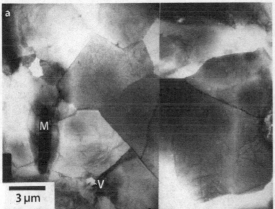
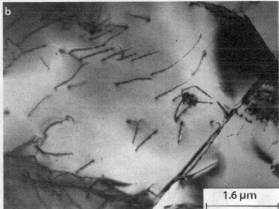

Fig. 3.36. a Low magnification TEM image of equidimensional grains in quartz ultramylonite. Note occurrence of voids (*marked V*) at triple points and grain boundary micas (*marked M*); **b** TEM image of free dislocations, visible as *dark lines*, inside small quartz grains from an ultramylonite. The homogeneous distribution of recovered dislocations indicates that dislocation creep was a significant deformation mechanism, but dominant grain boundary sliding is suggested by a random LPO in the quartz grains, measured by electron diffraction, and the voids at the grain boundaries. Quartz ultramylonite, Portugal. (Photographs courtesy Martyn Drury)

of recovery, and finally to combined GBM and SGR recrystallisation because of increasing ease of diffusion in the crystal lattice (Hirth and Tullis 1992). The same seems to apply to other minerals (Lafrance et al. 1996) but switches in accommodation mechanism will occur at other temperatures.

At high temperatures, diffusion processes may accompany or take over from dislocation climb and recrystallisation (Sect. 3.8).

The state of affairs during any stage of deformation and the final result that we observe in deformed rocks depend on the relative importance of the processes listed above and,

indirectly, on deformation parameters such as strain rate and temperature. In general, a high temperature and the presence of a fluid on grain boundaries promotes recovery and recrystallisation processes; high strain rate enhances crystal distortion. These facts have been known from the earliest age of metalworking; a sword or horseshoe can be shaped from a piece of metal by hammering if it is sufficiently heated. In thin section, only structures related to the last stages of the competing processes are normally preserved, formed shortly before temperature and/or strain rate fell below a critical value and the structures were 'frozen in'.

Box 3.8   Evidence for dynamic recrystallisation

Evidence for dynamic recrystallisation is usually more difficult to find than evidence for deformation or recovery. Two types of characteristic microstructures can be distinguished: partially and completely recrystallised fabrics.

In partially recrystallised fabrics a bimodal grain size distribution is characteristic, with aggregates of small new grains of approximately uniform size between large old grains with undulose extinction (Figs. 3.27–3.29, 3.37, ⊘Video 3.28a,b, ⊘Photo 3.23). The uniform size of new grains is due to deformation and recrystallisation at a specific differential stress (Sect. 9.6.2). The three mechanisms of dynamic recrystallisation can be distinguished as follows.

In the case of bulging (BLG) recrystallisation old grains can have patchy undulose extinction, kinks, deformation lamellae and evidence for brittle fracturing while grain boundaries are irregular and loboid with lobes on the scale of the new grains. New grains form at the expense of old grains along grain boundaries, and therefore form aggregates of small equally sized grains between the old grains (Figs. 3.27–3.29).

In the case of subgrain rotation (SGR) recrystallisation the transition from old to new grain is less abrupt. Old grains are flattened, show sweeping undulose extinction and contain subgrains the size of new grains, and gradual transitions in orientation from subgrains to new grains occur (Fig. 3.30, 3.31). Sub-grain boundaries can be seen to change laterally into grain boundaries. In the TEM, BLG recrystallisation is characterised by grains with a strongly variable dislocation density, while for SGR recrystallisation all grains have approximately similar dislocation density (Fig. 3.26; Tullis et al. 1990). A special lattice-preferred orientation may occur in recrystallised aggregates in the form of *orientation families* of grains, which may derive from large single parent grains that were completely substituted by SGR recrystallisation (see also domain shape preferred orientation, Box 4.2).

In the case of high-temperature grain boundary migration (GBM) recrystallisation, the distinction between old and new grains is difficult. Characteristic are large new grains with interlobate to amoeboid grain boundaries, internally subdivided in smaller subgrains (Figs. 3.32, 3.33). In quartz, chessboard-type subgrains are typical. Jessell (1987) proposed microstructures that can be used to recognise GBM and to establish the migration direction of a grain boundary. Grains of a second mineral such as micas can pin a grain boundary and cause 'pinning', 'window' or 'dragging' microstructures (Fig. 3.34, ⊘Video 3.34a,b,c). If a grain is almost completely replaced by a neighbour, 'left-over' grains with identical orientation may indicate the presence of an originally larger grain (Urai 1983; Jessell 1986; Fig. 3.34, ⊘Video 3.34d).

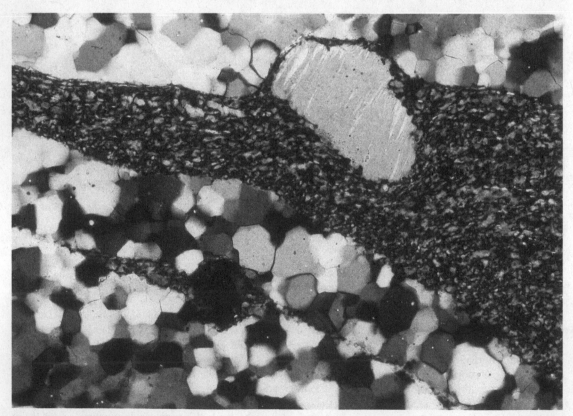

Fig. 3.37. Layer of fine-grained K-feldspar in quartz, both dynamically recrystallised. A perthitic fragment of a K-feldspar porphyroclast with flame-shaped albite lamellae is present in the recrystallised feldspar layer. Notice the difference in grain size of recrystallised quartz (coarse) and feldspar (fine). Granite mylonite. Qin Ling Mountains, China. Width of view 0.8 mm. CPL

Box 3.8   Continued

In practice, characteristic features of different recrystallisation mechanisms can be found together in one sample, since temperatures may change during deformation.

An aggregate of small, dynamically recrystallised grains around a crystal core with the same mineral composition is known as a *core-and-mantle structure*, provided that evidence (as mentioned above) exists that the structure developed by dynamic recrystallisation of the core mineral along its rim (Sect. 3.13; White 1976; Figs. 3.29, 5.22–5.25). If the mantle is extremely fine-grained and the mechanism by which it formed is uncertain, the term *mortar structure* has been used instead (Spry 1969). However, this term has a genetic implication as "mechanically crushed rock" and its use is therefore not recommended.

A completely recrystallised fabric may be difficult to distinguish from a non-recrystallised equigranular fabric. However, in an aggregate of grains formed by complete dynamic recrystallisation, the grains will show evidence of internal deformation, a lattice-preferred orientation (Sect. 4.4.2) and a relatively uniform grain size (Fig. 3.31).

Most arguments given above are based on optical microscopy. SEM observations also promise to become important to distinguish the effect of different mechanisms by precise and quantitative characterisation of grain and subgrain size and the frequency and distribution of boundary misorientation (Trimby et al. 1998).

3.10
Grain Boundary Area Reduction (GBAR)

Lattice defects are not the only structures that contribute towards the internal free energy of a volume of rock; grain boundaries can be considered as planar defects with considerable associated internal free energy. A decrease in the total surface area of grain boundaries in a rock can reduce this internal free energy (Vernon 1976; Poirier 1985; Humphreys and Hatherley 1995; Humphreys 1997; Kruhl 2001; Evans et al. 2001). Straight grain boundaries and large grains are therefore favoured and any polycrystalline material will strive towards a fabric with large, polygonal grains with straight boundaries to reduce the internal free energy (Figs. 3.38, 11.6b,c; ⊘Videos 3.38, 11.6a,b). We call this process of grain boundary migration resulting in grain growth and straightening of grain boundaries *grain boundary area reduction* (abbreviated GBAR). The reduction in internal free energy gained by GBAR is generally much less than that gained by GBM or SGR recrystallisation. Therefore, although GBAR occurs during deformation its effect is more obvious and may become dominant after deformation ceased, especially at high temperature (Sect. 3.11; Bons and Urai 1992).

The free energy represented by a grain boundary may depend on the orientation of the boundary with respect to the crystal lattice (Vernon 1976; Kruhl 2001). If the dependence of grain boundary energy on the crystal lattice is weak for a certain mineral, GBAR in a monomineralic

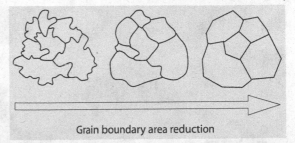

Grain boundary area reduction

Fig. 3.38. Illustration of the process of grain boundary area reduction (GBAR) through grain boundary adjustment and grain growth, resulting in a decrease in grain boundary energy. Irregular grain boundaries formed during deformation and dynamic recrystallisation are straightened to a polygonal shape, and some small grains are eliminated

rock will lead to the approach of an 'equilibrium-fabric' of polygonal crystals with contacts tending to make triple junctions with *interfacial angles* of approximately 120° in three dimensions (Figs. 3.39, 3.40a; ⊘Video 11.10b–d). Obviously, this angle can be smaller in oblique cross-sections (Fig. 12.2). Since similar structures form in foam, e.g. in a beer bottle, the fabric is often referred to as a *foam-structure* (Fig. 3.39; ⊘Video 11.10b–d). Large grains with many sides tend to increase in size while small grains with few sides shrink and eventually disappear during GBAR (⊘Video 11.10b–e).

Many aggregates where GBAR has been active show slightly curved grain boundaries. Small grains may have strongly outward curving boundaries (Fig. 3.39, ⊘Video 11.10b–e). On close inspection these may consist of many differently oriented straight segments parallel to crystallographic planes (Kruhl 2001; Kruhl and Peternell 2002). This curvature may be due to migration of the grain boundary in the direction of the centre of curvature during GBAR (Vernon 1976; Shelley 1993). However, care must be taken when applying this principle to deformed rocks since in SGR and GBM recrystallisation, new grains can have curved boundaries that migrate *away* from the centre of curvature (Figs. 3.26, 3.34).

If there is a correlation between grain boundary energy and the orientation of the crystal lattice, minerals are said to be anisotropic with respect to grain boundary energy (Vernon 1976). Minerals like quartz, olivine, feldspars, cordierite, garnet, carbonates, anhydrite and sulphides are weakly anisotropic; the effect is hardly visible in thin section but interfacial angles between grain boundaries in an equilibrium fabric commonly deviate from 120° (Fig. 3.40a; ⊘Video 11.10e). Minerals like hornblende and pyroxene are moderately anisotropic and many grain boundaries are parallel to {110} planes (Fig. 3.40b). Micas, sillimanite and tourmaline are strongly anisotropic and show a strong dominance of certain crystallographic planes as grain boundaries (Figs. 3.40c, 4.28c); in micas, (001) is dominant.

3.10

Fig. 3.39. Polygonal fabric of scapolite grains formed by static recrystallisation. Mt. Isa, Australia. Width of view 4 mm. CPL

Fig. 3.40. Effects of the anisotropy of minerals for grain boundary energy on grain boundary orientation. **a** If grain boundaries all have similar internal free energy, grains will be equidimensional and boundaries are not preferentially associated with specific crystallographic planes (indicated by *lines* in the crystal). **b** In the case of hornblende, some grain boundaries ({110} planes) have relatively lower internal free energy, and may be dominant in the aggregate. **c** In the case of micas, grain boundaries parallel to (001) are favoured over all others and idiomorphic grains are commonly abundant. **d** Typical shape of a quartz-mica aggregate where the low-energy (001) planes of micas dominate. **e** Illustration of the dihedral angle between minerals *A* and *B*, where the boundary between like minerals has a lower (*top*) or higher (*bottom*) energy than that between unlike minerals

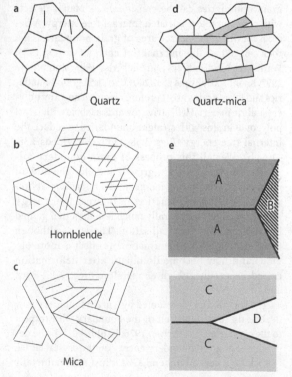

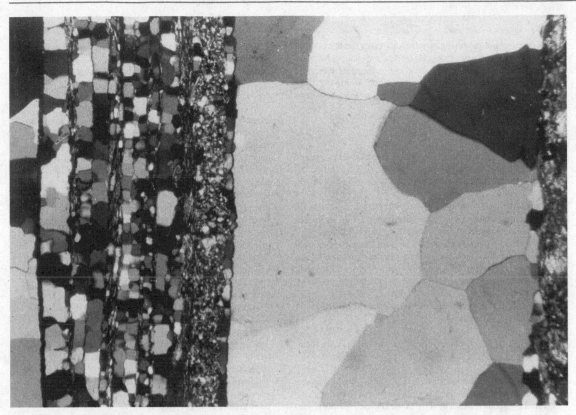

Fig. 3.41. Statically recrystallised quartz in a fabric with alternating quartz and feldspar layers. Feldspar is recrystallised and very fine-grained (e.g. in layers *left* of the *centre* and at *right*). Grain size of quartz depends on the width of the quartz layer; in thin layers, quartz grains are limited in their growth, leading to a clear dependence of statically recrystallised grain size on layer width. It is possible that quartz grain size was similar in all layers at the end of the deformation that formed the layering and before static recrystallisation started. Deformed quartz vein. Yilgarn Craton, Australia. Width of view 4 mm. CPL

Besides the anisotropy of individual minerals which influences interfacial angles in monomineralic aggregates, the nature of different minerals in contact is also of importance. In polymineralic aggregates where weakly and strongly anisotropic minerals are in contact, the grain boundaries tend to be defined by the more strongly anisotropic phase. For example, mica or tourmaline grains included in quartz can be idiomorphic (Figs. 3.35, 3.40d, 4.9). The anisotropy of minerals is also evident in the shape of included grains in rocks that underwent GBAR; grains of sillimanite in quartz, for example, usually show a strong predominance of favoured crystallographic directions for their boundaries. Notice, however, that this does not apply for inclusions in low to medium-grade rocks where inclusion boundaries have been relatively immobile after the growth of the host grain (Sect. 7.3).

In aggregates with phases of low anisotropy, another effect may be visible. The boundaries between grains of the same mineral can have another (commonly higher) grain boundary energy than those between grains of different minerals (Vernon 1976). In general, there is a tendency for high energy boundaries to decrease, and for low energy boundaries to increase in length. Consequently, the interfacial angle between the boundaries separating unlike minerals (also known as the *dihedral angle*; Hunter 1987; Fig. 3.40e) deviates from 120°.

The process of grain growth tends to lower the internal free energy of a grain aggregate even after a foam structure has been established, although grain growth becomes slower with increasing grain size (Olgaard and Evans 1988; Kruhl 2001). The grain size that is finally reached after GBAR depends on temperature, but also on the presence of other solid or liquid phases in grains and grain boundaries, variation in mineral chemistry and crystallographic preferred orientation (Evans et al. 2001). Of these factors, the possibility of grains to grow without obstruction by grains of other minerals seems most important (Masuda et al. 1991; Evans et al. 2001); consequently GBAR in layered rocks results in relatively coarse grains in wide monomineralic layers, and small grains in thin or polymineralic layers (Fig. 3.41, ⊘Video 3.41).

Box 3.9   Fabric nomenclature

An extensive and confusing terminology exists for the description of the geometry of grains and fabrics in metamorphic rocks (see also Box 1.1). Below, we give some of the most important terms, their meaning and their mutual relation (Moore 1970; Best 1982; Shelley 1993). The suffix 'blastic' refers to solid-state crystallisation during metamorphism.

Shape of individual grains

The following terms describe the shape of individual grains and can be used as prefix for 'grain' or 'crystal', e.g. euhedral crystal shape, anhedral grains (Fig. B.3.1):

- **euhedral** – with fully developed crystal faces. Less commonly, the term **idiomorphic** is used, mainly in igneous rocks. The term **automorphic** has similar meaning but is little used.
- **subhedral** – with irregular crystal form but with some well developed crystal faces. Less commonly, the term **hypidiomorphic** is used, mainly in igneous rocks (Fig. 3.9). The term **hypautomorphic** has an equivalent meaning but is little used.
- **anhedral** – without crystal faces. Less commonly, the terms **allotriomorphic, xenomorphic** and **xenoblastic** are used.
- **acicular** – needle-shaped.

Three terms are commonly used for large grains with inclusions:

- **poikiloblastic** – with numerous, randomly oriented inclusions of other minerals. The term **poikilitic** refers to a similar structure in igneous rocks. The term is mainly used for porphyroblasts.
- **skeletal** – refers to a spongy shape of a grain that occurs in thin seams between grains of other minerals that are nearly in contact (Fig. 7.6).

Shape of grain aggregates

The following terms can be used as a prefix for fabric, e.g. polygonal fabric, decussate fabric (Fig. B.3.1):

Grain boundary geometry

- **polygonal** – with straight grain boundaries and consisting of anhedral or subhedral grains (e.g. Fig. 3.39).
- **interlobate** – with irregular, lobate grain boundaries (e.g. Figs. 3.30, 4.9).
- **amoeboid** – with strongly curved and lobate, interlocking grain boundaries; like an amoeba.

Size distribution of grains

- **equigranular** – all grains with roughly equal size.
- **inequigranular** – non-gradational distribution of different grain size; an example is a bimodal distribution, with large grains of approximately equal size in a fine-grained equigranular matrix.
- **seriate** – a complete gradation of fine- to coarse-grained.

Special terms for the shape of grain aggregates

- **granoblastic** (less common crystalloblastic) – a mosaic of approximately equidimensional subhedral or anhedral grains. Inequant grains, if present, are randomly oriented (Fig. 3.37). The term **equigranular** has a similar meaning but is not restricted to metamorphic rocks. Many granoblastic fabrics exhibit a foam-structure (see main text; Fig. 3.39).
- **lepidoblastic** – a predominance of tabular mineral grains with strong planar preferred dimensional orientation (Fig. 4.8). This term is now generally substituted by a description of the foliation (Fig. 4.7; compare the first and second editions of Williams et al. 1954, 1982).
- **decussate** – an arrangement of randomly oriented elongate grains (such as mica) in a metamorphic rock.
- **reticular** – arranged in lozenges with two common directions, as in a fishing net
- **granolepidoblastic** – a combination of granoblastic and lepidoblastic fabric in the same rock. The term has become obsolete.
- **nematoblastic** – a predominance of acicular or elongate grains displaying a linear preferred dimensional orientation. This term has become obsolete as well, substituted by a description of the mineral lineation.
- **porphyroblastic** – inequigranular fabric, with large grains that grew during metamorphism and which are embedded in a finer-grained matrix (Chap. 7; Fig. 7.5).
- **mylonitic** – see Chap. 5 for a detailed description of mylonitic fabrics.
- **flaser** – a type of mylonitic fabric in which elliptical porphyroclasts lie in a finer mylonitic matrix. Since most mylonitic rocks exhibit this kind of fabric, the term is not particularly informative and is therefore not recommended for metamorphic rocks. (In sedimentary rocks the term flaser structure refers to the presence of small lenses of pelite in sandstone, indicative of a particular sedimentary environment).
- **clustered** or **anticlustered** distribution of grains of a certain phase in a polymineralic aggregate refer to the tendency of grains of one phase to group together (clustered) or to be spread out with minimum number of grains of that phase touching each other (anticlustered) – fields on a chess-board are perfectly anticlustered. Notice that anticustered is not the same as random (Kretz 1969; Kroustrup et al. 1988; Kruse and Stünitz 1999)

Grain shape can be quantified using the *PARIS factor* (Panozzo and Hürlimann 1983). This factor quantifies the irregularity of the grain boundary and is defined as the ratio of the actual length of a grain boundary divided by the length of the outline of the grain projection (imagined as a rubber band tied around the grain). A PARIS factor of 1 is a smooth round grain; values are progressively higher for interlobate and amoeboid grains.

This process is also known as *Zener pinning* (Nes et al. 1985; Evans et al. 2001). Especially the presence of small graphite grains in a rock may hamper the growth of other minerals (Krabbendam et al. 2003). This is the reason why many graphitic schists are fine-grained, even at high metamorphic grade. Similarly, in micaceous quartzites, pure quartzite layers are usually much coarser than quartz-mica layers (Fig. 3.41, ⊘Video 3.41).

A process similar to GBAR is *Ostwald ripening* or *liquid-assisted static recrystallisation* (Lifshitz and Slyozov

Box 3.9   *Continued*

Shape of grains

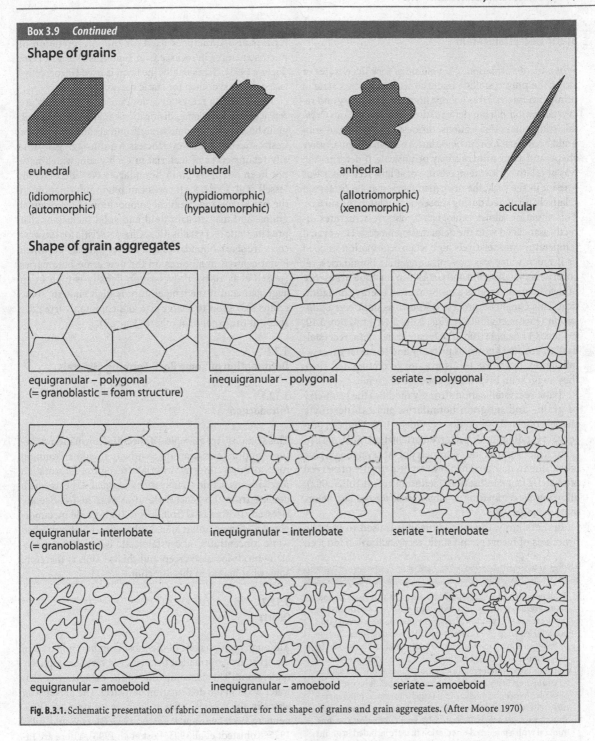

euhedral

(idiomorphic)
(automorphic)

subhedral

(hypidiomorphic)
(hypautomorphic)

anhedral

(allotriomorphic)
(xenomorphic)

acicular

Shape of grain aggregates

equigranular – polygonal
(= granoblastic = foam structure)

inequigranular – polygonal

seriate – polygonal

equigranular – interlobate
(= granoblastic)

inequigranular – interlobate

seriate – interlobate

equigranular – amoeboid

inequigranular – amoeboid

seriate – amoeboid

Fig. B.3.1. Schematic presentation of fabric nomenclature for the shape of grains and grain aggregates. (After Moore 1970)

1961; Evans et al. 2001). If grains of different size are surrounded by a fluid, small grains with a highly curved surface tend to dissolve while large grains with a smaller curvature of the surface grow (⊘Videos 11.6c–e, 11.10f). In this way, smaller grains tend to disappear while larger ones increase. The theory of Ostwald ripening is mostly applied in igneous petrology for development of phenocrysts (Park and Hanson 1999), but also for growth of porphyroblasts in metamorphic rocks (Miyazaki 1991; Carlson 1989).

3.11
Static Recrystallisation

When the deformation of a volume of rock decelerates or stops, the polycrystalline material will not be in a state of minimum internal free energy, not even if recovery and recrystallisation during deformation were important. Crystals still contain dislocations, dislocation tangles and subgrain boundaries. Grain boundaries have an irregular, wavy shape, and some minerals may be unstable. If deformation was at relatively low temperature or if little free water was present in the rock, the deformed fabric may be preserved relatively unaltered during subsequent uplift to the surface. This situation allows geologists to observe structures directly associated with the deformation process. However, if temperature was relatively high when deformation stopped or if much water was present along grain boundaries, recovery, recrystallisation and GBAR can continue in absence of deformation towards a lower internal energy configuration. This combined process is known as *static recrystallisation* (Evans et al. 2001; Figs. 3.38, 3.41, 11.6, Box 3.10, ⊘Videos 11.6a,b, 11.10b–d). Dynamic and static recrystallisation are also known as primary and secondary recrystallisation, but these terms are not recommended since they suggest an invariable sequence of events.

Static recrystallisation strongly modifies the geometry of grain- and subgrain boundaries and can destroy a shape-preferred orientation (⊘Video 11.10c) but can preserve crystallographic preferred orientation (Sect. 4.4). In quartz, for example, the asymmetry of the crystallographic fabric due to non-coaxial flow can be perfectly preserved after static recrystallisation (Heilbronner and Tullis 2002), allowing determination of shear sense for the last deformation stage.

In metallurgy, the term *annealing* is used to indicate processes of recovery and static recrystallisation induced

by passive heating of a previously deformed material. The term is also sometimes used for the interpretation of microstructures in rocks, e.g. in xenoliths (Vernon 1976; Shelley 1993). Occasionally, the term is used (incorrectly) as a general synonym for static recrystallisation.

During static recrystallisation, unstable minerals are replaced by stable ones, dislocation tangles are removed, grain boundaries become straight and grains tend to grow in size due to GBAR. Nevertheless, a grain aggregate usually retains cores of material in each grain, which have not been swept by grain boundaries (⊘Video 11.10d; Jessell et al. 2003). Such cores can retain information on the size, shape and chemical composition of the original grains, and may contain fluid and solid inclusions that predate static recrystallisation. Characteristic for *unswept cores* are sharp boundaries and an irregular shape, which is not centred in all cases on the new grain boundaries (Jessell et al. 2003). If dislocation density is high in an aggregate and if the temperature is high enough, some grains may grow to a large size and commonly irregular shape at the expense of others (Fig. 4.9).

3.12
Deformation of Some Rock-Forming Minerals

3.12.1
Introduction

This section gives examples of specific deformation structures and deformation mechanisms in some common rock-forming minerals. Criteria to recognise deformation mechanisms in thin section are mentioned. Aspects which deviate from the general trend as sketched above are stressed. Treatment is from low to high-grade metamorphic conditions unless stated otherwise. Most published work concentrates on crystalplastic deformation, especially on dislocation creep and this section is therefore somewhat biased in this direction.

3.12.2
Quartz

Although quartz is one of the most common minerals in the crust, its deformation behaviour is very incompletely understood. This is mainly due to the complex role that water plays in the deformation of quartz. The presence of water in the crystal lattice influences its strength (Kronenberg 1994; Luan and Paterson 1992; Gleason and Tullis 1995; Kohlstedt et al. 1995; Post et al. 1996). There are indications that with increasing water pressure in the pore space dislocation creep strength of quartz decreases, probably through an increase in water fugacity in the quartz grains (Luan and Paterson 1992; Post et al. 1996).

At very low-grade conditions (below 300 °C) brittle fracturing, pressure solution and solution transfer of

---

Box 3.10   Evidence for static recrystallisation

Evidence for static recrystallisation and its principal mechanism, grain boundary area reduction (GBAR), is provided by the presence of crystals with straight or smoothly curved grain boundaries (Figs. 3.39, 11.6) which lack undulose extinction or subgrains in a rock that was strongly deformed as shown by the presence of folds in layering, relict augen or the presence of a strong lattice-preferred orientation. Such grains are said to be *strain-free*. In a statically recrystallised fabric it is commonly possible to recognise relics of a largely destroyed older structure; relics of a foliation or porphyroclasts may be preserved. Static grain growth is indicated by small grains of a second mineral with a preferred orientation that are included in grains of the main mineral (Fig. 4.9), and by elongate strain-free crystals that define a foliation; these may have grown in a rock with an older foliation where they were hampered in their growth by grains of a second mineral (Figs. 3.41, 5.11, 5.12). Static recrystallisation may be followed once more by deformation inducing undulose extinction and dynamic recrystallisation, starting a new cycle.

material are dominant deformation mechanisms (Dunlap et al. 1997; van Daalen et al. 1999; Stipp et al. 2002). Characteristic structures are fractures in grains, undulose extinction, kink bands (Nishikawa and Takeshita 1999) and evidence for pressure solution and redeposition of material, sometimes in veins. Healed fractures are common, usually aligned with fluid and solid inclusions. BLG recrystallisation may locally occur at very low-grade conditions in strongly deformed quartz (Wu and Groshong 1991a).

At low-grade conditions (300–400 °C) dislocation glide and creep become important, mainly on basal glide planes in the (c)<a> direction. Characteristic structures are patchy and, at higher temperature, 'sweeping' undulose extinction (Fig. 3.17) and deformation lamellae (Fig. 3.18) occur. A dominant dynamic recrystallisation mechanism under these conditions is BLG recrystallisation (Stipp et al. 2002). Dauphiné deformation twinning is possible in quartz at low-grade conditions but also at higher temperature (Tullis 1970; Barber and Wenk 1991; Lloyd et al. 1992; Heidelbach et al. 2000; Lloyd 2000).

At medium temperatures (400–500 °C), dislocation creep is dominant, and prism {m}<a> slip becomes important. Characteristic are relatively strongly flattened old crystals and abundant recovery and recrystallisation structures (Fig. 3.41). Pressure solution may still play a role under these conditions (den Brok 1992). The dominant recrystallisation mechanism here is SGR recrystallisation (Lloyd and Freeman 1994; Stipp et al. 2002). Old grains may be completely replaced by recrystallised material. (Hirth and Tullis 1992; Stipp et al. 2002). Oblique foliations (Box 4.2) probably develop mainly in the combined SGR and GBM recrystallisation regime.

At 500–700 °C, recrystallisation is mostly by GBM recrystallisation, grain boundaries are lobate, and pinning- or migration microstructures are common (Jessell 1987; Stipp et al. 2002) at lower temperature ranges. Above 700 °C, prism-slip {m}<c> becomes important (Blumenfeld et al. 1986; Mainprice et al. 1986) and rapid recrystallisation and recovery cause most grains to have a strain-free appearance. Grain boundaries are lobate or amoeboid in shape (Fig. 4.9). A special type of approximately square subgrain structure occurs at these high grade conditions, known as *chessboard extinction* or *chessboard subgrains* (Fig. 3.23) which may be due to combined basal <a> and prism <c> slip (Blumenfeld et al. 1986; Mainprice et al. 1986; Stipp et al. 2002) or the α–β transition in quartz (Kruhl 1996). Under these metamorphic conditions strain-free monomineralic quartz ribbons can form (Box 4.2; Figs. 5.11, 5.12).

Temperature is an important, but not unique factor determining quartz deformation behaviour; this also depends strongly on strain rate, differential stress and the presence of water in the lattice and along grain boundaries. With increasing differential stress, more slip systems may become active since the critical resolved shear stress

of other slip systems is reduced. For example, at low temperature, with increasing differential stress the system (c)<a> is followed by {m}<a> and finally {r}<a>. At high temperature, the sequence is (m)<c>, {m}<a>, (c)<a> and {r}<a> (Hobbs 1985).

3.12.3
Calcite and Dolomite

At very low-grade conditions calcite deforms by fracturing and cataclastic flow (Kennedy and Logan 1998). The coarser grained fragments are heavily twinned and show undulose extinction, and are cut by veins and stylolites while small matrix grains can be strain- and twin free. Brittle deformation is apparently assisted by solution transfer, twinning and, especially in the fine-grained matrix, dislocation glide and BLG recrystallisation (Wojtal and Mitra 1986; Kennedy and Logan 1998).

At low-grade conditions and if water is present, pressure solution is dominant in calcite and leads to *stylolite* development (Box 4.3) although other mechanisms may also contribute (Burkhard 1990; Kennedy and Logan 1997, 1998). Calcite is special in that deformation twinning becomes important from diagenetic conditions onwards (Schmid et al. 1981; Sects. 9.6.2, 9.9). Twinning occurs along three {e}-planes inclined to the c-axis and is initiated at very low critical resolved shear stress (between 2 and 12 MPa, depending on temperature and mean stress; Turner et al. 1954; Wenk et al. 1986a; Burkhard 1993). However, the amount of strain that can be achieved by twinning is limited and must be accommodated at grain boundaries by pressure solution, grain boundary migration or grain boundary sliding. Evidence for the activity of these accommodating mechanisms in thin section are partly dissolved twins at grain boundaries, or twins that end before the grain boundary is reached, left behind by the migrating boundary. Twins can be used as indicators of temperature, strain and stress (Sects. 9.2, 9.5.1 and 9.6.3).

At low- to medium-grade metamorphic conditions, dislocation glide on r- and f-planes becomes important besides deformation twinning: {f}<r∩f> (six systems) at low temperature and {f}<a∩f> (three systems) at higher temperature (Takeshita et al. 1987; de Bresser and Spiers 1997). In addition, c<a> slip may become important at high temperature (Schmid et al. 1987; de Bresser and Spiers 1993, 1997; Barnhoorn et al. 2004). BLG recrystallisation is active under low-grade conditions and increases in importance with increasing temperature. SGR recrystallisation is active under a range of conditions (de Bresser et al. 2002; Ulrich et al. 2002; Bestmann and Prior 2003). Grain boundary sliding and 'superplastic' behaviour may be important in calcite if the grain size is very small (Schmid 1982; Schmid et al. 1987; Walker et al. 1990; Casey et al. 1998; Brodie and Rutter 2000; Bestmann and Prior 2003).

Dolomite behaves differently from calcite (Barber and Wenk 2001). It deforms by basal <a> slip at low to moderate temperatures and deformation twinning on f-planes at moderate to high temperatures. Twinning apparently does not develop below 300 °C, in contrast to calcite, which can even twin at room temperature. Notice that twinning occurs on different planes in calcite and dolomite. At low-grade conditions, dolomite is usually stronger than calcite, which causes commonly observed boudinage of dolomite layers in a calcite matrix.

3.12.4
Feldspars

Deformation behaviour of plagioclase and K-feldspar is rather similar and therefore the feldspars are treated together. Laboratory experiments and observation of naturally deformed feldspar have shown that feldspar deformation is strongly dependent on metamorphic conditions. The behaviour as observed by several authors (Tullis and Yund 1980, 1985, 1987, 1991, 1998; Hanmer 1982; Tullis 1983; Dell'Angelo and Tullis 1989; Tullis et al. 1990; Pryer 1993; Lafrance et al. 1996; Rybacki and Dresen 2000; Rosenberg and Stünitz 2003) is described below, according to increasing temperature and decreasing strain rate. Indicated temperatures are for average crustal strain rates. Notice, however, that these temperatures are only valid in case of chemical equilibrium between old and new grains; if new grains have another composition than old grains, e.g. more albite rich, other temperatures will apply (Vernon 1975; White 1975; Stünitz 1998; Rosenberg and Stünitz 2003).

At low metamorphic grade (below 400 °C) feldspar deforms mainly by brittle fracturing and cataclastic flow. Characteristic structures in the resulting cataclasite are angular grain fragments with a wide range of grain size. The grain fragments show strong intracrystalline deformation including grain scale faults and bent cleavage planes and twins. Patchy undulose extinction and subgrains with vague boundaries are normally present. TEM study of such structures has shown that they are not due to dislocation tangles or networks, but to very small-scale brittle fractures (Tullis and Yund 1987). In plagioclase, deformation twinning on albite and pericline law planes is important (Seifert 1964; Vernon 1965; Borg and Heard 1969, 1970; Lawrence 1970; Kronenberg and Shelton 1980; Passchier 1982a; Jensen and Starkey 1985; Smith and Brown 1988; Egydio-Silva and Mainprice 1999). Albite twins may form at the tips of microfaults and *vice versa* (McLaren and Pryer 2001).

At low-medium grade conditions (400–500 °C) feldspar still deforms mainly by internal microfracturing but is assisted by minor dislocation glide. Tapering deformation twins, bent twins, undulose extinction, deformation bands and kink bands with sharp boundaries may be present (Pryer 1993; Ji 1998a,b). Clearly separable augen

and matrix, or core-and-mantle structures are absent. BLG recrystallisation may occur (Shigematsu 1999). *Flame-perthite* (Sect. 9.5.4), a perthite with tapering 'flame-shaped' albite lamellae may be present in K-feldspar, especially at grain boundaries and high stress sites (Figs. 3.37, 7.28; Spry 1969; Augustithitis 1973; Debat et al. 1978; Passchier 1982a; Pryer 1993; Pryer and Robin 1995). Such perthite is thought to develop by albite replacement of K-feldspar driven by breakdown of plagioclase and sericite growth (Pryer and Robin 1995); replacement proceeds preferentially at sites of intracrystalline deformation such as where two feldspar grains are touching (Passchier 1982a; Pryer and Robin 1996). 'Bookshelf' microfracturing in feldspar is common at low-grade conditions, splitting the grains up into elongate 'book-shaped' fragments (Passchier 1982a; Pryer 1993; Sect. 5.6.12). Pryer (1993) claims that antithetic fracture sets are more common in the low temperature range, and synthetic fractures at higher temperature.

At medium-grade conditions (450–600 °C) dislocation climb becomes possible in feldspars and recrystallisation starts to be important, especially along the edge of feldspar grains. Recrystallisation is mainly BLG by nucleation and growth of new grains (cf. Borges and White 1980; Gapais 1989; Gates and Glover 1989; Tullis and Yund 1991). This is visible in thin section by the development of mantles of fine-grained feldspar with a sharp boundary around cores of old grains, without transitional zones with subgrain structures; typical core-and-mantle structures develop (Fig. 5.20) and micro-shear zones of recrystallised grains may occur inside the feldspar cores (Passchier 1982a).

Fracturing in feldspar becomes less prominent under these conditions but microkinking is abundant, probably associated with cataclastic failure at sites of dislocation tangles (Tullis and Yund 1987; Altenberger and Wilhelm 2000). If large kink-bands occur, they have unsharp boundaries (Pryer 1993). Fine-grained recrystallised material may resemble feldspar cataclasite described above, but has a uniform grain size and polygonal grains. Grain boundary sliding has been proposed as a deformation mechanism in this fine-grained feldspar (Vernon and Flood 1987; Tullis et al. 1990), but this is difficult to assess by optical means, and even by TEM. Optically, the only useful criteria are lack of a lattice-preferred orientation and unusual homogeneous mixing of feldspar grains and other minerals in the fine-grained aggregates. According to Tullis et al. (1990), microscopic gouge zones can undergo recrystallisation and develop into small ductile shear zones, destroying most evidence for earlier brittle faulting.

Towards higher temperature, deformation twinning is less abundant. Myrmekite growth becomes important along the boundaries of K-feldspar porphyroclasts (Sects. 5.6.9, 7.8.3). Myrmekite occurs mainly along crystal faces parallel to the foliation (Simpson 1985; Simpson and Wintsch 1989). Flame-perthite is abundant in K-feldspar (Pryer 1993).

At high-grade conditions (above 600 °C), dislocation climb and recovery are relatively easy in feldspar and real subgrain structures form (Vidal et al. 1980; Olsen and Kohlstedt 1985; Pryer 1993; Kruse and Stünitz 1999; Altenberger and Wilhelm 2000). Both SGR and BLG recrystallisation occur (Fig. 5.12). Core-and-mantle structures still occur, but the boundary between the core and the mantle is less pronounced than at lower temperature. Myrmekite along foliation planes is abundant. At low and intermediate pressure, feldspar grains are strain-free, with isolated micro-kink bands while flame-perthite is absent. Fracturing of grains can still be common (Berger and Stünitz 1996; Kruse et al. 2001). At high-pressure conditions, Altenberger and Wilhelm (2000) report microfractures, kinkbands, deformation bands, undulose extinction and flame perthite in K-feldspar and recrystallisation by SGR, or by BLG at high strain rate.

At ultra high-grade conditions (>850 °C), GBM recrystallisation has been reported for plagioclase in the presence of a melt phase (Lafrance et al. 1996, 1998; Rosenberg and Stünitz 2003), indicated by strain-free grains with interlobate grain boundaries and left-over grains (Fig. 3.34). However, compositional effects are again very important for such microstructures (Rosenberg and Stünitz 2003).

Several dislocation slip systems can be active in feldspars, especially at high temperature. In plagioclase, slip on (010)[001] and (001)<110> seems to dominate at medium to high-grade metamorphic conditions (Olsen and Kohlstedt 1984, 1985; Montardi and Mainprice 1987; Kruhl 1987a; Ji et al. 1988; Ji and Mainprice 1990; Kruse and Stünitz 1999; Heidelbach et al. 2000). Slip on {001}<100>, (010)[100] and {111}<110> is reported as well (Montardi and Mainprice 1987; Ji and Mainprice 1988; Dornbush et al. 1994; Ullemeyer et al. 1994; Marshall and McLaren 1977a,b; Olsen and Kohlstedt 1984, 1985; Ji and Mainprice 1987, 1988, 1990 and Stünitz et al. 2003). For K-feldspar, activity of (010)[100] has also been reported by Gandais and Willaime (1984). At high-grade metamorphic conditions, diffusion creep may be important in feldspar deformation (Gower and Simpson 1992; Selverstone 1993; Martelat et al. 1999). A deformation mechanism map for feldspar was constructed by Rybacki and Dresen (2004).

The limited number of active slip systems in feldspars leads to dynamic recrystallisation and core-and-mantle structures. At low temperature, BLG recrystallisation may nucleate on small brittle fragments in crush zones (Stünitz et al. 2003). Two types of mantled porphyroclasts (Sects. 5.6.5, 5.6.6) may develop in plagioclase at high temperature: relatively little deformed 'globular' porphyroclasts, similar to those at low temperature, which have (010)[001] slip systems in an unfavourable orientation for slip, and ribbon plagioclase grains which were in a favourable orientation for slip on (010)[001] (Ji and Mainprice 1990; Kruse et al. 2001; Brodie 1998; Olesen 1998; Box 4.2). Deformation twins, undulose extinction and deformation bands are common in such ribbons.

3.12.5
Micas

Micas deform mainly by slip on either (001)<110> or (001)[100], and therefore show abundant evidence for accommodation mechanisms such as pressure solution and fracturing (Kronenberg et al. 1990; Shea and Kronenberg 1992; Mares and Kronenberg 1993), undulose extinction, kinking and folding (Wilson 1980; Lister and Snoke 1984; Bell et al. 1986b). Folds and kinks are particularly common in mica; commonly, folding occurs on the outside and pressure solution or kinking in the core of a folded crystal. Fractures are commonly associated with deflection of basal planes and lead to barrel or fish-shaped boudinaged grains (Sect. 5.6.7). Grain boundary migration recrystallisation becomes important at medium to high grade (Bell 1998).

In the brittle domain, biotite may show crude kinking or layer parallel slip to develop 'cleavage steps' or mica fish (Sects. 5.6.7, 5.7.3; Kanaori et al. 1991). Biotite behaves ductilely at temperatures above 250 °C (Stesky et al. 1974; Stesky 1978). Muscovite is generally more resistant to deformation than biotite and therefore commonly forms mica fish in mylonite (Sect. 5.6.7).

3.12.6
Olivine

Different slip systems operate in olivine at different temperatures in the mantle (Nicolas and Christensen 1987; Mainprice and Nicolas 1989; Suhr 1993). At 'low' temperature (700–1000 °C), slip systems (010)[001] (Nicolas and Christen-sen 1987) or {110}[001] (Carter and Avé Lallemant 1970) have been reported, and additional slip on several planes that intersect along the [100] direction. The latter is called pencil glide on (0kl)[100]. Old grains of olivine show strong undulose extinction and subgrain boundaries. Olivine recrystallises to fine-grained crystals that are concentrated in shear zones by flow partitioning (Suhr 1993). At medium temperature around 1 000 °C, pencil glide on (0kl)[100] is dominant. At high temperature (T > 1000 °C), only (010)[100] dominates and at very high temperature (T > 1 250 °C), (010)[100] is dominant and (001)[100] may be active (Nicolas and Christensen 1987; Mainprice and Nicolas 1989). A polygonal granoblastic fabric of coarse-grained, strain-free olivine develops. A strong lattice preferred orientation of olivine and trails or bands of other minerals in olivine might be the only indication that the rock was deformed. The relatively coarse grain size of olivine (0.4–1 mm; Suhr 1993) corresponds to low flow stresses in the mantle at these levels (Sect. 9.6.2).

Besides temperature, water may influence slip system activity in olivine (Jung and Karato 2001). At high water content [001] slip may become predominant over [100] slip in olivine. Therefore, a LPO with an [001] maximum parallel to the stretching lineation could be related to hydration rather than low temperature (Jung and Karato 2001).

3.12.7
Orthopyroxene

In orthopyroxene, dislocation glide is dominant on (100)[001] (Coe and Kirby 1975; McLaren and Etheridge 1976; Mercier 1985; Dornbush et al. 1994). Other slip systems that have been found are (100)[010] and (010)[001] (Nazé et al. 1987; Dornbush et al. 1994). Optically visible subgrain boundaries are usually parallel to (100), (010) and (001).

(100)[001] dislocations in orthopyroxene are usually split into partial dislocations, separated by a stacking fault along which the crystal lattice is transformed into that of a clinopyroxene (Coe and Kirby 1975; McLaren and Etheridge 1976); as a consequence, exsolution lamellae of clinopyroxene can easily develop parallel to (100) and are therefore common in deformed orthopyroxene (Suhr 1993).

Under upper mantle conditions (up to 1 000 °C) orthopyroxene may form ribbon grains with aspect ratios up to 100:1 (Etheridge 1975; Nicolas and Poirier 1976; Mainprice and Nicolas 1989; Suhr 1993; Ishii and Sawaguchi 2002; Sawaguchi and Ishii 2003), or equidimensional porphyroclasts if grains had an orientation that was unsuitable for slip (Etchecopar and Vasseur 1987). The old grains may be surrounded by a mantle of fine recrystallised orthopyroxene. Ribbon grains probably form due to the dominant operation of the (100)[001] slip system (Dornbush et al. 1994). Garnet, spinel, plagioclase or quartz can form exsolution lamellae in orthopyroxene.

3.12.8
Clinopyroxene

In clinopyroxene, the unit cell is half the length of that of orthopyroxene in the a-direction. Burgers vectors in that direction are therefore shorter, and since the activation energy of a dislocation is proportional to the length of the Burgers vector, more active slip systems can be expected in clinopyroxene than in orthopyroxene.

At low temperature and/or high strain rate, deformation occurs by (100) and (001) twinning in combination with (100)[001] slip, but in nature this is mainly restricted to meteorites due to the breakdown of clinopyroxene at low temperature (Avé Lallemant 1978; Ashworth 1980, 1985). At high temperature (>500 °C) and/or low strain rate multiple slip occurs, mainly on $\{\bar{1}10\}1/2{<}110{>}$, $\{110\}[001]$ and (100)[001], and rarely on (010)[100] (van Roermund and Boland 1981; Phillipot and van Roermund 1992; van Roermund 1983; Buatier et al. 1991; Ingrin et al. 1991; Ratterron et al. 1994; Godard and van Roermund 1995; Bascou et al. 2001, 2002). Dislocation creep may be assisted by diffusive mass transfer and dynamic recrystallisation (Godard and van Roermund 1995; Mauler et al. 2000a,b; Bystricky and Mackwell 2001). Optically visible subgrain boundaries are usually parallel to $\{110\}$, (100), (010) and (001). Clinopyroxene does not easily form ribbons such as orthopyroxene at high temperature.

Garnet, spinel, plagioclase, hornblende (Fig. 7.29) or quartz can form exsolution lamellae in clinopyroxene. Exsolution can occur parallel to (100) and (001), but at temperatures above 700–750 °C only along (100).

3.12.9
Garnet

Although garnet behaves as a rigid mineral at low grade metamorphic conditions, several studies have presented evidence for ductile deformation of garnets such as lattice bending (Dalziel and Bailey 1968; Ross 1973) and dislocation substructures revealed by etching (Carstens 1969, 1971) and electron microscope studies (Allen et al. 1987; Ando et al. 1993; Doukhan et al. 1994; Ji and Martignole 1994; Chen et al. 1996; Voegelé et al. 1998b; Kleinschrodt and McGrew 2000; Prior et al. 2000, 2002). Elongate lensoid and folded shapes of garnet crystals parallel to the deformation fabric (Kleinschrodt and Duyster 2002; Ji and Martignole 1994), subgrain structures and a LPO are found in some garnets and can also be used as evidence for crystalplastic deformation (Prior et al. 2000, 2002; Kleinschrodt and McGrew 2000; Mainprice et al. 2004). The transition from brittle to crystalplastic deformation seems to lie at 600–800 °C (Voegelé et al. 1998b; Wang and Ji 1999). At low and medium grade conditions, garnet is much stronger than quartz and feldspar and does not deform when isolated in a quartzo-feldspathic matrix. At higher temperatures, the difference in strength decreases to an extent that all three minerals can deform together (Ji and Martignole 1996; den Brok and Kruhl 1996; Kleinschrodt and McGrew 2000). TEM studies give evidence for dislocation slip (Ando et al. 1993; Doukhan et al. 1994; Voegelé et al. 1998a,b; Ji et al. 2003). Since garnet has a cubic crystal structure, many slip systems can theoretically be activated. Dislocation glide of <100> dislocations in $\{011\}$ and $\{010\}$ planes and $1/2{<}111{>}$ dislocations that glide in $\{110\}$, $\{112\}$ and $\{123\}$ planes have been observed (Voegelé et al. 1998a), providing 66 possible slip systems. Of these, slip on the $1/2{<}111{>}\{110\}$ system seems to dominate. However, microstructures in garnet which are interpreted as an effect of crystalplastic deformation may also have formed by other, so far little investigated processes such as fracturing (Prior 1993; Austrheim et al. 1996) multiple nucleation and growth (Spiess et al. 2001), and diffusion mechanisms (Ji and Martignole 1994, 1996; den Brok and Kruhl 1996; Wang and Ji 1999; Ji et al. 2003). Ductile deformation of garnet can produce a lattice-preferred orientation but garnet seems to have weak preferred orientation in deformed rocks (Mainprice et al. 2004).

3.12.10
Amphiboles

The deformation behaviour of amphiboles is as yet poorly understood. In amphiboles, the crystal unit cell in the direction of the b-axis has more than twice the length of that in pyroxenes. Theoretically, due to the increased Burgers vector length, amphiboles should therefore be stronger in ductile deformation than clinopyroxenes. In practice, the opposite is commonly observed.

Presently available evidence on deformation of hornblende suggests that below 650–700 °C, amphiboles mostly deform by brittle deformation and dissolution-precipitation, and aggregates of fine-grained hornblende probably form by fracturing rather than dynamic recrystallisation (Allison and LaTour 1977; Brodie and Rutter 1985; Nyman et al. 1992; Stünitz 1993; Lafrance and Vernon 1993; Babaie and LaTour 1994; Berger and Stünitz 1996; Wintsch and Yi 2002; Imon et al. 2002, 2004). Dissolution of hornblende is probably balanced by deposition of amphibole of a different composition (Imon et al. 2004) or of other phases such as epidote, albite and biotite elsewhere in the rock (Berger and Stünitz 1996). Core-and-mantle structures on hornblende formed below 650–700 °C may also be due to fracturing (Nyman et al. 1992), but where recrystallisation is involved (Cumbest et al. 1989), it is probably driven by a difference in chemical composition rather than strain energy (Fitz Gerald and Stünitz 1993; Stünitz 1993). The main reason for this dominant brittle behaviour seems to be the excellent cleavage on {110} planes. At low temperature and/or high strain rate, amphiboles also deform by deformation twinning on ($\bar{1}$01) or (100) (Buck 1970; Rooney et al. 1975; Morrison-Smith 1976; Dollinger and Blacic 1975; Biermann 1981; Hacker and Christie 1990) and slip on (100)[001]. As in micas, slip on (100)[001] can lead to development of kinks.

At high temperature, above 700 °C and in dry rocks hornblende can apparently deform by crystalplastic deformation, and shows strain energy driven dynamic recrystallisation (Boullier and Gueguen 1998a; Kruse and Stünitz 1999; Fig. 3.24). At high temperature and/or low strain rate, several slip systems have been documented, mainly (hk0)[001] and (100)[001] but also {110}1/2<$\bar{1}$10> and (010)[100] (Rooney et al. 1975; Dollinger and Blacic 1975; Biermann and van Roermund 1983; Olsen and Kohlstedt 1984; Montardi and Mainprice 1987; Cumbest et al. 1989; Reynard et al. 1989; Skrotsky 1992; Kruse and Stünitz 1999). Subgrains are elongated parallel to the c-axis and subgrain boundaries consist of simple arrays of [001], [100] or <110> dislocations and are parallel to {110}, (100) or (010) (Biermann and van Roermund 1983; Reynard et al. 1989).

A characteristic structure in hornblende schists is that of 'garben' (German for stack), bundles of elongate hornblende crystals that are oriented in fan-like arrangements usually parallel to the foliation plane. Such 'garben' may develop by growth of subgrains in the direction of the c-axis in previously deformed hornblende crystals (Biermann 1979).

3.13
Deformation of Polymineralic Rocks

3.13.1
Introduction

Since most rocks are composed of more than one mineral, it is interesting to see how individual minerals behave in a polymineralic rock. Minerals do not always show the same dependence in behaviour on temperature and strain rate as in monomineralic aggregates, and may even behave in an entirely different way. The behaviour of polymineralic rocks is remarkably complex (Jordan 1987, 1988; Handy 1989, 1992; Bons 1993; Handy et al. 1999; Stünitz and Tullis 2001). The concept of a stress-supporting network is important; if 'hard' and 'soft' minerals coexist, the strength of an aggregate does not increase linearly with the amount of the hard mineral present. If few hard grains are present, the strength of the aggregate is similar to that of a monomineralic aggregate of the soft mineral; the hard minerals may rotate in the flow of the soft material, and may form core-and-mantle structures if they recrystallise on the outside. The strength of the aggregate increases suddenly when the grains of the hard mineral are so common and large that they touch and start to support the imposed differential stress. Obviously, the original shape of the grains is also important here. When the hard mineral is dominant, the strength of the aggregate will approach that of the pure hard mineral, but at higher strain the pockets of the soft mineral may interconnect and form shear zones that weaken the aggregate (Jordan 1987). The contrast in rheology between two minerals may change and even reverse with changing external conditions. Below, we discuss the behaviour of quartz-feldspar aggregates as an example of a polymineralic rock.

3.13.2
Quartz-Feldspar Aggregates

The study of deformed quartzofeldspathic rocks such as granites shows an interesting dependence of structure on metamorphic grade (Vernon and Flood 1987; Tullis et al. 1990, 2000). At very low-grade conditions feldspar and quartz deform both by brittle fracturing (Fig. 3.42). Microstructural observations suggest that feldspar is actually weaker than quartz at these conditions (Chester and Logan 1987; Evans 1988). This is probably due to the fact that feldspar grains have cleavage planes that reduce their strength. As a result, aggregates of elongate cataclased feldspar and quartz develop (Fig. 3.42) where part of the feldspar (especially K-feldspar) is transformed to kaolinite and sericite. A cataclastic foliation of fragmented grain clusters with fractures and preferred orientation of sheet silicates commonly develops (Evans 1988).

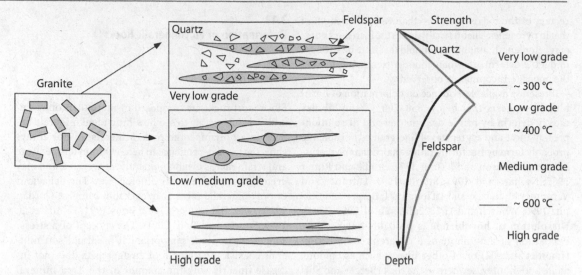

Fig. 3.42. Changes in the deformation behaviour of quartz-feldspar aggregates with depth. At *right*, a depth-strength graph with brittle (*straight line*) and ductile (*curved line*) segments of quartz and feldspar is shown. At very low grade, both quartz and feldspar are brittle, but feldspar is the weaker mineral. At low to medium-grade conditions, quartz deforms by dislocation creep and feldspar is the stronger mineral, developing core-and-mantle structure and mantled porphyroclasts (Figs. 5.12, 5.22–5.25). At high grade, feldspar and quartz deform by dislocation creep and have similar strength (Fig. 5.11)

Under low-grade conditions, feldspar is still brittle while quartz deforms ductilely by dislocation glide and creep (Fig. 3.42; Tullis and Yund 1977; Simpson 1985; Gapais 1989; Gates and Glover 1989; Fitz Gerald and Stünitz 1993; Stünitz and Fitz Gerald 1993). However, the strength contrast is now reversed, and quartz is the weaker mineral; feldspar porphyroclasts deform by fracturing and may develop to core-and-mantle structures as a result of neocrystallisation due to compositional disequilibrium. Cores show abundant evidence of brittle faulting and patchy undulose extinction. Stretched mantled porphyroclasts may form elongated wings that eventually define a compositional layering (Fig. 3.42). Quartz aggregates are elongate to ribbon-shaped and may consist of tightly folded crystals which have recrystallised to some extent (Passchier 1985; Hongn and Hippertt 2001). These low temperature ribbons tend to have grain boundaries and subgrain boundaries parallel to the long axis of the ribbons. They usually wrap around feldspar aggregates and deform much more homogeneously; cores of old quartz grains show abundant subgrains that laterally pass into recrystallised (new) grains. At high strain, 'augen' (German for eyes) of feldspar develop, separated by finely laminated aggregates of fine-grained quartz and feldspar.

At medium to high-grade conditions, both feldspar and quartz deform by dislocation creep assisted by diffusion and recrystallisation. Both minerals may form monomineralic and polymineralic ribbons that give the rock a banded appearance (Culshaw and Fyson 1984; McLelland 1984; Mackinnon et al. 1997; Hippertt et al. 2001; Box 4.2; Fig. 3.42). These ribbon grains may form by stretching of single crystals or crystal aggregates, but also by coalescence of grains (Hippertt et al. 2001). Such high temperature ribbons tend to have grain and subgrain boundaries oblique to the long axis of the ribbons. Both have subgrains in old grain cores and a gradual transition from the core to a recrystallised mantle. Feldspar augen are rare. Feldspar and quartz show similar deformation intensity and seem to have a relatively small contrast in strength.

At high-grade conditions, grain boundaries between quartz and feldspars are commonly strongly curved, with lobate and cuspate and even amoeboid shapes (Fig. 3.33; Passchier 1982a; Gower and Simpson 1992). This geometry may be due to deformation at high-grade conditions, possibly with a large component of solid-state diffusive mass transfer such as Coble or Nabarro-Herring creep (Gower and Simpson 1992).

One of the characteristic differences in behaviour of feldspar and quartz at low temperature and high strain rate is the development of core-and-mantle structures in feldspar, and more homogeneous deformation in quartz. This has been explained by Tullis et al. (1990) as a result of different deformation mechanisms of feldspar and quartz at these conditions; in feldspar dislocation climb is difficult and deformation occurs by BLG recrystallisation-accommodated dislocation creep (Dell'Angelo and Tullis 1989). The newly produced grains of feldspar are free of dislocations and relatively soft, and grain boundary migration can easily replace them by new grains once they develop dislocation tangles. As a result, the mantle of recrystallised feldspar grains surrounding feldspar cores is much softer than the core, and deformation is concentrated in the mantle, which grows in volume as the core shrinks with progressive deformation (Dell'Angelo and Tullis 1989). Diffusion processes may also play a role in the recrystallised mantles. In quartz, dislocation creep is accommodated by dislocation climb and SGR recrystallisation dominates. The new grains have the same dislocation density as old subgrains, and the new aggregate is equally strong as the old grains; consequently, no core and mantle structure develops and quartz deforms relatively homogeneously.

3.13.3
Deformed Rhyolites – an Exception

Some deformed rhyolites and ignimbrites are an interesting exception to the rule that in quartz-feldspar aggregates, deformed at low- to medium-grade conditions, quartz is the weaker mineral. In these rocks, quartz phenocrysts may survive as porphyroclasts (Figs. 3.7, 3.9, 3.10; Williams and Burr 1994). Probably, the fine-grained polymineralic matrix of a rhyolite can deform by grain boundary sliding, or pressure solution and precipitation at such a low differential stress that limited intracrystalline deformation is induced in quartz. Deformed rhyolites and ignimbrites can be recognised by the presence of euhedral to subhedral quartz phenocrysts with typical wriggly embayments (Figs. 3.9, 3.10). Obviously, the behaviour of quartz and feldspar in an aggregate is dependent not only on external conditions, but also on the original geometry of the aggregate before deformation.

3.14
Flow Laws and Deformation Mechanism Maps

In order to establish under which conditions deformation mechanisms as described in this chapter are active, data from experimental deformation are used in combination with observations on rocks deformed at known metamorphic conditions. Experimental deformation of rocks at a range of pressure and temperature conditions can give us some idea of the activity of deformation, recovery and recrystallisation processes at specific conditions. One drawback of experimental work is that geologically realistic strain rates in the order of 10^{-12} to 10^{-14} s^{-1} cannot be reproduced in experiments. Nearly all our data on deformation mechanisms are from experiments at much higher strain rate. However, for many deformation mechanisms, increase in temperature has an effect similar to a decrease in strain rate. Therefore, extrapolation of experimental results to geologically realistic strain rates is possible by 'projection' of data from experiments carried out at higher temperature.

The rheological behaviour of minerals and rocks is usually expressed in *flow laws* (Poirier 1985; Hirth et al.

2001; Mainprice et al. 2003). Some important and commonly quoted types of flow laws are given in Box 3.11. In the equation for dislocation creep given here, strain rate is independent of grain size but has a strong non-linear (power law) dependence of strain rate on stress. In the equations for diffusion creep, strain rate has a linear dependence on stress but a non-linear dependence on grain size. Flow laws have been proposed on the basis of experiments and theoretical considerations.

The parameters in the equations have been determined experimentally for a range of conditions. In many cases, these data are incomplete or difficult to compare because of differences in confining pressure, sample preparation etc. However, if a suitable set of data on rheology of a particular mineral can be found, it is possible to integrate the data to determine which mechanisms are expected to operate under particular conditions; in general, the mechanism that operates at the lowest differential stress for a particular strain rate is thought to be dominant. Conditions at which specific deformation mechanisms are dominant can be shown in a *deformation mechanism map*. Such a map shows fields in which certain deformation mechanisms are dominantly, although not exclusively, active. Also shown are projected curves for several strain rates, which give an indication of the relationship of stress and strain rate for a specific temperature. Cataclasis occurs only above a certain differential stress level, which is dependent on fluid pressure (Sibson 1977a) and temperature (Griggs et al. 1960). Since grain size plays a major role in determining which deformation mechanism will be active, several maps for different grain sizes are usually given.

Figure 3.43 shows an example of a deformation mechanism map for quartz, and the way in which it is constructed. Parameters that have been used are given in the inset. Using the equations given in Box 3.11, graphs are first made which plot shear stress against temperature at given strain rates for each of the deformation mechanisms (Fig. 3.43a–c). In such graphs, *normalised units are plotted*. This is done to obtain numbers that are dimensionless since this allows easy comparison of different materials (Box 2.5). For example, homologous temperature T_h ($T_h = T / T_m$ where T_m is the melting temperature of a mineral in K) is used on the horizontal scale instead of absolute temperature; $T_h = 0$ at 0 K and $T_h = 1$ at the melting temperature of the mineral. In this way, deformation behaviour of ice can be compared with that of steel if both are at the same T_h value. Similarly, normalised shear stress (σ / μ) is used instead of shear stress.

At any point in the graphs of Fig. 3.43a–c, a single strain rate is defined at a certain stress and temperature if other parameters are constant. When these diagrams are combined in pairs (Fig. 3.43d,e), each point will be attributed two strain rate values, one for each of the possible deformation mechanisms; the mechanism with the highest

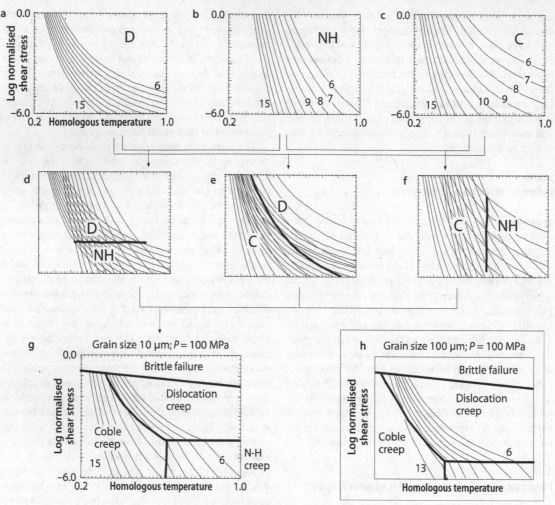

Fig. 3.43a–h. Method to construct a deformation mechanism map from experimental data. All graphs show contours of strain rate (10^{-6} to 10^{-15}, only exponents shown) in plots of normalised shear stress against homologous temperature (see text). *D* Dislocation creep; *C* Coble creep; *NH* Nabarro-Herring creep. **a–c** graphs for the three types of deformation mechanisms. If these graphs for single mechanisms are combined **d–f** fields of dominant deformation mechanism can be defined. Combination of the graphs for two mechanisms gives a deformation mechanism map **g**. The constructed map is for a grain size of 10 μm of single crystals at 100 MPa confining pressure in a dry environment. A similar map for 100 μm grain size **h** is shown for comparison

strain rate is thought to be dominant at that point. The boundaries between fields of dominant deformation mechanisms can be found by simply joining the intersection points of the strain rate curves when two diagrams are overlapping. In a similar way, the combined data define fields for each deformation mechanism, the deformation mechanism map Fig. 3.43g. Strain rate contours for each dominant mechanism alone are shown in the fields of Fig. 3.43g; these show sharp kinks on the bounding lines. Above a certain stress level, the material is thought to deform by brittle failure; the curve for brittle failure was taken from Griggs et al. (1960).

Figure 3.43g shows the deformation mechanism map for quartz with a grain size of 10 μm, and Fig. 3.43h with

a grain size of 100 μm. The transition between Coble and Nabarro-Herring creep is mainly influenced by temperature; the transitions of Nabarro-Herring creep to dislocation creep, and dislocation creep to brittle failure are mainly an effect of shear stress. The transition from Coble creep to dislocation creep is influenced by both temperature and shear stress. Temperature and strain rate have conflicting influence, as mentioned above, and this allows extrapolation of experimental data to geological strain rates. Notice that with increasing grain size, the dislocation creep field increases in size and the diffusion creep fields shrink (Fig. 3.43g,h).

Deformation mechanism maps are most useful for the prediction and comparison of experimental results. For

Box 3.11 Flow laws

The following flow laws are commonly quoted in the literature, and have been used to construct deformation mechanism maps in Fig. 3.43.

For one of the simplest models, bulk diffusion-controlled dislocation creep (also known as Weertman creep) the flow law is:

$$\dot{\varepsilon} = \frac{\mu b D_L}{kT} \cdot (\sigma/\mu)^3 \cdot e^{(-H_L/RT)}$$

For Coble creep:

$$\dot{\varepsilon} = \frac{A_C \mu V D_G W}{RTd^3} \cdot (\sigma/\mu) \cdot e^{(-H_G/RT)}$$

and for Nabarro-Herring creep:

$$\dot{\varepsilon} = \frac{A_{NH} \mu V D_L}{RTd^2} \cdot (\sigma/\mu) \cdot e^{(-H_L/RT)}$$

Notice that dislocation creep is non-Newtonian and that the diffusion creep flow types are Newtonian (Sect. 2.12). The symbols in the equation have the following significance (units in square brackets):

e   – exponential number (2.718281)
$\dot{\varepsilon}$ – shear strain rate [s^{-1}]
σ – shear stress [Nm^{-2}]
T – temperature [K]
b   – Burgers vector [m]
A_c – numerical factor for Coble creep depending on grain shape and boundary conditions
A_{NH} – numerical factor for Nabarro Herring-creep depending on grain shape and boundary conditions
H   – molar activation enthalpy for self-diffusion [J mol^{-1}]
D   – diffusion constant for self-diffusion [m^2 s^{-1}]

R   – gas constant [J mol^{-1} K^{-1}]
k   – Boltzmann constant [J K^{-1}]
μ – shear modulus [Nm^{-2}]
d   – grain size [m]
W   – grain boundary thickness [m]
V   – molar volume of the solid [m^3 mol^{-1}]
σ/μ – normalised shear stress [dimensionless number]

Parameters for Fig. 3.43

For the deformation mechanism maps in Fig. 3.43 the following parameters have been used:

T_m = 1550 K (melting temperature of quartz in the presence of water)
R   = 8.3143 J mol^{-1} K^{-1}
k   = 1.38062 × 10^{-23} J mol^{-1} K^{-1}
V   = 2.6 × 10^{-5} m^3 mol^{-1}
b   = 5 × 10^{-10} m
μ = 42 × 10^9 Nm^{-2} (Sosman 1927)
A_c = 141 (grain boundary sliding possible)
A_{NH} = 16 (grain boundary sliding impossible)
H_L = 243 × 10^3 J mol^{-1} for grain boundary diffusion used in the flow laws for Weertman creep and Nabarro-Herring creep at 450–590 °C and a mean stress of 100 MPa (Farver and Yund 1991a)
D_L = 2.9 × 10^{-5} m^2 s^{-1} Bulk oxygen self-diffusion in the presence of water for Weertman creep and Nabarro-Herring creep at 450–590 °C and mean stress = 100 MPa (Farver and Yund 1991a)
H_G = 113 × 10^3 J mol^{-1} for grain boundary diffusion used in the flow law for Coble creep (Farver and Yund 1991b)
$D_G W$ = 3 × 10^{-17} m^3 s^{-1} Bulk oxygen self-diffusion in the presence of water for Coble creep at 450–800 °C and 100 MPa mean stress (Farver and Yund 1991b)

example, they are defined for only one mean stress value (100 MPa in the case of Fig. 3.43). This is useful for experimental purposes where mean stress is usually kept constant, while temperature and strain rate are varied. However, in nature, mean stress and temperature increase together with increasing depth and this effect is usually not shown in an ordinary deformation mechanism map. We should not imagine a situation where one deformation mechanism takes over abruptly from another at a set temperature, pressure or other variable. In this sense, the fields in a deformation mechanism map can be slightly misleading; they indicate *dominant* deformation mechanisms – other deformation mechanisms may also be active in these fields, and towards a boundary one mechanism will gradually take over from the other.

Like most geological diagrams, deformation mechanism maps suffer from the disadvantage that too many parameters must be shown in just two dimensions. The effect of grain size on rheology is also strong and has to be shown on separate maps (Fig. 3.43g,h). Another disadvantage of deformation mechanism maps is that they cannot show all deformation mechanisms to advantage. Pressure solution, a very important mechanism in quartz (and probably in feldspar and other minerals; Wintsch and Yi 2002) is difficult to include because a deformation mechanism map is only valid for a specific mean stress. Since fluid pressure is important in pressure solution (but possibly also in dislocation creep; Tullis and Yund 1991), it is difficult to show exact boundaries for pressure solution in deformation mechanism maps. However, a field of pressure solution should plot on the low stress and temperature side, in the lower left-hand corner of Fig. 3.43g,h where it replaces the Coble creep field for quartz, calcite, feldspar and micas in most geological situations where water is present. Finally, flow laws and consequently deformation mechanism maps are valid for steady state flow; deformation of a recrystallising aggregate with porphyroclasts can therefore not be shown on deformation mechanism maps.

An alternative diagram to deformation mechanism maps is the *depth-strength diagram* commonly used to show a strength profile of the lithosphere (Figs. 3.42, 3.44)

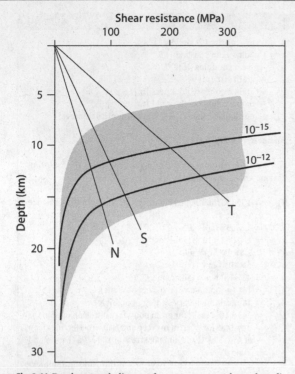

Fig. 3.44. Depth-strength diagram for quartz at a geothermal gradient of 30 °C km^{-1} and hydrostatic fluid pressure in brittle faults (based on Sibson 1983; Gleason and Tullis 1995). *Straight lines* are strength of brittle fracturing for thrust (*T*), strike slip (*S*) and normal faults (*N*). *Curved lines* are for dislocation creep at strain rates of 10^{-15} and 10^{-12} s^{-1}. Ornamented field shows the estimated error in the position of the dislocation creep curves due to uncertainty about flow laws to be used, and in estimations of rheological parameters (Gleason and Tullis 1995)

(Sibson 1983; Kohlstedt et al. 1995; Gleason and Tullis 1995). Such diagrams are valid for a specific mineral, grain size, strain rate, geothermal gradient, orientation of brittle faults and fluid pressure on faults. Differential stress or strength of the material is plotted against depth. There are two sets of intersecting curves. The straight curves show strength of a brittle fault; with increasing depth, the blocks on both sides of a fault are pressed together and thereby increase the differential stress that is needed to make the blocks move along the fault. The lower curves are for dislocation creep and represent a decrease in strength with increasing temperature at increasing depth. High in the crust, brittle faulting is therefore favoured and at deep levels dislocation creep. The crossover point is known as the *brittle-ductile transition*. Notice that this is not a simple surface in the crust; it lies at a different depth for different minerals (Fig. 3.42) and depends on bulk strain rate, fault orientation, geothermal gradient, grain size and probably other factors. In practice, a wide transitional zone where both mechanisms are active is usually present. The depth-strength diagram has other disadvantages; no deformation mechanisms other than just brittle faulting and dislocation creep have been considered. Pressure solution, for example, could considerably flatten the stress peak at the brittle-ductile transition. If a rock deforms by ductile flow and develops a crystallographic fabric, this can cause hardening and transition to brittle deformation without a change in external conditions. As shown above, in polymineralic rocks the situation is much more complex. Obviously, there is still some scope for development of diagrams to show distribution of deformation mechanism activity in geological applications.

Foliations, Lineations and Lattice Preferred Orientation

The main fabric elements, present in most deformed metamorphic rocks are discussed in this chapter.

Foliations are subdivided into primary and secondary ones, the first being of sedimentary or igneous origin and the second formed by deformational processes. They are described according to their morphological characteristics, and classified into two main groups: continuous foliations and spaced foliations. Several mechanisms thought to be responsible for foliation development are explained and discussed. These are: mechanical rotation of pre-existing grains, solution transfer, crystalplastic deformation, dynamic recrystallisation, static recrystallisation, mimetic growth, oriented growth in a differential stress field and micro folding.

Subsequently, the relation between secondary foliations and axial planes of folds, the XY-plane of tectonic strain and volume change is discussed. The dependence of foliation development from lithotype and metamorphic conditions is also treated.

Special attention is given to the practical use of foliations. Since they are present in the large majority of metamorphic rocks they are important reference structures that can commonly be traced between outcrops. They are especially useful for the establishment of overprinting relations of successive deformation phases.

Lineations are another fabric element of major interest. A new subdivision into object and trace lineations is followed in this book. Object lineations can be further subdivided into grain and aggregate lineations, and trace lineations include crenulation and intersection lineations. The development of lineations is to a large extent similar to that of foliations, but some differences are discussed.

The third fabric element treated in this chapter is lattice preferred orientation (LPO) of minerals. Especially in quartz and calcite this fabric element is not always obvious and special techniques may be required to determine whether an LPO is present. Several factors that influence the LPO pattern are discussed. LPO patterns in quartz are treated in detail and then compared to LPO patterns in other minerals.

4.1
Introduction

Many microstructures in rocks are defined by a preferred orientation of minerals or fabric elements. We distinguish foliations, lineations and lattice-preferred orientation.

The word *foliation* (Fig. 4.1) is used here as a general term to describe any planar feature that occurs penetratively in a body of rock. It may refer to thin rhythmic bedding in a sedimentary rock, to compositional layering in igneous rocks or to cleavage, schistosity, or other planar structures in metamorphic rocks (Sect. 4.5). Joints are normally excluded for not being sufficiently penetrative. We prefer this broad use of a descriptive term to genetic terms since it is often difficult to decide what the origin of a planar structure in a deformed rock is. Foliations may be defined by a spatial variation in mineral composition or grain size (Fig. 4.1a), by a preferred orientation of elongate or platy grains or aggregates of grains (Fig. 4.1b–f), by planar discontinuities such as microfractures (Fig. 4.1g), or by any combination of these elements (Fig. 4.1h).

Stylolites are irregular surfaces, usually in metasedimentary rocks that can define a crude foliation but may also occur isolated or crosscutting. They are described in Box 4.3.

A *lineation* is defined as any linear feature that occurs penetratively in a body of rock (Fig. 4.2). Fibres and striations on fault planes are not lineations since they only occur on specific planar surfaces in the rock, not penetratively. Two main types of lineations can be distinguished; *object lineations* and *trace lineations* (Piazolo and Passchier 2002a). Object lineations are defined by constituting elements that have a specific volume. Trace lineations are intersections of planes or microfolds on foliation planes that lack a distinct volume (Sect. 4.3). *Platelet lineations* are defined by planar minerals such as micas that share a common axis. They have elements of object- and trace lineations.

In three dimensions many foliations show an associated linear element, that is, the fabric elements defining the foliation may appear stronger in some sections normal to the foliation than in others. A complete transition from pure S tectonites (only a foliation) to LS tectonites (both a foliation and a lineation) to L tectonites (only a lineation) can be imagined. In practice, such transitions may actually exist within a single outcrop. Some lineations may develop from or into foliations with time. An example of the latter is the progressive development of slaty cleavage at a high angle to bedding (Box 4.6, ⊘Video B.4.5).

It is important in the description of a foliation to give the relation with a lineation, if present. The linear elements that are of the same age as foliations in a volume of rock are important in tectonic studies because they may furnish information on the direction of tectonic transport (Fig. 5.10). Such lineations must definitely been taken into consideration when deciding how to cut a thin sec-

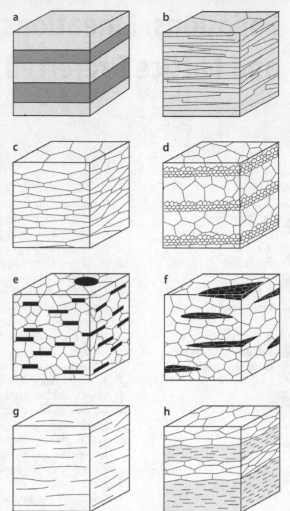

Fig. 4.1. Diagrammatic presentation of various fabric elements that may define a foliation. (After Fig. 5.1 in Hobbs et al. 1976). **a** Compositional layering. **b** Preferred orientation of platy minerals (e.g. mica). **c** Preferred orientation of grain boundaries and shape of recrystallised grains (e.g. quartz, carbonate) in a grain shape preferred orientation. **d** Grain-size variation. **e** Preferred orientation of platy minerals in a matrix without preferred orientation (e.g. mica in micaceous quartzite or gneiss). **f** Preferred orientation of lenticular mineral aggregates. **g** Preferred orientation of fractures or microfaults (e.g. in low-grade quartzites). **h** Combination of fabric elements **a**, **b** and **c**; such combinations are common in metamorphic rocks

tion from a rock sample (Sect. 12.5). Lineations are treated in Sect. 4.3.

Many rocks have a *lattice-preferred orientation* (LPO), a non-random orientation of the crystallographic axes of constituent minerals. Some foliations or lineations are defined by a LPO. However, we use the term here in a more restricted sense for minerals with an equant shape like quartz and calcite, the LPO of which cannot be seen in the field or in thin section without the aid of special techniques. LPO is treated in Sect. 4.4.

Fig. 4.2.
Diagrammatic representation of
various types of fabric elements
that may define a lineation

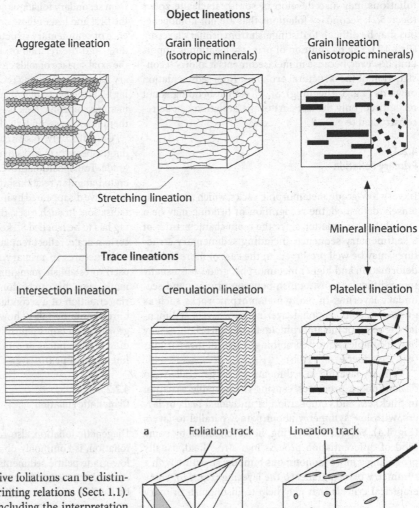

Aggregate lineation

Object lineations

Grain lineation
(isotropic minerals)

Grain lineation
(anisotropic minerals)

Stretching lineation

Mineral lineations

Trace lineations

Intersection lineation

Crenulation lineation

Platelet lineation

4.2
Foliations

In many areas several successive foliations can be distinguished in the field by overprinting relations (Sect. 1.1). Their study in thin section, including the interpretation of the metamorphic and deformational conditions during their formation, is an important tool to unravel the tectonic and metamorphic evolution of an area. Foliations are also used as reference structures to establish the relative growth periods of metamorphic minerals, especially porphyroblasts (Sect. 7.4). Foliations and lineations are generally more penetratively developed in any volume of rock than folds and are therefore better reference structures for the definition of deformation phases (Sect. 1.1).

Primary foliations are structures related to the original rock-forming process. Bedding in a sedimentary rock and magmatic layering in igneous rocks are the most common examples. A *diagenetic foliation* may be formed by diagenetic compaction. *Secondary foliations* are generated later (in the case of sediments: after lithification) as a result of deformation and metamorphism. This group includes cleavage, schistosity, differentiated compositional layering, mylonitic foliation (Box 4.4), etc.

Development of secondary foliations is usually seen as evidence that the rock deformed in a ductile way, but

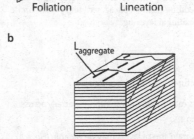

a   Foliation track   Lineation track

Foliation   Lineation

b

$L_{aggregate}$

Fig. 4.3. a Foliations are visible on outcrop surfaces as foliation tracks; these should not be confused with lineations. Object lineations are visible on outcrop surfaces as lineation tracks; these should not be confused with lineations, since they generally have another orientation. **b** Section through a mylonite with mylonitic foliation (*horizontal*) and shear band cleavage (*inclined*). The intersection lineation of the two cleavages is normal to the aggregate lineation on the foliation surfaces. This is an example of two lineations of approximately the same age that have different orientations and tectonic significance

4.2

foliations may also develop in some cataclasite zones (Sect. 5.2). Secondary foliations that are not homogeneous may be difficult to distinguish from primary layering (Box 4.1). The recognition of primary foliation is important, however, because in metasediments it allows reconstruction of the structural evolution from sedimentation onwards (e.g. S_0 (bedding), S_1, S_2, etc.). If bedding is not recognised, only the last part of the evolution can be reconstructed (e.g. S_n, S_{n+1}, S_{n+2}, etc.).

4.2.1
Primary Foliation

In very low-grade metamorphic rocks, which are not intensely deformed, the recognition of bedding may be a straightforward matter, since the main characteristics of a sedimentary sequence, including sedimentary structures, may be well preserved. In the case of more intense deformation and higher metamorphic grade, it is usually more difficult to distinguish between primary and secondary layering. In many metamorphic rocks such as gneisses a compositional layering may have a sedimentary, igneous or metamorphic/deformation origin, or may have a complex nature combining several of these origins (e.g. Passchier et al. 1990b).

Primary layering in sediments results generally from discontinuous processes, causing considerable variation in thickness and composition of individual beds or layers, with low symmetry about planes parallel to layers (Fig. 4.4). Secondary layering, however, forms by some kind of differentiation process in a stress field, usually producing a more monotonous bimodal structure with a symmetry plane parallel to the layering (Fig. 4.5). Some empirical criteria that may help to distinguish primary

from secondary foliations are listed in Box 4.1. In fact, only the first and last of these criteria are conclusive: the presence of sedimentary structures is good evidence for bedding, and the relation of a compositional layering with the axial surface of folds clearly demonstrates the secondary nature of a layering (contemporaneous with the folding). The presence of two crosscutting layering structures in a metamorphic rock is also good evidence that one of them must be secondary (Fig. 4.5).

Unfortunately, it is often impossible to recognise bedding, especially in rocks of medium to high metamorphic grade. Transposition processes (Box 4.9) may have obliterated angular relationships, or sedimentary structures may have disappeared by intense deformation and recrystallisation. In such cases, the oldest compositional layering has to be labelled S_n, keeping in mind that it may, at a certain scale, reflect remnants of bedding.

Structures in primary sedimentary layering can be used to establish younging direction in thin section. In many cases this can be done by recognition of asymmetric refraction of a secondary foliation through bedding. Care should be taken, however, since in some cases the growth of metamorphic minerals may invert graded bedding if large micas or other minerals grow in originally fine-grained pelitic layers.

4.2.2
Diagenetic Foliation

Diagenetic foliation, also referred to as bedding-parallel foliation, is commonly observed in very low-grade and low-grade pelitic sediments, which have undergone little or no deformation (Borradaile et al. 1982). It is defined by parallel orientation of thin elongate mica grains with

Box 4.1 Criteria to distinguish between primary and secondary foliations	
The arguments given below are useful to distinguish a primary foliation such as bedding from a spaced secondary foliation developed as a compositional layering.	
Primary foliation	**Secondary foliation**
Sedimentary structures may be recognised	No sedimentary structures present
Thickness of layers may show any variation, especially across strike (Fig. 4.4)	Little variation in thickness, usually two alternating approximately constant thicknesses (Fig. 4.5)
Composition and grain size of layers may be variable (Fig. 4.4)	Composition of layers usually bimodal (Figs. 4.5, 4.13)
Layering usually planar	Layering commonly lensoid or anastomosing (Fig. 4.5)
Rarely a symmetry plane parallel to layering	Normally a symmetry plane parallel to layering (Fig. 4.5)
Foliation never parallel to the axial plane of folds (however do not mistake folds for refraction)	Foliation (sub)parallel to the axial plane of folds of an earlier foliation; commonly developed by differentiation of fold limbs (Figs. 4.12, 4.13)

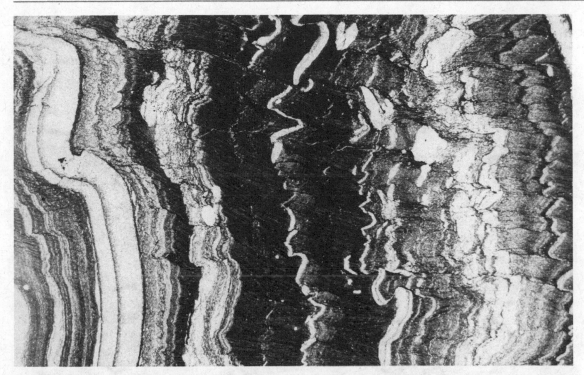

Fig. 4.4. Folded bedding S_0 (primary foliation – subvertical) with spaced cleavage S_2 (secondary foliation – close to horizontal), developed in *dark layers*. An older slaty cleavage (S_1) is present subparallel to S_0 but not visible at this magnification. Note variation in thickness and composition of bedding. (A detail of the *central upper part* is presented in Fig. 4.21). Pyrenees, Spain. Width of view 7 mm. PPL

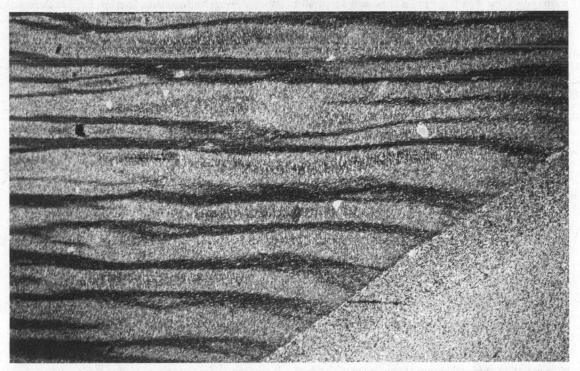

Fig. 4.5. Secondary foliation (S_2) (*horizontal*) developed by differentiation in limbs of crenulations. Remnant bedding (S_0) is visible in *lower right corner*. S_1 is parallel to S_0. The secondary compositional layering (S_2) has a monotonous bimodal character with a horizontal symmetry plane. Leiden Collection. Width of view 15 mm. Polars at 45°

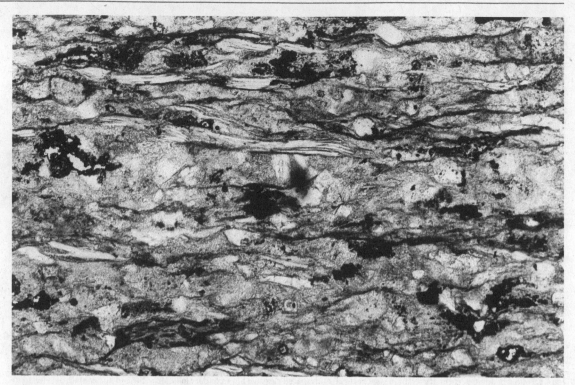

Fig. 4.6. Bedding-parallel diagenetic foliation defined by elongate detrital micas. Collio Formation. Southern Alps, Italy. Width of view 1.8 mm. PPL

frayed edges (Fig. 4.6). These micas are usually subparallel to bedding. Diagenetic foliation is thought to be the result of diagenetic compaction of sediment that contains detrital micas (Williams 1972a; Borradaile et al. 1982; Sintubin 1994a). The micas have rotated passively into an orientation parallel to bedding during compaction. Diagenetic foliation is an example of a foliation defined by the preferred orientation of micas that is not associated with folds. It is thought to precede and play an important role in development of secondary foliations in pelitic rocks (see below).

Maxwell (1962) and Roy (1978) have postulated that diagenetic or dewatering foliations may also be oblique to bedding and associated with synsedimentary folding, and may even be the initial stage of slaty cleavage. However, this idea is now largely abandoned since such foliations can usually be shown to have formed after the rock lithified. Oblique synsedimentary foliations do occur but seem to be extremely rare.

4.2.3
Secondary Foliations

Below, we present a morphological classification of secondary foliations and discuss the main processes involved in their development. Secondary foliations may show a large variation of morphological features. On the basis of these characteristics, a number of more or less descriptive names have been used such as slaty cleavage, crenu-lation cleavage, differentiated layering, fracture cleavage, schistosity etc. (see definitions in the glossary). Unfortunately, the use of these names is not uniform and some have been used with genetic implications. For example, the name *fracture cleavage* has been used for a discontinuous foliation with finely spaced compositional layering that possibly originated by preferential dissolution along fractures that are no longer visible; other interpretations of such structures that do not involve fractures are possible and the use of such genetic names should therefore be avoided. For this reason, we aim to use purely descriptive terms.

The concepts of *cleavage* and *schistosity* are so widely used that we maintain them as general terms for foliations thought to be of secondary origin. Cleavage is generally used for fine-grained rocks up to the scale where individual cleavage forming minerals (e.g. micas) can be distinguished with the naked eye; schistosity for more coarse-grained secondary foliations. Cleavage and schistosity therefore cover the complete range of secondary foliations, especially in field descriptions. Notice, however, that this distinction by grain size is not expanded consistently to all parts of foliation terminology; terms like crenulation cleavage, shear band cleavage, cleavage domain and cleavage lamellae (Sect. 4.2.6; Fig. 4.7) are generally used regardless of grain size. Another term that is occasionally used is *gneissosity*, for a coarse-grained secondary foliation in gneiss; use of this term is not recommended because of possible confusion with compositional layering in gneiss.

Morphological classification of foliations
(using an optical microscope)

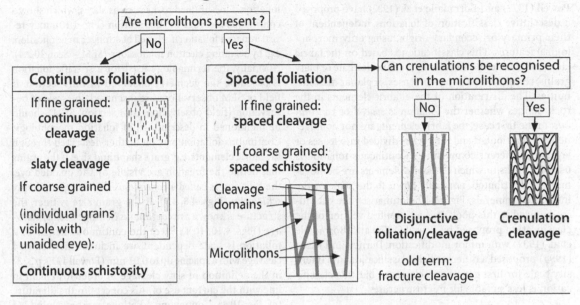

Are microlithons present ?

No → **Continuous foliation**

Yes → **Spaced foliation** → Can crenulations be recognised in the microlithons?

Continuous foliation

If fine grained:
continuous cleavage
or
slaty cleavage

If coarse grained

(individual grains visible with unaided eye):

Continuous schistosity

Spaced foliation

If fine grained:
spaced cleavage

If coarse grained:
spaced schistosity

Cleavage domains

Microlithons

Can crenulations be recognised in the microlithons?

No → **Disjunctive foliation/cleavage**
old term:
fracture cleavage

Yes → **Crenulation cleavage**

Useful criteria to describe spaced foliations:

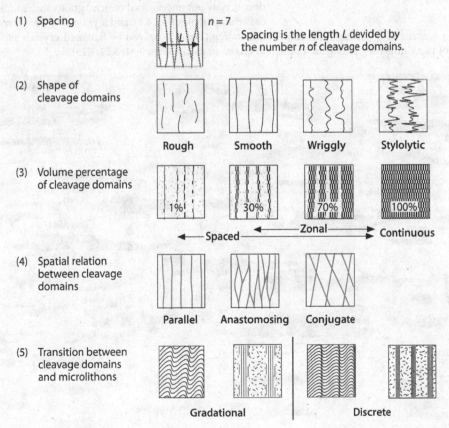

(1) Spacing

$n = 7$

Spacing is the length L devided by the number n of cleavage domains.

(2) Shape of cleavage domains

Rough   Smooth   Wriggly   Stylolytic

(3) Volume percentage of cleavage domains

1%   30%   70%   100%

← Spaced   ← Zonal →   Continuous →

(4) Spatial relation between cleavage domains

Parallel   Anastomosing   Conjugate

(5) Transition between cleavage domains and microlithons

Gradational   Discrete

Fig. 4.7. Morphological classification of foliations using an optical microscope. (After Powell 1979 and Borradaile et al. 1982)

4.2.4
Morphology of Foliations

Powell (1979) and Borradaile et al. (1982) have proposed a descriptive classification of foliations, independent of their primary or secondary origin, using only morphological features. This classification is based on the fabric elements that define the foliation such as elongate or platy grains, compositional layers or lenses, or planar discontinuities. The distribution of these fabric elements in the rock defines whether the foliation is *spaced* or *continuous*. In the first case, the fabric elements are not homogeneously distributed and the rock is divided into lenses or layers of different composition. Continuous foliation is used for rocks in which the fabric elements are homogeneously distributed, normally down to the scale of the individual minerals. Figure 4.7 summarises the classification used in this book. It is a simplified version of the classification proposed by Powell (1979) and Borradaile et al. (1982) with minor modification. Durney and Kisch (1994) proposed a different field classification and intensity scale for first generation cleavages, but this classification is less suitable for microstructures.

4.2.5
Continuous Foliation

A continuous foliation consists of a non-layered homogeneous distribution of platy mineral grains with a pre-

ferred orientation. Most common are minerals such as mica or amphibole (Fig. 4.8), but quartz (Fig. 3.31) or other minerals (Fig. 3.39) may also define a continuous foliation. Fine-grained rocks such as slates, which show a continuous cleavage in thin section (Fig. 4.10), may reveal a spaced foliation if studied at stronger magnification, e.g. by scanning electron microscope (SEM – Sect. 10.2.4). Normally, the terminology used for a specific foliation is based on the geometry observed in thin section. If field or SEM observations are discussed, the scale of observation (field observation, thin section, SEM) should be mentioned in descriptions of foliation morphology. Continuous foliation may be further described through the fabric elements, e.g. grain shape and size. If the grains that define the foliation are visible by the unaided eye, the foliation is called a *continuous schistosity* or simply *schistosity* (Figs. 4.8, 4.9). If the grain size is finer, the structure is known as a *continuous cleavage* or *slaty cleavage* (Figs. 4.10, 10.4). Since the continuous nature of a foliation is scale-dependent, we include finely spaced cleavage with a spacing up to 0.01 mm (Powell 1979, p 333) in the definition of slaty cleavage in order to remain in line with the current use of this concept in the literature (cf. Fig. 10.4). Continuous foliations can be subdivided into mineral foliations, defined by the preferred orientation of platy but undeformed mineral grains such as micas or amphiboles (Fig. 4.8), and a *grain shape preferred orientation* (Box 4.2) defined by flattened crystals such as quartz or calcite (Figs. 3.31, 3.39, 4.25).

Fig. 4.8. Continuous schistosity defined by parallel crystals of biotite, muscovite and quartz. Mt Isa, Australia. Width of view 1.8 mm. PPL

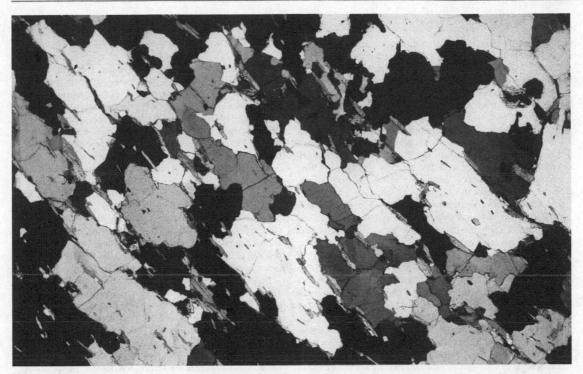

Fig. 4.9. Continuous foliation (schistosity) in an upper-amphibolite facies micaceous quartzite, mainly defined by subparallel micas. Note the irregular shape of quartz crystals as a result of secondary grain growth (Sect. 3.12.2). Undulose extinction and subgrains are probably due to late deformation after grain growth. Ribeira Belt, Rio de Janeiro State, Brazil. Width of view 18 mm. CPL

Fig. 4.10. Continuous cleavage on thin section scale (slaty cleavage) in a slate, defined by fine-grained micaceous material. Fabrics like this may be spaced foliations that contain domains without fabric elements or with folded older fabric elements when studied by SEM (Fig. 10.4) or TEM. Castellbo, Pyrenees, Spain. Width of view 1.8 mm. PPL

Box 4.2   Shape fabrics

In deformed rocks it is common to find a fabric composed of elongate or disc-like grains or grain aggregates that define planes, or lineations if they share a linear direction. In general, this type of fabric is known as a shape preferred orientation (SPO).

If the fabric is composed of elongate or disc-like single grains of minerals, which normally form equidimensional grains in an undeformed rock, such as quartz or calcite, the fabric is known as a grain shape fabric or grain shape preferred orientation (GSPO). A GSPO can be planar, linear of both. A linear GSPO is known as a grain lineation (Fig. B.4.2; Sect. 4.3). The term shape preferred orientation or GSPO is *not* used for a preferred orientation of platy minerals such as micas or amphiboles that also have an elongate shape in undeformed rocks.

GSPO can be developed in primary grains such as sand grains in a quartzite or oolites. Monocrystalline ribbons (Boullier and Bouchez 1978) can be regarded as an extreme case of such GSPO. These ribbons form mostly in minerals where only a single slip system operates such as orthopyroxene but can also form in quartz or feldspar under certain metamorphic conditions (Sect. 3.12). Monocrystalline ribbons do not only form by deformation; they can also develop from polycrystalline ribbons that are bounded by other minerals due to grain boundary migration, e.g. in GBM or static recrystallisation (Sect. 3.10, 3.11).

More commonly, GSPO develops in aggregates of secondary, recrystallised grains (Means 1981; Lister and Snoke 1984; Figs. B.4.2, 4.31c, 5.10f). Examples are shown in Figs. 4.25, 5.24, 5.30, 5.31. GSPO can develop by crystalplastic processes such as dislocation creep or solid-state diffusion (Sect. 3.4, 3.8) but solution transfer may also play a role. In the case of dislocation creep, the deformation intensity of each individual grain depends on its lattice orientation, since the activity of slip systems is a function of their orientation with respect to the kinematic frame (see below) (Fig. 4.24). This can explain why some quartz grains in a deformed quartzite may be much less deformed than others (Fig. 4.24); however, other reasons may be a considerable difference in original grain shape or late preferential grain growth of some crystals. At high homologous temperatures (Sect. 3.14), diffusion of ions through a crystal lattice becomes increasingly important (Nabarro-Herring creep). Grains can be flattened in this case without activity of slip systems or the presence of an intergranular fluid. This process may aid development of a grain shape-preferred orientation in high-grade rocks, but its importance is uncertain since the number of active slip systems also increases with temperature.

If subgrains obtain an elongate shape, they may define a weak foliation on thin section scale, which is named a subgrain shape preferred orientation (SSPO). SSPO and GSPO grade into each other where SGR recrystallisation transforms subgrains into new grains (Sect. 3.7.2).

The strength and orientation of a GSPO depends on finite strain, but there is no simple relationship. In an ideal case, a GSPO would form by deformation of a set of spheres with isotropic rheological properties. In this case, the GSPO would exactly mimic the geometry of the strain ellipsoid; a planar shape fabric would be parallel to the *XY*-plane of finite strain, and a linear shape fabric with the *X*-axis. However, in nature an older SPO may be overprinted and the resulting shape will not reflect finite strain of the latest deformation phase. Also, grains are not passive spheres, especially if they deform by dislocation creep. In this case, they only deform along certain slip planes and, depending on the overall flow field, some will deform more strongly than others (Chap 4.2.7.4; Fig. 4.24; Wilson 1984). If a rock consists of equidimensional grains that have an older lattice preferred orientation, e.g. by static recrystallisation

of a deformed rock, deformation by dislocation creep will form a GSPO that can be oblique to finite strain axes for the last deformation phase. If the fabric consists of an alternation of grains of different minerals, as in a granite, some minerals will deform more strongly than others. Moreover, grains do not always deform up to very high strain, but may be affected by recrystallisation to form new grains with a low aspect ratio, so that aggregates consist of grains of different shapes, depending on when they formed during the deformation process. The mean aspect ratio of all these grains will only reflect part of the finite strain. GBAR and static recrystallisation can also change the shape of grains during and after deformation.

If flow is coaxial a GSPO will at least lie approximately parallel to finite strain axes, even if the aspect ratio of the grains is not the same as the finite strain ratio. However, in non-coaxial flow the finite strain ellipse rotates away from the orientation of ISA with progressive deformation (Sect. 2.7). If grains recrystallise to form during this deformation process, their mean orientation will only reflect part of the finite strain, and the GSPO will lie somewhere between the orientation of the flow ISA and finite strain axes.

Besides a preferred orientation of single grains or subgrains, aggregates of grains can also have a preferred orientation, which is visible, if individual aggregates are bounded by grains or aggregates of other minerals. This type of fabric could be named an aggregate shape preferred orientation (ASPO; Fig. B.4.2). This kind of fabric is also generally referred to as a shape fabric, either a planar shape fabric or a linear shape fabric. The term polycrystalline ribbon is also occasionally used in thin section descriptions. Another name for a linear shape fabric is an aggregate lineation (Sect. 4.3).

ASPO most commonly forms by deformation of older aggregates of polycrystals such as conglomerates, or by deformation and recrystallisation of large grains (Fig. B.4.1; Sect. 4.2.7.5). Piazolo and Passchier (2002a) demonstrated that, even if an original fabric is undeformed, the strength of an ASPO depends not only on strain intensity, but also on the initial mineral distribution and grain size of the rock (Fig. B.4.1). Since the size of dynamically recrystallised grains depends on differential stress (Sect. 9.6.2), a fine-grained poly- or monomineralic rock will flatten but may recrystallise to grains of the same size. In such cases, no ASPO can form. Only if a rock is polymineralic, and if the original grain size or aggregate size exceeds that of the new recrystallised grains a new ASPO will form. An example of this influence is shown in Fig. B.4.1 where the effect of original grain size and fabric on development of an ASPO during deformation and dynamic recrystallisation is shown. Other possible mechanisms to form an ASPO are breakdown of large grains to other phases such as the common reaction of garnet to plagioclase upon decompression, followed by deformation (e.g. garnet to plagioclase), and by boudinage of a layer into rods or discs.

As for GSPO, there is a relation between ASPO and finite strain. ASPO is not easily reset by recrystallisation and therefore has a tendency to lie close to the *XY*-plane of the strain ellipsoid, provided it formed from equidimensional older elements, and is not overprinting an older ASPO. However, there is usually no good correlation between the 3D aspect ratio of aggregates and 3D finite strain geometry. Freeman and Lisle (1987) have shown that viscous spheres of a certain rheology embedded in a material of an other rheology do not mimic the shape of the strain ellipsoid, but tend to be more linear if the viscosity is higher and more planar if the viscosity is lower than that of the matrix. In conclusion, ASPO can in some cases be used to find the orientation of finite strain axes, but has to be interpreted with great care.

A special type of shape fabric can be defined by domains composed of grains that share a certain crystallographic preferred ori-

Box 4.2   *Continued*

entation, but not necessarily a GSPO (Eisbacher 1970; Garcia Celma 1982; Knipe and Law 1987; Lloyd et al. 1992; Law et al. 1990; Herwegh and Handy 1998). Such domains can have an elongate shape and define a domain shape preferred orientation (DSPO) (Figs. B.4.1, B.4.2) that can form a weak foliation oblique to other fabric elements. DSPO have also been referred to as orientation families of grains. Individual domains can also change shape by recrystallisation or rotation of grains. DSPO is usually inclined in the same direction as GSPO in the rock, but at a smaller angle to the fabric

attractor (Pauli et al. 1996; Herwegh and Handy 1998). DSPO can be an active foliation (Sect. 4.2.9.2) if it is defined by shear bands, but DSPO can also be strain-insensitive. It may form by SGR recrystallisation of larger grains.

All types of shape fabrics can occur combined in a single rock fabric, and they may also combine with other foliation and lineation types. A special name for a GSPO oblique to ASPO or mylonitic foliation is *oblique foliation* (Sect. 5.6). Three possible combinations of LPO are shown in Fig. B.4.2.

Fig. B.4.1.
Diagram depicting the effect of an initial fabric on the development of a shape preferred orientation (Fig. B.4.2) with respect to dynamic recrystallization. The diagram is highly simplified. Represented polymineralic rocks are limited to rocks with two mineral species. Grain sizes of recrystallised grains are assumed to be the same as initially fine grained phases. (After Piazolo and Passchier 2002a)

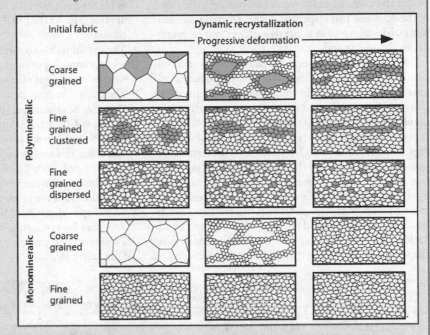

Fig. B.4.2.
Different types of shape fabrics. **a**, **b** and **c** show single shape fabric types, **d**, **e** and **f** show combinations of shape fabrics

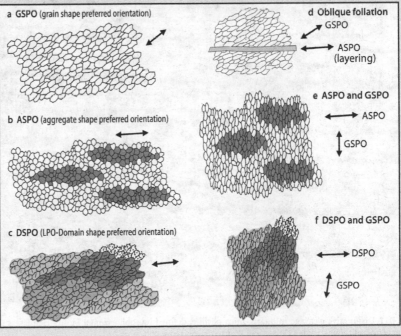

4.2.6
Spaced Foliation

Rocks with spaced foliation consist of two types of domains, *cleavage domains* (also known as *cleavage lamellae*) and *microlithons* (e.g. Fig. 4.5). As an alternative, the terms *M domain* (mica-rich), or *P domain* (phyllosilicate-rich) and *Q domain* (quartz-rich) have been used in micaschist or phyllite (Shelley 1993). Cleavage domains are planar and contain fabric elements subparallel to the trend of the domains. In metapelites, cleavage domains are usually rich in mica and in minerals such as ilmenite, graphite, rutile, apatite and zircon.

Microlithons lie between cleavage domains and contain fabric elements that have a weak or no preferred orientation, or which contain fabric elements oblique to the cleavage domains. Spaced foliations may be further subdivided according to the structure in the microlithons. If these contain microfolds of an earlier foliation (e.g. Figs. 4.5, 4.12, ⊘Video 4.12, ⊘Photo 4.12) the term *crenulation cleavage* is applied (Rickard 1961). If not, the structure is known as *disjunctive foliation* (or *disjunctive cleavage* if fine-grained, e.g. Fig. 4.11). The more general terms *spaced cleavage* and *spaced schistosity* are also used to describe fine-grained and coarse-grained disjunctive foliation (Fig. 4.7). Some spaced foliations contain lens-shaped microlithons and may be called *domainal spaced*

foliation (Fig. 4.14) or, if the spacing is sufficiently narrow, *domainal slaty cleavage*. Other morphological features of spaced foliations that may be considered in their description are (Fig. 4.7):

- The spacing of the cleavage domains.
- The shape of cleavage domains: rough (Gray 1978), smooth (e.g. Fig. 4.5), wriggly or stylolytic.
- The percentage of cleavage domains in the rock; if this is higher than 30%, the term *zonal foliation* may be applied (Fig. 4.13). At 100% the foliation becomes continuous.
- The spatial relation between cleavage domains: parallel, anastomosing or conjugate (two intersecting directions without signs of overprinting).
- The transition from cleavage domain to microlithon. This may be gradational (Figs. 4.12, 4.13, ⊘Video 4.12, ⊘Photos 4.12, 4.13) or discrete (e.g. Figs. 4.14, 4.15, 4.20). Note, however, some discrepancy in the literature about this use of *gradational*; some authors (Gray 1977; Powell 1979; Kisch 1998) use the term zonal crenulation cleavage to describe these gradational structures. Our use follows Borradaile et al. (1982).
- The shape of microfolds in crenulation cleavage. This may be symmetric (e.g. Fig. 4.12, ⊘Video 4.12, ⊘Photo 4.12), asymmetric (e.g. Fig. 4.13, ⊘Photo 4.13), tight, open, etc.

Fig. 4.11. Disjunctive cleavage in quartz-mica phyllite, defined by subhorizontal biotite-rich layers (cleavage domains) and quartz-mica layers (microlithons). Leiden Collection. Width of view 4 mm. PPL

Fig. 4.12. Differentiated crenulation cleavage in phyllite with symmetric microfolds: the foliation is defined by cleavage domains (*flanks of microfolds*) and microlithons (*fold hinge areas*). Note the difference in composition of the two domains and the gradual transition between both. Cordillera Real, Ecuador. Width of view 4 mm. PPL

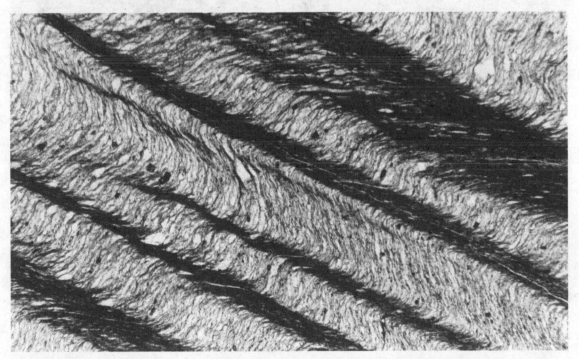

Fig. 4.13. Zonal crenulation cleavage with a percentage of cleavage domains variable from about 25% in the *lower left part* to almost 50% in the *upper right part* of the photograph; note the gradual transition between cleavage domains and microlithons, and the asymmetric character of microfolds resulting in relative mica enrichment predominantly in one of two alternating fold limbs (cf. Fig. 4.12, where both limbs are identical). Leiden Collection. Width of view 4 mm. PPL

Fig. 4.14. Domainal spaced cleavage with chlorite stacks (Sect. 7.4.2) in microlithons. Collio Formation. Southern Alps, Italy. Width of view 1.8 mm. PPL

Fig. 4.15. Discrete crenulation cleavage (S_2 *subhorizontal*) overprinting a slaty cleavage (S_1 *steep*) that is subparallel to bedding (S_0 *white layer at left*). The crenulation cleavage is selectively developed in more pelitic material and changes abruptly over a lithological boundary (*white layer*). Note the apparent offset of the white layer in the *lower left corner* that may reflect fault movement or removal of a flexure by pressure solution (cf. Fig. 4.21). Leiden Collection. Width of view 4 mm. PPL

The morphology of crenulation cleavages may show a vast array of variation (Fig. 4.12, ⊘Video 4.12, ⊘Photo 4.12, Fig. 4.13, ⊘Photo 4.13; Figs. 4.19–4.21, 4.35, 4.37); important factors that influence the final morphology, apart from the lithotype, are temperature and deformation intensity. Figures 4.18 and 4.19 (⊘Photo 4.19b1–7) show the inferred range of stages in crenulation cleavage development according to these two parameters (however, see also Box 4.5).

A special type of spaced foliation is *compositional layering*, where microlithons and cleavage domains are wide and continuous enough to justify the use of the term layering. Normally, this term is applied if the layering is visible to the unaided eye in a hand specimen.

Many transitional forms between foliation types as defined above occur in nature. In fact, the variation in morphology is almost infinite and we should realise that the proposed classification is meant as a way to facilitate communication between geologists and not as an objective in itself. For this reason, we have not tried to define strict boundaries between categories, and we advocate the use of a minimum of terminology. Where necessary, a good photograph or detailed drawing can supplement a description.

A foliation may change its morphology drastically within a single thin section (Fig. 4.15, ⊘Photo 4.15), or even disappear completely. This is generally related with the transition from one lithotype to another; foliation development is strongly dependent on lithotype. However, local strain distribution around fold hinges has its influence on foliation development too, and may produce a remarkable variation in foliation morphology along a single layer.

It is generally difficult to quantify the intensity or strength of foliations. However, relative strength of foliations can be compared in samples with a continuous foliation and similar grain size and mineral content using X-ray texture goniometry (Sect. 10.3.5; van der Pluijm et al. 1994; Ho et al. 2001).

4.2.7
Mechanisms of Foliation Development

4.2.7.1
Introduction

Secondary foliations develop in response to permanent rock deformation. The main controlling factors on their

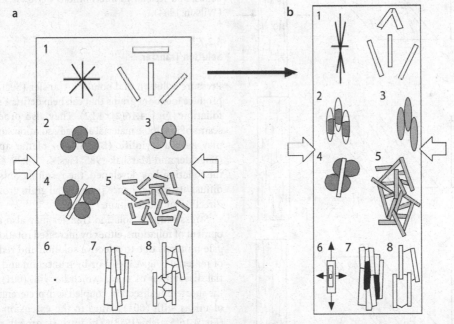

Fig. 4.16. Schematic diagram of some important mechanisms contributing to development of secondary foliations in rocks. **a** Fabrics at the onset of deformation. **b** Fabric elements after deformation. *1* Elongate crystals (*open rectangles*) rotate in response to deformation in a way similar to theoretical passive markers (*solid lines*) but there are differences; minerals may fold when normal to the shortening direction and thus strengthen a preferred orientation, or rotate at slower rate than material lines when highly oblique to the shortening direction. *2* Mineral grains change shape by stress-induced solution transfer; *grey* is original material, *white* are overgrowths. *3* Mineral grains change shape by crystalplastic deformation such as dislocation creep or solid-state volume diffusion. *4* Polymineralic aggregates develop foliations by processes *1 + 2* when assisted by stress-induced solution transfer. *5* Grain growth of micas parallel to (001) during or after shortening leads to an increase of foliation intensity because grains oriented in the direction of the foliation can grow to greater length than those in oblique orientations. *6* Oriented nucleation and growth of a mineral in a stress field. *7* Mimetic growth of elongate grains due to restrictions in growth direction imposed by an existing foliation. *8* Restricted growth parallel to platy minerals

development are rock composition, stress orientation and magnitude, metamorphic conditions including temperature, lithostatic and fluid pressure, and fluid composition. The relation between morphology and genetic processes is usually complex, and the description of foliation morphology should therefore be separated from the interpretation of processes involved. Our knowledge of these processes is still incomplete, although research over the last 30 years has increased our understanding considerably (e.g. Siddans 1972; Wood 1974; Means 1977; Oertel 1983; Skrotzki 1994; Durney and Kisch 1994; Worley et al. 1997; Ho et al. 1996, 2001; Williams et al. 2001; Fueten et al. 2002).

This section gives a list of the main processes that are currently thought to play a role during formation of secondary foliations (Figs. 4.16, 4.17). In a number of cases, like the examples cited below, it may be possible to indicate which processes have been important, but in general several of the processes probably operate together.

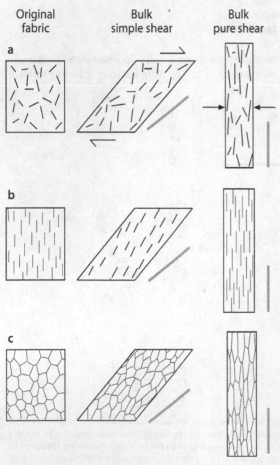

Fig. 4.17. Development of some foliations by progressive simple shear and pure shear of: **a** a random initial orientation of isolated elongate or planar minerals; **b** an initial preferred orientation of isolated elongate or planar minerals; **c** originally equidimensional grains. In **c** a grain shape preferred orientation is formed. *Grey bars* indicate the direction of the *XY*-plane of finite strain for the deformation shown

4.2.7.2
Mechanical Rotation of Tabular or Elongate Grains

During homogeneous ductile deformation, a set of randomly oriented planes will tend to rotate in such a way that their mean orientation will trace the direction of the *XY*-plane of finite strain (Figs. 4.16(1), 4.17a; Jeffery 1922; March 1932). A similar effect is thought to apply to tabular or elongate grains with a high aspect ratio such as micas or amphiboles in deforming rocks (Gay 1968; Oertel 1970; Tullis and Wood 1975; Tullis 1976; Wood et al. 1976; Willis 1977; Wood and Oertel 1980; Means et al. 1984; Lee et al. 1986; Ho et al. 1995, 1996; Sintubin 1994b, 1996, 1998; Sintubin et al. 1995; Fig. 4.16(1)). If an earlier preferred orientation was present, the foliation will not trace the *XY*-plane (Fig. 4.17b) in the case of bulk simple shear; deformed originally equidimensional grains will trace the *XY*-plane in this case.

If deformation in a rock with random tabular or elongate grains, such as a mica-bearing granite, occurs along spaced shear zones, rotation of fabric elements in these shear zones can develop a spaced foliation in a homogeneous parent rock (Wilson 1984); micas will tend to become parallel and relatively closely spaced in the shear zones, and less so in microlithons between shear zones (Wilson 1984).

4.2.7.3
Solution Transfer

Pressure solution and solution transfer (Sect. 3.3) may produce inequant grains that can help define a secondary foliation (Figs. 4.16(2), 4.22). They also produce dark seams of insoluble material along dissolution surfaces that may have a stylolitic (Box 4.3) or planar appearance (Engelder and Marshak 1985; Figs. 4.15, 4.20, 4.21). After the foliation has developed, the resulting anisotropy of diffusivity may enhance preferential grain growth in the direction of the foliation.

Stress-induced solution transfer may also aid development of foliations, either by increased rotation of elongate minerals due to selective solution and redeposition of material (Fig. 4.16(4)) or by truncation and preferential dissolution of micas which lie with (001) planes in the shortening direction, coupled with preferential growth of micas with (001) planes in the extension direction (Fig. 4.16(5); Ishii 1988). The intrinsic growth rate of micas is anisotropic and fastest parallel to (001) planes (Etheridge et al. 1974; Rosenfeld 1985). Solution transfer including micas will therefore lead to a preferred orientation, even in the absence of rotation (Ishii 1988; see mimetic growth, Sect. 4.2.7.6).

Solution transfer is very important for the formation of spaced foliations, especially younger foliations overprinting older ones (Sect. 4.2.10.2).

Box 4.3   Stylolites

Pressure solution is common in low-grade deformation and is usually active throughout a rock volume on the grain scale, leading to development of foliations and grain-scale dissolution and deposition features. Pressure solution can locally be enhanced, for example in strain caps aside a rigid object. Localised pressure solution is often concentrated along surfaces that may originate as joints or fractures, particularly in limestone (Stockdale 1922; Dunne and Hancock 1994; Petit and Matthauer 1995; Renard et al. 2004), but also in other macroscopically homogenous, fine grained rocks (Dewers and Ortoleva 1990; Railsback and Andrews 1995; Railsback 1998; Karcz and Scholz 2003; Gratier et al. 2005). Such surfaces are normally highly indented and consist in three dimensions of interlocking teeth of wall rock. These surfaces are therefore known as stylolites (from Latin *stylus*, a stake or pen). Stylolites can be subdivided into bedding parallel and transverse stylolites. Teeth in the stylolite surface commonly have secondary phases such as mica grains along the crowns, while teeth walls are commonly parallel so that the stylolite might be pulled apart without breaking the teeth. The indented shape of stylolites forms by preferred pressure solution along one side of the surface, usually due to a concentration of non-soluble phases on the opposite side (Fig. B.4.3a). A difference between stylolites and amoeboid grain boundaries (Box 3.9) formed by grain boundary migration is the concentration of material on the crowns of the teeth, and protrusions with inward sloping walls (Fig. B.4.3b). Stylolites are generally enriched in insoluble material such as opaques and mica with respect to the wall rock. Where the wall rock is locally rich in insoluble material, the stylolite is also locally enriched in such material (Borradaile et al. 1982). Stylolites are most common in carbonate rocks with a certain clay content, but can also occur in some sandstones. Bedding parallel stylolites are commonly considered to be diagenetic (Andrews and Railsback 1997). Transverse stylolites occur in rocks with insoluble residue material of 2–20 wt%.

In many texts, the word stylolite is used for pressure solution generated surfaces with teeth normal to the stylolite surface, while slickolites have teeth oblique to the surface (Fig. B.4.3c). A classification of stylolites was presented by Guzetta (1984) and Andrews and Railsback (1997).

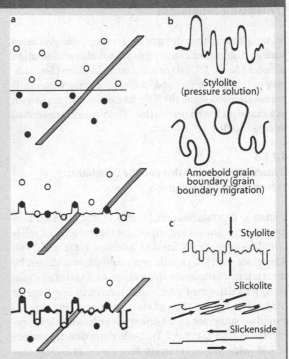

Fig. B.4.3. a Development of a stylolite in rocks with insoluble inclusions; material behind inclusions may be protected from solution and form interlocking *teeth*. **b** Stylolites formed by pressure solution differ from amoeboid grain boundaries formed by grain boundary migration in that they have teeth with parallel sides that allow the two halves to be "pulled apart". **c** Explanation of the terms stylolite, slickolite and slickenside. In a stylolite, teeth and inferred shortening direction are normal to the plane, in slickolites oblique and in slickensides parallel

Fig. B.4.4.
Stylolite in limestone. A late calcite-filled vein transects part of the stylolite. Width of view 2 mm. PPL. (Courtesy Daniel Köhn)

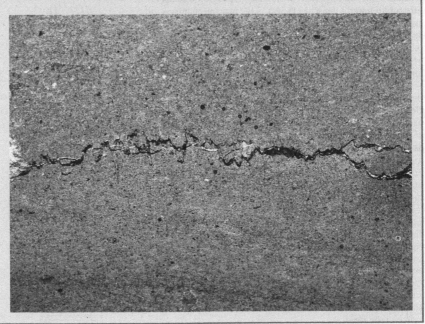

4.2.7.4
Crystalplastic Deformation

Mineral grains that deform by crystalplastic processes such as dislocation creep, pressure solution and solution transfer (Sects. 3.3, 3.4) or solid state diffusion (Sect. 3.8) may obtain a flattened and/or elongate shape with maximum extension along the *XY*-plane of finite strain known as a shape preferred orientation. This process is described in detail in Box 4.2.

4.2.7.5
Dynamic Recrystallisation and the Orientation of New Grains and Subgrains

Dynamic recrystallisation (Sect. 3.7) and oriented new growth of, e.g. mica are important mechanisms of foliation development (White and Johnston 1981; Ishii 1988; Kanagawa 1991). Dynamic recrystallisation is driven by the tendency to decrease free energy, such as stored strain energy in deformed grains and interfacial free energy. Kinking or tight folding of existing mica grains may accumulate sufficient strain energy to enhance bulging recrystallisation (Sect. 3.7.2). Little deformed fragments of old mica grains or strain-free nuclei can grow into the damaged crystal lattice with a preferred orientation that contributes to the secondary foliation (Fig. 4.18b).

Fig. 4.18. Inferred range of stages in crenulation cleavage development with increasing deformation (*vertical axis*) and temperature (*horizontal axis*; cf. Bell and Rubenach 1983). Figure 4.19 illustrates this same sequence with photographs. At low temperature (up to lower greenschist facies), **a**, the main mechanisms for crenulation cleavage formation are thought to be differentiation by solution transfer and rotation, whereas at higher temperatures, **b**, recrystallisation and grain growth (including new minerals) are probably dominant factors. At stage *1* gentle crenulations have formed in the original foliation S_1, but no S_2 cleavage is apparent yet. Some recrystallisation may occur in D_2 fold hinges. At stage *2* the crenulations are somewhat tighter and a discrete S_2 crenulation cleavage is visible. S_1 is still the dominant fabric. At stage *3* the new cleavage has developed to such an extent that S_1 and S_2 are of approximately equivalent importance in the rock. Recrystallised microfolds known as polygonal arcs may be visible at the higher temperature range, especially in **b***3*. At stage *4* S_2 clearly predominates and S_1 is only recognisable in some relic fold hinges. In stage **b***4* new grains grown along S_2 dominate the fabric. Finally, stage *5* shows the end product of the process where S_1 is completely transposed and not recognisable any more. Most rocks will follow some path from the *upper left* to the *lower right corner* of the diagram during development of a crenulation cleavage (compare Fig. 4.28). Other factors that influence the development of crenulation cleavage are the presence and activity of a fluid phase, the presence of soluble minerals and the growth of new minerals. The step to complete transposition at low temperature (**a***4*–**a***5*) seems to be difficult without recrystallisation and grain growth. This may be the reason that old foliations are often better preserved in low-grade rocks

In the case of quartz and feldspars, recovery may lead to subdivision of equant grains into elongate subgrains (e.g. Fig. 4.26). If further deformation leads to SGR recrystallisation (Sect. 3.7.3), the subgrains may become new independent grains that, by their shape, define a foliation (subgrain shape preferred orientation; Box 4.2). Recrystallisation is associated with reequilibration of the chemical composition of minerals in the rock to metamorphic conditions during cleavage development (White and Knipe 1978; Gray 1981; Knipe 1981; White and Johnston 1981; Ishii 1988; Williams et al. 2001). In many cases the minerals in cleavage domains reflect metamorphic conditions during cleavage development, and those in the microlithons older, even diagenetic conditions (Knipe 1979, 1981; White and Johnston 1981; Lee et al. 1984, 1986).

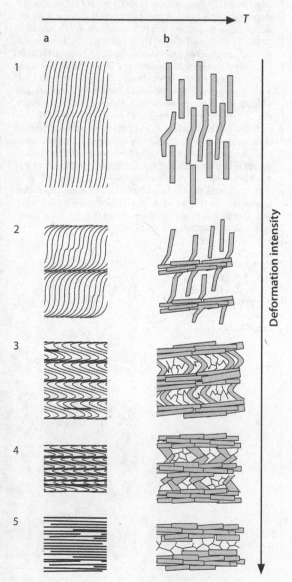

Fig. 4.19.
Natural examples of foliations from various areas, which are thought to represent stages of the foliation development sequence shown in Fig. 4.18. Leiden Collection. Width of view **a***1* 2 mm; **a***2* 2 mm; **a***3* 2.5 mm; **a***4* 2 mm; **a***5* 2 mm; **b***1* 1 mm; **b***2* 2 mm; **b***3* 2 mm; **b***4* 1 mm; **b***5* 1 mm. PPL (see also ⊘Photo 4.19b1–7)

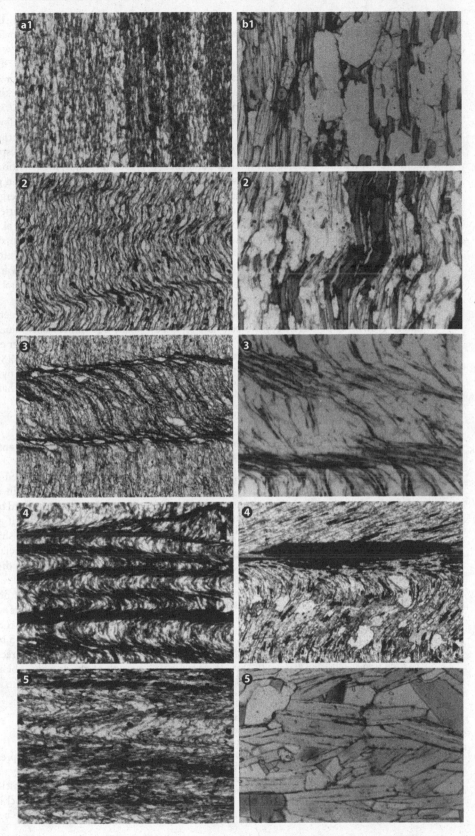

4.2.7.6
Static Recrystallisation and Mimetic Growth

Foliations can be modified in several ways after deformation ceases. If low-grade foliation is subjected to considerable temperature increase in the absence of deformation, as in contact aureoles, the strength of this foliation normally decreases due to nucleation and growth of new minerals over the foliated fabric in random orientation, changing a foliated rock into a *hornfels*. Limited heating, however, without a change in mineral paragenesis, can also strengthen a foliation by growth of micas that are approximately parallel to the foliation and preferred dissolution of grains in unfavourable orientations (Ho et al. 2001). The latter probably occurs because of stored strain energy in grains with (001) planes oblique to the original shortening direction.

In some rocks, elongate crystals that help define a secondary foliation may actually have grown in the direction of the foliation after the deformation phase responsible for that foliation ceased. This process is known as *mimetic growth*. The elongate crystals may have replaced

---

Box 4.4   Mylonitic foliation and monocrystalline ribbons

A foliation in mylonite is usually referred to as *mylonitic foliation*; it is generally a spaced foliation composed of alternating layers and lenses with different mineral composition or grain size, in which more or less strongly deformed porphyroclasts are embedded; the mylonitic foliation wraps around these porphyroclasts (Sect. 5.3). Some lenses are single crystals with an unusual planar or linear shape that define or strengthen a foliation in the rock. Such lenses are known as monocrystalline *ribbons* (Sect. 5.3.5). Common examples are quartz ribbons, but ribbons of mica, feldspar and orthopyroxene are also known (Sect. 3.12). In low to medium-grade mylonites, quartz ribbons are strongly elongate and show strong undulose extinction, deformation lamellae, subgrain structures and dynamic recrystallisation, mainly along the rim of the ribbons. Commonly, such ribbons show extinction banding parallel to their long axis, which may be due to folding of the crystal lattice (Boullier and Bouchez 1978; Passchier 1982a). Most ribbons probably form by extreme flattening and/or stretching of large single crystals.

In high-grade gneiss, quartz ribbons consist of single crystals with an elongate shape, which lack intracrystalline deformation structures (Figs. 5.11, 5.12). Such monocrystalline quartz ribbons are also known as *platy quartz* (Behr 1965; Frejvald 1970; Boullier and Bouchez 1978) and commonly include equidimensional or elongate feldspar grains. The quartz may contain rutile needles that have a preferred orientation or show boudinage, indicating that these ribbons have been subject to strong deformation. Monocrystalline quartz ribbons in high-grade gneiss are probably formed by strong deformation followed by recovery and significant grain boundary migration that removed most older grain boundaries and intracrystalline deformation structures (Sect. 3.12.2). In this case, static recrystallisation leads to elongate single crystals of quartz because other minerals hamper grain growth in directions normal to the ribbons.

---

existing minerals inheriting their shape (Fig. 4.27a); they may have nucleated and grown within a fabric with strong preferred orientation, following to some extent this orientation (Figs. 4.16(7), 4.27b); or they may have grown along layers rich in components necessary for their growth, in this way mimicking the layered structure in their shape fabric (Sect. 7.3; Fig. 4.27c). Some monocrystalline ribbons may develop in this way. Mimetic growth is probably an important process in the later stages of foliation development, especially at medium to high-grade metamorphic conditions. Since micas grow fastest in the (001) direction, grain growth catalysed by reduction of interfacial grain energy can lead to strengthening of an existing preferred orientation (Figs. 4.16(5), 4.28, ⊕Photo 4.28; Etheridge et al. 1974; Ishii 1988). Crenulation cleavage may be progressively destroyed by this process transforming itself into an irregular schistosity (Fig. 4.28, ⊕Photo 4.28). Partly recrystallised relicts of crenulation cleavage microfolds as in Fig. 4.28c are known as *polygonal arcs*.

An effect similar to mimetic growth is growth of normally equidimensional minerals such as quartz or calcite between micas or other elongate crystals with a preferred orientation (Fig. 4.16(8)). Due to restriction in their growth direction imposed by the micas, such grains may obtain an elongate shape that strengthens the pre-existing foliation.

4.2.7.7
Oriented Growth in a Differential Stress Field

The possibility of oriented nucleation and growth of metamorphic minerals in a differential stress field (Fig. 4.16(6)) was suggested by Kamb (1959) and is thermodynamically possible; it may produce a strong preferred orientation of both shape and crystal habit without necessarily being associated with high strain. However, rocks subject to high differential stress are usually deformed, and it is difficult to prove that a mineral-preferred orientation did not develop by one of the processes outlined above. Some well developed schistosities in medium to high-grade rocks with undeformed crystal habit and straight grain boundaries may be a result of this process, but static recrystallisation and mimetic growth of grains which obtained their preferred orientation by rotation may form a similar fabric (Fig. 4.28).

4.2.7.8
Microfolding

If an older planar fabric is present in the rock, the associated mechanical anisotropy may give rise to a harmonic, regularly spaced folding which produces some of the most intriguing structures in rocks, crenulation cleavage. The limbs of the folds may line up to form a crude foliation,

Fig. 4.20. Discrete crenulation cleavage (S$_2$ *subhorizontal*) overprinting a slaty cleavage (S$_1$ trending from *top left* to *bottom right*). The crenulation cleavage is defined by *horizontal dark seams* with wriggly to smooth appearance. The seams are interpreted as accumulations of insoluble material along dissolution surfaces. Concepción, Chile. Width of view 1.8 mm. PPL

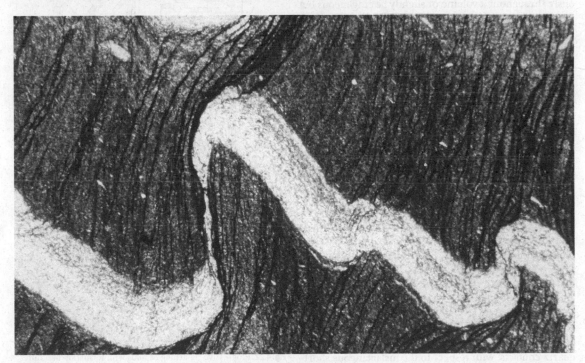

Fig. 4.21. Crenulation cleavage (S$_2$ *subvertical*) overprinting a slaty cleavage (S$_1$) that is parallel to bedding (S$_o$). Development of the crenulation cleavage was accompanied by solution effects. The extreme attenuation of the vertical fold limb in the quartz-rich (*light-coloured*) layer coincides with the presence of accentuated *dark seams* along the S$_2$ plane in adjacent micaceous layers. Both are interpreted as the result of preferred dissolution enhanced by the orientation of the fold limb, as explained in the text. This figure is a detail of Fig. 4.4. Pyrenees, Spain. Width of view 5 mm. PPL

but in many cases solution transfer or oriented crystallisation or recrystallisation of new grains (Gray and Durney 1979a,b) become important after the folds have reached a certain amplitude, and develop a spaced foliation along limbs of microfolds (Figs. 4.12, 4.13, 4.18, 4.19; White and Johnston 1981; Williams et al. 2001). Spaced foliations can, however, also form without folding of the older fabric (Fig. 4.20; Sect. 4.2.7.3; Durney 1972; Engelder and Marshak 1985). Besides harmonic microfolding of a foliation, disharmonic microfolding or kinking of individual micas can also increase mica-preferred orientation by rotation of mica segments away from the shortening direction (Fig. 4.16(1); Engelder and Marshak 1985).

4.2.8
Development of Spaced Foliations

Spaced foliations and tectonic layering have a marked uniformity in the spacing between cleavage planes and several ideas have been postulated on how this develops (Williams 1990). In most cases, some form of dissolution-precipitation and transport of material through a fluid phase in combination with a mechanical interaction is postulated for the development of spaced foliations. Three groups of mechanisms have been postulated:

One option is that the periodicity develops spontaneously throughout a volume of slightly heterogeneous but unfoliated rock in which compaction localizes by a self-organisation mechanism due to the interaction of stress and chemical gradients (Dewers and Ortoleva 1990). Macroscopic patterns of alternating cementation and compaction result which represent cleavage seams or stylolites and microlithons.

A second possibility is that foliation develops as single cleavage plane "seeds", which develop into cleavage planes, while new planes are initiated on both sides at a regular distance. This could happen if strong quartz rich domains form next to developing cleavage domains. These quartz rich domains will then initiate new cleavage domains at their margins, thus gradually filling the rock volume with a spaced fabric (Robin 1979; Fueten and Robin 1992; Fueten et al. 2002). Such mechanisms, however, are difficult to prove.

A third possible mechanism is through the development of microfolds in an older foliation (Trouw 1973; Cosgrove 1976; Gray 1979; Gray and Durney 1979a; Beutner 1980; Wright and Platt 1982; Woodland 1985; Southwick 1987; Ho et al. 1995, 1996; Worley et al. 1997; Stewart 1997; van der Pluym et al. 1998; Williams et al. 2001; Fueten et al. 2002). The folding of an earlier foliation produces a difference in orientation of planar elements, such as mica-quartz contacts, with respect to the instantaneous shortening direction. This may enhance preferred dissolution in fold limbs, which produces a secondary foliation in the form of a differentiated crenulation cleavage (Figs. 4.12, 4.22) and eventually a compositional layering in which

fold hinges may have been erased (Fig. 4.18). Figures 4.18 and 19 show a progressive sequence of development of crenulation cleavage with increasing pressure and temperature. This sequence can be understood as an example of progressive development of many spaced foliations (see also ⊙Photo 4.19b1–7 and Box 4.5).

The efficiency of differentiation by solution transfer depends on the abundance of a fluid phase and is therefore most active under diagenetic and low-grade metamorphic conditions. The mechanism is also dependent on the presence of one or more soluble minerals. Gray and Durney (1979a) published the following mineral sequence according to decreasing mobility by solution transfer: calcite > quartz > feldspar > chlorite > biotite > muscovite > opaques. In quartz- or carbonate-bearing phyllites, solution transfer seems to operate quite well:

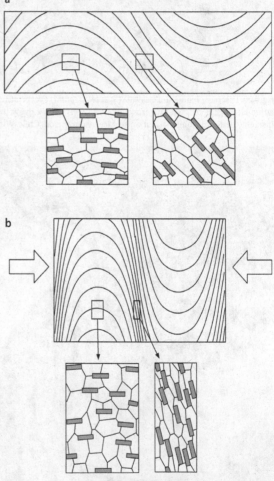

Fig. 4.22. Progressive tightening of folds with formation of a differentiated crenulation cleavage (S_2) by preferential dissolution of quartz in fold limbs caused by the orientation of quartz-mica contacts with respect to the σ_1 direction; resolved normal stress over these contacts is higher in fold limbs than in hinges. **a** and **b** are two stages in progressive deformation (cf. Figs. 4.12, 4.13)

Box 4.5   Fabric gradients

One of the problems in tectonics is that the evolution of structures cannot be directly observed in nature. As a result, there has been a tendency to look for intermediate stages or gradients in the geometry of structures, here referred to as *fabric gradients*. Fabric gradients are gradual changes in the fabric of a rock over a certain distance in the field (e.g. Fig. 1.5) or in thin section (e.g. Fig. 4.15). Examples are increasing tightness of folds, a decreasing grain size in a mylonite (Fig. 5.9), a decrease in angle between two foliations, an increase in amplitude of crenulations and gradual appearance of a second foliation (Fig. 4.19). If such fabric gradients are associated with changes in strain or metamorphic grade, it is tentative to interpret them as evolutionary stages in the development of the most evolved fabric. As far as can be determined with experiments, this assumption commonly holds. This is fortunate, since it allows us to reconstruct and study fabric evolution processes, which would otherwise remain inaccessible. It is dangerous, however, to assume that such fabric gradients always and in all aspects represent a sequence of evolutionary stages. The simple fact that fabric gradients are found at the surface implies that intermediate stages of the fabric gradient cannot be regarded as intermediate stages on a *P-T-t* loop. For example, in a fabric gradient of increasingly complex foliations with euhedral micas, the grains may have been subhedral during the evolution of every part of the fabric gradient, but micas across the gradient obtained a euhedral shape by late static recrystallisation.

certain minerals concentrate commonly in the fold hinges (quartz, calcite, feldspar, chlorite) and others (biotite, white mica, opaque minerals) in the limbs. This may be due to the high solubility of quartz and calcite, and the effect of enhanced permeability where micas are present (Gray and Durney 1979a,b; Engelder and Marshak 1985; Schweitzer and Simpson 1986). As a consequence, differentiation is not common in pure mica phyllites.

Examples are also known where ion exchange takes place between developing microlithons and cleavage domains. White mica and chlorite may be redistributed in this way, chlorite concentrating in the microlithons, and white mica in the cleavage domains (Waldron and Sandiford 1988; Price and Cosgrove 1990). High-resolution compositional mapping of minerals in cleavage and microlithon domains is a powerful tool to recognise newly grown minerals. Williams et al. (2001) give an example from the Moretown Formation, western Massachusetts, where most newly grown plagioclase grew in hinge (microlithon) domains and a large amount of phengitic muscovite in limb (cleavage) domains.

Some spaced foliations which have mainly formed by solution transfer processes may occur as *cleavage bundles* (Fig. 4.23a; Southwick 1987; Fueten and Robin 1992) centred on thin parts of layers, fold closures or other objects that may have acted as stress concentrators, or as continuous '*mica films*' in psammites (Fig. 4.23b; Gregg 1985). Such foliations probably nucleated near the stress concentration site, and grew out into the surrounding medium normal to the shortening direction (Fletcher and Pollard 1981; Gregg 1985; Tapp and Wickham 1987).

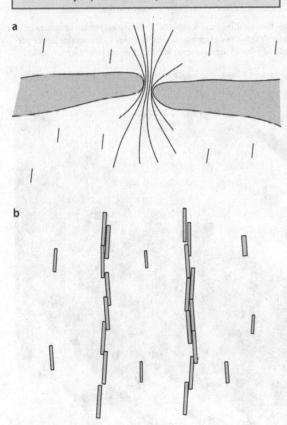

Fig. 4.23. a Cleavage bundle nucleated on a gap in a bedding plane probably related to strain concentration. **b** Mica films developed in psammite as a result of solution transfer

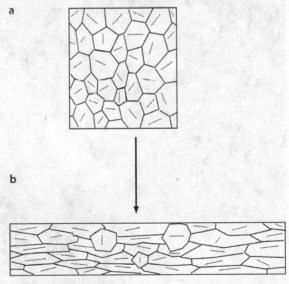

Fig. 4.24a,b. Schematic diagram of development of a foliation by crystalplastic deformation illustrating the role of lattice orientation. *Tracing* in grains indicates active slip planes for dislocations in quartz. The grains with *horizontal* and *vertical slip planes* do not deform because of their special orientation

Fig. 4.25. Low-grade metaconglomerate with highly elongated, mainly monocrystalline quartz pebbles. Undulose extinction of quartz indicates that crystalplastic deformation was important. Some pebbles are more deformed than others, probably due to favourable lattice preferred orientation for easy flattening (cf. Fig. 4.24b). Western Alps. Width of view 13 mm. Polars at 45°. (Sample courtesy Gordon Lister)

Fig. 4.26. Elongate subgrains in a deformed quartz crystal. Note formation of new, elongate grains in *lower right* and *upper left corners*. St. Barthélemy, Pyrenees, France. Width of view 1.8 mm. CPL

Solution transfer plays probably also a major role in the development of disjunctive cleavages that evolve by preferred dissolution along sets of parallel fractures; the fractures may act as channelways for the fluids with enhanced dissolution along them, causing accumulation of residual material that results in the formation of cleavage domains.

4.2.8.1
Development of Spaced Foliation without Dissolution-Precipitation

If the original rock is coarse grained, development of a shape fabric may be sufficient to create a spaced cleavage. Alternatively, a contrast between domains of different mineralogy, including individual mineral grains, is likely to produce mechanical instabilities during deformation that may result in the nucleation of micro shear bands (Goodwin and Tikoff 2002). The less competent mineral or material tends to become elongated along these shear bands, leading to a compositional layering defined by subparallel lenses of this less competent material. Although the initial shear bands may make an angle up to 60° with the XY-plane of finite strain, they tend to rotate progressively towards this plane during subsequent deformation, resulting in an anastomosing network, roughly parallel to the XY-plane (Jordan 1987; Goodwin and Tikoff 2002). This mechanism is essentially mechanical and may occur in any bi- or polymineralic medium, from poorly lithified sediment under diagenetic conditions up to granulite facies gneisses. Even in monomineralic rocks the crystal lattice orientation of individual grains may cause gradients in competency, related to the orientation of slip systems (Fig. 4.24).

This mechanism is capable to produce compositional layering at various scales, and is not necessarily accompanied by dissolution and precipitation or other diffusional mass transfer mechanisms. However, the change in shape of the less competent domains must involve cataclasis, grain boundary sliding, crystal plastic deformation or a combination of these mechanisms, treated above (e.g. Sects. 4.2.7.2, 4.2.7.4).

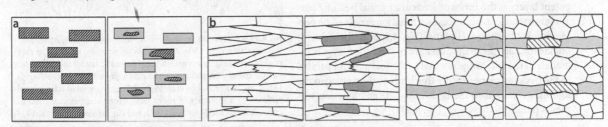

Fig. 4.27. Three examples to show how mimetic growth may play a role in the formation of secondary foliation. **a** A foliation defining mineral may be substituted, after deformation has ceased, by another mineral that inherits its shape and so continues to define the older foliation. **b** A new mineral may grow in a fabric with strong preferred orientation, mimicking this preferred orientation to a certain extent (e.g. biotite in a muscovite fabric). **c** Certain minerals may follow pre-existing compositional banding because of limited mobility of ions (Sect. 7.3; e.g. cordierite or staurolite may follow pelitic bands because of availability of Al³⁺ ions)

Fig. 4.28. Progressive obliteration of crenulation cleavage structure by grain growth of micas. Many somewhat irregular schistosities may be the result of such a process (cf. Figs. 4.18, 4.19, 4.22). **a** Fine-grained phyllite with vertical crenulation cleavage (lower greenschist facies). Pyrenees, Spain. Width of view 1.2 mm. PPL. **b** Coarse phyllite with micas that grew at least partially after crenulation, lower amphibolite facies. Carrancas, Southern Minas Gerais, Brazil. Width of view 3 mm. PPL. **c** Schist with coarse micas showing a fabric in which 'ghost' folds or polygonal arcs are just recognisable (amphibolite facies). Marsfjällen, Sweden. Width of view 5 mm. CPL

4.2.9
Geological Context of Foliation Development

4.2.9.1
Foliations and Folds

Commonly, secondary foliations are referred to as *axial planar foliations* (e.g. Hobbs et al. 1976), i.e. they show a consistent geometrical relationship with the axial planes of folds (Figs. 4.4, 4.29, 4.30, ⊘Photo 4.30). This relation was recognised as early as Sedgwick (1835) and Darwin (1846), and is generally accepted to indicate that folds and foliation developed during the same deformation phase. Commonly, foliations are not perfectly parallel to axial planes of folds, but symmetrically arranged with respect to the axial plane (Fig. 4.30a). This effect is known as *foliation fanning*. A foliation may also *refract* where it passes from one lithology to another. Foliation fanning and refraction can be due to strain partitioning generally related to viscosity contrast (Treagus 1983, 1999), or to passive rotation of relatively competent layers in the limbs of folds after initial bedding parallel shortening. In some cases, a foliation may even be perpendicular to the axial plane (Fig. 4.30b). This is due to buckle folding where the outer arc of the folded layer is extended in the fold hinge normal to the fold axis but parallel to the bedding plane; locally this leads to shortening in the fold hinge *normal* to the bedding plane and foliation segments parallel to bedding (Fig. 4.30b).

The intersection lineation of foliation and a folded surface is usually parallel to the fold axis if folds and foliation are of the same age. If the intersection lineation is oblique to the fold axis, the structures are known as *foliation-transected folds* (Johnson 1991). Foliation transected folds may form if the vorticity vector of non-coaxial flow was oblique to the fold axis, or if folds and foliation are of different age. Some foliations, such as diagenetic foliation and foliations in shear zones, need not be associated with folds at all.

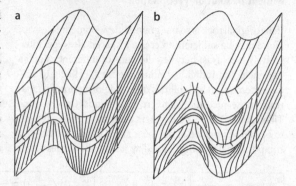

Fig. 4.30. a Refraction of foliation in competent layers. Apart from the refraction defined by a change in orientation, a change in morphology commonly occurs: in the psammitic layers the foliation is usually disjunctive whereas in the pelitic ones it may be continuous. (After Fig. 5.3 in Hobbs et al. 1976). **b** Highly variable foliation orientation in a sequence of rocks with strong competency contrast. (After Roberts and Strömgård 1972, and Fig. 5.16 in Hobbs et al. 1976)

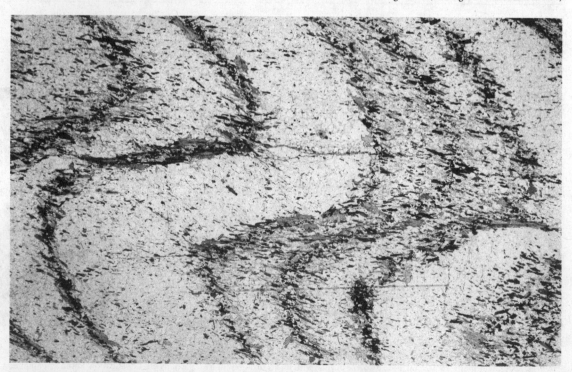

Fig. 4.29. Secondary foliation (*subhorizontal*) defined by preferred orientation of micas parallel to the axial plane of folds. Quartz mica schist. São Felix de Cavalcante, Goiás, Brazil. Width of view 17 mm. PPL

Box 4.6   Geometric development of foliations

Irrespective of the processes involved in foliation development, the geometry of a developing fabric in rocks may change dramatically during its evolution. The simplest possible situation is deformation of a random fabric into a foliation or lineation, where fabric development simply reflects increasing strain. If an older fabric exists, its deformation may lead to inhomogeneities such as folding or boudinage, but may also lead to homogeneous deformation at some scales; such homogeneous deformation can show drastic changes in the geometry of the strain ellipsoid and the associated fabric. A well-documented example is the development of slaty cleavage at a high angle to diagenetic foliation and bedding (Reks and Gray 1982; Ramsay and Huber 1983). After deposition, a pelitic sediment will undergo diagenetic compaction that may lead to a significant volume loss, associated with expulsion of part of the pore fluid. This causes development of a diagenetic foliation parallel to bedding (Sect. 4.2.2; Fig. B.4.5b, ⊘Video B.4.5). Subsequent superposition of a tectonic strain usually causes development of a new foliation oblique to the diagenetic foliation. At small tectonic strain, the tectonic and diagenetic strains may produce the same degree of anisotropy, and result in an effectively linear fabric. If such rocks are uplifted and eroded, the result has been called a *pencil cleavage* (Graham 1978; Reks and Gray 1982, 1983; Ramsay and Huber 1983; Fig. B.4.5c, ⊘Video B.4.5). If tectonic strain increases beyond development of a linear fabric, a new foliation is formed, usually with relics of the diagenetic foliation in microlithons (Figs. B.4.5d, 10.4, ⊘Video B.4.5). This sequence of fabrics seems to be common in development of a first slaty cleavage in pelitic sediments (Reks and Gray 1982; Ramsay and Huber 1983). It also illustrates the relevance of the distinction between diagenetic, tectonic and total strain (Box 4.7). It should be noted that linear fabrics similar to pencil cleavage can also form by constrictional strains under some circumstances, rather than by foliation overprint (Ramsay 1981).

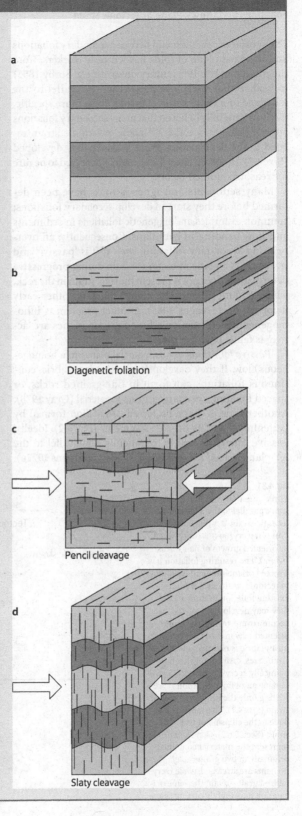

Diagenetic foliation

Pencil cleavage

Slaty cleavage

Fig. B.4.5. Development of slaty cleavage. **a, b** During diagenesis, the rock is vertically compacted and water is expelled, leading to considerable volume loss and a bedding parallel diagenetic foliation (Fig. 4.6). **c** With onset of tectonic shortening, a foliation starts to develop oblique to the diagenetic foliation. The interference of both gives rise to a linear fabric, pencil cleavage. **d** With increasing tectonic strain, the new foliation increases in strength and a slaty cleavage develops. In microlithons of the slaty cleavage, relics of the diagenetic foliation may be preserved (e.g. Figs. 4.14, 10.4; see also ⊘Video B.4.5)

4.2.9.2
Foliations and the *XY*-Plane of Tectonic Strain

The symmetric relationship between secondary foliations and the axial plane of folds has led early workers, from the middle of the 19th century onwards (e.g. Sorby 1853) to realise that such foliations may be parallel to the *XY*-plane of a finite strain ellipsoid. To be more specific, it is now generally believed that many secondary foliations approximately trace the *XY*-plane of *tectonic* strain related to the deformation phase in which they developed (Box 4.7). However, exact parallelism is expected to be rare for reasons outlined below.

Many sediments and igneous rocks have been deformed before they start to develop secondary foliations; common examples are diagenetic foliations in sediments and flow banding in batholiths. Consequently, an overprinting secondary foliation, even if it is 'passive' and traces the *XY*-plane of tectonic strain during progressive deformation, does not represent the total strain in the rock, which includes diagenetic compaction and other early deformation (Treagus 1985). Another problem is inhomogeneous deformation where foliation planes are 'active' as faults or shear zones.

'*Passive foliations*' act as material planes in a homogeneous flow. If they develop from a random fabric, continuous foliations can form in fine-grained rocks, or spaced foliations in coarse-grained material (Gray 1978). Another type is shape preferred orientation formed by flattening of grains or rock fragments (Box 4.2). Ideally, passive foliations will be parallel or subparallel to the *XY*-plane of tectonic strain (Fig. 4.17a,c; Williams 1972a).

> **Box 4.7   Strain nomenclature**
>
> Strain may be subdivided into parts related to periods of the progressive strain history. The following terms are currently in use:
>
> - **Diagenetic strain** – strain resulting from diagenetic processes such as compaction and dewatering.
> - **Tectonic strain** – strain induced by tectonic deformation, usually after diagenesis.
> - **Incremental strain** – (infinitesimally small) increment of strain.
> - **Finite strain** – part of the tectonic strain, i.e. strain accumulated over a specific period of time. It may, for example, refer to the strain of the D_1 deformation episode in comparison to the combined strain acquired during D_2 and D_3, or even to the tectonic strain. The term 'finite' (accumulated over a measurable period of time) is also used as a contrast to 'incremental'.
> - **Total strain** – normally this term refers to the total accumulated strain of a rock, including diagenetic- and tectonic strain.
>
> A more detailed description of this terminology is given in Means (1979).

However, if an older anisotropy existed, several paths can be followed. If an old foliation lies at a high angle to the shortening direction, it may rotate towards a new orientation without development of folds or new foliation planes (Fig. 4.17b). If the older anisotropy plane is oblique to the shortening direction, a new foliation may develop oblique to the previous one, gradually replacing it (Fig. 4.31a); this is the case for many disjunctive foliations. Alternatively, the earlier anisotropy may cause microfold-

Fig. 4.31.
Three situations where a foliation is not parallel to the *XY*-plane of tectonic strain. **a** A diagenetic foliation (*grey* crystals) is overprinted by oriented growth of new micas (*white*). The resulting foliation has a mixed orientation and is oblique to tectonic strain axes. **b** A pre-existing foliation in non-coaxial flow may develop microfolds that become overgrown in the hinges by oriented new micas. The resulting mean fabric is oblique to tectonic strain axes. **c** An aggregate of dynamically recrystallising grains obtains an oblique foliation representing only the last increments of strain (Box 4.7). This fabric is oblique to the ellipsoid of tectonic strain (Sect. 5.6.2). *Grey domains* represent the material contained originally in two grains at *left*: these domains are stretched, while recrystallised grains retain the same orientation and slightly oblong shape

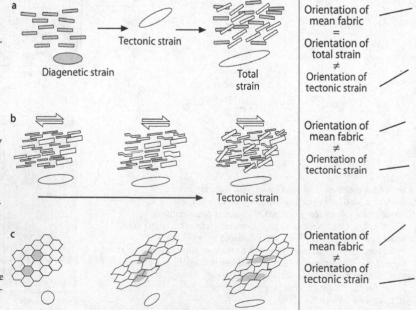

ing or micro-shear zone development, and the new foliation follows the axial planes of folds, or the shear zones (Hobbs et al. 1982; Fig. 4.31b). Mawer and Williams (1991) describe a situation where fold hinges develop in a continuous foliation deformed in non-coaxial progressive deformation; new micas overgrow newly formed fold hinges, these become unrecognisable and a mixed foliation is formed with an orientation oblique to the XY-plane of tectonic strain (Fig. 4.31b; Mawer and Williams 1991). Even ordinary slaty cleavage normally replaces a diagenetic foliation and is therefore not necessarily exactly parallel to the XY-plane of tectonic strain (Figs. B.4.5, 4.31a, ⊘Video B.4.5). In most of the cases mentioned above, the foliation is oblique to the XY-plane of tectonic strain, except in the case of very high strain values.

Some foliations are active as fold limbs or micro-shear zones. These 'active foliations' will never be parallel to tectonic strain axes, unless they become passive by rotation. Examples are some constrictional crenulation cleavages (Rajlich 1991), and shear band cleavages (Sect. 5.6.3). Care is needed even in assessment of apparently 'passive' foliations because foliation planes, once formed, are easily mobilised as planes of shear movement (Bell 1986). In many practical examples there is evidence of such 'reactivation', resulting in shear movement along foliation planes during deformation post-dating their formation.

Finally, there are 'oblique foliations' (Box 4.2; Figs. 5.10, 5.30), which represent only the last part of the tectonic strain. These foliations are not normally parallel to the XY-plane of tectonic strain (Fig. 4.31c; Ree 1991) but form wherever some process such as recrystallisation or grain boundary sliding resets the shape of elongated grains formed by dislocation creep. As a result, the foliation will only represent the last part of the deformation history (Box 4.2; Fig. 4.31c).

4.2.9.3
Foliations, Strain and Volume Change

It is presently unclear to what extent solution transfer associated with foliation development leads to bulk volume change. Shortening values normal to the foliation up to 70% are mentioned in the literature, but most observations are in the range of 30% (Gray 1979; Southwick 1987). Bulk volume loss of up to 80% has been reported, especially for slaty cleavage development at very low- and low-grade metamorphic conditions (Ramsay and Wood 1973; Wright and Platt 1982; Etheridge et al. 1983; Beutner and Charles 1985; Ellis 1986; Wright and Henderson 1992; Goldstein et al. 1995, 1998). On the other hand, many studies concluded that little or no significant bulk volume change accompanied cleavage formation (Waldron and Sandiford 1988; Wintsch et al. 1991; Tan et al. 1995; Saha 1998; Davidson et al. 1998). On theoretical grounds, bulk volume loss on a large scale is expected to be of minor

importance at deeper crustal levels since large volumes of fluid would necessarily have to flux through the rock to remove material in solution (Engelder 1984; Bhagat and Marshak 1990). However, Goldstein et al. (1998) argue that in accretionary complexes large volumes of water are passing through the rocks and that in such settings large volume losses are to be expected. The difficulty is that volume loss during foliation development can rarely be directly measured in deformed rocks (Sect. 9.2). Well-pre-

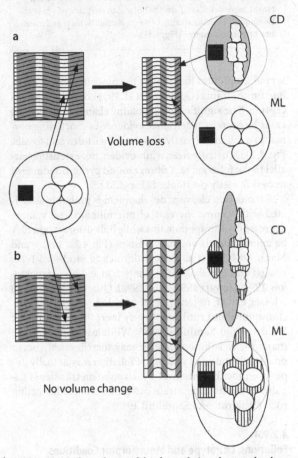

Fig. 4.32. Two end-member models of crenulation cleavage development in plane strain. The onset of crenulation cleavage development is shown in the *squares* at *left*. Schematic enlargements of an aggregate of four quartz grains (*white*), a pyrite cube (*black*) and a passive *marker circle* are given. Deformed situations in cleavage domains (*CD grey*) and microlithons (*ML white*) are shown in *rectangles* at *right*. Local strain and volume loss in both situations are indicated schematically (not to scale) by the elliptical shape of the deformed marker circle and the outline of the original circle. **a** Significant volume loss in cleavage domains while microlithons are undeformed. Quartz grains are partly dissolved in cleavage domains but no fibres form near the quartz or pyrite grains. **b** Volume-constant deformation where volume loss in cleavage domains is compensated by volume increase of microlithons. Quartz grains are partly dissolved in cleavage domains but have fibrous overgrowths in cleavage domains and microlithons (*vertical striping*); fibres also occur next to pyrite cubes. If no pyrite cubes or similar objects are present, and if overgrowths on quartz are not clear, situations **a** and **b** are difficult to distinguish

---

Box 4.8   Area and volume change

In geological practice, it is easy to confuse area change and volume change. Area change is a component of two-dimensional strain and is measured in a plane, e.g. from stretch values; it causes a change in the cross-sectional area of a structure (e.g. a fossil). Volume change is a component of three-dimensional strain. Area change is not a direct measure for volume change. For example, even if a thin section shows evidence for area increase, bulk volume loss may occur if shortening is significant in the direction normal to the thin section. Only if strain is two-dimensional, i.e. if stretch normal to the plane of observation equals 1 (plane strain), can area change be used as a measure of volume change (Fig. 4.32).

---

served graptolites that permit the measurement of absolute finite strains (Goldstein et al. 1998) constitute an exception. The significance of volume change may in some cases be overestimated since evidence of shortening normal to a foliation (partly dissolved structures and fossils; Fig. 4.21) is usually clear, while evidence of extension parallel to the foliation (e.g. fibres around pyrite, boudinaged micas) is easily overlooked (Fig. 4.32).

Crenulation cleavage development is probably associated with volume increase of microlithons and volume decrease of cleavage domains while bulk deformation may be approximately volume-constant (Fig. 4.32; Erslev and Mann 1984; Lee et al. 1986; Waldron and Sandiford 1988; Bhagat and Marshak 1990; Wintsch et al. 1991; Mancktelow 1994; Stewart 1997; Saha 1998). Quartz, albite and, to a lesser extent, micas are exchanged in pelites while zircon, apatite and rutile are largely inert (Southwick 1987; Waldron and Sandiford 1988; Williams et al. 2001). In many rocks, solution transfer may therefore only occur on a small scale and spacing of foliation may actually depend on the distance over which solution transfer is capable of maintaining strain compatibility in a deforming rock (Waldron and Sandiford 1988).

4.2.9.4
Foliations, Lithotype and Metamorphic Conditions

Secondary foliations develop by processes mentioned in Sect. 4.2.7, but in different lithotypes and under different metamorphic conditions, these processes operate to different extents. A brief outline of present ideas is given below.

In pelites, mechanical rotation, pressure solution transfer, crystallisation, recrystallisation and oriented nucleation are all competing processes. In many cases, a diagenetic foliation may have been present before onset of foliation development. In some cases, at very low-grade or nonmetamorphic conditions, cleavage domains develop oblique to the diagenetic fabric by stress-induced solution transfer or development and rotation of micro shear zones (Goodwin and Tikoff 2002) with no- or minimal folding,

leading to spaced foliation (e.g. Fig. 4.14). In most cases, microfolds (mechanical rotation) develop in the diagenetic foliation and this initial stage is followed by solution transfer of material between hinges and limbs, usually quartz from limbs to hinges (Williams 1972a; Cosgrove 1976; Gray 1979; Waldron and Sandiford 1988), and/or syntectonic crystallisation or recrystallisation of micas in cleavage domains (Tullis 1976; White and Knipe 1978; Knipe 1981; White and Johnston 1981; Lee et al. 1986; Kisch 1991). These effects are thought to be mainly temperature-dependent, solution transfer occurring at lower grade than syntectonic crystallisation and recrystallisation (Kanagawa 1991; Kisch 1991). Consequently, solution transfer may be followed by syntectonic crystallisation (Weber 1981). With increasing temperature in the absence of deformation, a preferred orientation may even be strengthened further by mimetic mica growth (Siddans 1977; Weber 1981; Ishii 1988). In some slates, the stage of folding and rotation may be absent and the foliation develops by syntectonic crystal growth without mechanical rotation (Woodland 1982; Gregg 1985; Ishii 1988).

After a first foliation is developed, renewed shortening at a low angle to the existing foliation may cause development of a second foliation; again, the early foliation may be folded or truncated by developing new cleavage domains, and either solution transfer or new growth of mica and possibly other minerals such as plagioclase (Williams et al. 2001) may dominate. This leads to disjunctive or crenulation cleavage. If differentiation is strong and accompanied by recrystallisation, evidence of early foliations may be obscured and a *compositional layering* develops. The term *differentiated layering* is also commonly used for such structures, but since it can be difficult to distinguish sedimentary layering from secondary layering, the non-genetic term compositional layering is preferred.

In psammites, continuous foliation can form in fine-grained rocks, or spaced foliations in coarse-grained material (Gray 1978). In the second case, *mica films* (Fig. 4.23b) may develop by solution transfer and mica growth (Gregg 1985) and/or by the development of micro shear zones (Goodwin and Tikoff 2002).

In limestones, foliation development is strongly dependent on temperature and mica-content. Solution transfer and twinning are important at low temperature (Sect. 3.12.3; Davidson et al. 1998) and can lead to a grain shape preferred orientation defined by elongated carbonate grains, or a coarse spaced foliation (stylolites). A primary high mica content of limestone may cause development of slaty cleavage and cleavage bundles. In one case the growth of illite + kaolinite + quartz + anatase in cleavage domains was reported to accompany the removal of calcite by dissolution (Davidson et al. 1998). Passive rotation of micas is mainly responsible for mica-preferred orientation in limestone at low temperature (Alvarez et al. 1976; Mitra and Yonkee 1985; Kreutzberger and Peacor 1988). At higher temperature, crystalplastic flow and twinning are important, and a foliation is mostly formed by elongate grains. In all cases, foliations in limestones are less well developed than in pelites.

Metabasites deformed at low-grade conditions give rise to continuous or spaced foliations defined by preferred orientation of amphiboles, chlorite, epidote, micas and lenses of different composition. Mechanical rotation and oriented growth of new minerals is more important then solution transfer. At medium to high-grade conditions, oriented mineral growth and crystalplastic deformation are the main mechanisms of foliation development. Metabasites can be equally suitable to determine metamorphic grade as pelites, especially at low-grade metamorphic conditions.

In many mountain belts the onset of cleavage is marked by a *cleavage front*, that separates rocks with cleavage from rocks without cleavage (Mattauer 1973; Holl and Anastasio 1995). Although this front is strongly dependent on deformation intensity and also lithologically controlled, a minimum pressure of about 200 MPa corresponding to an overburden of 5–7 km (Engelder and Marshak 1985) and a minimum temperature of about 200 °C for pelitic siliciclastic rocks and 175 °C for limestones (Groshong et al. 1984a) can be estimated. Criteria for incipient cleavage development are given in Kisch (1998).

4.2.10
Practical Use of Foliations

4.2.10.1
Introduction

Foliations can be used to obtain information on strain, metamorphic conditions and overprinting relations. In Sect. 4.2.9.2 it is shown how secondary 'passive foliations' can be used to find the approximate orientation of the XY-plane of tectonic strain related to the formation of that particular foliation (not the total strain), provided the problems mentioned in Sect. 4.2.9.3 are kept in mind. Especially continuous foliations may be used to a first approximation to identify the XY-plane of strain. Foliations, which are thought to have developed by mechanical rotation of fabric elements only, can in principle be used to estimate the magnitude of finite strain (Sect. 9.2).

If a foliation is defined by parallel minerals that show a "growth fabric" (e.g. Fig. 4.8) without signs of post-crystalline deformation, the foliation must have formed under metamorphic circumstances during which these minerals were stable. A foliation defined by euhedral amphiboles must have formed under circumstances where these amphiboles were stable. However, care must be taken to distinguish cases where a pre-existing foliation has undergone an increase in metamorphic conditions after deformation, producing new minerals that may have grown mimetically over the existing foliation (Figs. 4.27, 4.28). Relicts of older mineral assemblages may help to recognise these situations.

In many areas where several overprinting foliations can be recognised, a "main foliation" may have formed under peak metamorphic circumstances, whereas later events are characterised by deformation of this main foliation to produce weaker and less penetrative foliations under lower metamorphic or drier circumstances. This may be related to the fact that during progressive metamorphism, water is generally released by mineral reactions favouring complete recrystallisation during deformation. After the peak of metamorphism, under retrograde circumstances when the rock has lost most of its water, recrystallisation is difficult and deformation results mainly in folding, distortion or mylonitisation of earlier fabrics.

4.2.10.2
Overprinting Relations

Overprinting relations between foliations are probably the most useful tools to furnish a reference frame for the study of the tectonic evolution of a body of rock. They are particularly appropriate for study in thin section because of their penetrative nature and because of the usually small size of the fabric elements. The principle for establishing a sequence of foliation planes is quite simple: if microfolds are visible the folded surfaces are always older than the fabric elements developed along the axial surface, or cutting the folds. Any surface associated

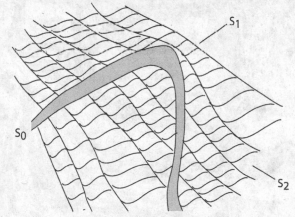

Fig. 4.33. D_1 fold, folding bedding (S_0) with development of S_1 foliation along the axial surface of the fold. Later D_2 deformation folded S_1 to produce an S_2 crenulation cleavage in pelitic layers that cut the D_1 fold through both limbs. Note the deviation of S_2 around the more resistant fold hinge

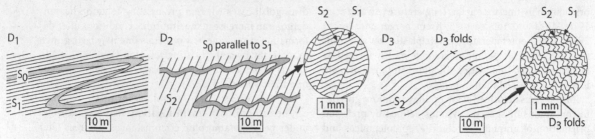

Fig. 4.34. Schematic presentation of a common sequence of foliation development in slate and schist belts. Enlargements of typical microstructures are shown at *right*. Further explanation in text

Fig. 4.35.
a Crenulation cleavage S_3 (*vertical*) with microlithons in which the folded cleavage (*subhorizontal*) is an older crenulation cleavage S_2. S_2 and S_3 correspond respectively to stages *4* and *3* in the scheme in Fig. 4.18. Eastern Alps, Austria. Width of view 1.5 mm. PPL. **b** Detail of **a** showing folded S_1 in microlithons of S_2. Width of view 0.3 mm. PPL. (Photographs courtesy S. Wallis)

with the axial planes of folds is genetically related with those folds, but foliation planes that cut folds obliquely are younger than the folds (Fig. 4.33).

A general outline of a common sequence of events in slate and schist belts may serve to illustrate how the analysis of overprinting relations works (Fig. 4.34; cf. Hobbs et al. 1976, their Chap. 9; Williams 1985). During a first deformation phase (D_1), a penetrative slaty cleavage is developed at varying angles with bedding, according to the position in large D_1 folds that are commonly asymmetric. In the long limbs the angle between S_1 and S_0 may become so small that it is not visible any more in the field or even in thin sections (Fig. 4.34). The slaty cleavage (S_1) may be spaced or continuous, but is generally not a crenulation cleavage as analysed under normal microscopic amplification. However, if analysed by SEM, it may show crenulation cleavage features, folding a bedding-parallel foliation of diagenetic origin.

A second phase of deformation (D_2) commonly produces a crenulation cleavage, folding S_1 (Fig. 4.34). Various stages or morphologies may be present depending on the intensity of deformation (cf. Bell and Rubenach 1983) and according to grain growth in response to metamorphic circumstances (Figs. 4.18, 4.19; ☉Photo 4.19b1–7).

A third phase of deformation may be recognised by folding of the S_2 crenulation cleavage (Figs. 4.34, 4.35). This may in some cases result in interesting structures, since according to their orientation certain limbs may be refolded and others straightened out (Figs. 4.36, 4.37, ☉Photo 4.37a–c). Later phases of deformation may be recognised in a similar way by overprinting (folding) of earlier foliations.

The main problem of this analysis is to establish how to correlate foliations from one thin section to another, from one outcrop to another, or even from one analysed area to another. This is a matter that is hard to solve with general rules, but the following suggestions may be of help (see also Williams 1985). Deformation may be quite heterogeneously distributed through a rock body, especially the deformation that post-dates peak metamorphic conditions. It is, for instance, common to find D_3 or D_4 deformation features concentrated in narrow zones, leaving other areas without visible effects. Shear zones are, of course, the most spectacular example of this local concentration of deformation. On the other hand, foliations induced during peak metamorphic conditions are normally widespread and remarkably continuous over large areas. These may, however, vary abruptly because of lithological variation (e.g. strong foliations may disappear abruptly at the contact of a calc-silicate rock because of the lack of platy minerals to define a foliation). It is important in the correlation of foliations to pay attention to their relation with metamorphism, since metamorphic conditions usually do

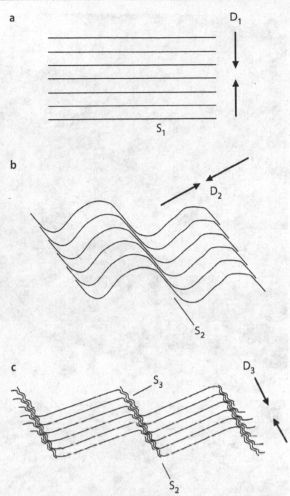

Fig. 4.36. Sequence of events leading to selective refolding of a second foliation (S_2) by D_3 while the older foliation (S_1) seems unaffected. **a** S_1 is formed by vertical compression. **b** Oblique lateral compression by D_2 caused a steep S_2 differentiated crenulation cleavage. **c** Oblique D_3 compression is applied, resulting in selective refolding of differentiated limbs of D_2 folds because of their orientation. The other limbs are progressively unfolded until S_1 becomes approximately parallel to the axial plane of D_3 folds

not change much from one outcrop to the next, unless post-metamorphic faulting is involved. In rare cases (e.g. Lüneburg and Lebit 1998) successive deformation phases produced only a single cleavage, reflecting the total strain ellipsoid.

Especially in the field, intrusive veins or dykes can be important to distinguish phases of deformation and their associated foliations of different age. These bodies may have intruded over a relatively short period of time and may be recognised over a large area by their similar composition and orientation. Structures cut by the veins are older, whereas younger structures affect the veins by folding, shearing or other deformation.

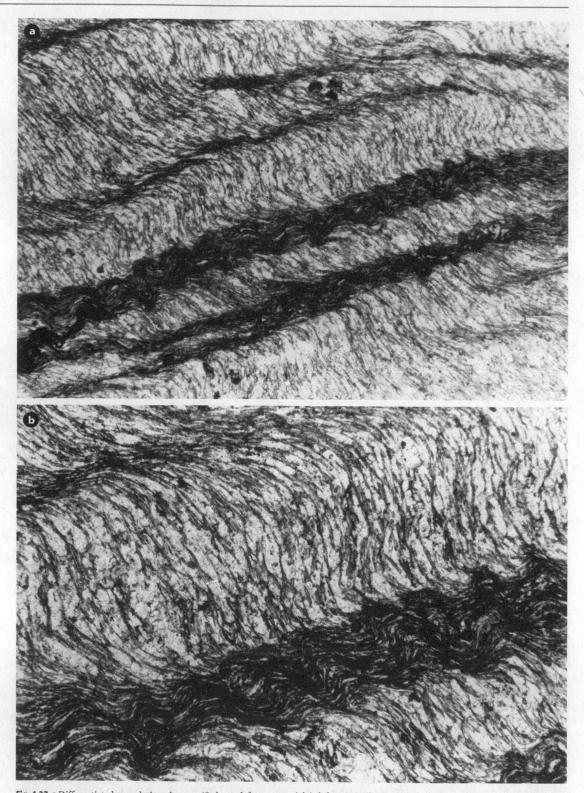

Fig. 4.37. a Differentiated crenulation cleavage (S_2 *lower left* to *upper right*) deforming a finely spaced disjunctive cleavage (S_1 *upper left* to *lower right*). S_2 is folded during D_3. This is an example of a structure thought to have developed according to the scheme explained in Fig. 4.36. Açunguí, São Paulo, Brazil. Width of view 4 mm. PPL. **b** is a detail of **a**. Width of view 1.8 mm. PPL

4.3

In many metamorphic terrains it is difficult or impossible to use bedding as a reference plane in outcrop. The distribution of lithotypes on the map may allow establishment of approximate contacts between stratigraphic units, but these may not coincide with lithologic contacts in outcrop. The latter contacts are usually parallel to "the main foliation" and may be difficult to follow along strike. In such areas, *transposition* of one or more foliations (including bedding) has occurred. Transposition is usually defined as the progressive erasure of a reference surface (S_0, S_1, S_n, etc.) due to tight folding accompanied by some differentiation process. However, it can also be used in a more general sense for erasure of an older structure by strong younger deformation. Turner and Weiss (1963; see also Davis 1984) have given some good examples of bedding transposition on the outcrop scale. The concept is clearly scale-dependent; a number of en-echelon disrupted bedding lenses may be mistaken for real bedding if seen in an outcrop smaller than these lenses. In a large outcrop where a number of lenses are visible, the oblique position of the enveloping surface of bedding may still be recognisable.

Transposition may also occur on the scale of a thin section. The sequence of crenulation cleavage development (Figs. 4.18, 4.19) is a good example of transposition of S_1 by S_2. Figure 4.38 shows a natural example of a D_2 fold where S_2 is clearly distinct from S_0/S_1 in the fold hinges, whereas in the limbs transposition has occurred and all three planes, S_0, S_1 and S_2, have become parallel. The parallelism of S_0 and S_1 probably indicates that a similar process occurred during D_1.

4.3
Lineations

4.3.1
Terminology of Lineations

Terminology of lineations was redefined by Piazolo and Passchier (2002a), as follows: Object lineations can be subdivided into *grain lineations* and *aggregate lineations* (Fig. 4.2). Grain lineations are defined by parallel oriented elongate single crystals. These can be deformed single crystals of normally equidimensional shape such as quartz or calcite (grain shape preferred orientation – Box 4.2), or of euhedral or subhedral mineral grains with an elongate shape such as amphibole, tourmaline or sillimanite. An aggregate lineation is a type of shape preferred orientation defined by elongate aggregates of equidimensional or slightly elongate grains (Box 4.2). Common examples are aggregates of dynamically recrystallised grains replacing a large deformed older crystal. It is possible and even common that only part of the rock volume defines an object lineation (Fig. 4.2). Trace lineations include *crenulation lineations* and *intersection lineations* (Fig. 4.2). Intersection lineations are formed by intersecting foliations (Fig. 4.2) while crenulation lineations are defined by hinge lines of microfolds in a foliation surface (Fig. 4.2).

Several other terms are used in the geological literature for lineations. The word *stretching lineation* is commonly used as a general term for aggregate lineation and grain lineation if the constituting grains are defined by deformed aggregates or single crystals. However, the term stretching lineation has genetic implications and can be misleading. Elongate crystals or aggregates can form by stretching, but also by boudinage into thin strips normal to the stretching direction, or by vein formation. Therefore, the term stretching lineation should only be used if it is clear that aggregates lie in the direction of the X-axis of the finite strain ellipsoid. Grain lineations made up of deformed large crystals can also be classified as *stretching lineations* for this reason (Fig. 4.2). The term *mineral lineation* has been used for the preferred orientation of non- or little deformed euhedral or subhedral mineral grains with an elongate shape such as amphibole, tourmaline or sillimanite. Mineral lineation is a special type of grain lineation. *Cataclastic lineation* (Tanakà 1992) consists of elongate fragments and aggregates of fragments in the more fine-grained matrix of a cataclasite (Sect. 5.2).

Since lineations are defined as linear structures that occur penetratively in a volume of rock, they do not include linear features that only occur on certain surfaces in the rock. For example, *slickenside striations* and similar structures that occur restrictedly on slickensides (e.g. Means 1987) or other fault planes are not considered to be lineations and are not treated here, since they can rarely be studied in thin section. Another common mistake is to

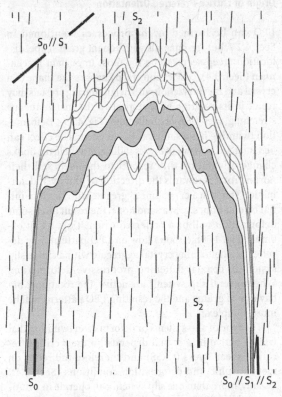

Fig. 4.38. Isoclinal D_2 fold showing the parallel orientation of S_0, S_1 and S_2 in fold limbs and S_2 oblique to S_0/S_1 in the hinges

4.4

refer to the intersection of a foliation and a flat outcrop surface as a lineation (Fig. 4.3); this could be referred to as a *foliation track*. By analogy, intersections of a flat outcrop surface with an object lineation give rise to a *lineation track* (Fig. 4.3).

4.3.2
Development of Lineations

The development of lineations is complex and treated only briefly in this book.

Aggregate lineations can form by stretching of equidimensional aggregates of grains to a linear shape, or by similar deformation of large single crystals and subsequent recrystallisation (Box 4.2). Grain lineations either form by deformation of equidimensional grains without recrystallisation or by dissolution and growth (similar to Fig. 4.16(2, 3); Box 4.2). Minerals with a typical elongate shape such as sillimanite, tourmaline and amphiboles may either rotate or grow in a preferred direction creating a grain (or mineral) lineation. As a result, most aggregate and grain lineations are parallel to the X direction of the finite strain ellipsoid. However, it has to be investigated with care whether these lineations really represent a direction of extension, or whether they formed by boudinage or pressure solution.

Trace lineations form by polyphase deformation, or by changes in the deformation regime as e.g. in the case of shear band cleavage development (Sect. 5.6.3). These structures do not have a simple relationship with strain axes or directions of tectonic transport, although they are commonly oblique to object lineations of the same age (Fig. 4.3). Trace lineations are commonly parallel to buckle fold axes, and at a high angle to the shortening direction in a volume of rock.

There is no simple relationship between types of lineations and metamorphic conditions. As for foliations, the nature of minerals that constitute aggregate of grain lineations gives information on metamorphic conditions during development of the lineation.

Aggregate and grain lineations that represent the longest strain axis (X) can be used to find the "direction of tectonic transport" if they form in ductile shear zones with approximately simple shear flow (Passchier 1998). Trace lineations can be used to find the orientation of fold axes in buckle folds, and to obtain information on the nature of polyphase deformation; in principle each intersection lineation represents a deformation phase, but for instance a third deformation phase, D_3, may produce different intersection lineations between S_3 and S_2, S_3 and S_1 and S_3 and S_0.

Object lineations are commonly overprinted by trace lineations of different age, or even of the same age. It is also common that foliation planes contain more than one generation of trace lineations, each representing a deformation phase. Overprinting object lineations of different age on one foliation plane are less common, but do occur.

4.4
Lattice-Preferred Orientation (LPO)

4.4.1
Introduction

In many deformed rocks, the lattice orientation of crystals is not randomly distributed, but arranged in a systematic way. Such rocks have a *lattice-preferred orientation* (LPO) for a specific mineral. In the case of crystals with a planar or elongate shape in a particular crystallographic direction such as micas and amphiboles, an LPO is easy to recognise as a foliation or lineation. However, for minerals such as quartz and calcite this is more difficult. In the case of quartz, the presence of an LPO can be checked by inserting a gypsum plate under crossed polars; when the microscope table is turned, a dominant blue or yellow colour for a quartz aggregate in different orientations is an indication for an LPO. In other minerals with higher birefringence, special techniques are required to determine if an LPO is present. Several processes can contribute to development of an LPO (Skrotzki 1994). LPO patterns and LPO development in quartz is treated in some detail as an example.

4.4.2
Origin of Lattice-Preferred Orientation

LPO can be formed by the processes mentioned in Sect. 4.2.7, but for minerals with equant grain shape, dislocation creep seems to be the most important mechanism (Sect. 3.4). Dislocation creep changes the shape of a crystal and the interaction with neighbouring crystals may result in its rotation with respect to the instantaneous stretching axes (ISA) of bulk flow (Fig. 4.39). Deformation twinning has a similar rotation effect. The effect can be visualised by a pile of books sliding on a shelf; the books change orientation with respect to the shelf and their normal rotates towards the direction of gravity. If deformation starts in a crystalline aggregate with random initial orientation, e.g. in a sandstone, the result after some deformation will be a preferred orientation. As an example, Fig. 4.39b,c shows how an LPO pattern may develop in a deforming crystal aggregate with a single slip system in coaxial flattening progressive deformation. When several slip systems are active, the rotational behaviour of grains and the resulting LPO patterns will be more complex.

The type of slip systems or deformation twinning that will be active in a crystal depends on their critical resolved shear stress (CRSS) and therefore indirectly on metamorphic and deformation conditions (Sect. 3.4). Usually, more than one slip system can operate in a mineral and the CRSS of each slip system changes with temperature and chemical activity of certain components,

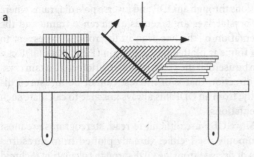

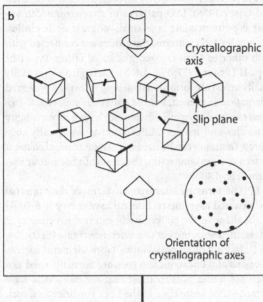

Crystallographic axis

Slip plane

Orientation of crystallographic axes

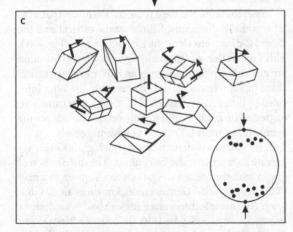

Fig. 4.39. a Reorientation of a pile of books by slip: an axis normal to the books (*bold line*) rotates towards the direction of gravity. Development of LPO in crystals due to dislocation glide on slip systems operates in a similar way. **b** Flattening of an aggregate of crystals with a single slip system normal to a crystallographic axis (*bold line*). **c** All crystal axes rotate towards the compression direction except those parallel or normal to this direction. Those parallel to the compression direction may deform by kinking or twinning with rotation of the segments

and may even 'overtake' that of other slip systems. At low differential stress, only one slip system may be active, but at higher differential stress, several slip systems can operate simultaneously. In fact, for maintenance of cohesion between grains, five independent slip systems should be operating (Lister 1977). In silicates, however, which usually have low crystal symmetry, fewer slip systems are active and space problems are accommodated at low temperature by lattice bending, kinking, fracturing and, at high temperature, by dynamic recrystallisation or grain boundary sliding.

The type of LPO pattern that is formed in a rock depends on many factors, the most important of which are (Schmid 1994):

1. The slip systems that are operating and the amount of activity on each slip system.
2. The ratio of stretching rates along the ISA of the flow, i.e. plane strain, flattening or constrictional flow. These rates determine in which direction crystals rotate and thereby the shape of the fabric (Fig. 4.41).
3. The finite strain. Usually, if the flow pattern does not change during deformation, the LPO pattern increases in strength and sharpness with increasing strain but undergoes only slight changes in geometry (Sect. 4.4.4.2).
4. The kinematic vorticity number. In initially isotropic materials, non-coaxial progressive deformation leads to LPO patterns with monoclinic symmetry, and coaxial progressive deformation to patterns with higher symmetry.
5. The activity of bulging and grain boundary migration dynamic recrystallisation. Recrystallisation may influence an LPO pattern in several ways but the effect is difficult to predict; it may weaken an existing pattern by generation of new, randomly oriented grains; or it may strengthen a pattern or part of a pattern by removing (consuming) certain grains with a relatively high dislocation density. Grains that are unfavourably oriented for slip may be removed by this process if they developed a high dislocation density because of constriction by neighbours (Jessell 1987; Ree 1990). However, the reverse is also possible; such grains may have low dislocation density, since all deformation is taken up in softer neighbours, and therefore consume grains favourably oriented for slip (Gleason et al. 1993). Evidence for both processes has been found in experiments. Static recrystallisation may also affect LPO patterns, but the effect is uncertain (Humphreys and Hatherley 1995; Heilbronner and Tullis 2002; Park et al. 2001).
6. Growth of grains from solution. The growth rate in many minerals is dependent on crystallographic direction, and growth of minerals from solution can therefore produce a preferred orientation (Shelley 1979, 1989, 1994).

Theoretically, it should be possible to use LPO patterns as a source of information on the six parameters mentioned above. However, our understanding of the development of LPO is unfortunately still sketchy. Most successful has been the application of LPO patterns with monoclinic symmetry to determine sense of shear (Sect. 4.4.4.3).

The study of the development of LPO proceeds through several angles of approach. Observation of natural LPO patterns and comparison with known temperature, strain geometry and vorticity of the progressive deformation can give an indication of the influence of these parameters on LPO development. However, in natural LPO, the deformation history is usually unknown and may have been more complex than is assumed; early parts of the development are most likely erased. Slip systems may be identified by observation of lattice defects in naturally deformed crystals by TEM (Blacic and Christie 1984; Hobbs 1985). However, lattice defects in natural deformed rocks may be formed late, after the LPO was developed (White 1979a; Ord and Christie 1984). Theoretical and numerical modelling of fabric development using a pre-set choice of slip systems have been very successful in modelling LPO patterns (Etchecopar 1977; Lister 1977; Lister and Price 1978; Lister et al. 1978; Lister and Paterson 1979; Lister and Hobbs 1980; Etchecopar and Vasseur 1987; Jessell 1988b), but theoretical studies suffer from assumptions that may be wrong and simplifications necessary to operate computer models. Furthermore, only monomineralic aggregates have been simulated, while most of the interesting fabrics in rocks occur in polymineralic aggregates. The most successful, but possibly also most laborious approach to study LPO development, is experimental deformation of rocks at high pressure and temperature and subsequent analysis of the LPO patterns in deformed samples, in combination with TEM analysis of lattice defects (Green et al. 1970; Tullis et al. 1973; Dell'Angelo and Tullis 1989).

4.4.3
Presentation of LPO Data

The orientation of a crystal in a reference frame is only completely defined if the orientation of three crystal axes is known; this means that three numbers are needed to represent the orientation of a single crystal in a reference frame. However, if an LPO is to be presented in this way, it can only be done as points in a three-dimensional diagram. Such a diagram is known as an orientation distribution function diagram or *ODF* (Fig. 4.40a). In practice, it may be difficult for the inexperienced to read such diagrams. Geologists usually rely on polar diagrams such as stereograms to plot the orientation of crystals (Fig. 4.40a); however, these are only useful if just one crystallographic direction, such as the c-axis of quartz, is plotted. In this way, only part of the LPO pattern of a crystalline aggregate is presented. Other methods of presentation are cross-

sections through an ODF, and inverse pole diagrams where the crystal axes are taken as a reference frame and the orientation of the lineation in the rock with respect to this frame is plotted for each grain (Fig. 4.40a). ODF can also be useful if the preferred orientation of a certain crystal direction that is of interest cannot be measured directly; from an ODF it is always possible to calculate such orientations.

Since ODF are difficult to read, stereograms are most commonly used, either directly plotted from measured data or derived from the ODF through calculation (Schmid and Casey 1986). LPO patterns in stereograms can appear as point maxima or as small- or great circle girdles. In complex LPO patterns, the girdles are connected with each other to form crossed girdles of either Type I or Type II (Lister 1977; Fig. 4.40b). Cleft girdles (actually small circles) are formed in flattening strain. If a preferred orientation is present, but the pattern is vague, *pole-free areas* can be distinguished. In order to enhance visibility of girdles and maxima, LPO patterns are usually contoured. Contours can be used to derive a *fabric skeleton*, a pattern of lines connecting the crests of the contour diagram (Fig. 4.40b).

LPO patterns are interpreted in terms of their internal and external asymmetry. Internal asymmetry is defined by the shape of the pattern itself; external asymmetry is determined with respect to a reference frame (Sects. 2.4, 5.6.1); lacking other possibilities, fabric elements such as foliations and lineations in a rock are normally used as a reference frame, notably those that are thought to have formed at the same time as the LPO. For briefness, such reference foliations and lineations are given in this chapter as S_r and L_r.

In stereograms, standard presentation of LPO patterns is with the Y-direction of finite strain vertical and the X- and Z-directions along the EW and NS axes (Fig. 4.41). This implies that a corresponding foliation and lineation are presented in the diagram as an E-W-trending vertical plane (S_r) and horizontal line respectively, the latter indicated by dots on the circle (L_r; Fig. 4.40b). L_r is usually an aggregate or grain lineation. This orientation shows the symmetry of most LPO patterns advantageously.

It is commonly useful to show which grains in an aggregate have a particular orientation. The distribution of grains with particular orientations can be given in a map of the sample under consideration, known as an *AVA diagram* (German: 'Achsenverteilungsanalyse' – analysis of orientation of axes; Sander 1950; Heilbronner-Panozzo and Pauli 1993). In practice, AVA diagrams are made for the LPO pattern of a single crystal axis, such as c-axes. An AVA diagram can be presented by plunge direction of c-axes for each grain, presented as lines (Fig. 4.24) or, more advanced, by colours representing different orientations. AVA can be of great help for the interpretation of LPO patterns and of the way in which they develop (Sect. 10.3).

Fig. 4.40.
a Orientation of a quartz crystal in a reference frame defined by a foliation (S_r), lineation (L_r) and foliation pole. The full crystal orientation is given by Eulerian angles ϕ, ψ and θ. Orientation of the c-axis is given by angles α and β. Three diagrams that are commonly used to present LPO patterns are shown. In an ODF diagram the full orientation of the crystal is represented. In a pole diagram the orientation of individual axes of the crystal can be plotted; in this case, only the c-axis. In an inverse-pole diagram the orientation of L_r is plotted with respect to crystallographic axes. **b** Examples of pole diagrams with contours of pole density showing two types of crossed girdles (Lister 1977) of quartz c-axes. The shape of the girdles is highlighted by use of a fabric skeleton that traces the crests of the contour diagram

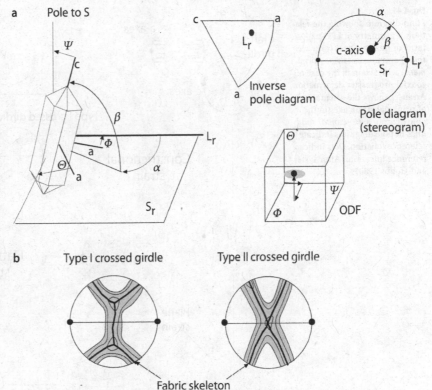

4.4.4
LPO Patterns of Quartz

4.4.4.1
Introduction

Figure 4.41 shows the influence of flow type and finite strain on the geometry of c-axis LPO patterns of quartz that accumulated by coaxial progressive deformation at low- to medium-grade metamorphic conditions (Tullis 1977; Lister and Hobbs 1980; Schmid and Casey 1986; Law 1990; Heilbronner and Tullis 2002; Takeshita et al. 1999; Okudaira et al. 1995). Small circle girdles are most common but in plane strain, small circle girdles are connected by a central girdle to produce Type I crossed girdles (Fig. 4.40b). Other c-axis LPO patterns that develop in coaxial progressive deformation are Type II crossed girdles, which seem to form in constriction (Fig. 4.40b; Bouchez 1978), and point maxima around the Y-axis of strain. Both patterns seem to form at higher temperature than the patterns shown in Fig. 4.41 (Schmid and Casey 1986; Law 1990). Increasing temperature also seems to cause an increase in the opening angle of the small circle girdles (Kruhl 1998).

In the case of non-coaxial progressive plane strain deformation, other c-axis patterns develop (Fig. 4.42) (Behrmann and Platt 1982; Bouchez et al. 1983; Platt and Behrmann 1986). Most common are slightly asymmetric Type I crossed girdles, and single girdles inclined to S_r and L_r (Burg and Laurent 1978; Lister and Hobbs 1980; Schmid and Casey 1986). At medium to high-grade conditions, single maxima around the Y-axis are common, while at high grade (>650 °C), point maxima in a direction close to the aggregate lineation L_r occur (Mainprice et al. 1986). c-axis patterns as shown in Figs. 4.41 and 42 represent only a small part of the full LPO of quartz and the orientation of other directions, such as <a>-axes, should also be known to allow interpretation of LPO development; in Figs. 4.41 and 4.42, patterns for <a>-axes are therefore shown beside c-axes. Nevertheless, c-axis patterns are most commonly represented in the literature since they can easily be measured on a U-stage; for other crystallographic directions more advanced equipment such as a goniometer (Sect. 10.3.5) is needed.

The patterns in Fig. 4.41 can be explained as an effect of the activity of slip planes in quartz; at conditions below 650 °C, slip in <a> directions on basal, prism and rhomb planes is dominant in quartz. As a result, <a>-axes tend to cluster close to planes and directions of maximum incremental shear strain (at 45° to ISA; Fig. 4.43a). In flattening, <a>-axes cluster in small circles around the shortening direction, similar to the situation in Fig. 4.39c. In constriction, a small circle girdle of <a>-axes around the extension direction forms and in plane strain there are two directions in the XY-plane. Slip on basal planes con-

Fig. 4.41.
Flinn diagram showing the relation of geometry of LPO patterns of quartz c-axes (*grey contours*) and a-axes (*striped ornament*) with strain in the case of coaxial progressive deformation. An inset shows the orientation of principal strain axes in the pole diagrams. *Horizontal solid lines* in pole diagrams indicate reference foliation. *Dots* indicate reference lineation. (After Lister and Hobbs 1980)

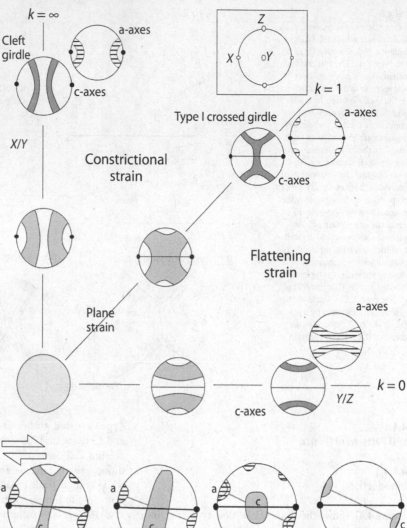

Fig. 4.42.
Pole diagrams showing four types of contoured LPO patterns of quartz c-axes (*grey*) and a-axes (*striped*) such as develop with increasing metamorphic grade in non-coaxial progressive deformation. The variation is due to a change in the dominant slip systems. Explanation in text

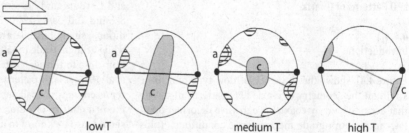

low T   medium T   high T

tributes mainly to c-axes in the periphery of the diagram, slip on prism planes to those in the centre, and slip on rhomb planes between both (Fig. 4.43a). Type II crossed-girdle c-axis patterns probably develop in constriction when rhomb slip is dominant over prism slip (Bouchez 1978; Schmid and Casey 1986).

In non-coaxial progressive deformation, domains of material line rotation are not of equal size as in coaxial progressive deformation (Sect. 2.7). As a result, one of the <a>-axes maxima is favoured and the c-axis patterns may be similar to those in Fig. 4.41 but one part will be better developed than the other. Consequently, the pattern of <a>- and c-axes obtains a monoclinic symmetry. For example, at high strain accumulated by simple shear at low to medium-grade metamorphic conditions, the Type I crossed girdle and double <a>-axes maxima are replaced

by a single <a>-axes maximum parallel to the movement direction (the fabric attractor) and a single girdle of c-axes normal to the flow plane (Figs. 4.42, 4.43b; Sect. 2.9). The c-axes from the periphery to the centre of the girdle stem from c-axes of grains deformed by basal, rhomb and prism slip respectively (Fig. 4.43). At low temperature, basal <a> slip is most important and the girdles may have a strong cluster of c-axes in the periphery. With increasing temperature, prism <a> slip becomes more important (Wilson 1975; Bouchez 1977; Lister and Dornsiepen 1982; Law 1990) and the girdle tends to a maximum around the Y-axis (Figs. 4.42, 4.43b). At very high temperature and hydrous conditions, prism <c> slip operates (Lister and Dornsiepen 1982; Blumenfeld et al. 1985; Mainprice et al. 1986), and causes a c-axis maximum subparallel to the attractor (Figs. 4.42, 4.43), and <a> axes normal to it.

a Coaxial

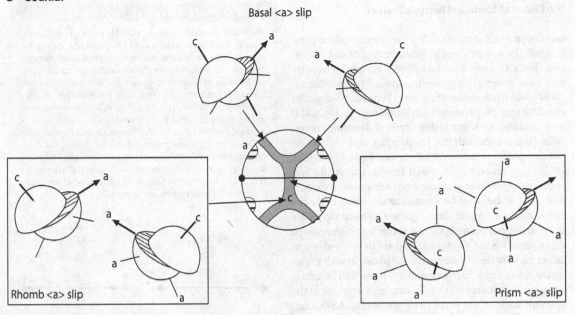

b Non-coaxial

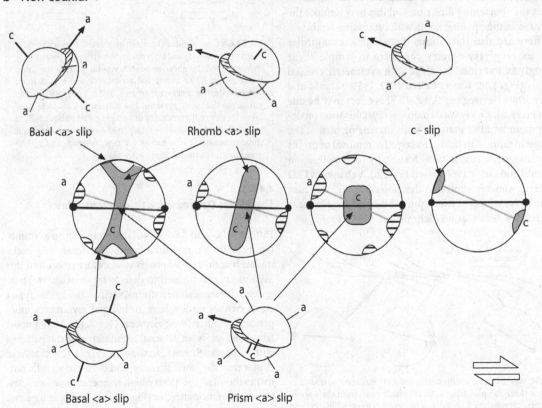

Fig. 4.43. a Illustration of the contribution of equidimensional quartz crystals with aligned <a>-axes and basal, rhomb or prism slip planes to a Type I crossed girdle pattern formed in coaxial progressive deformation. Ornamentation of contoured patterns of c- and a-axes in polar diagrams as in Fig. 4.42. **b** The same for several patterns that develop in non-coaxial progressive deformation. At *right* a pattern that developed by slip in direction of the c-axis

4.4.4.2
The Effects of Strain and Recrystallisation

Increasing strain at constant flow parameters and temperature will theoretically lead to strengthening of an LPO pattern (Fig. 4.41; Lister and Hobbs 1980), but fabric geometry may also change with increasing strain. The opening angle of small circle girdles of c-axes in Type I crossed girdles of a flattening progressive deformation regime (Fig. 4.41a) may decrease with increasing strain if dynamic recrystallisation is subdued; low temperature and high strain rate also produce relatively small opening angles (Tullis et al. 1973; Marjoribanks 1976; Jessell 1988b). This is an effect of the competition of grain rotation in response to dislocation glide and dynamic recrystallisation. Small circle girdles of c-axes are mostly due to grains with basal slip planes at 45° to the shortening ISA; these grains have high resolved shear stress on the slip system and are in optimal orientation for 'easy slip'. However, the slip planes in such grains rotate away from this orientation towards the fabric attractor with increasing strain, causing a decrease in the opening angle of the small circle girdle (Fig. 4.39). High temperature and low strain rate promote grain boundary migration that may consume grains that have rotated towards the shortening direction and this may hamper the decrease in the opening angle with increasing strain.

There are also indications that single c-axis girdles with external asymmetry can form in simple shear through an intermediate stage with symmetric crossed girdles (Fig. 4.44; Garcia Celma 1982, 1983; Schmid and Casey 1986; Herwegh et al. 1997). This effect may be due to an increasing effect of dynamic recrystallisation on development of LPO patterns with increasing strain; the change in fabric can be due to selective removal of grains in unfavourable orientations for slip by recrystallisation (Schmid and Casey 1986; Jessell 1988a,b). A change in LPO pattern geometry similar to that shown in Fig. 4.44 may occur in transition between pure shear and simple shear progressive deformation (Schmid and Casey 1986).

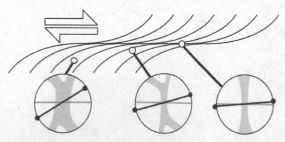

Fig. 4.44. Pole diagrams with contoured LPO patterns of quartz c-axes for three samples from a ductile shear zone, from the edge to the centre. *Black solid line* and *dots* represent reference foliation and lineation. The *horizontal grey line* marks the fabric attractor plane. The pattern changes from a Type I crossed girdle to a single girdle and becomes sharper with increasing strain to the centre of the zone but does not rotate with respect to the fabric attractor

---

Box 4.10 Problems with the interpretation of shear sense using quartz LPO

The reader should be aware of possible pitfalls in the assessment of shear sense using quartz LPO. First, the shear sense observed may not be associated with the other fabric elements visible in the rock. For example, where thin quartz seams occur between other mineral grains, the flow kinematics may have been completely different from the bulk flow. Another possible error occurs at high strain, if fabrics are vague or incomplete, e.g. due to recrystallisation or the presence of grains of a second mineral. In this case, the orientation of the LPO pattern with respect to a reference foliation S_r that is approximately parallel to the fabric attractor will give the wrong shear sense if it is interpreted as an external asymmetry (Fig. B.4.6). It is therefore wise not to rely exclusively on quartz c-axis fabrics to determine shear sense.

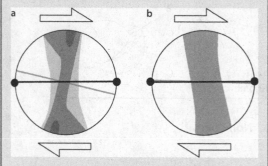

Fig. B.4.6. a Contoured LPO pattern of quartz c-axes with a clear internal asymmetry and external asymmetry with respect to S_r indicating dextral shear sense. **b** A similar pattern but more vague so that it lost its internal asymmetry, and at high finite strain for dextral shear sense when S_r and L_r are approximately parallel to the fabric attractor. The pattern in **b** could be mistaken for external asymmetry of a single girdle oblique to S_r at low strain, and erroneously interpreted to represent a sinistral shear sense. *Black solid line* and *dots* represent S_r and L_r. The *grey line* marks the fabric attractor plane

---

4.4.4.3
Shear Sense Determination Using Quartz Fabrics

Patterns of c- and <a>-axes of quartz obtain an asymmetry (actually a monoclinic symmetry) when they accumulate by non-coaxial progressive deformation, and this asymmetry can be used to deduce sense of shear. There are two elements; some patterns, such as the skewed Type I c-axis crossed girdles, have an internal asymmetry, independent of other fabric elements (Fig. 4.42). More important, however, is an external asymmetry of the patterns with respect to S_r and L_r in the rock (Fig. 4.42). <a>-Axes cluster near the fabric attractor, and c-axes in a girdle normal to the attractor. At very high temperature, c-axes cluster near the fabric attractor (Fig. 4.42); S_r and L_r in the same rock rotate towards the fabric attractor but will lie in the extension quadrant of the flow; consequently, there will be an angle between S_r or L_r and elements of the LPO fabric which can be used to determine sense of shear (Fig. 4.42).

4.4.5
LPO Patterns of Other Minerals

As in quartz, LPO in other minerals strongly depends on active slip systems and on the geometry and symmetry of the flow pattern, resulting in a similarity of LPO geometry and strain geometry. As for quartz, slip systems and fabric elements tend to rotate towards the fabric attractor, resulting in a common subparallel orientation of components of the LPO, and S_r and L_r in the rock.

Calcite c-axis LPO patterns show a similar influence of flow symmetry on pattern geometry. At low temperature (<300 °C) coaxial progressive flattening produces a maximum around the Z-axis of strain. In constriction a girdle through Y and Z develops and in coaxial plane strain a maximum around Z and a minor girdle through Y (Wenk et al. 1986a; Shelley 1993). In simple shear flow, a c-axis girdle develops as for quartz, but *dipping in the opposite direction* (in the direction of the shortening ISA). This is due to the fact that in calcite at low temperature, e-twinning is largely responsible for development of the LPO fabric (Schmid 1982; Behrmann 1983; Schmid et al. 1987; Wenk et al. 1987; Rutter et al. 1994; Lafrance et al. 1994; Rutter 1995; Khazanehdari et al. 1998; Casey et al. 1998; Pieri et al. 2001).

Experimental and computer models indicate that at high temperature in plane strain, split maxima may develop around the shortening axis (Wagner et al. 1982; Wenk et al. 1986a, 1987; Schmid et al. 1987; Takeshita et al. 1987; de Bresser 1991). Natural examples of such a high temperature LPO pattern in calcite were reported by de Bresser (1989), Burlini et al. (1998), Kurz et al. (2000), Leiss and Molli (2003) and Barnhoorn et al. (2004).

In *plagioclase*, (010) trends to parallelism with S_r and [001] with L_r at medium to high-grade metamorphic conditions (Olsen and Kohlstedt 1985; Montardi and Mainprice 1987; Ji and Mainprice 1988; Shaosheng and Mainprice 1988; Mainprice and Nicolas 1989; Dornbush et al. 1994; Egydio-Silva and Mainprice 1999). At very high grade, however, [100] trends to parallelism with L_r (Kruhl 1987b; Dornbush et al. 1994). In shear zones, the pole to (010) tends to show an external asymmetry with respect to a planar and linear shape preferred orientation, which can be used to determine shear sense as in the case of quartz c-axes.

Olivine has a complex behaviour that is strongly dependent on temperature (Mainprice and Nicolas 1989). At relatively 'low' temperatures (700–1 000 °C) corresponding to lower crustal or upper mantle levels where olivine is at the limit of crystalplastic behaviour, diffuse girdles of principal axes may occur. At medium temperature (≈1 000 °C), a girdle of [0kl] normal to L_r and a point maximum of [100] parallel to L_r may form. At high temperature (>1 100 °C) a point maximum of [010] normal to S_r and [100] parallel to L_r develops; and at hypersolidus conditions (>1 250 °C) a point maximum of [100] parallel to L_r and partial girdles of [010] and [001] are formed. Fabrics with external asymmetry do occur and can be used to determine shear sense (Avé Lallemant and Carter 1970; Mercier 1985).

In *clinopyroxene* three main types of LPO have been described; (a) [100] normal to S_r and [001] parallel to L_r or S_r (Mainprice and Nicolas 1989; Phillipot and van Roermund 1992); (b) [010] normal to S_r and [001] parallel to L_r or S_r (Mainprice and Nicolas 1989); (c) [010] normal to L_r and [001] parallel to L_r (Helmstead et al. 1972; van Roermund 1983, 1992). (a) is mainly found in peridotite massifs with pyroxenite layers and may either be formed at low temperature and high strain rate or, if this is unlikely, by post-tectonic crystal growth (van Roermund 1992). (b) and (c) form by crystalplastic deformation by multiple slip in medium to high temperature eclogites by activity of dislocations with predominantly [001] and <110> Burgers vectors (van Roermund 1983, 1992; Buatier et al. 1991). Their difference seems to reflect different strain types, constriction for Type (b) and flattening for Type (c) (Helmstead et al. 1972; van Roermund 1992). Asymmetric fabrics have not been reported for clinopyroxene in the literature.

In *orthopyroxene* deformed at high-grade metamorphic conditions, LPO of [100] has been reported at a high angle to S_r and of [001] gently inclined with respect to L_r (Dornbush et al. 1994). This preferred orientation is associated with the dominance of (100)[001] as a slip system. The LPO reported by Dornbush et al. (1994) is slightly asymmetric and can be used to determine sense of shear.

Little is known about LPO in *amphiboles*, but the available data indicate a strong similarity with clinopyroxene. [001] is commonly parallel to L_r (Gapais and Brun 1981; Rousell 1981; Shelley 1994) and either (100) (Mainprice and Nicolas 1989) or (110) (Gapais and Brun 1981) parallel to S_r. Preferred orientations at medium to high grade may be due to crystalplastic deformation, but LPO in amphiboles at low-grade conditions may be due to rigid body rotation, dissolution-precipitation or oriented growth (Ildefonse et al. 1990; Shelley 1993, 1994; Imon et al. 2004). Hornblende fabrics reflect the strain symmetry (Gapais and Brun 1981).

Shear Zones

A subject that has fascinated many geologists over the last decades is treated in this Chapter. Concentration of deformation along shear zones, producing mylonites or brittle fault rocks is discussed as related to depth and therefore metamorphic environment. The different fault rocks are treated from low to high grade or from fast to slow strain rates. The first part of the chapter deals with a description of brittle fault rocks that may be cohesive or incohesive, and of pseudotachylytes that form by seismic events.

Mylonites are produced predominantly by ductile deformation and usually show significant recrystallisation of the matrix. Characteristic fabric elements are porphyroclasts in a finer grained matrix. Attention is given to the dynamics of mylonite development and to the influence of different metamorphic conditions. Many shear zones show evidence of repeated activity under different metamorphic conditions or strain rates, producing e.g. narrow cataclasites cutting mylonite zones.

The determination of shear sense in shear zones is of crucial importance for the reconstruction of the tectonic evolution of e.g. colliding continents, crustal strike-slip movements or escape tectonics. A number of well established shear sense indicators is presented and discussed. These include: displacement of markers, foliation curvature, shear band cleavage including C/S fabrics, mantled porphyroclasts, mineral fish, quarter structures and lattice preferred orientation. Other more problematic shear sense indicators are also evaluated. Special attention is dedicated to a new topic called flanking structures, in which crosscutting elements such as faults or veins transect the mylonitic foliation at an angle. The last section is on shear sense indicators in the brittle regime.

5.1
Introduction

In general, deformation in rocks is not homogeneously distributed. One of the most common patterns of heterogeneous deformation is the concentration of deformation in planar zones that accommodate movement of relatively rigid wall-rock blocks. Deformation in such high-strain zones usually contains a rotation component, reflecting lateral displacement of wall rock segments with respect to each other; this type of high-strain zone is known as a *shear zone*. Deformation in a shear zone causes development of characteristic fabrics and mineral assemblages that reflect *P-T* conditions, flow type, movement sense and deformation history in the shear zone. As such, shear zones are an important source of geological information.

Shear zones can be subdivided into brittle zones or faults, and ductile zones (Chap. 3). Ductile shear zones are usually active at higher metamorphic conditions than brittle shear zones (Figs. 3.44, 5.2). Major shear zones which transect the crust or upper mantle have both brittle and ductile segments. The depth of the transition between dominantly brittle and ductile behaviour depends on many factors such as bulk strain rate, geothermal gradient, grain size, lithotype, fluid pressure, orientation of the stress field and pre-existing fabrics (Sect. 3.14). Ductile shear zones may develop in marbles at metamorphic conditions where quartzites would deform by brittle fracturing, and different minerals in a small volume of rock can show contemporaneous brittle and ductile deformation (Fig. 3.42).

Major shear zones can be active for considerable periods of time, and material in the shear zone may be transported upwards or downwards in the crust. Consequently, rocks in major shear zones commonly show evidence of several overprinting stages of activity at different metamorphic conditions. Minor shear zones may also show several overprinting stages since shear zones, once formed, are easily reactivated.

A special terminology is used for rocks that have been deformed in shear zones, partly independent of their lithology (Sibson 1977b). They are usually referred to as *fault rocks* or *deformation zone rocks* (Sibson 1977b; Scholz 2002; Schmid and Handy 1991; Blenkinsop 2000), even if deformed in ductile shear zones. The most common types are *brittle fault rocks*, *mylonites* and *striped gneiss*. An excellent detailed treatment of fault rocks and structures in shear zones is given in the fault related rocks atlas edited by Snoke et al. (1998).

5.2
Brittle Fault Rocks

5.2.1
Introduction

Brittle fault rocks form by fault propagation through intact rock, commonly along some older plane of weakness, and formation of a volume of *brittle fault rock* in a *fault zone* along the active fault. Many processes influence the ability of a fault to propagate and slip such as the regional stress field and geometry of rock units, fluid pressure in the fault and wall rock and the interaction of the brittle fault rock with fluids. Faults can show *velocity weakening* or *velocity strengthening* behaviour (Tse and Rice 1986; Chester et al. 1993). In the first case, resistance to sliding decreases with increasing velocity and the fault can produce earthquakes. In the second case, faults decelerate and sliding is asysmic and stable. A higher temperature seems to promote velocity strengthening behaviour (Shimamoto 1989; Chester et al. 1993). Presently, there are no ways to distinguish brittle fault rocks that form in velocity weakening and strengthening segments of faults, except for pseudotachylyte which forms exclusively on rapidly moving faults (Sect. 5.2.5). Fluids that infiltrate a fault strongly influence its mechanical behaviour, but in a complex way. An increased fluid pressure decreases the strength of the fault by decreasing the effective normal stress over the fault (Sect. 3.2). Fluids may also cause weakening by reaction of stronger phases to weaker minerals in fault rocks, or by stress corrosion in the fault process zone (Chap. 3). Fluids may cause fault rock strengthening by precipitation of vein material such as quartz, calcite or even K-feldspar, cementing fault rock fragments together (Fredrich and Evans 1992; Wintsch et al. 1995; Wintsch 1998). Precipitation of vein material may even cause a decrease in permeability of the fault zone and thereby an increase in fluid pressure and fault rupture, after which permeability is increased and fluid pressure falls until renewed precipitation causes the next cycle of fault activity (Sibson 1990; Cox et al. 1991; Cox 1995).

5.2.2
Incohesive Fault Rocks

Brittle fault rocks can be subdivided into incohesive and cohesive types. Incohesive brittle fault rocks are usually found in faults, which have been active at shallow crustal levels. They occur in fault zones of variable thickness and can be subdivided into *incohesive breccia*, *incohesive cataclasite* and *fault gouge*. Incohesive breccia consists for more than 30 vol-% of angular fragments of the wall rock or of fractured veins, separated by a fine-grained matrix. In cataclasite, less than 30 vol-% fragments are present in the fine-grained matrix. In fault gouge, few large fragments occur isolated in the matrix. This matrix may be foliated, and fragments commonly have a lensoid shape (Chester et al. 1985; Chester and Logan 1987; Evans 1988; Kano and Sato 1988; Lin 1996, 1997; Takagi 1998). The wall rock and included fragments in incohesive cataclasite and fault gouge commonly show polished surfaces (slickensides) with striations or fibres (slickenfibres – Sect. 6.2.5) that can be used to determine movement direction and shear sense (Sect. 5.7.2) along the fault zone.

Fig. 5.1.
Optical (**a**) and cathodoluminescence (**b**) image of a dilatational
brittle fault zone in a quartzite
host rock of the Muth Formation
(India; sample of Erich Draganits, Vienna). Sense of shear is
dextral (see rotated quartz chips
that flake off from the larger
grains). Created voids are ce-
mented with quartz that is dark
in the cathodoluminescence
image. Thus fluids have to be
involved in the cementation of
the brittle fault. The brittle particles acted as precipitation nuclei. Notice microcracks and
transgranular fractures in the
quartz grains outside the fault
zone. Width of view 4 mm.
(Photographs courtesy of
Erich Draganits and Michel
Bestmann)

5.2.3
Deformation Bands

Deformation bands are mm-wide planar brittle shear
zones in undeformed, porous, quartz-rich, clay-poor sedimentary rocks (Aydin 1978; Aydin and Johnson 1978, 1983;
Antonellini et al. 1994; Davis 1999; Mair et al. 2000, 2002;
Davatzes et al. 2003; Main et al. 2001). Notice, however,
that the term *deformation band* is also used for a special
type of undulose extinction (Sect. 3.6).

Deformation bands are brittle fault zones that develop very
close to the Earth's surface in poorly or even unconsolidated
porous sediment (Fig. 5.1; Underhill and Woodcock 1987;
Antonellini et al. 1994; Cashman and Cashman 2000). They
are normally planar and up to tens of metres long, with
occasional ramp or branching structures. They typically
have a small to very small displacement homogeneously
distributed over the band, without localisation as on faults.

Most deformation bands have significantly lower porosity than the undeformed wall rock associated with cataclasis.

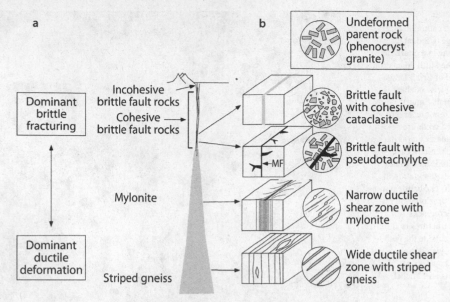

Fig. 5.2. Distribution of the main types of fault rocks with depth in the crust. **a** Schematic cross-section through a transcurrent shear zone. The zone may widen, and changes in geometry and dominant type of fault rock occur with increasing depth and metamorphic grade. **b** Schematic representation of four typical fault rocks (out of scale) and the local geometry of the shear zone in a 1-m-wide block, such as would develop from a phenocryst granite. Inclined (normal or reverse) shear zones show a similar distribution of fault rocks and shear zone geometry with depth. No vertical scale is given since the depth of the transition between dominant ductile deformation and brittle fracturing depends on rock composition, geothermal gradient, bulk strain rate and other factors (Sect. 3.14). *MF:* Main fault vein

Deformation bands form in high porosity rocks or at high differential stress, high mean stress and high strain (*critical state theory:* Schofield and Wroth 1968). In such bands, cataclasis is associated with collapse of the pore space, grain rotation and cataclastic flow (Menéndez et al. 1996), especially in rocks with high porosity and good sorting. Good sorting implies high stress concentration points where grains touch, while poor sorting effects more equal stress distribution over grains. In such cases, polycrystalline grains and feldspar grains are fragmented while monocrystalline quartz grains are more resistant. Deformation bands with decrease in porosity are also known as *compaction bands* (Mollema and Antonellini 1996). Some deformation bands show no significant change in porosity, but are defined by preferred orientation of grains in the band. There are also deformation bands with an increase in porosity. These form mainly at low strain in low porosity rocks without much cataclasis, and probably at very low mean stress close to the Earth surface (Antonellini et al. 1994). Such bands are also known as *dilatation bands* (Du Bernard et al. 2002). Clay content of the rock also influences deformation behaviour.

Deformation bands can occur single or, more commonly, in bundles or *zones* of subparallel bands which taken together can lead to significant displacement (Main et al. 2001). These zones probably develop due to strain hardening in individual deformation bands, leading to development of new ones. Cataclasis and hardening in a deformation band may eventually be followed by localisation of motion on a brittle slip plane, and softening.

Deformation bands that are associated with a change in porosity are of great economic importance since they influence rock permeability and the shape of water and hydrocarbon reservoirs in rocks (e.g. Aydin 2000; Fisher and Knipe 2001; Ogilvie and Glover 2001).

5.2.4
Cohesive Fault Rocks

Cohesive fault rocks can be subdivided into *cohesive breccia* (Fig. 5.3), *cohesive cataclasite* and *pseudotachylyte*. The distinction between breccia and cataclasite is as discussed for incohesive fault rocks. The cohesive nature of the rock is due to precipitation crystallisation of minerals such as quartz, calcite, epidote, chlorite or K-feldspar from a fluid. K-feldspar only precipitates if the solution is highly alkaline, which could occur if fluid infiltration into a freshly crushed rock is limited (Wintsch et al. 1995; Wintsch 1998).

Cohesive breccia and cataclasite are less easily identifiable in outcrop than incohesive fault rocks; for example, incohesive cataclasite in quartzite is obvious because of weathering contrasts, but cohesive cataclasite may differ from undeformed host rock only by a darker colour. Cohesive breccia and cataclasite can be formed in any rock type. Usually, fragments of all sizes occur hampering a clear distinction between matrix material and fragments (Figs. 3.5, 5.3). The contact between the fault rock and the intact wall rock is usually a gradual transition of decreasing brittle deformation intensity. Cohesive cataclasite and breccia

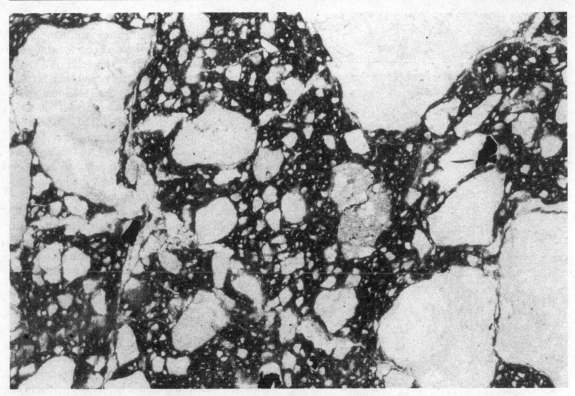

Fig. 5.3. Cohesive fault breccia in quartzite. Angular fragments of variable size are present. Orobic Alps, Italy. Width of view 8 mm. PPL

commonly show evidence for abundant pressure solution and precipitation effects. Rock fragments are transected by healed cracks aligned with fluid inclusions. Veins of quartz, calcite, epidote or chlorite, and in ultramafic rocks, serpentine, are common. These veins form during and after brittle deformation, since they have commonly been fractured. Although most cataclasites have random fabrics, *foliated cataclasite* does occur, especially where the host rock is rich in micas (Chester et al. 1993; Evans and Chester 1995; Lin 1996, 1997, 1999; Lin et al. 1998; Chester and Chester 1998; Evans 1998; Mitra 1998). Such rocks may contain a compositional layering (Kanaori et al. 1991) and a preferred orientation of mica fragments, elongate grains or new-grown micas (Evans 1988) wrapping around large grains of resistant minerals, e.g. quartz and feldspar in granitic rocks. The foliation can also be spaced and result from parallel alignment of minor shear fractures (Chester et al. 1985) or of dissolution planes filled with opaque material (Mitra 1998). Shear band cleavage structures (Sect. 5.6.3) are common in such foliated cataclasites (Lin 1999, 2001). Deformation mechanisms in cataclasite are mainly cataclastic flow in and between grains, grain boundary sliding and pressure solution. Cohesive breccia and cataclasite are thought to develop at greater crustal depth than incohesive ones. Unless stated otherwise, cataclasite and breccia are understood to mean the cohesive form in the following sections.

5.2.5
Pseudotachylyte

Pseudotachylyte is a cohesive glassy or very fine-grained fault-rock with a very distinct fabric (Magloughlin and Spray 1992). Its curious name derives from its resemblance to tachylyte, a mafic volcanic glass, while the material is obviously not of volcanic origin (Shand 1916). Pseudotachylyte has a number of characteristic geometric features that usually allow its distinction from other brittle fault rock types. It is composed of a dark matrix material with minor inclusions of mineral or wall rock fragments (Figs. 5.4–5.6). It usually occurs in a characteristic geometrical setting of a planar *main fault vein* (Sibson 1975; Spray 1992) or *generation surface*, up to a few mm thick and irregular *injection veins* which branch from the main fault vein into the wall rock (Figs. 5.2, 5.4, 5.5). Main fault veins are usually planar, up to a few centimetres wide and occur as straight bands in outcrop. Occasionally, they occur as pairs of subparallel surfaces, connected by injection veins. The volume of rock containing injection veins is known as the *reservoir zone* (Magloughlin and Spray 1992). Main fault veins can be difficult to spot in the field, especially if they lie parallel to older layering or foliation; injection veins are more conspicuous and allow recognition of pseudotachylyte in outcrop. Pseudotachylyte veins have distinct, sharp and

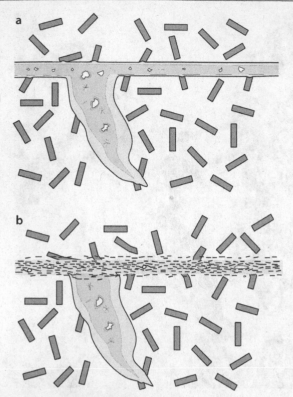

Fig. 5.4. a Schematic drawing of a typical pseudotachylyte with main fault vein, injection vein, internal compositional banding and typical inclusions. The boundary with the wall rock is sharp. Mica grains in the wall rock show corrosion along the contact with pseudotachylyte. **b** Pseudotachylyte in which the main fault vein has been reactivated as a mylonite zone. The mylonite can be recognised as a former pseudotachylyte by its fine-grained homogeneous nature and the presence of injection vein relicts

(for main fault veins) straight boundaries with the wall rock. They never show transitional zones of decreasing brittle deformation intensity towards the wall rock as is usually the case for cataclasite or breccia. The wall rock can be cataclased or faulted, but these structures are normally transected by the younger pseudotachylyte. Pseudotachylyte is thought to form by local melting of the rock along a brittle fault plane due to heat generated by rapid frictional sliding (10^{-2} to 1 m s^{-1}; Philpotts 1964; Sibson 1975, 1977a,b; Grocott 1981; Maddock 1986; Maddock et al. 1987; Spray 1987, 1992, 1995, 1997; Shimamoto and Nagahama 1992; O'Hara 1992; Swanson 1992; Lin 1994; Legros et al. 2000; Bjørnerud and Magloughlin 2004).

Pseudotachylyte occurs associated with events such as meteorite impact (Martini 1992; Thompson and Spray 1994; Spray et al. 1995; Hisada 2004), crater collapse, caldera collapse and giant landslides on superficial *superfaults* (Masch et al. 1985; Reimold 1995; Spray and Thompson 1995; Spray 1997; Legros et al. 2000), in veins thicker than 1 cm, but occurrences of veins less than 1 cm wide are more common and are thought to be associated with

seismic activity on brittle faults in common tectonic settings. Most pseudotachylytes therefore form in the upper to middle crust. However, some occurrences from the deep crust, which apparently formed at granulite or eclogite facies, have been reported (Austrheim and Boundy 1994; Boundy and Austrheim 1998; Clarke and Norman 1993).

Melting at temperatures between 750–1 600 °C is thought to occur on the main fault vein of a pseudotachylyte (Austrheim and Boundy 1994; Camacho et al. 1995; Lin and Shimamoto 1998; O'Hara 2001; Di Toro and Pennacchioni 2004). Some of the melt may intrude minor faults, which branch from the main fault vein into the wall rock, and form injection veins (Figs. 5.2, 5.4a, 5.5). The small volume of melt formed in this way cools rapidly to the temperature of the host rock. As a result, the melt quenches to a glass or very fine-grained, aphanitic material that occurs along fault planes and adjacent branching injection veins (Figs. 5.4a, 5.5). There is some evidence that rock crushing may precede the melting stage in some pseudotachylytes. Pseudotachylyte is normally not associated with growth of quartz- or calcite veins and generally occurs in massive, dry, low-porosity rocks such as granite, gneiss, granulite, gabbro and amphibolite. This is because the fluid present in porous rocks lowers the effective normal stress over a fault plane upon heating; consequently, not enough frictional heat can be produced to cause local melting. Therefore, pseudotachylyte is not normally found in porous sedimentary rocks (for a possible exception see Killick 1990). It is not found in marble because of the dissociation of carbonates at high temperature and the resulting decrease in normal stress over a fault, and the ductile flow in carbonates, which inhibits build-up of high differential stress. It may seem curious that pseudotachylyte is a product of high temperature (melt generation) related to low temperature brittle fault zones, while such local melting is rare in higher-grade ductile shear zones. In brittle fault zones, however, elastic strain energy may be stored for a long period of time and is released in a matter of seconds in a small volume of rock along faults; in ductile shear zones, heat is dissipated continuously over a larger volume of rock and is therefore usually insufficient to cause a significant rise in temperature.

The matrix of pseudotachylyte is commonly black, dark brown, green or red and relatively homogeneous, but may contain a compositional layering of irregular thickness, which follows the contours of the vein (Fig. 5.4a). This layering is commonly of a different colour along the vein wall and in the interior, and is interpreted to result from selective melting of the wall rock. The layering may be folded and folds are interpreted to have formed by fluid flow in the melt. Even sheath folds (Sect. 5.3.2) parallel to the displacement direction of the wall rock have been observed in the layering (Berlenbach and Roering 1992). Amygdules derived from gas bubbles are sometimes present in the matrix (Maddock et al. 1987; Magloughlin 1989, 1998a,b).

Fig. 5.5. Pseudotachylyte in a foliated amphibolite. The foliation is cut by the pseudotachylyte. A main fault vein transects the rock from *top right* to *lower left*. An injection vein occurs at *left*. Isolated fragments lie in a dark pseudotachylyte matrix. Boundaries between pseudotachylyte and wall rock are sharp. Paine, South Chile. Width of view 14 mm. PPL

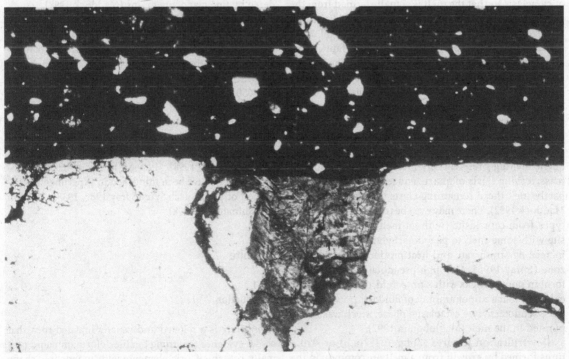

Fig. 5.6. Contact of a pseudotachylyte vein (*top*) and the wall rock (*bottom*). The boundary is sharp where quartz and feldspar grains (*white*) are in contact with the pseudotachylyte matrix, but an embayment exists where a biotite grain (*centre*) is in contact. This structure is attributed to preferential corrosion of the biotite crystal by pseudotachylyte melt along an originally straight fracture surface. The biotite grain is strongly kinked, probably due to brittle deformation preceding pseudotachylyte generation. Vestfold Hills, Antarctica. Width of view 5 mm. PPL

The mineral composition of inclusions in pseudotachylyte is commonly disproportional to the mineral composition of the wall rock; quartz and to a lesser extent feldspar are common as inclusions, while Fe-Mg rich Al-silicates are under-represented (Figs. 5.4a, 5.5, 5.6; Maddock 1986; Maddock et al. 1987; Lin 1994). Micas are rarely present as inclusions. Quartz fragments have angular outlines with numerous internal fractures and fluid inclusion planes while fragments of feldspar, hornblende or pyroxene tend to be rounded. Where the contact of pseudotachylyte and wall rock is a straight fracture, as along main fault veins, embayments of pseudotachylyte may exist where micas or amphiboles were in contact with the pseudotachylyte matrix. The matrix of a pseudotachylyte differs from that of cataclasite or breccia in that the smallest size fragments are lacking and isolated fragments are contained in a relatively homogeneous matrix (cf. Figs. 5.3, 5.4a, 5.5, 5.6). All these features are attributed to preferential dissolution of Fe-Mg Al silicates and feldspar in the pseudotachylyte melt (O'Hara 1992; Lin 1994). Another typical microstructure of pseudotachylyte is the presence of aggregates of small sulphide particles in larger quartz fragments. These sulphide droplets may have formed from a sulphide bearing melt (Magloughlin 1992).

Microstructural and experimental evidence suggests that most pseudotachylyte forms through an initial stage of cataclasis so that the melt is actually formed from the crushed rock rather than from the intact wall rock (Swanson 1992; Spray 1995; Ray 1999; Fabbri et al. 2000; Ray 2004). In some pseudotachylyte main fault veins, cataclasite occurs in isolated pockets along the contact with the wall rock, but not in injection veins (Ermanovics et al. 1972; Killick et al. 1988; Magloughlin 1989, 1992; Curewitz and Karson 1999). Hydrated ferromagnesian minerals were preferentially fragmented into a fine-grained cataclasite groundmass with included larger fragments of feldspar and quartz (Allen 1979); subsequent melting preferentially incorporates this groundmass, leaving clasts of quartz and feldspar, but may also partly melt these remaining clasts (Magloughlin 1989; Maddock 1992). There may even be transitional fault rock types from cataclasite (without melt) through cataclasite with some melt to pseudotachylyte, which form by increasing strain rate and heat production in the fault zone (Spray 1995). Although pseudotachylyte does not form in porous rocks with a pore fluid, there is chemical evidence that a minor amount of fluid was present in the pre-pseudotachylyte cataclasite phase, which was incorporated in the melt (Magloughlin 1992).

Devitrification features (Lofgren 1971a,b) or structures formed by growth from a melt are common in the matrix of pseudotachylytes (Maddock 1986; Maddock et al. 1987; Lin 1994). They are similar to those observed in obsidian and consist of idiomorphic acicular grains

> **Box 5.1   Misidentification of pseudotachylyte**
>
> Some dark cataclasites and layers or veins filled with dark minerals such as chlorite or tourmaline resemble pseudotachylyte in the field and even in thin section. Pseudotachylyte differs from these rocks by (1) the sharp boundaries with the wall rock; (2) the occurrence of injection veins; (3) evidence for melting such as a relative scarceness of micas, pyroxene and hornblende as inclusions in the matrix and the corrosion of such minerals along vein contacts; (4) presence of spherulites and devitrification structures and (5) the absence of contemporaneous quartz or calcite veins. Most pseudotachylyte has a chemical composition almost identical to the host rock, while other veins or cataclasite zones will usually show a different composition.

of feldspar, biotite, amphibole or orthopyroxene known as *microlites* (Lofgren 1971a,b; Toyoshima 1990; Macaudière et al. 1985; Magloughlin 1992; Di Toro and Pennacchioni 2004). Microlites may occur as simple acicular grains, as skeletal and dendritic shapes, or be arranged into *spherulites* (Lofgren 1974; Doherty 1980; Clarke 1990; Lin 1994, 1998; Di Toro and Pennacchioni 2004). There may be a sequence of increasingly complex shape from acicular, skeletal, dendritic to spherulitic from margin to core of a pseudotachylyte vein, which may be due to differences in cooling rate (Lin 1994, 1998). Spherulites of biotite or feldspar are commonly nucleated on inclusions. Microlites in pseudotachylyte are commonly powdered by fine magnetite grains (Maddock 1998).

Melting temperature of pseudotachylytes is hard to determine since they do not form by equilibrium melting. Information can be obtained from microlite composition, or from the presence or absence of certain minerals with different melting points in the matrix, between 550–650 °C for micas, through 1 100–1 500 °C for feldspars and pyroxenes, to 1 700 °C for dry quartz (Toyoshima 1990; Lin 1994). Estimates on melting temperatures using these tools range from 750 °C to exceeding 1 700 °C. (Wallace 1976; Sibson 1975; Maddock et al. 1987; O'Hara 1992; Lin and Shimamoto 1998). Many of the microstructures observed above have been mimicked in experimental generation of pseudotachylyte (Spray 1987, 1988, 1995; Lin and Shimamoto 1998).

5.3
Mylonite

5.3.1
Introduction

A *mylonite* is a foliated and usually lineated rock that shows evidence for strong ductile deformation and normally contains fabric elements with monoclinic shape symmetry (Figs. 5.2, 5.7–5.9, ⊚Photos 5.8, 5.9a,b; Bell and Etheridge 1973; Hobbs et al. 1976; White et al. 1980; Tullis et al. 1982; Hanmer and Passchier 1991). Mylonite is a

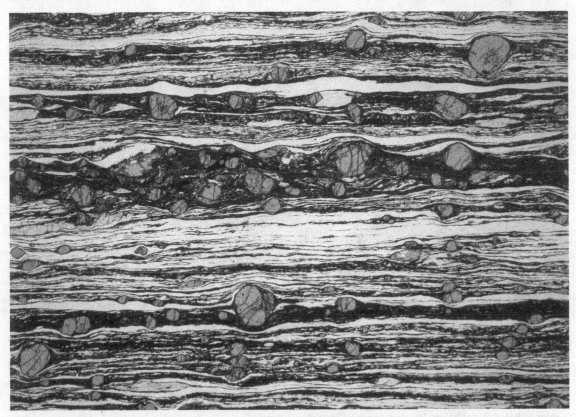

Fig. 5.7. Mylonite derived from pelitic gneiss with quartz, feldspar, garnet and micas in a section parallel to the aggregate lineation and normal to the foliation. Alternating layers rich in quartz (*clear*) and feldspar (*grey*), with porphyroclasts of garnet define the mylonitic foliation. Sense of shear indicators are poorly developed in this section but subtle stair-stepping (Sects. 5.6.5–5.6.7) of wings on porphyroclasts and small C'-type shear bands (Sect. 5.6.3) indicate a dextral shear sense. Marsfjällen, Sweden. Width of view 13 mm. PPL

strictly structural term that refers only to the fabric of the rock and does not give information on the mineral composition. Mylonite should therefore not be used as a rock name in a stratigraphic sequence.

Mylonite occurs in high-strain zones known as *mylonite zones*, interpreted as exhumed, 'fossil' ductile shear zones. The contact of a mylonite zone and unaffected wall rock tends to be a gradual fabric transition. Grain size in the mylonite is usually smaller than that in the wall rock. (Fig. 5.9). Mylonite zones can occur in any rock type and have been described from a sub-millimetric scale to zones several km wide. (Bak et al. 1975; Hanmer 1988). The intensity of deformation may vary from one mylonite zone to another but is always high. The word 'mylonite' derives from the Greek 'μυλων' (a mill) since the original opinion on these rocks was that they formed by brittle 'milling' of the rock (Lapworth 1885). However, present use of the word mylonite refers to rocks dominantly deformed by ductile flow, while brittle deformation may play a minor role in isolated included lenses or grains (Bell and Etheridge 1973; Tullis et al. 1982); in other words, the stress-supporting network is affected by crystalplastic deformation (Sect. 3.8).

5.3.2
Characteristic Fabric Elements

Mylonites can be recognised in the field by their small grain size and strongly developed, unusually regular and planar foliations (Figs. 5.7–5.9) and straight lineations. Lenses and layers of fine-grained material that are common in mylonites are thought to derive from a more coarse-grained parent rock by intracrystalline deformation and recrystallisation. Such deformed lenses usually have a 'surf-board' shape defining both a planar and linear fabric element. This shape may be explained by common development of mylonites in an approximately plane strain regime such as in simple shear.

Many mylonites contain *porphyroclasts* (Figs. 5.6, 5.7; Box 5.2) which are remnants of resistant mineral grains of a size larger than grains in the matrix. The foliation in the matrix wraps around porphyroclasts (Figs. 5.7, 5.8, 5.10). Porphyroclasts develop because of a difference in rheology between constituent minerals; relatively 'hard' minerals will form porphyroclasts, while relatively soft ones form part of the matrix. However, porphyroclasts do not always form in the

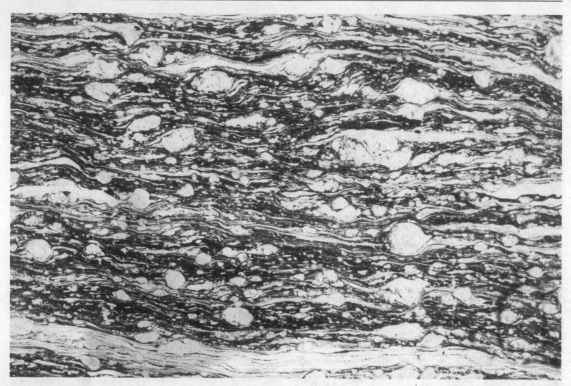

Fig. 5.8. Quartz-feldspar mylonite. Section parallel to the aggregate lineation and normal to the foliation. Lenses of recrystallised quartz and feldspar define the mylonitic foliation. The foliation wraps around feldspar porphyroclasts. Minor shear bands (Sect. 5.6.3) define the sense of shear as dextral. St. Barthélemy, Pyrenees, France. Width of view 10 mm. PPL

Box 5.2   Porphyroclasts and porphyroblasts

Porphyroclasts and porphyroblasts are relatively large, single crystals in a fine-grained matrix. The word porphyroclast is also used for a rounded polycrystalline rock fragment in a more fine-grained matrix. Porphyroclasts (from 'clasis' – breaking) are inferred to have formed by diminution of the grain size in the matrix. They are therefore typical for mylonites and cataclasites; they are relic structures of a more coarse-grained original fabric. The word *clast* is often used as a short equivalent commonly with the constituent mineral as a prefix (feldspar clast). Common minerals that form porphyroclasts are feldspar, garnet, muscovite, hornblende and pyroxenes. Quartz forms porphyroclasts only in very special cases (Sect. 3.13.3; Figs. 3.9, 3.10). Porphyroclasts should not be confused with detrital clasts in sediments. Porphyroblasts (from 'blasis' – growth) are inferred to have formed by growth of crystals of specific mineral species, while crystals in the matrix did not grow to the same extent (cf. Chap. 7). The word *blast* is commonly used as a short equivalent. They are common in non-mylonitic phyllites and schists. In some cases, original porphyroblasts in a schist or gneiss may become porphyroclasts when the schist or gneiss is mylonitised.

same minerals, since rheological properties of minerals depend on metamorphic conditions and initial grain size (Sect. 3.12).

The planar fabric element of mylonites is known as a *planar shape fabric, shape preferred orientation* (Box 4.2) or more specifically as a *mylonitic foliation* (Box 4.4); the linear fabric element is known as a *linear shape fabric* or *aggregate lineation* (Sect. 4.3, Box 4.2; Fig. 5.10). Low-strain lenses around which the shape fabric anastomoses are common in mylonites (Fig. 5.10), from lozenge-shaped single feldspar crystals (Figs. 5.8, 5.12) to km-scale lenses.

Well-developed aggregate lineations are mainly found in polymineralic rocks where grain size reduction has taken place (Piazolo and Passchier 2002b; Sect. 4.3). In originally fine-grained rocks, especially if they were monomineralic and no grain size reduction took place, aggregate lineations may be absent even if strain is high. At high-grade deformation conditions, grain lineations dominate.

Mylonites commonly contain two or even three foliations, inclined to each other at small angle, that are thought to have developed contemporaneously (Fig. 5.10). These are further explained in Sect. 5.6. Quartz, calcite and feldspar commonly also show evidence of lattice preferred orientation (Sect. 4.4.5). Foliations in mylonite are locally subject to tight or isoclinal folding (Figs. 1.2, 5.10, ⊘Video 1.2). In most cases, the axial planar foliation in these folds cannot be distinguished from the main

Fig. 5.9. Zone of ultramylonite with straight internal layering (*left*) in a coarse-grained host rock composed of quartz, feldspar and biotite. Section parallel to the aggregate lineation and normal to the foliation. A minor mylonite zone transects the *centre* of the photograph. Along this zone deflection of the foliation in the wall rock indicates dextral sense of shear. Pernambuco, Brazil. Width of view 8 mm. CPL

foliation in the mylonite; this indicates that such folds should not be regarded as the effect of a separate phase of deformation affecting an older mylonitic fabric, but as the result of a local distortion in the flow field during mylonite genesis (Sect. 2.5; Cobbold and Quinquis 1980). Some of these folds are *sheath folds*, that is, they have a tubular shape parallel to the aggregate or grain lineation (Fig. 5.10; Cobbold and Quinquis 1980; Lacassin and Mattauer 1985; Alsop and Holdsworth 2004). Others are non-tubular but cylindrical with a straight fold axis parallel to the lineation; these are known as *curtain folds* (Fig. 5.10; Hartwig 1925; Lotze 1957; Passchier 1986a). Curtain folds commonly decrease in amplitude and fade out laterally (Fig. 5.10).

An important characteristic of many mylonites is a clear difference in geometry of structures in thin sections cut normal and parallel to the aggregate or grain lineation (Fig. 5.10). In sections normal to the lineation, the rock may seem relatively undeformed or structures have orthorhombic symmetry (Fig. 3.29); in sections parallel to the lineation, the deformation fabric is usually much stronger, and structures with monoclinic symmetry that may be used as shear sense indicators appear (Sects. 5.5, 5.6). Characteristic is the curved shape and decreasing intensity of mylonitic foliations away from

the core of the shear zone (Fig. 5.10, top left). The overall monoclinic symmetry of mylonite zones and of fabric elements in them reflects the monoclinic symmetry of non-coaxial flow in a shear zone.

It is sometimes difficult to decide if a certain strongly deformed rock in an isolated outcrop should be called a mylonite or not. In such cases it is important to use good illustrations in publications.

5.3.3
Mylonite Classification

Mylonites are classified according to the metamorphic grade at which deformation took place (e.g. high-grade mylonite) or according to the lithotype or mineralogy in which they are developed (e.g. quartzite-mylonite, granodiorite-mylonite, quartz-feldspar mylonite). If mylonite develops in a monomineralic rock it is referred to as calcite-mylonite, quartz-mylonite etc. (Burlini and Kunze 2000). Another commonly used classification of mylonites is based on the percentage of matrix as compared to porphyroclasts (e.g. Spry 1969; Sibson 1977b; Scholz 1990; Schmid and Handy 1991). Rocks with 10–50% matrix are classified as *protomylonites* (Fig. 5.9, right hand side); with 50–90% matrix as mylonites (or mesomylonites; e.g.

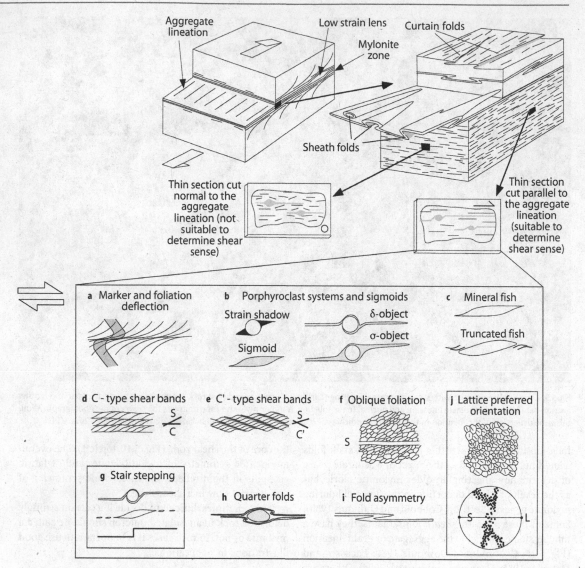

Fig. 5.10. Schematic diagram showing the geometry of a mylonite zone and the nomenclature used. For thin sections parallel to the aggregate lineation, the most common types of shear sense indicators are shown. Further explanation in text. This figure is schematic and does not show all possible geometries. Other figures in this chapter show more detail

Figs. 5.7, 5.8) and rocks with over 90% matrix as *ultra-mylonites* (Fig. 5.9 at left). The problem with this classification is that an arbitrary limit has to be defined between matrix grain size and porphyroclast grain size. Another problem is that mylonites developed at high metamorphic grade or in fine-grained or monomineralic parent rocks do not normally develop porphyroclasts; for this reason, ultramylonite does not necessarily represent a higher strain than mylonite or protomylonite. Other commonly used terminology is *blastomylonite* for a mylonite with significant static recrystallisation and *phyllonite* for a fine-grained mica-rich mylonite (resembling a phyllite). Some authors use the term phyllonite as a synonym for ultramylonite.

5.3.4
Dynamics of Mylonite Development

The relatively high finite strain values reached in mylonites imply that strain rate in the mylonite zone must have exceeded that in the wall rock for some time, and that the material in the zone must have been 'softer' than the wall rock. Nevertheless, many mylonites have the same chemical and mineral composition as the wall rock. Apparently, changes occur in the rheology of material in a ductile shear zone after its nucleation. This effect is known as *softening* or *strain-softening* (Sect. 2.12). The most important mechanisms that contribute to softening are (White et al. 1980; Tullis et al. 1990):

1. A decrease in grain size, which enhances activity of grain size-dependent deformation mechanisms such as diffusion creep and grain boundary sliding (Sect. 3.9; Fig. 3.43, ⊘Video 11.10a; Allison et al. 1979; White et al. 1980; Schmid et al. 1977; Behrmann and Mainprice 1987; Fliervoet et al. 1997; Ji et al. 2004). This decrease in grain size is caused by the fact that the size of new grains formed by dynamic recrystallisation is a function of differential stress (Sect. 9.6.2). However, de Bresser et al. (1998) suggest that this mechanism may not be very efficient.

2. GBM recrystallisation, which replaces hardened crystals by new, easily deformable crystals without dislocation tangles (Fig. 3.26a). Notice that SGR recrystallisation (Fig. 3.26b) will not lead directly to softening since new grains have the same dislocation density as the old ones (Tullis et al. 1990).

3. Growth of new minerals, which are more easily deformable than minerals of the host rock (reaction softening; Mitra 1978; White et al. 1980; Hippertt and Hongn 1998). The replacement of feldspars by aggregates of white mica and quartz is an example.

4. Transformation of large grains of the host rock to new phases in a fine-grained aggregate such as in symplectite formation. Such a newly formed aggregate of minerals may be softer that the original grains not because its individual phases are more easily deformable than the old grains, but because it is more fine grained, and therefore favours another deformation mechanism (Furusho and Kanagawa 1999; Kruse and Stünitz 1999; Newman et al. 1999). An example is the transformation of large K-feldspar grains to myrmekite (Tsurumi et al. 2003).

5. Development of a lattice-preferred orientation of mineral grains which places them in a position for easy dislocation glide (geometric softening; Ji et al. 2004).

6. Enhanced pressure solution due to decrease in grain size and opening of voids and cracks (Rutter 1976; Stel 1981).

7. 'Hydrolytic' weakening of minerals due to diffusion of water into the lattice (Sect. 3.12.2; Luan and Paterson 1992; Kronenberg 1994; Post and Tullis 1998). Quartz at high-grade metamorphic conditions contains little intragranular water and is relatively strong (e.g. Nakashima et al. 1995). If such dry quartz is brought to amphibolite facies conditions and subject to water in the pore fluid, it may be weakened rapidly by infiltration of water into the lattice, probably through crystal defects (Kronenberg et al. 1990; Post et al. 1996; Post and Tullis 1998). Under greenschist facies conditions, however, water infiltration into the quartz lattice is slow and may only affect quartz rheology if grain size is small, or if aided by fracturing of the grains or by grain boundary migration (Post and Tullis 1998).

8. Development of shear bands or shear band cleavage (Ji et al. 2004).

5.3.5
Mylonite Development at Different Metamorphic Conditions

Although the fabric of mylonites is strongly dependent on the lithotype and original structure of the rock in which it develops, a general fabric gradient exists for all rock types with increasing metamorphic grade, depending on the rheology and melt temperature of constituent minerals (e.g. structures in granite mylonite formed at 400 °C may resemble those in peridotite mylonite formed at 800 °C). As an example, consider the effect of metamorphic grade on mylonitisation of a bimineralic rock with a mineral A that is 'hard' and a mineral B that is 'soft' at low-grade conditions due to a different number of active slip systems with different critical resolved shear stress (Sect. 2.3.4; compare feldspar-quartz aggregates in Sect. 3.13.2).

At very low grade, A and B deform by brittle fracturing and a brittle fault rock forms.

At low-grade conditions A deforms in a brittle manner and B by dislocation creep (Handy et al. 1999). Differential stresses are high (Figs. 3.42, 5.2) and mylonites are therefore fine-grained with fragmented, angular porphyroclasts of A embedded in ductilely deformed grains of B that wrap around the porphyroclasts. Foliations and lineations are usually well developed. Mylonite zones tend to be narrow with sharp boundaries.

At medium grade, A and B both deform by crystal-plastic processes, but A is still stronger than B. As a result, well-developed mylonites form with a mylonitic foliation containing fragments of partly recrystallised porphyroclasts of A. Most of the shear sense indicators mentioned in Sect. 5.6 may be recognised in mylonites formed under such conditions. Foliations and lineations are well developed.

At high grade, shear zones tend to be wider than at lower grade, since softening and localisation mechanisms are less efficient than at lower metamorphic grade (Hanmer et al. 1995; Whitmeyer and Simpson 2003). Under these conditions, the difference in rheology between A and B decreases, diffusion becomes more important and differential stresses are low (Figs. 3.42, 5.2). At low strain rate, the result can be a layered rock with few porphyroclasts and a relatively coarse grain size. Grains in the matrix may have a reticular shape. Except from the compositional layering, foliations and lineations tend to be weakly developed. The rock may appear to be weakly deformed, but isoclinal folds in layering may show the intensity of strain. Such high-grade mylonites may be recognised by elongate recrystallised ribbons of B (Box 4.4) and by few large porphyroclasts of A (e.g. Fig. 5.12), which are usually symmetric. They are known as *ribbon mylonite* (McLelland 1984; Hanmer et al. 1995; Hippertt et al. 2001) or, if quartzo-feldspathic and relatively coarse

Fig. 5.11. Striped gneiss composed of alternating layers of recrystallised feldspar and quartz ribbons. Some of the quartz ribbons contain strongly elongated single crystals, probably formed by grain boundary migration within the ribbon. Section parallel to the aggregate lineation and normal to the foliation. Três Rios, Rio de Janeiro State, Brazil. Width of view 2.3 mm. CPL

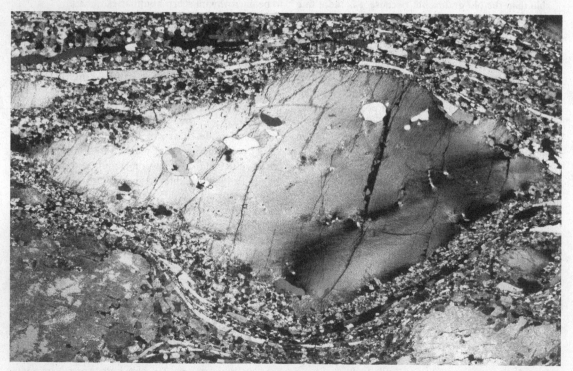

Fig. 5.12. K-feldspar porphyroclast with undulose extinction. A mantle of recrystallised feldspar with isolated polycrystalline quartz ribbons surrounds the porphyroclast. Notice the strongly elongated single crystals of quartz in the ribbons. Section parallel to the aggregate lineation and normal to the foliation. Três Rios, Rio de Janeiro State, Brazil. Width of view 10 mm. CPL

grained, *striped gneiss* (Fig. 5.11, ⊘Photo 5.11). In granitoid rocks, monocrystalline quartz ribbons and polycrystalline feldspar ribbons are common in such striped gneisses (Fig. 5.12). Some high grade mylonites contain elongate porphyroclasts of pyroxene or garnet (Hanmer 2000) which form by intracrystalline deformation (Chap. 3) and possibly by fracturing of kinked pyroxene crystals (Hanmer 2000).

Even at high metamorphic grade, ultramylonites may still form, probably at high strain rate (Whitmeyer and Simpson 2003). Such ultramylonites may contain porphyroclasts and notably mineral fish (Fig. 5.33) with thin or no mantles, possibly due to limited cohesion between clasts and matrix (Kenkmann 2000), or to low differential stresses and high recovery rate which limit recrystallisation in such clasts (Pennacchioni et al. 2001).

The fabric gradient sketched above is generally valid for polymineralic rocks but metamorphic conditions of transitions depend on mineral composition of the parent rock. However, fabric is only a rough indicator and cannot be used alone to determine metamorphic grade in mylonites; this should be done using minerals, which have grown or recrystallised during the deformation. Since mylonite zones may have a long history of reactivation, relicts of older fabrics may be present in low strain lenses. It is tempting to use these low strain lenses to determine the metamorphic conditions of mylonite genesis because of the large, weakly deformed crystals they contain, but the results may indicate metamorphic conditions prior to mylonitisation. Another factor that has to be taken into account is static recrystallisation, which may re-equilibrate minerals in mylonites after deformation.

5.4
Complex Fault Rocks

Since many shear zones have a long period of activity or can be reactivated, several fault rock types can overprint each other in a single shear zone. Most common are brittle fault rocks, which transect mylonite, since mylonite forms at depth and has to pass the field of brittle fracturing before it reaches the surface (Grocott 1977; Strehlau 1986; Scholz 1988; Passchier et al. 1990a). Such overprinting by brittle structures is usually easy to recognise. However, it may be difficult to differentiate a low-grade mylonite where some minerals were deformed by brittle fracturing from a mylonite overprinted by cataclasite formation. These situations can be distinguished because cataclasite will transect all minerals in the mylonite, usually along narrow zones.

A less common type of superposition is ductile deformation of brittle fault rocks. Ductilely deformed cataclasite or breccia does occur (Guermani and Pennacchioni 1998) but can be difficult to recognise if the ductile overprint is strong. Some pseudotachylyte veins have undergone ductile deformation after their solidification; these are easier to recognise, even after strong ductile overprint. In fact, pseudotachylyte veins seem to act as the preferred nucleation sites of mylonite zones in many locations (Fig. 5.4b; Allen 1979; Sibson 1980; Passchier 1982b, 1984; Passchier et al. 1990a; Takagi et al. 2000). The recognition of ductilely deformed pseudotachylyte is important, since the presence of a brittle deformation phase is an indication for either deformation at shallow crustal depth, or unusually high strain rates (Passchier et al. 1990a).

Evidence for weak ductile deformation of pseudotachylyte is the presence of flattened inclusions and a mica-preferred orientation in the matrix (Passchier 1982b, 1984). Strongly deformed pseudotachylyte veins are difficult to distinguish from thin ultramylonite zones, which lack a brittle predecessor. Indications may be an unusual ultra fine-grained (<5 µm) homogeneous matrix of biotite, quartz and feldspar with isolated porphyroclasts of quartz, but fewer or no clasts of other minerals, the presence of sulphide aggregates in quartz inclusions, and sharp boundaries of ultramylonite with the wall rock (Passchier 1982a, 1984). Ductile deformation of pseudotachylyte is usually restricted to the main fault veins, and injection veins may be less deformed and still recognisable; if a suspicion exists that a mylonite may have a pseudotachylyte predecessor, the presence of injection veins should be investigated in the field or in hand specimen. Any narrow, dark mylonite zone in metamorphic rocks should be checked for relicts of pseudotachylyte structures.

Ductile deformation of pseudotachylyte may be caused by a separate tectonic event after increase in metamorphic conditions to the depth where the rock deforms ductilely (Passchier et al. 1990a), or by ductile deformation at the level where the pseudotachylyte formed; the latter could happen in the transition zone between dominant ductile deformation and brittle fracturing (Fig. 5.2; Sibson 1980; Passchier 1982b, 1984). A pseudotachylyte vein may rapidly crystallise into a fine-grained aggregate under such conditions. At the high differential stress level sustained at these conditions, a fine-grained aggregate such as a crystallised pseudotachylyte may deform ductilely by diffusion-assisted grain boundary sliding while the coarse-grained wall rock is rigid (Sect. 3.14; Sibson 1980; Passchier 1982b, 1984).

It has to be kept in mind that mylonites, like other metamorphic rocks, record mainly peak and retrograde metamorphic conditions. Brittle faulting or early mylonitisation that predates these conditions may be completely obliterated by recrystallisation. Important early thrust structures in many metamorphic terrains do not contain any brittle fault rocks due to this process.

5.4

5.5
Sense of Shear

5.5.1
Introduction

The direction of movement on a shear zone can usually be shown to lie subparallel to striations, slickenfibres or aggregate and grain lineations. Once this direction is established, it is necessary to determine the sense of displacement (sinistral or dextral, normal or reverse) or *sense of shear*. Traditionally, this was mainly done using markers in the wall rock such as displaced layering and dykes or deflection of layering or foliation into a shear zone (Sects. 5.5.2, 5.5.3). Additionally, the geometry of structures in the zone can be used to determine sense of shear. This means that it is possible to determine shear sense for a shear zone in thin section, even without seeing the zone in the field. Microstructures that can be used to determine shear sense in mylonite are described in Sect. 5.6. Some microstructures can also be used to determine shear sense for brittle fault rocks in thin section. These are discussed in Sect. 5.7.

5.5.2
Displacement and Deflection of Markers

The simplest and best known sense of shear indicator is the displacement of markers such as dykes, veins, xenoliths and bedding along a shear zone. Commonly, there is also a deflection of markers near the zone, due to a strain gradient. Interpretation of both structures is straightforward in the case of a linear marker, or if the

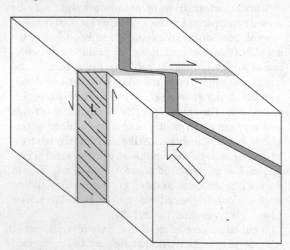

Fig. 5.13. Schematic presentation of a shear zone illustrating how deflection of a marker can give the wrong sense of shear if not used in a section parallel to the movement direction (here marked by a lineation *L*); the frontal block moves up and to the left, but the deflected layer indicates an apparent sinistral displacement on a horizontal surface

movement direction is normal to a planar marker (Fig. 5.10a). Notice, however, that in other cases the apparent offset on an outcrop surface or in a thin section may give the wrong sense of shear (Fig. 5.13, ⊘Video 5.13, ⊘Photo 5.13; Wheeler 1987b).

5.5.3
Foliation Curvature

Foliated ductile shear zones may show a gradient in foliation development from an undeformed wall rock towards the core of the zone; the foliation has a characteristic curved shape that can be used to determine sense of shear (Fig. 5.10a). This structure develops if the foliation is 'passive' (Sect. 4.2.9.2) and reflects the orientation of the finite strain axes (Sect. 4.2.9.2); the curvature reflects a gradient in finite strain from its peak in the core of a shear zone outwards (Sect. 9.2). It develops because foliations rotate from a position between the instantaneous extension axis and the fabric attractor towards the latter with increasing strain in non-coaxial progressive deformation (Ramsay and Graham 1970; Ramsay 1980a).

Foliation curvature is a reliable shear sense indicator if the movement direction, indicated by a lineation, is normal to the axis of curvature of the foliation. However, care should be taken not to confuse it with deflection of an older foliation, where the movement direction may be oblique to the axis of curvature of the foliation, in which case some sections through the structure may give the wrong shear sense (Sect. 5.5.2).

5.6
Microscopic Shear Sense Indicators in Mylonite

5.6.1
Introduction

One of the most characteristic properties of mylonites and some other rocks from ductile shear zones is that fabric elements and structures show monoclinic shape symmetry (Fig. 5.10). This effect is a result of the non-coaxial progressive deformation in shear zones due to relative displacement of the wall rocks. Lineations, finite strain axes and most foliations rotate towards the fabric attractor, which is oblique to the extensional ISA (Sect. 2.9) and their sense of rotation usually equals the sense of shear. Since different fabric elements track the fabric attractor to different degrees, complex structures with a distinct monoclinic shape symmetry (in the literature usually referred to as *asymmetry*) develop in mylonites. This asymmetry can be used to deduce sense of shear. We distinguish *internal* and *external asymmetry* for either single objects and fabric elements, or combinations of fabric elements with a characteristic monoclinic shape symmetry. A similar terminology is used for LPO patterns (Sect. 4.4.3). With internal asymmetry we mean that the object itself has an asymmetry without the need to involve other fabric elements or a reference frame; with external asymmetry we mean that the orientation of the object with respect to other fabric elements determines its asymmetry. This section describes the most common monoclinic microstructures, and explains how sense of shear can be derived from them.

In order to determine shear sense in a mylonite zone, thin sections should be properly oriented. Sections normal to the symmetry axis of shear sense markers give best results. This symmetry axis will be near the vorticity vector for flow in the shear zone. The plane normal to the vorticity vector is known as the vorticity profile plane (VPP). In most mylonite zones, the VPP lies normal to the intermediate strain axis. This means that a hand specimen should be sectioned parallel to the VPP, i.e. parallel to the aggregate or grain lineation and normal to the compositional layering or main foliation (Fig. 5.10). Problems may arise when several lineations and foliations are present: this is usually due to overprinting of several deformation phases. In this case, it will be necessary to reconstruct the deformation sequence first, then to decide which phase is of interest: usually, sense of shear can only be determined for the last phase or phases. Obviously, only oriented samples should be used to determine shear sense.

In crustal mylonite zones, which developed under low- to medium-grade metamorphic conditions, a large number of sense-of-shear markers are available, and most are empirically established. (Reviews in Bouchez et al. 1983; Simpson and Schmid 1983; Passchier 1986a; Hanmer and Passschier 1991). The most important ones visible in thin section are shown schematically in Fig. 5.10 and are discussed below.

All observations are for sections parallel to the aggregate lineation and normal to the foliation. In all other sections the structures either show a less pronounced monoclinic symmetry, or orthorhombic and higher symmetry.

5.6.2
Foliation Orientation

Many shear zones do not show foliation curvature (Sect. 5.5.3) as in Fig. 5.10a, but have sharp boundaries with the wall rock. The mylonite in the shear zone can have several foliations, which make a small angle with the wall rock and with each other. Such foliations can be mica-preferred orientation, a layering or a shape preferred orientation. If they develop during mylonite genesis, they can be good shear sense indicators. Mica preferred orientations, and aggregate shape preferred orientation (Box 4.2) are 'passive' foliations and are commonly slightly oblique to the shear zone boundary.

(Fig. 5.10a; Ramsay and Graham 1970; Simpson and Schmid 1983). This external asymmetry is due to rotation of the foliation towards the fabric attractor in noncoaxial flow, and has the same external asymmetry as the foliation gradient discussed in Sect. 5.5.2.

Another type of foliation in mylonites is defined by grain shape preferred orientations within monomineralic domains (Box 4.2). Aggregates of small grains in mylonites (usually formed by dynamic recrystallisation) can be characterised by a slightly elongate shape of most of the grains (Means 1981; Lister and Snoke 1984). This grain shape preferred orientation is usually oblique to compositional layering or mica-preferred orientation in a mylonite (Fig. 5.10f). The relation between shear sense and geometry of such an *oblique foliation* is as shown schematically in Figs. B.4.2, 4.31c and 5.10f (see also Box 4.2). Examples are shown in Figs. 5.22, 5.24, 5.31. Oblique foliations have been reported for quartz (Brunel 1980; Law et al. 1984; Law 1998; Knipe and Law 1987; Lister and Snoke 1984; Dell'Angelo and Tullis 1989), carbonates (Schmid et al. 1987; de Bresser 1989; Barnhoorn et al. 2004), olivine (van der Wal et al. 1992; Zhang and Karato 1995) and rock analogues such as ice (Burg et al. 1986), octachloropropane (Jessell 1986; Ree 1991) and norcamphor (Herwegh et al. 1997; Herwegh and Handy 1996, 1998). They are assumed to develop by an interplay of passive deformation and rotation of grains in noncoaxial flow resulting in increasingly elongate grains, and processes such as grain boundary migration and fracturing or microshearzone development which produce more equidimensional grain shapes (Fig. 4.31c; Means 1981; Ree 1991; Herwegh and Handy 1998). In this way, the foliation will remain fixed in orientation with respect to the kinematic frame of progressive deformation, usually at an angle of 20–40° to the fabric attractor (Dell'Angelo and Tullis 1989; Ree 1991; Fig. 4.31c). The actual angle probably depends on the vorticity number of flow, the recrystallisation mechanism and on the efficiency of fabric developing and fabric destroying processes (Hanmer 1984a; Herwegh and Handy 1998). For olivine, an alternative mechanism of kinking and grain boundary migration has been proposed (van der Wal et al. 1992). Such foliations are therefore to some extent *strain-insensitive*.

Oblique foliation occurs mostly as a grain shape preferred orientation (Box 4.2) in monomineralic layers of quartz or calcite in layered low- to medium-grade mylonites (Fig. 5.31); examples of polymineralic strain-insensitive foliations are less common, and occur mainly in medium- to high-grade mylonites (Hanmer and Passchier 1991); an example is the mica-preferred orientation (S_m; Fig. 5.24) oblique to the mylonitic compositional layering observed in some micaceous mylonites (Passchier 1982a). In ultramylonites such an oblique foliation is common and visible under crossed polars as a preferential extinction of the matrix at an angle of less than 5° to the layering. Such oblique foliations result from the fact that some fabric elements (elongated new quartz grains, micas) rotated through a smaller angle than the mylonitic foliation during non-coaxial progressive deformation.

Another type of strain-insensitive foliation that can occur in mylonites is a *domain shape preferred orientation* of grains that share a certain crystallographic preferred orientation (Box 4.2).

5.6.3
Shear Band Cleavage

A mica-preferred orientation or compositional layering may be transected at a small angle by sets of subparallel minor shear zones (Figs. 5.10d,e, 5.14–5.17, ⊘Photo 5.15). Such minor shear zones are known as *shear bands* and the complete structure is a *shear band cleavage* (Roper 1972; White 1979b; Gapais and White 1982). Shear band cleavage may superficially resemble crenulation cleavage but develops by extension along the older foliation rather than shortening. Some authors (e.g. Platt and Vissers 1980) therefore use the terms 'compressional' crenulation cleavage (as treated in Chap. 4), and *extensional crenulation cleavage* for shear band cleavage. Geometric differences between these two groups of structures are given in Box 5.3.

Two types of shear band cleavage are distinguished: *C-type* and *C'-type* (Figs. 5.10d,e, 5.14). In the literature, this distinction is not always made. Our C- and C'-type shear band cleavage correspond to C- and C'-bands of Berthé et al. (1979a,b). C'-type equals extensional crenulation cleavage of Platt and Vissers (1980); we favour the term C'-type because it is older and non-genetic.

C'-type shear band cleavage is oblique to shear zone boundaries and to the older foliation in micaceous mylonites (Figs. 5.15, 5.16, ⊘Photo 5.15; White 1979b; Platt and Vissers 1980). The angle between the shear bands and the shear zone margin is 15–35° (Dennis and Secor 1987; Passchier 1991b; Blenkinsop and Treloar 1995). C'-type shear band cleavage develops mainly in strongly foliated mylonites such as phyllonites and mylonitic micaschists; usually, shear bands fail to continue into more weakly foliated layers (e.g. quartz layers) in such rocks (Fig. 5.15). The shear bands are usually anastomosing, short and wavy (Figs. 5.14, 5.15, ⊘Photo 5.15). An older mylonitic foliation is cut by the shear bands and is deflected in the bands in the same way as foliation curvature in a large-scale shear zone (Sect. 5.5.3). Normally, the intersection of the older foliation and the shear bands is normal to aggregate or grain lineations on both these surfaces. This is taken to indicate that both the older foliation and the shear band cleavage formed during the same event of mylonitic deformation. However, the C'-type shear band cleavage may have a deviant mineral composition, usually indicative of retrograde metamorphic conditions (McCaig 1987; Norrell et al. 1989). Commonly, only one set of C'-type shear bands is developed in fabrics but occasionally a second, less developed set may be present almost orthogonal to the main set (Fig. 5.14; Harris and Cobbold 1985; Behrmann 1987). Also, a younger set of shear bands may overprint an older set that has a more gentle inclination to the shear zone boundary (Platt and Vissers 1980; Passchier 1991b).

Development of C'-type shear band cleavage is only partly understood. It seems to develop late during shear

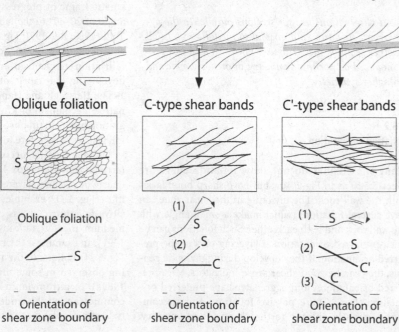

Fig. 5.14.
Three types of foliation pairs that are common in ductile shear zones. The shear zone is shown (*top*) with typical foliation curvature. The main differences in geometry between the foliation pairs are shown in the *centre*. Elements used to determine sense of shear are shown *below*. Further explanation in text

Oblique foliation

C-type shear bands

C'-type shear bands

Oblique foliation

Orientation of shear zone boundary

Orientation of shear zone boundary

Orientation of shear zone boundary

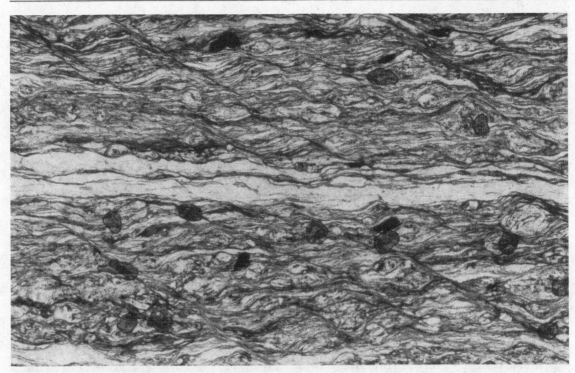

Fig. 5.15. C'-type shear band cleavage (from *upper left* to *lower right*) transecting the main foliation in a micaschist. Note that the cleavage does not continue into the quartz ribbons in the *centre* of the photograph. Dextral shear sense. Menderes Massif, Turkey. Section parallel to the aggregate lineation and normal to the foliation. Dark round patches are due to thin section preparation. Width of view 3.5 mm. PPL. (Photograph courtesy R. Hetzel)

Fig. 5.16. C'-type shear bands in a quartz-feldspar-biotite-garnet mylonite, indicating dextral shear sense. Section parallel to the aggregate lineation and normal to the foliation. Marsfjällen, Sweden. Width of view 4 mm. PPL

Fig. 5.17. Polished hand specimen showing a C/S fabric in granite. C-planes are *horizontal*, S-planes trend *upper right* to *lower left* between C-planes. Shear sense is dextral. Section parallel to the aggregate lineation and normal to the foliation. South Armorican Shear zone, France

Box 5.3	Difference between compressional (normal) crenulation cleavage and shear band cleavage (extensional crenulation cleavage)

Shear band cleavage (Fig. 5.14) and compressional crenulation cleavage (Figs. 4.12, 4.15, 4.18) have a similar geometry and could be confused by a casual observer. Nevertheless, they have a different morphology and kinematic significance. We outline some morphological and kinematic differences below.

Compressional crenulation cleavage (ccc)	Shear band cleavage (sbc)
Morphology	
Angles between older foliation and ccc generally between 45° and 90°	Angle between older foliation and sbc less than 45°
Folds in older foliation have a large amplitude with respect to ccc spacing	Folds in older foliation have a small amplitude with respect to sbc spacing
ccc-planes irregular but penetrative	sbc-planes smooth and short, commonly anastomosing
Host rock is usually a folded phyllite	Host rock is usually a mylonite or phyllonite
Kinematics	
ccc develops at a high angle (close to 90°) to the bulk shortening direction and represent a foliation approaching the fabric attractor	sbc develops oblique to the bulk shortening direction and represents a zone of enhanced non-coaxial flow
Usually a shortening component normal to ccc	Usually an extension component normal to sbc

zone activity after a strong mineral preferred orientation has already been established, and represents probably an energetically favourable flow partitioning in strongly anisotropic materials (Fig. 5.19, ⊘Video 11.7b; Platt and Vissers 1980; Platt 1984; Dennis and Secor 1987; Passchier 1991b; Hafner and Passchier 2000). In an isotropic rock, two sets of conjugate shear bands would be expected around the shortening ISA; since the foliation in a mylonite lies between the extensional ISA and the fabric attractor, one of these potential shear band orientations makes a smaller angle with the foliation than the other. Strongly foliated mylonites may have a strong mechanical anisotropy, which inhibits development of the steeper set (Fig. 5.19) and causes a decrease in the angle between the main set and the older foliation (Cobbold et al. 1971; Cobbold 1976; Platt and Vissers 1980).

The geometry of shear bands imposes a geometric limit on the possible flow regimes in mylonites during their

Fig. 5.18. Photomicrograph of mylonitic quartzite with mica fish arranged between C-type shear bands. Dextral shear sense. Varginha, Minas Gerais, Brazil. Width of view 4 mm. CPL. (Sample courtesy Rodrigo Peternel)

growth; the segments between shear bands may have rotated synthetically or antithetically, dependent on the vorticity and volume change in a shear zone (Passchier 1991b). If microlithons between C'-type shear bands were rigid during C'-type cleavage development, the microlithons and shear bands must have rotated antithetically while the shear zone wall rock must have extended parallel to the zone (Fig. 5.19). There are indications that C'-type shear band cleavage is especially well developed in such 'stretching' shear zones (Passchier 1991b; Hafner and Passchier 2000) and independent evidence from LPO analysis indicates that stretching shear zones may be common (Schmid 1994). If flow in a shear zone is simple shear when C'-type shear bands form, the bands must have rotated synthetically and microlithons must have been deforming.

C-type shear band cleavage is part of a so-called C/S fabric (Figs. 5.14, 5.17) that consists of S-planes (from French 'schistosité'), transected by planar distinct C-type shear bands or C-planes (from French 'cisaillement', meaning shear; Berthé et al. 1979a,b; Vernon et al. 1983; Lister and Snoke 1984; Krohe 1990; Toyoshima 1998). C/S fabric is also written as S-C or C-S fabric in the literature. The C-type shear bands in C/S fabric are parallel to shear zone boundaries (Figs. 5.10e, 5.14, 5.17, 5.18) and relatively straight and continuous, unlike C'-type shear bands. C/S-fabric forms in weakly foliated mylonites with a small percentage of micas. It is most common in medium-grade shear zones and especially in deformed granites, where C-type shear bands anastomose around feldspar porphyroclasts. C-type shear bands may nucleate at sites of high differential stress adjacent to feldspar porphyroclasts, and subsequently propagate and join up (Hanmer and Passchier 1991; Ildefonse and Mancktelow 1993).

C/S fabric probably reflects inhomogeneous simple shear. Contrary to C'-type shear band cleavage it may develop from the earliest stage of mylonite generation onward. The foliation in the microlithons probably continues developing while the shear bands grow, contrary to many C'-type shear band cleavages where the microlithons are probably rigid. C/S fabrics can be overprinted by C'-type shear band cleavage (Berthé et al. 1979b).

Lister and Snoke (1984) proposed a modification of the nomenclature and distinguished Type I and Type II C-S mylonites. Type I corresponds to C/S fabrics as described above, while Type II refers to sets of parallel shear bands without clear S-planes. The principle example of Type II C-S mylonites used by Lister and Snoke (1984), however, refers to a stair stepping of wings of small mica grains adjacent to mica-fish in quartzite mylonite (Sect. 5.6.7). They envisage that the wings represent C-type shear bands adjacent to the mica-fish. However, since such wings do not necessarily form by flow partitioning along shear bands (Sect. 5.6.7), we discourage the use of Type II C-S mylonite terminology.

a   Simple shear

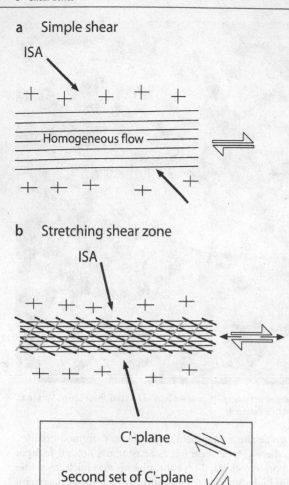

b   Stretching shear zone

Fig. 5.19a,b. Schematic diagram showing the development of C'-type shear band cleavage with relatively rigid microlithons. The C'-type shear bands develop at a late stage of the shear zone activity after a foliation is well established. Shear bands can theoretically develop in two orientations, but only one, at a small angle to the foliation, is realised. Notice that the shortening direction must be relatively steep with respect to the shear zone and that the shear zone must be extending

Shear band cleavages have both internal and external asymmetry that can be used as shear sense indicators (e.g. Malavieille and Cobb 1986; Davis et al. 1987; Saltzer and Hodges 1988). The internal asymmetry is a sigmoidal shape of the older foliation between shear bands (Fig. 5.14(1)). The external asymmetry is the angle between the enveloping surface of the older foliation and the shear bands (Fig. 5.14(2)). C'-type shear band cleavage has an additional external asymmetry element since the shear bands are inclined to the shear zone in a characteristic way (Figs. 5.14(3), 5.15, 5.16, ⊘Photo 5.15). Shear sense in both C- and C'-type shear bands is syn-

thetic with that of the main shear zone. The asymmetry of C'-type shear band cleavage is opposite to that of oblique foliation (Fig. 5.14).

5.6.4
Porphyroclast Systems in Mylonites – Introduction

Porphyroclasts are large single crystals in a more fine-grained matrix of different, usually polymineralic composition, which are common in mylonitic rocks (Fig. 5.20). If they are equidimensional and have a sharp boundary with the matrix, they are known as *naked clasts* or, if the porphyroclasts have an elongate shape with monoclinic symmetry, *mineral fish* (Fig. 5.20). In many cases, porphyroclasts have attached polycrystalline rims that differ in structure or composition from the matrix; such assemblages are known as *porphyroclast systems*. Porphyroclast systems can have a number of characteristic shapes that can be used to determine shear sense and other kinematic parameters. In many cases, the surrounding rim has a tapering shape on opposite sides of the porphyroclast. If the material in the rim is of the same composition as the porphyroclast, the rim is called a mantle and the structure is known as a *mantled porphyroclast* or *mantled clast* (Figs. 5.10b, 5.20). If the rim has a composition different from the porphyroclast, the adjacent tapering domains are known as *strain shadows* and the entire structure as a *porphyroclast with strain shadows* (Fig. 5.20; Sect. 6.3–6.5 and treated separately below). Such stain shadows are commonly composed of carbonate, quartz, mica or opaque minerals, which were apparently not formed by reaction with the porphyroclast, but by precipitation from solution. If it can be shown that the material in the rim is formed by transformation of the porphyroclast, the rim is known as a reaction rim (Sect. 7.8, cf. Lafrance and Vernon 1998).

Sigmoids are aggregates of grains of a mineral A in a matrix of another mineral, lacking a clear porphyroclast core (Fig. 5.20). They can have a similar shape to σ-type mantled clasts or mineral fish.

In geological practice, there is a tendency to call all structures in Fig. 5.20 "sigma-objects" indiscriminately. We discourage this attitude, since each of the structures in Fig. 5.20 forms by a different mechanism, as explained below.

Porphyroclasts, fish, strain shadows and sigmoids are observed in many geological environments including volcanic flows (Benn and Allard 1989; Luneau and Cruden 1998), cataclasite (Cladouhos 1999a,b) and mylonite (Passchier 1987; Hanmer 1990; Shelley 1995; Masuda et al. 1995c).

In the following sections, we first describe the shape of different types of porphyroclast systems (Sects. 5.6.5, 5.6.6), followed by a review of the present understanding

Fig. 5.20.
Schematic diagram of the principal types of objects encountered in the matrix of mylonites. This includes large single crystal shapes such as naked clasts and mineral fish, single crystal porphyroclasts with rims such as mantles, reaction rims or strain shadows, and polycrystalline aggregates such as sigmoids. *A*, *B*, etc. refer to mineral types

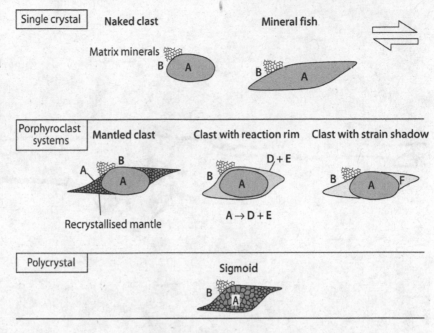

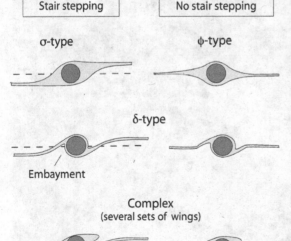

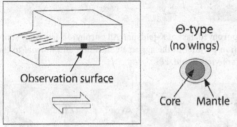

Fig. 5.21. Classification of mantled porphyroclasts. Dextral sense of shear

of their development (Sect. 5.6.7), and finally their usefulness as shear sense indicators is discussed (Sect. 5.6.8).

5.6.5
Mantled Porphyroclasts

Mantled porphyroclasts consist of a central single crystal and a fine-grained mantle of the same mineral. Common examples are porphyroclasts of feldspar in a matrix of quartz-feldspar-mica, of orthopyroxene in peridotite and of dolomite in a calcite matrix. The fine-grained soft mantle can be deformed into *wings* (or trails) that extend on both sides of the porphyroclast parallel to the shape preferred orientation in the mylonite (Passchier and Simpson 1986). Wings are thought to stretch and change shape while the porphyroclast core remains rigid or continues to recrystallise along the contact with the rim, shrinking in size (Sect. 5.6.7.2). Wing shape can be used as a shear sense indicator and contains information on rheology of the matrix and the matrix-clast coherence (Sect. 9.3.4).

Four types of mantled porphyroclasts have been distinguished in the literature based on the shape of the wings (Hanmer 1984b; Passchier and Simpson 1986; Hooper and Hatcher 1988): ϕ-type, σ-type, δ-type, and complex mantled clasts (Figs. 5.21, 5.25, ⊘Photos 5.9b, 5.23, ⊘Videos 5.22, 5.23). Θ-type mantled clasts lack wings but have a mantle with orthorhombic symmetry (Passchier 1994). σ-type-, δ-type and complex mantled clasts have monoclinic shape symmetry. σ-type mantled clasts have wide mantles near the porphyroclast with two planar faces and two curved faces that define an internal

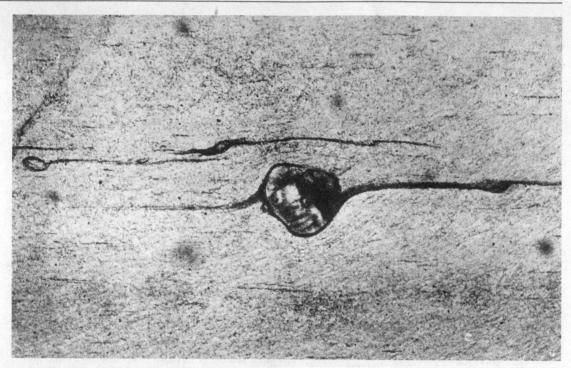

Fig. 5.22. Quartzite mylonite with δ-type (*centre*) and small σ-type mantled porphyroclast of K-feldspar. All porphyroclasts show stair-stepping. A weak oblique foliation trends from *top right* to *bottom left*. Section parallel to the aggregate lineation and normal to the foliation. Dextral shear sense. St. Barthélemy Massif, Pyrenees, France. Width of view 6 mm. PPL

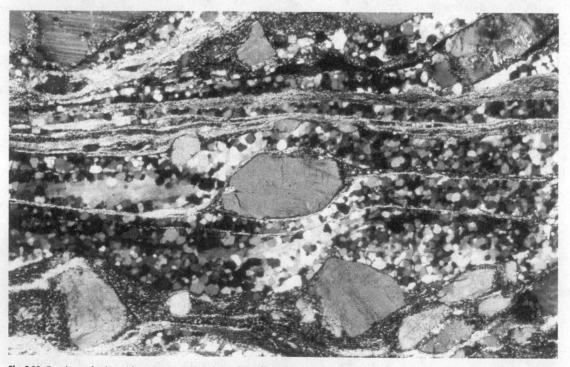

Fig. 5.23. Granite mylonite with a σ-type mantled porphyroclast of K-feldspar (*centre*) in a matrix of recrystallised quartz. Feldspar porphyroclasts at the top and bottom are mantled by fine-grained dynamically recrystallised feldspar. The grain aggregate of quartz probably formed by dynamic recrystallisation but shows effects of static recrystallisation. Section parallel to the aggregate lineation and normal to the foliation. Dextral shear sense. St. Barthélemy Massif, Pyrenees, France. Width of view 4 mm. CPL

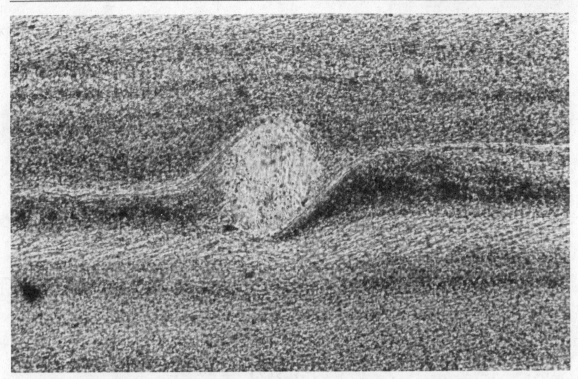

Fig. 5.24. δ-type mantled porphyroclast of K-feldspar in quartz-feldspar-mica ultramylonite. The object shows weak stair-stepping. A layer rich in white mica below the porphyroclast shows an oblique mica fabric. Section parallel to the aggregate lineation and normal to the foliation. Dextral shear sense. St. Barthélemy Massif, Pyrenees, France. Width of view 0.3 mm. PPL

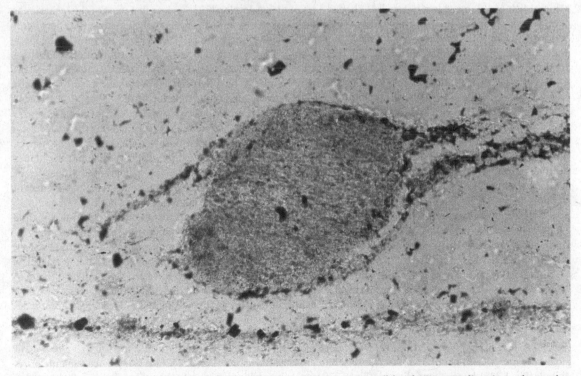

Fig. 5.25. Complex mantled porphyroclast of K-feldspar in quartzite mylonite. Section parallel to the aggregate lineation and normal to the foliation. Dextral shear sense. St. Barthélemy Massif, Pyrenees, France. Width of view 2 mm. PPL

asymmetry. The tips of the wings lie at different elevation on both sides. This difference in elevation is referred to as *stair-stepping* (Figs. 5.21–5.25, ⊘Photos 5.9b, 5.23, ⊘Video 5.22; Lister and Snoke 1984). Passchier and Simpson (1986) distinguished σ_a- and σ_b-type mantled clasts; the former occur isolated in a mylonitic matrix; the second as part of developing C/S fabrics. δ-type mantled clasts have narrow wings and characteristic bends in the wings adjacent to the porphyroclast. As a result, two embayments of matrix material occur adjacent to the porphyroclast. Not all δ-type mantled clasts have stair-stepping (Fig. 5.21). Complex mantled clasts have more than one set of wings (Fig. 5.21).

δ-type and complex mantled clasts mainly occur in high strain mylonites, while σ-type mantled clasts occur also at lower strain. Naked clasts occur commonly in ultramylonites, especially at high grade (Whitmeyer and Simpson 2003; Pennacchioni et al. 2001). ϕ-type mantled clasts are most common in high-grade relatively coarse-grained mylonites. σ-type mantled clasts should not be confused with fish or sigmoids, or with asymmetric strain shadows and strain fringes treated in Chap. 6 since each category is formed by different mechanisms (see Fig. 5.20 for differences).

5.6.6
Mineral Fish

Mineral fish are elongate lozenge or lens-shaped single crystals, which are common in mylonites. They characteristically lie with their longest dimension at a small angle to the mylonitic foliation (Figs. 5.30, 5.31, 5.33, 5.37). Most common are mineral fish of large single white mica crystals known as *mica fish* in micaceous quartzitic mylonites (Figs. 5.10c, 5.28, 5.30, 5.31, ⊘Photos 5.29a–c, 5.31; Eisbacher 1970; Choukroune and Lagarde 1977; Simpson and Schmid 1983; Lister and Snoke 1984; ten Grotenhuis et al. 2003; Sawaguchi and Ishii 2003). Commonly, trails of small mica fragments extend into the matrix from the tips of isolated mica fish (Figs. 5.30, 5.31; Lister and Snoke 1984) with well-defined stair stepping (Sect. 5.6.5). Ten Grotenhuis et al. (2003) proposed a morphological subdivision of mica fish into six groups (Fig. 5.28, ⊘Photos 5.29a–c, 5.31), based on a study of crystals from an upper greenschist facies mylonite from Brazil.

Besides white mica, a number of other minerals can develop as mineral fish with a similar orientation with respect to the mylonitic foliation, as observed in mylonites from a variety of locations and metamorphic grade. Presently known are examples of biotite (Fig. 5.33a), tourmaline (Fig. 5.34), K-feldspar (Fig. 5.33d), garnet (Fig. 5.33e), plagioclase (Fig. 5.33c, ⊘Photo 5.33c), staurolite, kyanite, amphibole (Fig. 5.33h), hypersthene, diopside (Fig. 5.33f), apatite, rutile, hematite, prehnite (Mancktelow et al. 2002; ten Grotenhuis et al. 2003), leucoxene (Oliver and Goodge 1996), sillimanite (Fig. 5.36; Pennacchioni et al. 2001; Mancktelow et al. 2002), olivine (Mancktelow et al. 2002) and quartz (Fig. 5.35; Bestmann et al. 2000, 2004; ten Grotenhuis et al. 2003).

Besides normal mineral fish, described as Type 1 by Mancktelow et al. (2002), Pennacchioni et al. (2001) described sillimanite fish of an unusual truncated shape in an amphibolite facies ultramylonite (Type 2 of Mancktelow et al. 2002). These sillimanite fish have the same backward inclination as normal fish with respect to the foliation, but with a mirror image lozenge-geometry to normal fish (Fig. 5.36, ⊘Photo 5.36a,b). The geometry is defined by short shear bands that trail off the truncated end of these sillimanite fish.

Polycrystalline mica layers in quartzite or marble can also have a lozenge shape and are known as *foliation fish*. Since they are commonly bordered by shear bands this structure grades into C/S fabric (Sect. 5.6.3). It can also be used as a reliable shear sense indicator.

5.6.7
The Development of Porphyroclast Systems

5.6.7.1
Introduction

It is as yet not completely clear how porphyroclast systems develop but it is clear that most of the scenarios as sketched in Box 5.4 can apply in nature. Real porphyroclasts have a complex 3D shape, while most of the modelling described in Sect. 5.6.7.2 is two-dimensional. However, 2D cross-sections along the vorticity profile plane through porphyroclast systems and mica-fish usually give a good approximation of the 3D shape (Jezek et al. 1994). Porphyroclasts may be rigid, deformable or weaker than their surrounding matrix material; they may have a mantle with similar or lower viscosity than the matrix, or bonding between the porphyroclast and the matrix may be reduced due to crystallographic properties or high fluid pressure, which is comparable to the effect of a thin weak zone close to a rigid object.

An isolated rigid porphyroclast without a mantle which has perfect bonding with the matrix and no recrystallisation nor erosion will rotate as a rigid object, or may become stationary in general non-coaxial flow (Figs. B.5.2, B.5.5), but this behaviour will be difficult to recognise in thin section. If differential stresses are high in the rim of the porphyroclast local crystalplastic deformation and storage of dislocation tangles in the rim of a porphyroclast can lead to dynamic recrystallisation in the rim of the porphyroclast to form a *core-and-mantle structure* (Box 3.8; White 1976). Once a mantle of recrystallised material exists, this will normally be weaker than the remaining porphyroclast and will deform with

Box 5.4   Rigid objects in a ductile matrix – theory

Research into development of porphyroclasts and mineral fish is still in progress. This section aims to give an overview of the factors that are presently thought to play a role in their development (Passchier and Simpson 1986; vandenDriessche and Brun 1987; Passchier et al. 1993; Passchier 1994; Pennacchioni et al. 2000; Biermeier et al. 2001; Ceriani et al. 2003).

A spherical rigid object with perfect bonding between object and a homogeneous extensive matrix in a non-coaxial flow will rotate with respect to flow-ISA with an angular velocity of half the shear strain rate, just as a paddle wheel inserted in a flowing river (Jeffery 1922; Fig. B.2.3). However, even if the strain rate and vorticity of the flow are constant, the angular velocity of a rigid object will be fluctuating if the object is not a sphere (Fig. B.5.1, ⊘Video B.5.1a,b; Ghosh and Ramberg 1976; Passchier 1987b). Elongate objects will accelerate and decelerate with changing orientation and, if the vorticity number is between that for pure and simple shear, may even become stationary in the flow when they exceed a critical aspect ratio (Figs. B.5.1, B.5.2, B.5.6, ⊘Video B.5.2; Ghosh and Ramberg 1976; Freeman 1985; Passchier 1987b; Jezek et al. 1994; ten Brink 1996; Arbaret et al. 2001). In simple shear, however, all rigid objects with perfect bonding to the matrix except material lines rotate permanently. The rotational behaviour

of rigid objects is mostly dependent on their length-width ratio, and much less on their actual rectangular, rhomboid or elliptical shape (Jeffery 1922; Ghost and Ramberg 1976; Arbaret et al. 2001).

Even if objects do not obtain a true stationary position, the variable rotation rate may cause a statistical preferred orientation of elongate objects with orientation and strength that depend on finite strain and vorticity number (Fig. B.5.3; Passchier 1987; Masuda et al. 1995c; Piazolo et al. 2002; Marques and Coelho 2003).

An additional complication arises when rigid objects are close together, or are close to a non-deforming wall rock. In that case, objects may rotate more slowly, become irrotational or even show antithetic rotation (Marques and Coelho 2001; Biermeier et al. 2001). Series of objects can also become stationary in a tiling arrangement (Fig. 5.48; Ildefonse et al. 1992a,b; Tikoff and Teyssier 1994; Mulchrone et al. 2005).

Another scenario is that the central object has higher viscosity than the matrix, but is not rigid. In such cases, the central object may be deforming while it rotates, which can lead to complex rotational behaviour and either permanent or pulsating deformation of the central object (Passchier and Sokoutis 1993; Piazolo and Passchier 2002a; Fig. B.5.4). The same applies to an object that is weaker than the matrix (Treagus and Lan 2003),

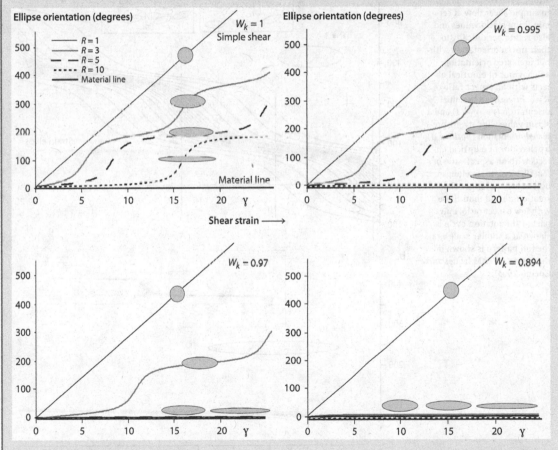

Fig. B.5.1. Rotational behaviour of rigid objects in simple shear flow and several types of general flow. Shear strain component of the deformation γ is given on the horizontal axis, orientation of the objects on the vertical axis. With increasing aspect ratio (R), objects show increasingly oscillating angular velocity in simple shear. In general flow, objects with an aspect ratio higher than a critical value can reach a stable position and rotate no further. W_k is the kinematic vorticity number (Sect. 2.5.2). (Calculations by Sara Coelho)

Box 5.4   Continued

Fig. B.5.2. Rotational behaviour of ellipses with aspect ratio $R = 3$ but different initial orientation in simple shear (*grey curves*) and in general flow with $W_k = 0.89$ (*black curves*). In simple shear, the ellipses rotate permanently irrespective of their initial orientation. At $W_k = 0.89$ such ellipses reach a stable position. Some ellipses can be seen to actually rotate backwards towards the stable position. (Calculations by Sara Coelho)

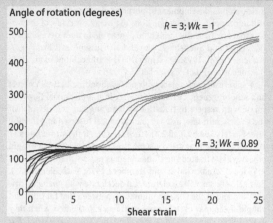

Fig. B.5.3.
Groups of rigid elliptical objects with different orientation can reach a periodic strong preferred orientation in simple shear flow at certain finite strain values and also rotate periodically to their initial orientation without preferred orientation.
a Behaviour of elliptical objects with an aspect ratio of $R = 3$. *Inset* shows the intial orientation (*position I*) and a strain value with strong preferred orientation (*position II*);
b behaviour of elliptical objects with an aspect ratio of $R = 10$. Since these elongate objects have a longer slow-rotation period than those with low aspect ratio, only part of the rotation cycle is shown at a similar scale as **a**; the full pattern is shown in the *inset*. (After Marques and Coelho 2003)

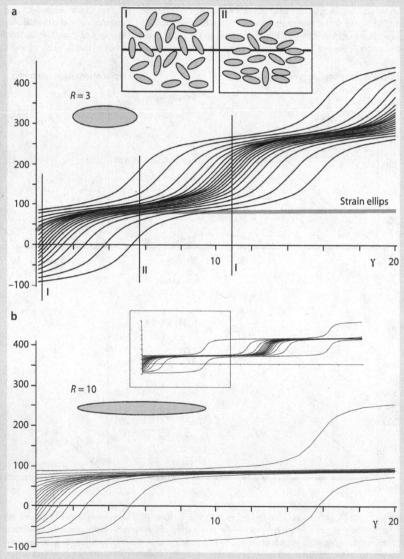

Box 5.4   Continued

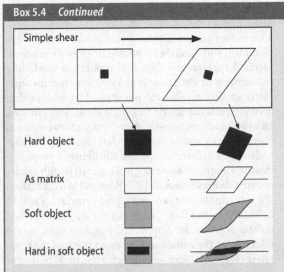

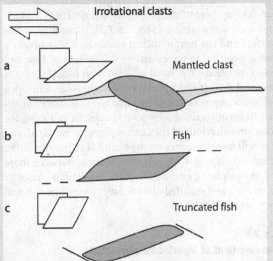

Fig. B.5.4. Behaviour of square deformable inclusions in simple shear flow. If an object is harder (higher viscosity) than the matrix, it will rotate and deform less than a cube of the matrix material would. If an object is softer (lower viscosity) than the matrix it will deform more strongly than a cube of the matrix material would. A hard object in a soft matrix can rotate against the shear sense because it is contained within the strongly deforming soft material. Object shapes based on Treagus and Lan (2000, 2003) and on finite difference numerical modelling by Passchier

Fig. B.5.5. Three types of porphyroclasts that lie in a stable, non-rotating orientation in a ductile flow. **a** Elongate clast in general flow with a critical aspect ratio; **b** fish and **c** truncated fish, which can be stable in simple shear

but here the rotational behaviour and deformation will be different again (Fig. B.5.4).

Another interesting and geologically relevant arrangement is that of a rigid central object, surrounded by a mantle that is *softer* than the matrix (Fig. B.5.4). In that case, the rotational behaviour of the rigid core object will strongly deviate from that described above and depends on object shape and orientation and on thickness and geometry of the mantle (e.g. Ildefonse and Mancktelow 1993; Odonne 1994; ten Brink and Passchier 1995; Bjørnerud and Zhang 1995; Pennacchioni et al. 2000; Kenkmann 2000; Ceriani et al. 2003; Schmid and Podladchikov 2004; Bose and Marques 2004). One interesting consequence of this arrangement is that elliptical or isolated rectangular objects will rotate in non-coaxial flow, but may show antithetic rotation, opposite to shear sense, if initially parallel to the flow plane and surrounded by a wide soft mantle (Fig. B.5.4; ten Grotenhuis et al. 2002; Mancktelow et al. 2002; Ceriani et al. 2003; Schmid and Podladchikov 2004). Monoclinic or lozenge-shaped objects show the same antithetic rotation in some cases, but can even obtain stable positions independent of finite strain (Pennacchioni et al. 2001; Mancktelow et al. 2002). This stable orientation is antithetic with respect to stable positions of objects with perfect bonding (Figs. B.5.1, B.5.5). In other orientations, the rotation rate of the rigid central object depends not only on its orientation and the vorticity of flow in the matrix, but also on width of the soft mantle (Ceriani et al. 2003).

Figure B.5.6 summarises the relation between object shape, object orientation with respect to ISA, and kinematic vorticity number of flow in yet another way. In this three-dimensional graph (Fig. B.5.6), a wing-shaped surface exists on which objects are irrotational; inside and outside of the wing, objects rotate in opposite direction. Only objects with perfect matrix bonding and an

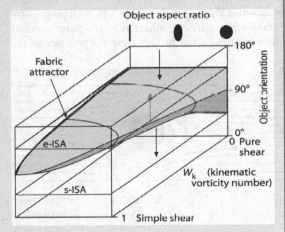

Fig. B.5.6. Diagram showing the orientation where rigid objects can be irrotational in two-dimensional flow without area change. The positions where objects do not rotate form a wing-shaped surface. Objects rotate in the direction of the *arrows* towards the *light grey* attractor surface, where they become irrotational. Objects are also stationary at the *dark grey* surface, but in a metastable position, which is not important in practice. The *solid curved line* at the *back* indicates the position of the attractor for material lines (fabric attractor). Notice that a rotation window exists for objects with a small aspect ratio at high vorticity flow. Two *horizontal planes* represent the orientations of the extensional (*e-*) and shortening (*s-*) ISA respectively

ellipsoidal shape are considered; those with a rectangular or irregular shape can show more complex rotational behaviour.

ongoing deformation (Passchier and Simpson 1986; Fig. 5.27). The resulting shape can give information about the development of the porphyroclast system as a whole and bulk deformation (Sect. 5.6.7.2). If bonding is imperfect and the porphyroclast elongate, it may obtain a stable position in the extensional quadrant of flow and erode to obtain the form of a mineral fish (Fig. B.5.5; Sect. 5.6.7.4). If fluid pressure is high, voids will open aside porphyroclasts and be filled with a different material to form strain shadows or fringes. In this case, the rotation behaviour of the central object is impaired, and this will have influence on the shape of structures in the fringes (Chap. 6). All kinds of transitions between these situations can be envisaged, where bonding increases or decreases, and mantled objects may change into fish and vice versa (Fig. 5.20).

5.6.7.2
Development of Mantled Porphyroclasts

If the mantle of an equidimensional porphyroclast has the same rheology as the matrix, the following behaviour can be predicted. A porphyroclast in a flowing fine-grained matrix will cause a perturbation of the flow field, as shown in Fig. 5.26 and ⊘Video 5.26. With progressive deformation, particles adjacent to the porphyroclast move in ellipses, but further away the presence of the

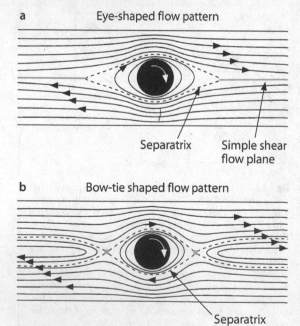

a   Eye-shaped flow pattern

Separatrix   Simple shear flow plane

b   Bow-tie shaped flow pattern

Separatrix

Fig. 5.26. Perturbation of a simple shear flow pattern around a spherical rigid object in simple shear flow, in the vorticity profile plane. Two types have been reported: **a** eye-shaped and **b** bow-tie-shaped flow perturbations. A separatrix surface lies between elliptical displacement paths near the rigid object and open displacement paths further away

porphyroclast only causes a deflection of the displacement paths (Passchier et al. 1993). In simple shear, a boundary can be defined between the far field displacement paths and the elliptical paths, known as a *separatrix*. Experimental data and numerical modelling (Passchier et al. 1993; Bons et al. 1997) show that the separatrix around a spherical porphyroclast with perfect bonding with the matrix in simple shear flow can have an 'eye-shape' or a 'bow-tie shape' in a section normal to the rotation axis of the porphyroclast (Fig. 5.26). Porphyroclasts of a different shape and slip between porphyroclast and matrix can give rise to a separatrix with a more complex shape (Bons et al. 1997; Bose and Marques 2004). If a soft mantle exists around a porphyroclast, it will be deformed in the flow. The geometry of the deformed mantle depends, for a spherical porphyroclast, on the thickness of the mantle and the exact shape of the separatrix (Fig. 5.27). Wide mantles give rise to ϕ- or σ-type clasts in eye or bow-tie shaped separatrices respectively (Figs. 5.23, 5.27c,f, ⊘Videos 5.27c,f,fs,fss); thinner mantles produce δ-type clasts by wrapping of the wings around the rotating central porphyroclast (Figs. 5.22, 5.24, 5.27b,e, ⊘Videos 5.27b,e,es,ess), and very thin mantles give no wings at all (Θ-type clasts, Fig. 5.27a,d, ⊘Videos 5.27a,d; Passchier et al. 1993). Experimental evidence and numerical modelling indicates that the shape of the separatrix depends on several factors such as the initial shape and orientation of the porphyroclast, the change of these factors with time, the bonding between clast and matrix, the rheology of the matrix, the flow vorticity in the matrix and finite strain (Passchier 1988a; Passchier and Sokoutis 1993; Passchier et al. 1993; Passchier 1994; Bjørnerud and Zhang 1995; Pennacchioni et al. 2000; Bose and Marques 2004). It probably also depends on an 'isolation factor' of the porphyroclast in the shear zone, which is high when the porphyroclast is isolated in a relatively wide shear zone, and low if it lies in a relatively narrow shear zone or if porphyroclasts are close together (Marques and Coelho 2001; Bose and Marques 2004). Only a bow tie shaped separatrix leads to significant stair stepping (Passchier 1994). If recrystallisation rate in the rim of the porphyroclasts is small, a δ-type clast can form since only part of the recrystallised mantle will cross the separatrix. If the porphyroclast is recrystallising rapidly and syntectonically, the separatrix will shrink and the mantle will extend over most of the separatrix. In this case, σ-type clasts will develop. If the porphyroclast has an elongate shape, the separatrix will change shape while the clast rotates and secondary wings may form, resulting in complex clasts (Fig. 5.21); complex clasts may also form where recrystallisation rate is irregular or if the porphyroclast starts recrystallising after a δ-type clast was formed (Passchier and Simpson 1986; Passchier 1994). This effect is caused by the fact, that once a δ-type wing is established, no new material

Fig. 5.27a–f.
Idealised development of mantled porphyroclasts around spherical core objects which do not shrink during progressive deformation for eye-shaped (*top*), and bow-tie shaped (*bottom*) separatrices and different initial mantle thickness. Note that δ-type mantled objects start their development as σ-type mantled objects

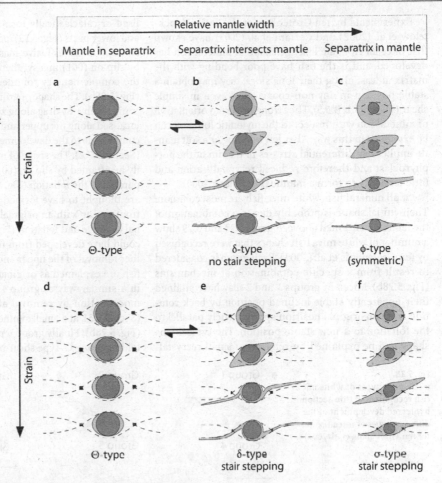

can be added from the porphyroclast to the wing; the material in the wing nearest to the porphyroclast is moving towards it, not away from it; if new recrystallised material is produced, new wings develop.

An important consequence of the model discussed above is that for exceptionally wide mantles and an eye-shaped separatrix, φ-type clasts with approximately orthorhombic symmetry will develop at high strain and high recrystallisation rate, even in simple shear. φ-type clasts therefore are not diagnostic for coaxial flow.

In general flow, elongate porphyroclasts can obtain a stable position, and formation and stretching of mantle material can continue in these cases. Such stable porphyroclasts have a "forward tilted" orientation and σ_a- or φ-type mantles (Fig. B.5.5). δ-type or complex mantle shapes are only possible around permanently rotating porphyroclasts (Passchier et al. 1993; Passchier 1994).

If the mantle is softer than the matrix, the behaviour of porphyroclasts changes. In the case of equidimensional clasts, the central clast may rotate faster than expected for passive mantles, and δ-objects can form. If objects are elongate, they may show the behaviour of fish, as explained below.

5.6.7.3
Development of Sigmoids

Sigmoids lack a rigid central clast, and usually show signs of internal deformation and recrystallisation in the entire object. They may have formed by boudinage and separation of σ-type asymmetric boudins of layers or veins (Sect. 5.6.12), by ductile deformation of rectangular grains which recrystallised completely (Treagus and Lan 2003, 2004), by complete recrystallisation of the core of a σ-type mantled clast, or inhomogeneous flow in the edge of a lens shaped aggregate.

5.6.7.4
Development of Mineral Fish

Characteristic of all mineral fish is the typical elongate lozenge or lens shape, a strong preferred orientation and lack of evidence for rotation. The lozenge shape may have developed by internal deformation (Treagus and Lan 2003), erosion by recrystallisation or pressure solution, and lateral growth by precipitation of dissolved material.

Experiments by ten Grotenhuis et al. (2002), Manck-telow et al. (2002) and Ceriani et al. (2003) have shown that once an elongate lozenge shape of mineral fish has developed, and if the fish have poor bonding with the matrix at least along their long sides, they may obtain a stable position in any non-coaxial flow, even in simple shear (Figs. B.5.1, B.5.5). The strong preferred orientation of mineral fish with respect to the mylonitic foliation can be explained in this way. Also, low bonding or a soft mantle implies low differential stresses in the rim of the porphyroclast and therefore reduced recrystallisation and little tendency to form a mantle (Kenkmann 2000).

Of all mineral fish, white mica fish are most common. Their initial shape is probably due to a combination of the factors mentioned above. Figures 5.28 and 5.29 show a number of white mica fish shapes that were recognised by ten Grotenhuis et al. (2003). Each group is considered to result from a specific combination of mechanisms (Fig. 5.28b). Micas of groups 1 and 2 may have attained their apparently stable inclined position by back rotation from an original position approximately parallel to the foliation to a new stable position. The typical lens shape can be explained by removal of small recrystal-lised or cataclastically torn-off grains along the upper and lower parts (Fig. 5.32), possibly accompanied by dissolution an/or diffusive mass transfer (Fig. 5.28b) and by slip on (001) in a synthetic sense, that could explain the complementary rounded parts of the lens shapes (Fig. 5.28b). The shape of group 2 fish is thought to evolve from group 1 by drag along zones of concentrated shear, localised along the upper and lower contacts (Fig. 5.28b), comparable to the development of σ-type mantled clasts (Sect. 5.6.7.2). The shape of mica fish of group 3 can easily be attained by slip on (001), starting from a position parallel to the foliation (Fig. 5.28b). Mica fish of group 4 are thought to have formed by antithetic slip on (001) from grains with an original high angle between internal cleavage and foliation (Fig. 5.28b). Alternatively they could have developed from fish of groups 1 or 2 by further removal of the upper and lower parts. Group 5 mica fish are explained as originated from short thick micas in a similar way as group 4 ones, but with additional modification by removal along C'-type shear bands (Sect. 5.6.3), at a small synthetic angle with the foliation (Fig. 5.28b). Finally group 6 micas may result from drag folding along C'-type shear bands (Fig. 5.28b).

Fig. 5.28.
a The main types of white mica fish recognised in thin section. **b** Inferred development of the different types of mica fish (after ten Grotenhuis et al. 2003)

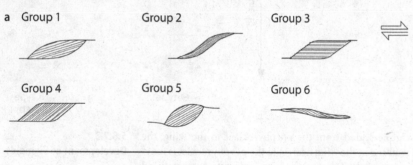

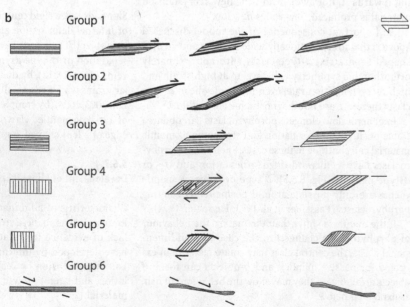

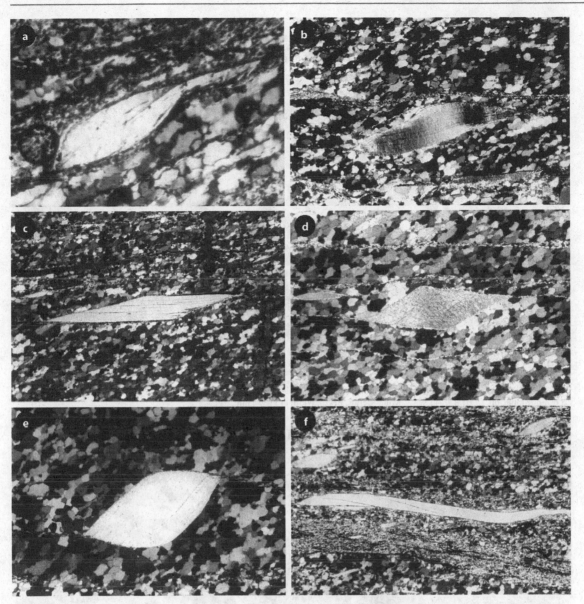

Fig. 5.29. Photomicrographs of different types of mica fish. **a** Lenticular mica fish with slightly inclined tips showing undulose extinction, transitional between groups 1 and 2; **b** lenticular fish with internal discontinuity in the right tip of the fish; **c** rhomboidal shaped fish with (001) parallel to longest side of the fish, group 3; **d** rhomboidal shaped fish with (001) parallel to the shortest side of the fish, group 4; **e** fish with small aspect ratio, group 5; **f** mica fish with high aspect ratio belonging to group 6. Samples are from Conceição do Rio Verde, Brazil. Shear sense in all photographs is dextral. Width of view **a** 3 mm, **b** 0.75 mm, **c, d** and **e** 3 mm, **f** 6 mm. CPL

Biotite fish (Fig. 5.33a; ten Grotenhuis et al. 2003) probably form in the same way as white mica fish but are less common. In most mylonites, biotite recrystallises more readily than white mica, possibly leading to the early destruction of biotite fish. K-feldspar fish in a quartz-feldspar matrix (ten Grotenhuis et al. 2003) in high-grade rocks may also form by internal deformation (Fig. 5.33d), possibly assisted by dynamic recrystallisation along the rim. Hyperstene fish in a quartz-feldspar matrix investigated by ten Grotenhuis et al. (2003) did not show evi-

dence of internal deformation and were interpreted to have formed essentially by grain size reducing mechanisms. By contrast, Ishii and Sawaguchi (2002) described a case where orthopyroxene porphyroclasts embedded in a fine-grained olivine matrix suffered strong crystal-plastic deformation. Tourmaline fish (ten Grotenhuis et al. 2003) show little or no intracrystalline deformation but seem to change shape exclusively by dissolution and precipitation, cutting the internal zoning patterns (Fig. 5.34). The same may apply to some feldspar fish, because of

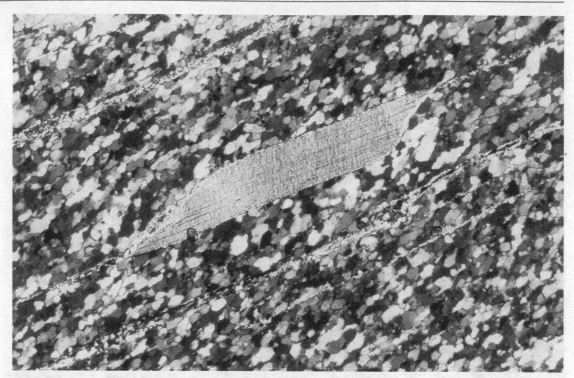

Fig. 5.30. Mica fish of group 3 in a quartzite mylonite. Note stair stepping. Quartz in the matrix is dynamically recrystallised and developed an oblique foliation. Dextral shear sense. Minas Gerais, Brazil. Width of view 4 mm. CPL

Fig. 5.31. Mica fish of group 5 as shown schematically in Fig. 5.28a from quartzite mylonite. Note stair-stepping of wings of mica fragments. Quartz in the matrix is dynamically recrystallised and developed an oblique foliation. Dextral shear sense. Minas Gerais, Brazil. Width of view 4 mm. CPL

Box 5.5   Misinterpretation of fault rocks

Misinterpretation of fault rocks is relatively common. It is therefore important to pay maximum attention to certain details that reveal metamorphic conditions and strain rates. In this box, we present a number of common errors and misconceptions and clues to avoid them. In thin section, the nature of fault rocks can usually be established without problems, unless there is overprint of one type on the other. In the field, however, the situation is more difficult.

Dark fine-grained rocks in a shear zone can be ultramylonite, cataclasite or pseudotachylyte. A planar dark vein in a low porosity rock such as a granite or gneiss with sharp boundaries to the wall rock and with injection veins is probably a pseudotachylyte. Characteristic of pseudotachylyte is also colour banding (Sect. 5.2.5) with different coloured core and edges, which extend into the injection veins.

If the dark fault material has a gradual transition to the wall rock, and occurs in an anastomosing geometry, or if it occurs in a sedimentary rock, it is more likely to be a cataclasite. An ultramylonite can have sharp boundaries as well, but lacks injection veins and colour banding. Characteristic is a good aggregate lineation and planar shape preferred orientation but this only develops in ultramylonite that is inhomogeneous, not in massive, homogeneous fine-grained ultramylonite (Box 4.4; Sect. 5.3). In thin section, the nature of vein-wall rock contacts and porphyroclasts can help to make a distinction. Sharp boundaries with corrosion of micas are typical for pseudotachylyte; ultramylonite and cataclasite have less sharp boundaries usually with recrystallisation in the adjacent large grains for ultramylonite, and fracturing and numerous fluid and solid inclusions in adjacent grains for cataclasite.

Coarser grained mylonite with porphyroclasts is occasionally confused with porphyry, rhyolitic ignimbrite or even with some finely layered sedimentary rock. In such cases, mylonites can be recognised in the field by the presence of aggregate lineations and in thin section by the presence of mantled porphyroclasts, mica fish and by a finely grained recrystallised matrix with an LPO in quartz domains.

Foliated cataclasite can also strongly resemble low-grade mylonite. Both have foliations and asymmetric structures such as porphyroclasts, mica fish and shear bands that can act as shear sense indicators. Foliated cataclasite is associated with polished fault planes (slickensides) and striations or fibres while low-grade mylonites have penetrative aggregate lineations. In foliated cataclasite, quartz is deformed to a fine fractured aggregate. In mylonite, quartz shows evidence for dynamic (usually BLG) recrystallisation.

In high-grade gneisses, it is difficult to recognise ductile shear zones. Usually, all rocks are deformed to some extent, and the difference between deformed and more deformed is small. High grade mylonites have stronger fabrics than their wall rocks, strong shape preferred orientations, grain lineations of constant orientation and often isoclinal folds. Shear sense indicators are, unfortunately, rare in high-grade ductile shear zones.

Faults that overprint each other within a reactivated shear zone are common; most common are brittle faults overprinting mylonites and occasionally mylonites overprinting a brittle fault rock. Different overprinting phases of brittle fault rocks can sometimes be recognised if they differ in direction of slip, and have striations or fibres in two superposed populations of different direction. Mylonites overprinting mylonites during a later deformation event can be hard to distinguish. However, if they differ in metamorphic grade, low-grade mylonites tend to occupy a smaller rock volume than high grade ones, so that the branches of low-grade mylonite will separate lenses of high-grade mylonite. These are easy to distinguish if the shape preferred orientation in both mylonites has a different orientation, especially the lineations; if *two* object lineations of different orientation are found in one mylonite zone this could be an indication for transpression (Passchier 1998), but it is more likely that the zone formed in two phases. However, if these are parallel, the two generations of mylonite can only be distinguished by the microstructure, e.g. the brittle or ductile behaviour of feldspar and the development of aggregate or grain lineations.

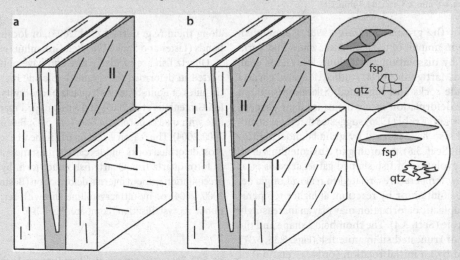

Fig. B.5.7. Mylonite zones may have several phases of activity. This image shows two shear zones with polyphase deformation. **a** An older mylonite (I) is cut by a younger narrow shear zone (II) in which the lineation has a different orientation than in (I). Because of this, the two phases can be distinguished in the field. **b** The younger low-grade mylonite (II) has a lineation with the same orientation as high-grade mylonite (I). In this case, it is more difficult to distinguish between the two phases in the field, but the microstructure can be used to resolve this problem. (I) was active at high-grade conditions, visible from ductile deformation of feldspar and GBM-recrystallisation of quartz. (II) was active at low-grade conditions since feldspar is brittle and recrystallised, and quartz recrystallised by SGR-recrystallisation. Grain size of quartz exaggerated

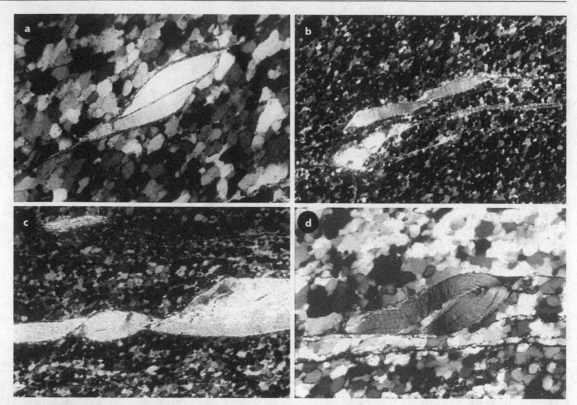

Fig. 5.32. Photomicrographs of microfaults separating mica fish in two or more smaller parts. **a, b** Different stages of a process in which a mica fish is divided in two parts along basal planes with synthetic sense of movement. **c** Micro-faults through a mica fish at a high angle to the basal planes, showing antithetic movement. **d** Folded mica fish with tight fold hinges. On the left hand side, the fish is split along the fold hinge. All samples are from Conceição do Rio Verde, Minas Gerais, Brazil. Shear sense in all photographs is dextral. Width of view **a** 1.5 mm, **b** 6 mm, **c** 3 mm, **d** 1.5 mm. CPL

zonation in the grains (McCaig 1998). Garnet fish (Fig. 5.33e) in amphibolite facies shear zones may form exclusively by dissolution and volume loss (Azor et al. 1997). Ji and Martignole (1994) studied elongated garnets in high grade rocks and suggested dislocation slip and recovery as deformation mechanisms for their garnets, but den Brok and Kruhl (1996) suggested that these structures could also have formed by grain boundary diffusional creep (Sect. 3.8). According to ten Grotenhuis et al. (2003), the evolution of fish shaped garnet crystals is, at relatively low temperatures, related to removal of garnet by pressure solution or by reaction; at higher temperatures crystalplastic deformation may play an increasingly important role (Sect. 3.4). The rhomboidal shape and the inclination of truncated sillimanite fish (Figs. 5.36, 5.37) is explained by an initial rotation (or back-rotation) of rectangular microboudins towards a stable position inclined to the mylonitic foliation (Mancktelow et al. 2002). The crystals then change their shape to a rhomboidal form by dissolution and/or reaction against C'-shear bands. It is still unclear if trails of particles that stretch out from the tips of most mineral fish into the matrix are passively stretched wings, in which case there is no displacement

along them (e.g. Ceriani et al. 2003), or localised shear bands (Lister and Snoke 1984; ten Grotenhuis et al. 2002).

Quartz fish are relatively rare and have only been reported in deformed fine-grained volcanic rocks such as rhyolites or ignimbrites with quartz phenocrysts (Fig. 3.10; ten Grotenhuis et al. 2003) and fine-grained marble mylonite with quartz porphyroclasts (Fig. 5.35; Bestmann et al. 2000, 2004). The reason is that quartz is the weaker phase in most deformed rocks, and special circumstances are needed to form quartz fish. Quartz fish either form by solution-precipitation without internal deformation (Bestmann et al. 2000, 2004) or by intracrystalline slip with recovery and minor recrystallisation (compare Fig. 3.10).

Fig. 5.33. Photomicrographs of different minerals showing fish ▶ shapes similar to mica fish. **a** Biotite fish with small recrystallised biotite grains along the rims, from Santa Rosa Mylonite zone, California. **b** Ilmenite fish from Conceição do Rio Verde, southern Minas Gerais, Brazil. PPL. **c** plagioclase fish; **d** K-feldspar fish; **e** garnet fish; **f** diopside fish; **g** plagioclase and biotite in a combined fish shape intergrowth; **h** hornblende fish. **c–h** in quartz matrix from high-grade gneiss. Varginha, southern Minas Gerais, Brazil. Width of view **a** 6 mm, **b–h** 1.5 mm. CPL

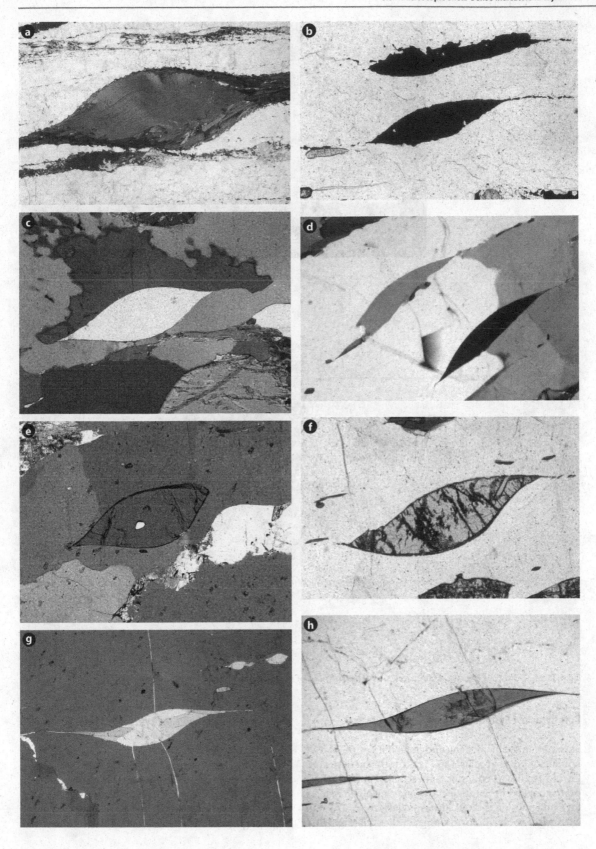

Fig. 5.34.
EBS-image of a tourmaline fish from Lambari, Brazil, showing zoning in the centre and new growth of tourmaline at the tips of the fish

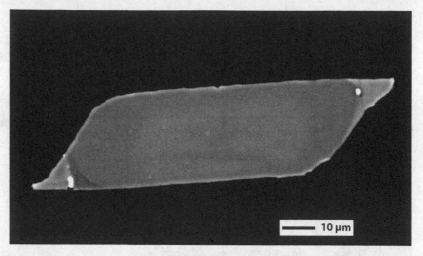

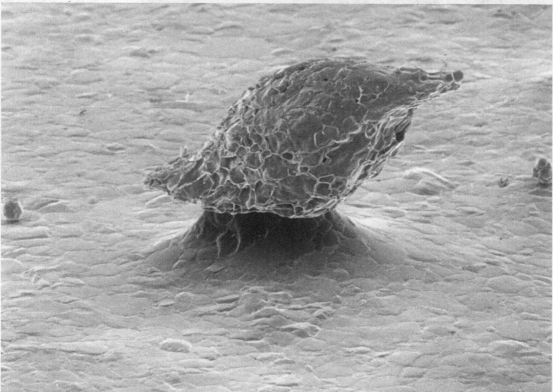

Fig. 5.35. SEM image of quartz fish in marble from Thasos. Width of view 0.8 mm. (Courtesy M. Bestmann)

5.6.8
Porphyroclast Systems as Shear Sense Indicators

Summarising our present understanding of the development of porphyroclasts and related structures, the following applies. If cohesion between clast and matrix is perfect, naked or mantled clasts and sigmoidal systems form; if cohesion is reduced, mineral fish form, or strain shadows if voids can form alongside the clast (Fig. 5.20).

Equidimensional naked or Θ-type clasts cannot be used as shear sense indicators. σ_a-type-, σ_b-type-, δ-type- and complex mantled clasts can be used as shear sense indicators using their internal asymmetry, and the stair-stepping of the wings. Stair stepping is a useful shear sense indicator; the wings step up in the direction of movement of the upper block (Figs. 5.21, 5.22–5.25). However, care should be taken in inhomogeneous mylonite with many large porphyroclasts or low strain lenses; if a mantled

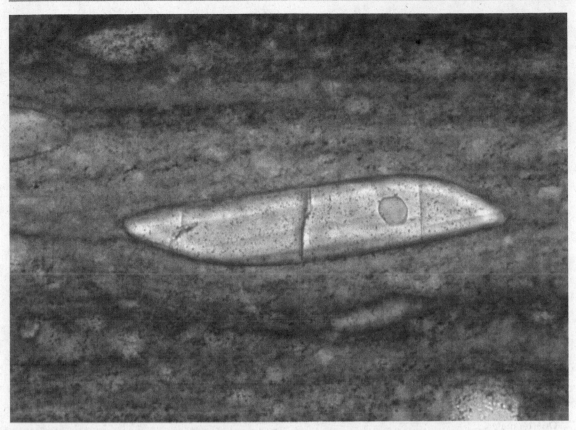

Fig. 5.36. Truncated sillimanite fish in ultramylonite. In the matrix fine-grained ilmenite outlines an oblique foliation inclined to the mylonitic foliation. The inclusion in the sillimanite is apatite. Dextral shear sense. Monte Mary, Italy. PPL. (Photograph courtesy Giorgio Pennacchioni)

porphyroclast lies between two large rigid objects, its asymmetry may reflect the relative movement of these two objects rather than the bulk shear sense in the shear zone. Therefore, mantled clasts are most reliable as shear sense indicators where they occur isolated in a fine-grained mylonitic matrix, e.g. in ultramylonite.

Sigmoids have the same shape and orientation as σ-type mantled clasts and, although their development is not completely understood, their shape can be used to determine shear sense in the same way (Fig. 5.20).

The geometry of porphyroclasts with pressure shadows and fringes is formed in a different way as mantled porphyroclasts, and both structures should not be confused. Mantled clasts occur exclusively in mylonites, while strain shadows can also be found in low strain rocks. Their use as shear sense indicators is explained in Chap. 6.

All mineral fish are good shear sense indicators, irrespective of their mechanisms of development. The inclination of the long axis of fish with respect to the foliation and a commonly developed monoclinic shape symmetry, with one curved and one planar side, can be used as shear sense indicators (Figs. 5.20, 5.28, 5.30, 5.31). If trains of small fragments trail away from the tips of the fish, these will show stair stepping which

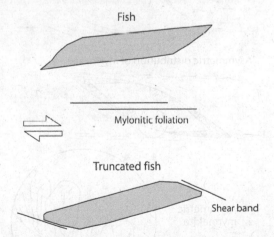

Fig. 5.37. Schematic diagram showing the geometry of normal mineral fish and truncated fish

can be used as a shear sense indicator as in the case of mantled clasts. An exception are mica fish of group 6 (Fig. 5.28), which are localised on shear bands, and truncated sillimanite fish (Fig. 5.37; Type 2 of Mancktelow et al. 2002) which have a different geometry from other types.

5.6.9
Quarter Structures

Porphyroclasts without mantles may show an asymmetric distribution of microstructures over the four quarters defined by the foliation and its normal; such structures have been named *quarter structures* (Fig. 5.38; Hanmer and Passchier 1991). Quarter structures are geometrical features that do not need to coincide with flow symmetry axes. Several types of quarter structures have been described and have been empirically established as shear sense indicators.

Quarter folds. Microfolds in the quarters that lie in the extensional direction are known as *quarter folds* (Fig. 5.38; Hanmer and Passchier 1991). Quarter folds probably develop by rotation of layering into the extension field of flow when passing the top of a porphyroclast during progressive deformation.

Quarter mats. Mica concentrations adjacent to a porphyroclast in the quarters that lie in the shortening direction are known as *quarter mats* (Hanmer and Passchier 1991). These probably form by preferential removal of quartz by solution transfer at stress concentration sites adjacent to a porphyroclast.

Asymmetric myrmekite. Myrmekite is sometimes concentrated in shortening quarters in the rim of K-feldspar crystals (Sect. 7.8.3; Simpson 1985; Simpson and Wintsch 1989). It probably forms by preferential proceeding of the K-feldspar breakdown reaction and associated volume loss at sites of high differential stress (Simpson and Wintsch 1989). The arrangement of quartz lamellae in myrmekite may also show an internal monoclinic symmetry (Fig. 5.38 inset) which can serve as an independent, internal shear sense indicator (Simpson and Wintsch 1989).

5.6.10
Lattice-Preferred Orientation

The lattice-preferred orientation (LPO) of minerals commonly shows a monoclinic symmetry. Crystals with elongate shape such as mica and amphiboles can develop a monoclinic oblique fabric with respect to another foliation such as a layering, as described above. Moreover, the deviation in orientation of such minerals around the mean can be skewed. Such skewness is difficult to measure optically, but has been detected in slates by X-ray goniometry (Sect. 10.3.5) and can be used to determine sense of shear (O'Brien et al. 1987).

Minerals with equant grain shape such as quartz, calcite, feldspar and olivine in mylonites commonly show a monoclinic symmetry of LPO with the symmetry axis normal to the aggregate lineation and parallel to the main mylonitic foliation. LPO patterns in pole diagrams for a single crystallographic axis such as c- or a-axes in quartz can have an internal monoclinic symmetry defined by the shape of the pattern, and an external asymmetry with respect to foliations in the rock. Both asymmetries are useful shear sense indicators. Further details on LPO development and interpretation are given in Sects. 4.4.3–4.4.5. A method to measure quartz c-axes is presented in Sect. 10.3.

5.6.11
Vergence of Asymmetric Fold Sections

Cross sections through folds in mylonites parallel to the lineation commonly show a dominant vergence (Fig. 5.10). If the folds are sheath folds generated during mylonite formation and the section is strictly parallel to the movement direction, the vergence may be reliable as a shear sense indicator. In most cases, the three-dimensional shape of the folds is unknown and it may also be unclear whether the folds are older or related to the mylonite formation. In those cases, asymmetric folds are unreliable as shear sense markers.

Quarter structures

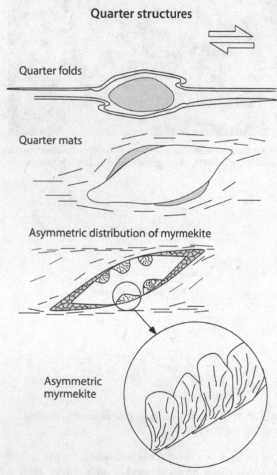

Quarter folds

Quarter mats

Asymmetric distribution of myrmekite

Asymmetric myrmekite

Fig. 5.38. Three types of quarter structures in mylonite that can be used to determine sense of shear. The structures are defined by an asymmetric distribution of fabric elements over the four quarters defined by the foliation and its normal. Myrmekite can also have an internal asymmetry that can be used to determine shear sense

5.6.12
Potential Shear Sense Markers

A number of structures in mylonites are potential shear sense markers, but they are either more cumbersome to use than the ones mentioned above, or the results are not consistent. This is probably mainly because their development is as yet incompletely understood. In time, these structures may also be useful, but we advise not to rely exclusively on one of them at present. The most common ones are:

Deformed Veins

If a large number of deformed veins of different orientation are present in a deformed rock, they can be used to obtain detailed information on the deformation process (Talbot 1970; Hutton 1982; Passchier 1990a; Fig. 5.39). This applies to veins in outcrop, but also to *microveins* that can be studied in thin section, formed

by fracturing and precipitation of material from solution (Chap. 6). Such microveins have been studied in deformed pseudotachylyte (Passchier 1986b) and metachert (Wallis 1992a).

Veins can shorten, extend, or shorten and extend in sequence depending on vein orientation in the flow, and on flow parameters such as vorticity and volume change. If sufficient viscosity contrast exists between veins and matrix, veins can suffer boudinage on extension, or they can fold on shortening. Figure 5.39 shows the distribution of domains of shortened (s), extended (e), first shortened-then-extended (se) and first extended-then-shortened (es) material line categories for different types of progressive deformation (see also Sect. 2.7; Fig. 2.8). Since veins do not start folding or enjoying boudinage at the onset of deformation, and may unfold when being stretched, the boundaries of material line categories do not correspond exactly to domains of folding (f), boudinage (b), folding-then-boudinage (fb) and boudinage-then-

Fig. 5.39.
Distribution of domains of deformed lines in space for four categories of progressive deformation. Note how the distribution of the domains is dependent on the type of progressive deformation; in non-coaxial progressive deformation the distribution has monoclinic symmetry and can be used to determine sense of shear. The largest field of first shortened-then-extended lines (*se*) indicates the dominant rotation direction of lines, and therefore sense of shear (dextral). Only in the case of polyphase deformation are domains of first extension-then-shortening possible

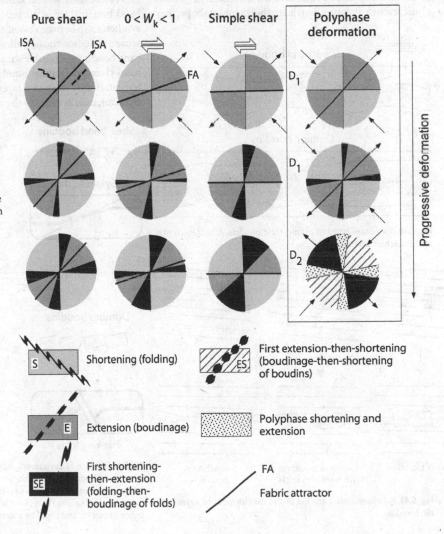

folding (bf) in veins. Nevertheless, it is useful to plot the orientation of deformed veins against deformation category (f, b, bf or fb) and to compare the symmetry with the theoretically predicted patterns. At least sense of shear should be deductible; if the pattern is asymmetric, the sense of asymmetry indicates shear sense (Fig. 5.39, ⊘Video 5.39). Complexly deformed veins are notably informative. Notice that first folded, then boudinaged veins (fb) can occur in any flow type, but veins with shortened or folded boudins (bf) are important: such a deformation sequence can *only* result from (1) a high rigid body rotation component of flow such as near rigid objects, or (2) polyphase deformation (Passchier 1990a). In most cases, the presence of folded or shortened boudins is due to polyphase deformation.

Asymmetric Boudins

Asymmetric boudins can occur in sedimentary or tectonic layering. If boudins have an asymmetric (monoclinic) rather than symmetric (orthorhombic) shape, they

can occasionally be used to determine sense of shear. Two types of asymmetric boudins can be distinguished in layering on mm–m scale, based on detailed analysis of field examples and experimental work (Goldstein 1988; Goscombe and Passchier 2003): shear band boudins and domino boudins (Fig. 5.40). *Shearband boudins* have a long, curved lenticular shape and large relative displacement and synthetic drag on an inter-boudin surface that is gently inclined to the boudin exterior surface. *Domino boudins* have an angular shape, an inter-boudin surface steeply inclined to the boudin exterior surface with small relative displacement and unique antithetic flanking folds instead of synthetic drag (Fig. 5.40). Development of either type of asymmetric boudins depends mainly on the initial boudin neck orientation, normal or oblique to the long axis of the boudinaged object, on the rigid or deformable character of such objects, and on the flow type (Fig. 5.41). If asymmetric microboudins can be identified by their shape as either shearband or domino boudins, they can be used as shear sense indicators (Fig. 5.42). Shear band boudins are in all reported cases associated with synthetic slip between boudins, i.e. relative displacement sense of boudins equals bulk shear sense. Therefore, shear band boudins are reliable shear sense indicators, like shear band cleavage. Domino boudins are more difficult to interpret. If they form in a layer parallel to the bulk fabric attractor, usually close to the mylonitic foliation in mylo-

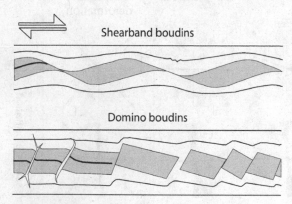

Fig. 5.40. Two main types of asymmetric boudins. After Goscombe and Passchier 2003

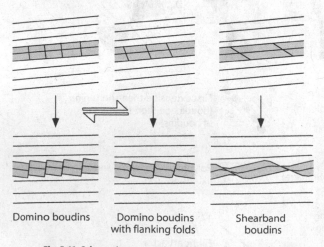

Fig. 5.41. Schematic presentation of the development of asymmetric boudins

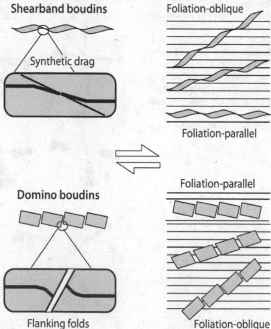

Fig. 5.42. The use of asymmetric boudins as shear sense indicators. Shear band boudins are reliable shear sense indicators in all orientations. Domino type boudins are reliable if parallel to the mylonitic foliation. If oblique to the mylonitic foliation they may have the opposite geometry and could be incorrectly interpreted

nites, inter-boudin slip is antithetic to bulk shear sense; however, if a layer or object that suffers boudinage is highly oblique to the fabric attractor, inter-boudin slip may be synthetic (Fig. 5.42). Since it is not always possible to determine the orientation on a train of domino boudins, they are to be used with caution.

Fragmented Porphyroclasts

Elongate rigid porphyroclasts of amphiboles, pyroxenes or feldspar grains can be separated into aggregates of fragments with a geometry similar to asymmetric boudins, separated by seams of cataclasite (Fig. 5.43, ⊘Photo 5.43; Duebendorfer and Christensen 1998; Babaie and LaTour 1998). As in the case of boudins, there are *domino type* and *shear band type fragmented porphyroclasts* (Fig. 5.44). The geometry depends on the bulk shear sense and the initial orientation of microfaults in the grains, which may be partly controlled by crystallographic directions in the porphyroclasts (Sect. 3.12.4) and by flow type. In porphyroclasts, fragment geometry also depends on the original grain shape. Simple sets of fractures tend to dominate when the porphyroclast suffered small internal strain, but at higher strain porphyroclasts split to form *mosaic fragmented porphyroclasts* (Duebendorfer and Christensen 1998; Babaie and LaTour 1998). Domino-type and mosaic fragmented porphyroclasts are more

Domino-type fragmented porphyroclast

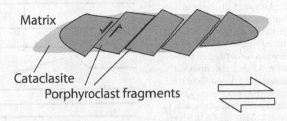

Matrix

Cataclasite

Porphyroclast fragments

Mozaik fragmented porphyroclast

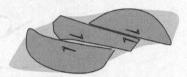

Shear band type fragmented porphyroclast

Fig. 5.44. Illustration of three types of fragmented porphyroclasts, and their interpretation in terms of bulk shear sense (*large arrows*)

Fig. 5.43. Microfaults transecting feldspar porphyroclasts in a section parallel to the aggregate lineation and normal to the foliation in granite mylonite. The faults do not continue into the mantle of recrystallised feldspar surrounding the porphyroclasts. Dextral shear sense. Both synthetic (*bottom left*) and antithetic (*top right*) microfaults are present. St. Barthélemy, Pyrenees, France. Width of view 9 mm. CPL

Fig. 5.45.
a Schematic presentation of a flanking fold structure in foliation or bedding (*host element-HE*) around a vein or fault (*crosscutting element-CE*). The geometry can be defined in terms of lift and slip. Lift defines the elevation of the far-field HE above the cutoff point against the CE; slip is displacement of the HE along the CE. Both are positive as shown in the *inset* for a structure in the hanging wall. **b** Simplified geometries of flanking structures along CE with different slip and lift. (After Coelho et al. 2005)

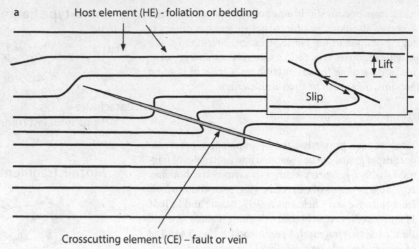

a   Host element (HE) - foliation or bedding

Lift

Slip

Crosscutting element (CE) – fault or vein

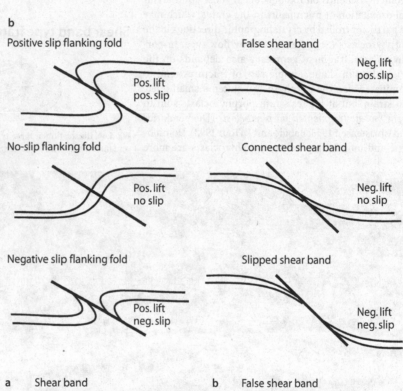

b

Positive slip flanking fold

Pos. lift
pos. slip

False shear band

Neg. lift
pos. slip

No-slip flanking fold

Pos. lift
no slip

Connected shear band

Neg. lift
no slip

Negative slip flanking fold

Pos. lift
neg. slip

Slipped shear band

Neg. lift
neg. slip

Fig. 5.46.
Shear bands (**a**) and false shear bands (**b**) have a different slip sense, but this is only obvious if displaced markers can be recognised. If multiple markers are displaced (**c** and **d**), the distinction may not be visible

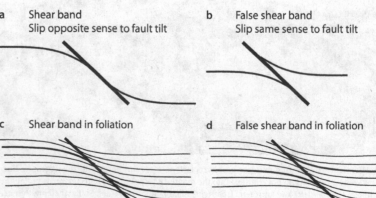

a   Shear band
Slip opposite sense to fault tilt

b   False shear band
Slip same sense to fault tilt

c   Shear band in foliation

d   False shear band in foliation

common than shear band type. It is possible that, at higher metamorphic grade, zones of recrystallised grains separate fragments instead of fracture zones.

Although attempts have been made to use fragmented porphyroclasts as shear sense markers (Simpson and Schmid 1983), they are relatively unreliable. Interpretation is usually as shown in Fig. 5.44. The overall shape of the aggregate is commonly that of a sigmoid, and this overall shape, together with the inferred domino or shear band type of the fracture system can be used to determine the shear sense.

Flanking Structures

In many mylonite zones, oblique veins or faults transect the mylonitic foliation or layering. Where such veins or faults (here referred to as *crosscutting elements* (CE); Passchier 2001; Coelho et al. 2005) show evidence that they were intruded or active during mylonite generation, their interrelationship with a foliation or other *host element* (HE) can give asymmetric structures that can occasionally be used as shear sense indicators. Shear band cleavage discussed in Sect. 5.6.3 is one example, but many other geometries have been observed known as *flanking structures* (Fig. 5.45). Critical elements of such structures are *slip* and *lift* of the host element with increasing distance from the crosscutting element (Coelho et al. 2005; Fig. 5.45; ⊘Video 11.10a–d). Many combinations are possible. In shear bands, normal slip can be accompanied by non-rotation or back-rotation of the crosscutting element

(⊘Video 11.10c). However, *false shear bands* with the reverse sense of slip also occur (Fig. 5.45). If these develop in a setting where individual layers can be traced, there is no problem, but if they occur in a foliation, they may be confused with normal shear bands (Fig. 5.46). Forward rolling of the host element produces common structures known as *flanking folds* (⊘Video 11.10a,b,d; Passchier 2001; Graseman and Stüwe 2001; Graseman et al. 2003; Exner et al. 2004; Wiesmayr and Grasemann 2005). Flanking folds can combine with reverse, normal or no slip (⊘Video 11.10a–d). In all cases, the geometry of these structures changes laterally along the crosscutting element, and dies out into the wall rock where the crosscutting element ends (Fig. 5.45; ⊘Video 11.10a–d).

Shear bands are straightforward in their interpretation as discussed above (Sect. 5.6.3) but false shear bands and flanking folds are not. Different mechanisms can develop structures of similar or slightly different geometry, and the use of flanking folds or false shear bands as shear sense indicators is therefore only possible if the exact evolution of the structure can be reconstructed (Graseman et al. 2003; Exner et al. 2004). Two major mechanisms can cause combinations of slip, lift, and rolling on crosscutting elements; (1) active slip along a fault in a ductile shear zone (⊘Video 11.10a–c), or presence of a planar competent body such as a vein (⊘Video 11.10d) creates a deviating flow type close to the fault which changes the geometry of the host element; (2) infiltration of fluids from a vein changes the rheology of the adjacent host rock and further deformation

Fig. 5.47.
a Flanking fold without slip in layering that is deflected against a vein with an alteration zone (*dotted*) can form in at least three different ways: by coaxial flattening **b**, by dextral shear **c** and by sinistral shear **d**

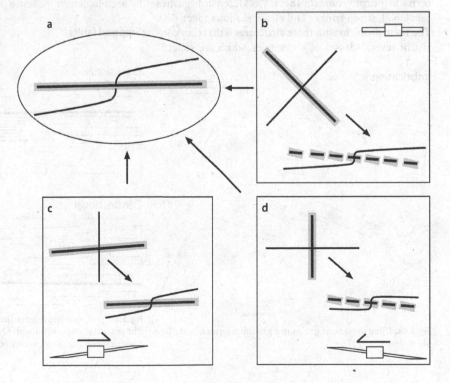

changes host element geometry in this domain. Reliable shear sense indicators are flanking folds without slip along veins where some kind of alteration may have caused forward rotation of the host element. This can be recognised since the crosscutting element is not stretched, but the host element is (Fig. 5.47).

Imbrication

The presence of a large number of rigid elongate crystals in a deforming material may result in their interference causing *tiling* or *imbrication* of the objects (Fig. 5.48; Fernandez et al. 1983; Arbaret et al. 1996). This type of structure seems to be common in phenocryst granites deformed in the liquid state (Blumenfeld and Bouchez 1988; Chap. 8), and has been described macroscopically by Blumenfeld (1983) for feldspar grains in granitoid gneiss, and for porphyroclasts in ultramylonite (Brunel 1986; Blumenfeld and Bouchez 1988; Pennacchioni et al. 2001). The asymmetry of imbrication can indicate shear sense, but a large number of observations of tiled grains in a single location are needed for a reliable analysis (Mulchrone et al. 2005).

Complex Indicators

Several microstructures with monoclinic shape symmetry do not exclusively occur in mylonites. These structures may be reliable shear sense indicators, but need close observation and classification before they can be applied. The most common examples are inclusion patterns in porphyroblasts (Sect. 7.6.8), tension gashes (Sect. 6.2), strain fringes and strain shadows (Sect. 6.3). The main problem with these structures is that they consist of several classes of geometries, which are superficially similar but formed in different ways, commonly in opposite sense of shear (e.g. Sect. 7.6.8). Examples of these complex shear sense indicators are treated in Chap. 6 and 7.

Tensional and Constrictional Stepover Sites

In mylonites with a strong mica foliation deformed at low grade or high strain rate, approaching the brittle deformation regime, small extensional fractures or microfolds may occur between foliation-parallel but 'overstepping' micro-shear bands or faults (Fig. 5.49). If such locations systematically show the same stair stepping of faults for constrictional or tensional stepover sites, they can be used as shear sense markers. Care should be taken, however, since slip on micro-shear bands may not be the only deformation in the sample, and need not be representative for bulk deformation; i.e. sense of shear on the micro-shear bands may be opposite to bulk shear sense.

'V'-pull-apart Microstructures

Hippertt (1993) introduced a new potential shear sense indicator in the form of pull-apart structures that occur in the rim of feldspar porphyroclasts at low metamorphic grade. Fractures in the edge of the porphyroclasts may open to a V-shape and are filled with quartz or another mineral. In some mylonites these V-pull-apart microstructures have a persistent asymmetry that can be used to determine sense of shear.

Stepped faults

Imbrication

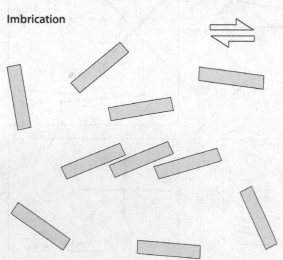

Fig. 5.48. Tiling of feldspar grains in a granite deformed in the liquid state

Extensional

Constrictional

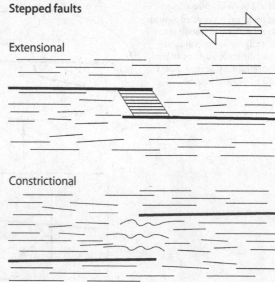

Fig. 5.49. Tensional and constrictional stepover sites in a mylonite. The tensional site is filled with fibrous material (Sect. 6.2.5). The constrictional site shows microfolding of the foliation

5.7
Shear Sense Indicators in the Brittle Regime

5.7.1
Introduction

The internal fabric of gouge, cataclasite or breccia may contain microstructures that can be used to determine shear sense, as in mylonites. A problem in gouges and cataclasites is that with some exceptions (Tanaka 1992), penetrative lineations that could be used to determine the movement direction are lacking. Instead, *slickensides* (polished fault surfaces) occur that may contain parallel ridges or grooves (known as *slickenlines* or *striations*), linear aggregates of cataclased material as isolated lenses or behind obstacles, or fibres (known as *slickenfibres*; Sect. 6.2.5) that are parallel or slightly inclined to the fault surface (Means 1987). Care should be taken when determining the movement direction along a fault in the field from striations or slickenfibres on slickensides, because they commonly only show the last movement stage on the fault (Tanaka 1992). In any case, specimens for establishment of shear sense should be oriented with respect to the macroscopically determined movement direction.

Sense of shear can be determined in the field from displacement of markers as for ductile shear zones (Sect. 5.5.2) or from the shape of striations, slickenfibres or minor faults on slickensides (Petit 1987). If slickenfibres consist of calcite, the orientation distribution of deformation e-twins can be used to determine sense of movement on the fault (Laurent 1987). Other shear sense criteria in thin section are given below.

5.7.2
Incohesive Brittle Fault Rocks

Although it is usually difficult to sample and cut gouge or incohesive cataclasite, thin sections of such materials may give information on shear sense. A gouge normally consists of rock and mineral fragments in a matrix rich in clay minerals. This matrix may show a uniform extinction under crossed polars due to preferred orientation of the clay minerals and may be layered or foliated (S in Fig. 5.50). The foliation is also referred to as a *P-foliation*. Shear bands (Sect. 5.6.3) may also be present in the matrix (Lin 2001). Fragments of mineral grains resistant to fracturing can develop asymmetric strain shadows (Lin 2001). Both the extinction direction and shear bands can be used to determine shear sense as described for oblique foliation and shear bands in mylonite (Sect. 5.6.3). Sets of subsidiary shear fractures with distinct orientation and movement sense, known as *Riedel shears* may also be present (Riedel 1929; Chester et al. 1985; Chester and Logan 1987, 1998; Rutter et al. 1986; Evans 1990; Lo-

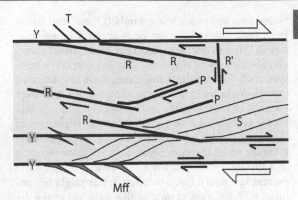

Fig. 5.50. Schematic diagram showing the characteristic geometry and shear sense of the most common types of Riedel shears (*R*-, *R'*-, *P*-, *Y*- and *T-shears*) in a brittle fault zone. Shear sense is mostly established from deflection of older shears or, in foliated cataclasite or gouge, deflection of the foliation (*S*). *Mff* – microscopic feather fractures

gan et al. 1992). They are subdivided into R-, R'-, P- and Y-shears, each with a characteristic orientation and shear sense (Fig. 5.50). Y-shears act as boundary faults for the brittle fault zone or are parallel to the boundary. Riedel shears resemble ductile shear bands but form by brittle fracturing. In some fault rocks, extensional fractures without displacement form at 20–50° to Y-shears or boundaries of the gouge zone. These are known as *T-fractures* (Petit 1987) or, if opened into wedge-shaped tensional veins, as *microfault induced microcracks* or *microscopic feather fractures* (Mff – Fig. 5.50; Friedman and Logan 1970; Conrad and Friedman 1976; Teufel 1981; Blenkinsop 2000).

Riedel shears are useful shear sense indicators. Since suitable displaced markers are usually rare, deflection of a foliation or of older Riedel shears by younger shears (e.g. P by R; Y by R as shown in Fig. 5.50) can be used to determine shear sense. Synkinematic veins of quartz and calcite are common in brittle fault rocks and occur in any stage of break-up from intact veins to rounded fragments (Chester and Logan 1987). Orientation of fibres in the veins (Sect. 6.2) or displacement of vein fragments can be used to determine shear sense. Additional shear sense indicators in some gouge zones are asymmetric boudins and folds, sigmoids (Takagi 1998) and mica-fish, with a geometry similar to those in mylonites.

Shear sense indicators in fault gouge as described above may form at very shallow depth, even within metres from the Earth's surface (Lin 2001).

5.7.3
Cohesive Brittle Fault Rocks

In most cohesive brittle fault rocks determination of sense of shear is difficult, except in foliated cataclasite. The most common shear sense indicators in foliated cohesive cataclasites are shear bands (Chester and Logan 1987; Petit 1987); these can be interpreted in a similar way as those in ductilely deformed rocks. Riedel shears may also be present (Fig. 5.50). Care has to be taken, however, with R-shears; in many cases, these are deflected towards Y-shears in a sense as shown in Fig. 5.50 (top; Logan et al. 1979; Strating and Vissers 1994). This deflection is probably due to a gradient in the orientation of the stress field near Y-shears during development of R-shears, which therefore obtain a curved shape (Strating and Vissers 1994). If many finely spaced R-shears are present in a non-foliated cataclasite, they might be confused with foliation planes; in that case, use of the deflection towards Y-shears as a shear sense indicator would give the wrong result (Fig. 5.50). However, R-shears can be recognised because their spacing is dependent on the spacing of Y-shears, which is not the case for foliation planes (Strating and Vissers 1994).

Foliated cataclasite can resemble low-temperature mylonite in many aspects. Although deformation in brittle fault rocks is by sliding on microfaults, fracturing and pressure solution, it creates a kind of coarse penetrative flow that can have similar effects on the development of asymmetric structures as ductile flow by dislocation creep and recrystallisation. With some care and cross-checking with other indicators, these asymmetric structures can be used as shear sense indicators in a way similar to those discussed above for ductile flow.

Porphyroclasts of minerals resistant to fracturing can form porphyroclasts with σ-type strain shadows of fragmented grains or newly grown phases (Bos et al. 2000). Foliation in the cataclasite can be deflected by C-type and C'-type shear bands on macro and micro scale (Lin 1999, 2001; Bos et al. 2000). These shear bands are filled with very fine-grained crushed material. Deformed inherited micas can be used as kinematic indicators in some foliated cataclasites. Inherited biotite grains in a cataclased granite can form cleavage steps or biotite fish similar to those in mylonites (Kanaori et al. 1991; Lin 1999, 2001). Such fish have trails of fine-grained crushed mica. Statistical analysis of the orientation of (001) in these micas against the main foliation trace allows determination of shear sense.

5.7.4
Pseudotachylyte

In pseudotachylyte, shear sense can rarely be determined in thin section. Displacement of markers is the best indicator, but this gives unreliable results (Fig. 5.13), especially since no slickensides are present as for gouge and cataclasite. In some pseudotachylyte veins, the orientation of injection veins or of fracture sets in the wall rock can be used to determine shear sense. Particularly useful are main fault veins that grade into faults with associated Riedel fractures (Swanson 1998).

Dilatation Sites –
Veins, Strain Shadows, Fringes and Boudins

Dilatation sites are voids in a rock filled with fluid, in which material may be deposited from solution. They include veins, strain shadows, fringes and microboudins. Voids can form in rocks at sites of high rheology contrast and/or where fluid pressure is high. The filling of such dilatation sites can produce interesting patterns that may contain a lot of information on the way the void opened, and indirectly on flow in the wall rock.

Veins can contain blocky, elongate or fibrous material and especially fibres carry much information on deformation in the rock. It is possible to obtain information on relative motion of vein walls, but even on the relative velocity of vein opening and filling using these structures.

Fringes are vein structures that occur on both sides of rigid objects such as pyrite crystals. Fringes belong to the geometrically most complex microstructures, and contain detailed information on the progress of deformation in a volume of rock. Unfortunately, they are difficult to interpret, but a lot of progress has been made on this subject in recent years. The final sections of this chapter deal with strain caps, strain shadows and microboudinage.

6.1
Introduction

Many deformed rocks contain sites with a deviant mineralogy and fabric, interpreted as an effect of rearrangement of material by local dilatation and precipitation during deformation. Such 'dilatation sites' can be isolated and elongate (*veins*), flanking rigid objects (*strain shadows*) or occur in the neck of boudinaged layers or elongate crystals (Fig. 6.1). Dilatation sites can be filled by solidified melt, or by precipitation of material from solution in an aqueous fluid, and it is the latter situation, which is discussed in this chapter. Dilatation sites are usually filled with polycrystalline material, which may be equigranular, but may also consist of parallel-oriented elongate or rod-shaped crystals (Figs. 6.1–6.3).

Veins and strain shadows are some of the most complex microstructures to be found in rocks, and contain much information about deformation and deformation history (Figs. 6.1–6.3; Box 6.1; Taber 1918; Mügge 1928; Pabst 1931; Zwart and Oele 1966; Choukroune 1971; Durney and Ramsay 1973; Beutner and Diegel 1985; Etchecopar and Malavieille 1987; Kanagawa 1996; Bons 2000; Köhn et al. 2000; Hilgers and Urai 2002). Strain shadows

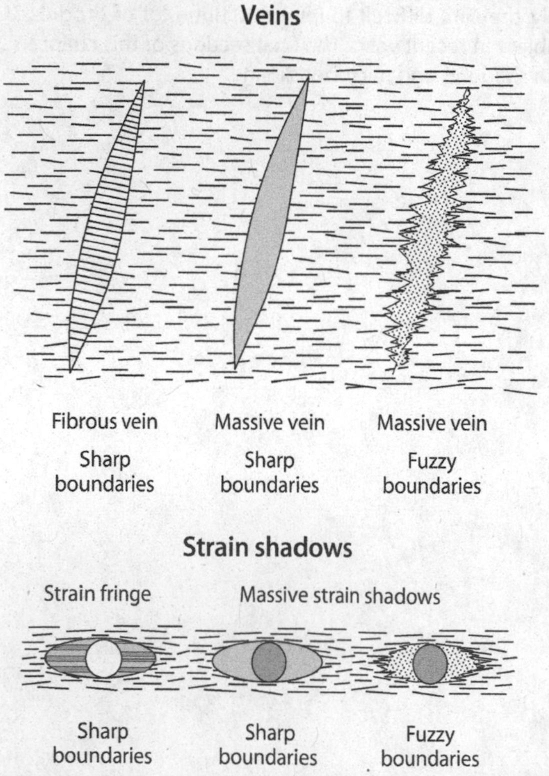

Veins

Fibrous vein — Sharp boundaries

Massive vein — Sharp boundaries

Massive vein — Fuzzy boundaries

Strain shadows

Strain fringe — Sharp boundaries

Massive strain shadows — Sharp boundaries — Fuzzy boundaries

Fig. 6.1. Schematic drawing of fibrous and massive veins, strain fringes and strain shadows as treated here. Massive veins and strain shadows are filled by a polycrystalline aggregate with a granoblastic fabric

are also known as *pressure shadows*. Since the shape of the aggregates gives primarily information on strain distribution around an object, and not on forces, we advocate the use of the term strain shadow. Some veins and strain shadows have fuzzy boundaries (Fig. 6.1). They may form by local alteration of the wall rock along a fracture (*replacement veins*) or rigid object, or by deformation and recrystallisation of veins with sharp boundaries (Sect. 6.5).

The development of veins and strain shadows is associated with the circulation of fluids in rocks, both for transport of material and for propagation and opening of the vein. Fluid pressure in rocks is usually between hydrostatic pressure (the pressure of a water column at a particular depth) and the smallest principal stress σ_3 in the rock (Sect. 2.11). Due to deformation, reduction of pore volume or metamorphic dewatering reactions, connectivity to the surface may be partly or completely lost, and pore fluid pressure P_f in rocks can increase and approach a critical value P_c:

$$P_f = P_c = \sigma_3 - T_s$$

where T_s is the tensile strength of the rock on pre-existing planes of weakness. If this value is reached, cracks filled with fluid can open at any depth and minerals can be precipitated in such cracks. Since T_s is usually small and σ_3 close to the overburden pressure, except in extensional tectonics, the critical value can also be expressed by the *pore-fluid factor* λ_v (Sibson 1990), a ratio of the pore fluid pressure to the vertical or lithostatic stress

$$\lambda_v = P_f / \sigma_v$$

If $\lambda_v \approx 0.4$, fluid pressure is hydrostatic, if $\lambda_v = 1$, it is near lithostatic. Fluid filled fractures form when $\lambda_v > 1$ in a compressive regime. If differential stress $(\sigma_1 - \sigma_3)$ is small in such cases, tensional fractures open but if differential stress exceeds a certain limit (a commonly used rule of thumb is $(\sigma_1 - \sigma_3) = 4\,T_s$; Jaeger 1963; Secor 1965), depending on rheological parameters of the rock, the opening fractures will have a shear component (Fig. 9.4).

The role of fluids in the formation and growth of veins and strain shadows is relatively complex. Fluid pressure may fluctuate in cycles of increasing pressure, leading to fracturing and vein opening, and subsequent falling pressure due to drainage during which mineral deposition commonly occurs (Vrolijk 1987; Cosgrove 1993; Ohlmacher and Aydin 1997; Bons 2000; Oliver and Bons 2001). Some veins, however, show evidence of persistent high fluid pressure during mineral growth (Henderson et al. 1990). In other vein segments, e.g. in strain shadows alongside a rigid object or in jogs along a shear

Fig. 6.2. Antitaxial vein of calcite fibres in a slate. A median line is visible. Appalachians, New York, USA. Width of view 10 mm. CPL. (Thin section courtesy Janos Urai)

Fig. 6.3. Fibrous strain fringe of quartz developed alongside a pyrite grain. Individual fibres are undeformed but show bends, which are interpreted to result from changes in the orientation of ISA with respect to the fabric in the wall rock. Lourdes, Pyrenees, France. Width of view 4 mm. CPL. (Photograph courtesy D. Aerden)

Box 6.1   Veins as a source of information on bulk deformation

The geometry of vein patterns is determined by the stress field in a deforming rock and fluid pressure, but its interpretation is beyond the scope of this book. On a microscopic scale, fibres or elongate crystals can be used to determine the relative motion of vein walls. As discussed above and shown in Fig. 6.15 crystals growing in a vein will only track the opening direction if the vein has suitably rough growth surfaces and will open in small increments. Real fibres of a normally non-fibrous mineral such as calcite or quartz are most reliable, since they can only form by tracking. Elongate crystals are more difficult to interpret, but if they are oblique to a grain wall, or have identical bends, they must be tracking to some extent; elongate crystals at right angle to a vein surface, however, are *not* indicative of straight opening; they can form by tracking or non-tracking growth. In veins with elongate or even blocky crystals, inclusion trails and ghost fibres are the most reliable tools to establish opening direction. If they are not present, inclusion bands may indicate movement direction by the lining-

up of jogs in the bands. If fibres, elongate crystals or inclusion patterns are not straight, it is important to establish whether veins are unitaxial, antitaxial or syntaxial before interpreting the motion history of vein walls. It is also important to establish if motion is in the plane of the thin section, or moves out of it; in the latter case, more sections or even a full 3D study are necessary. Once vein wall motion direction and history has been established, a simple interpretation in terms of regional tectonics is only possible for shear veins. Extension veins may themselves rotate with respect to bulk kinematic axes in non-coaxial flow, and even if the relative motion of the wall rocks can be established, this carries only indirect information on the far-field flow. In such cases, microstructural information must be combined with field evidence to reconstruct bulk deformation patterns. Fringe structures are even more difficult to interpret, since except for coaxial flow histories, fringes will always rotate with respect to the kinematic frame and the central object as outlined in Sect. 6.3.

fracture (Fig. 6.8), fluid pressure may actually fall during crack opening because of the increasing volume in the jog or strain shadow. Material deposited in a vein or strain shadow can be transported towards the dilatation site from outside in an open system along fracture networks, through the pore space, or along grain boundaries. This process is commonly referred to as *advection* and may cause changes in the chemical and isotope composition of a vein and its wall rock. Fluid may even move along veins in dislocation-fashion, without permanent opening of large segments (Cosgrove 1993; Oliver and Bons 2001) or be pumped in and out of a fracture network by occasional breaking and healing of mineral "seals", or by jog-opening (Vrolijk 1987; Cosgrove 1993; Ohlmacher and Aydin 1997; Oliver and Bons 2001). Material deposited in veins and strain shadows can also be derived from the surrounding wall rock in a closed system, e.g. by dissolution and precipitation of quartz or calcite. In this case, material can be transported in a circulating fluid, or in a stationary fluid by *diffusion* (Oliver and Bons 2001). It has been suggested that in some veins, pressure of the growing crystals on the wall rock in the case of super saturation may also open a crack ("force of crystallisation": Means and Li 2001; Wiltschko and Morse 2001).

Veins usually form in consolidated rocks, but there are indications that they may also form in unconsolidated sediment (Fisher and Bryne 1990; Orange et al. 1993; Vannucchi 2001). Criteria proposed to recognise such a setting includes deformation of the wall rock during vein growth, deformation of the wall rock by the tips of euhedral crystals in the vein and clouds of randomly oriented inclusions.

In most cases, crack opening and filling is no unique event, but occurs repeatedly or continuously during a deformation phase. In such cases, the resulting veins or

strain shadows contain much and detailed information on the deformation path, and are therefore some of the most useful structures to reconstruct tectonic events in thin section (e.g. Durney and Ramsay 1973; Ramsay and Huber 1983; Etchecopar and Malavieille 1987; Hilgers and Urai 2002).

In microtectonic analysis, veins and strain shadows have a number of important applications:

1. Veins indicate the transport of material by fluids over some distance. Pressure solution may have been important, and the possibility of volume change must be considered.
2. Veins can be used to unravel polyphase deformation by crosscutting relations, (Wallis 1992a); they can also give information on the earthquake cycle (Davison 1995) and hydrocarbon migration (Parnell et al. 2000).
3. The shape of veins and strain shadows and of the crystals in these structures can be used to determine shear sense and, in some cases, other deformation parameters.
4. The composition of veins, strain shadows and fluid inclusions in them allow assessment of the chemical composition of fluids that accompany deformation (Sect. 10.5).
5. Some veins and strain shadows can be dated (Sect. 9.8).

Although field observations and thin sections are the main source of information on the development of veins and strain shadows, it is now also possible to use two types of experimental modelling for the study of these structures; analogue experiments (Means and Li 2001; Hilgers et al. 1997; Bons and Jessell 1997; Köhn et al. 2003; Chap. 11) and numerical modelling (Bons 2001; Köhn et al. 2000, 2001a,b; Hilgers et al. 2001; ⊘Videos 6.23, 6.27).

6.2
Veins

6.2.1
Crystals in Veins

Many dilatation sites such as veins or strain shadows contain parallel oriented elongate crystals (Williams 1972b; Durney and Ramsay 1973; Ramsay 1980b; Machel 1985; Passchier 1982a; Bons 2000). This concerns naturally elongate minerals such as asbestos, actinolite or mica but more commonly minerals which do not normally have an elongate shape such as quartz, calcite, dolomite, feldspar, gypsum and anhydrite (Williams 1972b; Durney and Ramsay 1973; Ramsay 1980b; Machel 1985; Passchier 1982a). The elongate habit of minerals in such veins appears to be due to a special growth mechanism (Figs. 6.4, 6.5). Imagine a crack that forms in a crystalline aggregate of quartz or calcite. Immediately on opening, new crystalline material can be deposited on the existing crystals from solution in the fluid. If neighbouring crystals grow at similar rates, the overgrowth obtains an elongate shape; crystals that grow slowly because they have an unsuitable crystallographic orientation will become thin and end while their neighbours make contact[1]. A process of *growth competition* can lead to aggregates of few equidimensional grains, but if the growth competition is suppressed, many elongate grains will result. A gradient in shape can be envisaged between *fibres*, *elongate grains* and equidimensional or *blocky grains* (Fisher and Brantley 1992; Bons and Jessell 1997; Hilgers and Urai 2002; Fig. 6.4). Fibres are extremely elongate rod- or hose-shaped grains with parallel boundaries, a width of 10–350 μm and a length width ratio up to 100. Elongate grains have a smaller length-width ratio and are tapering (Fig. 6.4).

6.2.2
The internal Structure of Veins

In the case of an isolated crack, fibrous growth can occur into the widening void from both sides at the same rate. This process can be referred to as *bitaxial* growth (Bons 2000; Hilgers et al. 2001). The result is a symmetric vein with a central plane where deposition of new material takes place. In thin section this plane appears as a clear line known as the *median line* (Fig. 6.2). The median line is often marked by small opaque grains and a discontinuity in the elongate crystal fabric. Growth of elongate grains or fibres as described here from the vein wall towards the vein centre is known as *syntaxial growth*,

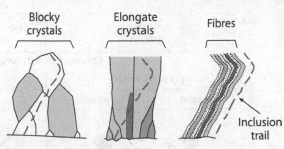

Fig. 6.4. Three types of common crystal shapes in veins. Only fibres follow the opening trajectory of a vein that is indicated in the crystals as a dotted trail of solid inclusions

and the veins as syntaxial fibre veins (Ramsay and Huber 1983) or *syntaxial veins* (Fig. 6.5a, ⊘Video 6.5ab). Syntaxial veins are commonly asymmetric, i.e. with an off-centred median line, and in extreme cases grow from one side of the vein only. Such *unitaxial veins* (Fisher and Bryne 1990; Köhn and Passchier 2000; Bons 2000; Hilgers et al. 2001; Oliver and Bons 2001) lack a median line (Fig. 6.5a).

Some veins are filled by growth of a mineral that is not the main constituent of the wall rock, e.g. a calcite vein in quartzite. In such cases, the growth usually occurs along the contact of elongate grains or fibres and the wall rock, i.e. on both sides of the material in the vein (Fig. 6.5b). A weak median line defined by small equidimensional grains or fragments of the wall rock is normally present in the centre of the fibrous aggregate, indicating the initial nucleation site of the vein filling. This type of growth towards the wall rock is termed *antitaxial growth*, and the veins are called antitaxial fibre veins (Ramsay and Huber 1983) or *antitaxial veins* (Figs. 6.2, 6.5b, ⊘Video 6.5ab). Material in antitaxial veins commonly consists of fibres rather than elongate crystals, and veins are commonly symmetric. Single fibres in antitaxial veins can be continuous over the median line, in contrast to elongate grains or fibres in syntaxial veins. *Composite veins*, in which an antitaxial vein segment is sandwiched between syntaxial vein rims are also possible (Fig. 6.5c, ⊘Video 6.5cd). Such a vein has three 'median' lines. In fact, most fibrous antitaxial calcite veins investigated to date have thin quartz-chlorite selvages along the vein wall contact and could therefore be classified as composite veins (Hilgers and Urai 2002).

Veins can be filled during a single event of gradual growth of crystals into a fluid-filled cavity, (either in a vein opened by a single event, or one that is opening faster than the crystal growth rate) or by periodic opening and filling of a narrow cavity, here referred to as *periodic growth*. If periodic growth happens by periodic sealing and fracturing, the process is known as *crack-seal growth* (Ramsay 1980b). If growth, either periodic or continuous, always occurs at the same site in the vein, e.g. the

---

[1] We have simplified most drawings in this section by omitting this tapering effect and imagine all fibres to grow at the same rate.

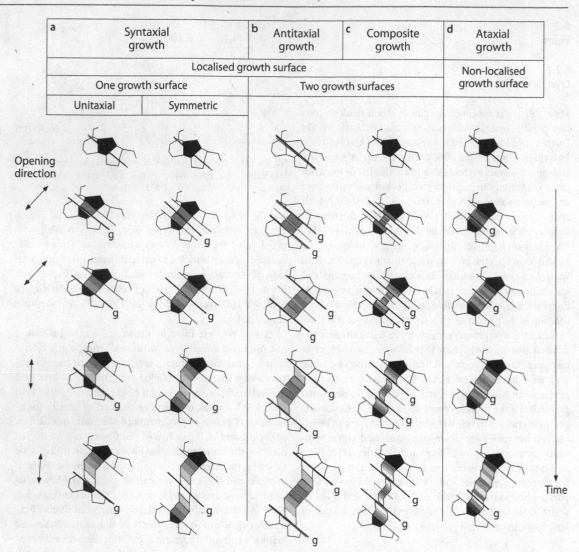

Fig. 6.5. Five types of fibre growth in veins common in nature. Youngest parts of the fibres are shown in the lightest colour. Growth surface is indicated by a *bold dark line* (*g*). **a** Syntaxial growth. Fibres are in continuity with the wall rock crystals; the oldest part of the fibres lies along the edge of the vein. A median line marks the final position of the growth surface if growth is from both vein walls. If growth is from one vein wall only, growth is unitaxial and a median line is lacking. **b** Antitaxial growth; fibres grow from the centre outwards. The vein usually consists of other material then the wall rock. The oldest part of the vein is along the median line. Two growth surfaces are localised along the contact with the wall rock. **c** Composite growth, with a syntaxial and antitaxial component. Two growth surfaces are present at the contact between the vein segments where the youngest parts of the vein are situated. **d** Ataxial growth. Fibres are in continuity with crystals in the wall rock but lack a localised growth surface; growth is by repetitive fracturing at different sites. Young and old parts of the fibres can be mixed throughout the vein. No median line is present

vein boundary, this leads to development of synthetic, antithetic or composite veins with median lines as discussed above (Fig. 6.5a–c, ⊘Videos 6.5ab, 6.5cd). Alternatively, veins may form by repeated fracturing and growth at alternating different sites in the vein (Fig. 6.5d, ⊘Video 6.5cd). Such non-localised, 'ataxial' cracking and growth produces veins with jogged or stepped and rarely smooth elongate grains without a median line that are in continuity with fragments of single crystals on both sides of the vein. Such elongate grains are known as

stretched crystals (Figs. 6.5–6.7; Durney and Ramsay 1973; Bons 2000, 2001) and are relatively common in nature (Hilgers and Urai 2002). We use the terms *ataxial veins* for veins with such crystals (Figs. 6.5d, 6.6, 6.7, ⊘Video 6.5cd). Note, that lack of a median line is a characteristic of both ataxial and unitaxial veins.

Some veins consist of pockets of elongate crystals, or columns of crystalline material with inclusion bands, and sections where the wall rock material collapsed into the veins; such structures may have formed where water

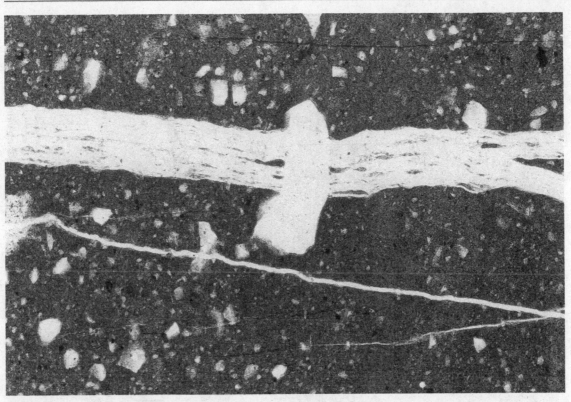

Fig. 6.6. Ataxial vein of quartz in fine-grained tourmalinite. A quartz grain in the tourmalinite (*centre*) has been cut by the vein and has been modified into a stretched crystal. Several planes of solid inclusions (tourmalinite fragments) occur in the vein and show that cracking was partly along the vein margins. Orobic Alps, Italy. Width of vein 4 mm. PPL

pockets were retained in the veins by high fluid pressure while the vein partially filled with crystalline material, followed by vein collapse when the water pressure decreased (Fisher and Bryne 1990; Henderson et al. 1990).

Many veins, as described above, lie at a high angle to their opening direction and are known as *extension veins* or *tension gashes* (Ramsay and Huber 1983; Figs. 6.8, 9.4; ⊘Video 6.8); however, veins can also form at a small angle to the opening direction, e.g. along bedding planes in the case of flexural slip. In that case they are referred to as *shear veins* (Ramsay and Huber 1983) (Figs. 5.49, 6.8, 6.9, 9.4; ⊘Video 6.8). Extension and shear veins are end members of a possible range of geometries.

Fluid and solid *inclusions* are commonly present in fibres at regular intervals. Solid inclusions are usually fragments of the wall rock, or small equidimensional mineral grains of a size considerably smaller than the main phase in the vein. Fluid and solid inclusions are generally concentrated at specific sites in a vein. They commonly occur in *inclusion bands*, irregularly shaped surfaces 2–30 μm apart that reflect the shape and orientation of the vein-wall rock contact (Figs. 6.9, 6.10; Ramsay 1980b; Cox and Etheridge 1983; Cox 1987; de Roo and Weber 1992; Fisher et al. 1995; Köhn and Passchier 2000). Compositional banding in veins which can be visible in

thin section or by cathodoluminescence (Sect. 10.2.1) is commonly parallel to inclusion bands. Inclusion bands and compositional banding may represent sudden changes in the opening rate or direction of a continuously or intermittently growing vein or in the composition of the fluid involved in vein growth (Wiltschko and Morse 2001; Means and Li 2001; Hilgers and Urai 2002). Alternatively, they may result from a crack-seal process in or at the edge of the vein (Ramsay 1980b; Figs. 6.10, 6.11; ⊘Video 6.11); the crack may temporarily seal and then reopen at approximately the same site, including fragments of the wall rock. In some veins, inclusion bands are composed of small mica grains that lie parallel to a foliation in the wall rock (Cox and Etheridge 1983; van der Pluijm 1984; Fisher and Bryne 1990); they are interpreted as overgrowths of micas on the vein wall rock, subsequently separated from the wall by cracking. Each mica-rich plane can be interpreted as representing a distinct growth and cracking event in a developing antitaxial vein (Cox and Etheridge 1983). From the spacing of such planes, individual crack openings in a vein have been estimated at 4 to 200 μm (Ramsay 1980b; van der Pluijm 1984; Cox 1987; Hilgers and Urai 2002). In some veins, inclusion bands are interrupted and occur only in some elongate crystals (Fig. 6.11, ⊘Video 6.11; Fisher and

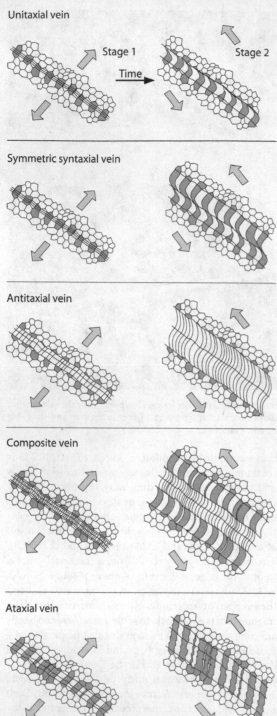

Fig. 6.7. Development of the five types of veins with tracking fibres described in the text. A change in the relative motion of the wall rocks (*arrows*) can cause curvature of the growing fibres in the veins. Notice that the sense of curvature of the fibres depends on the type of vein, and that ataxial veins develop straight fibres which indicate a mean displacement direction of the vein wall rocks

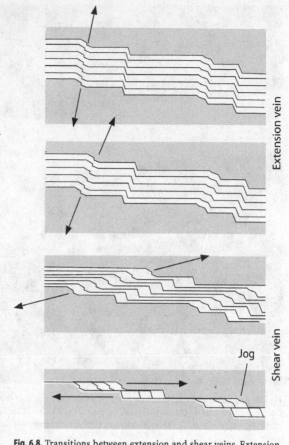

Fig. 6.8. Transitions between extension and shear veins. Extension veins open normal to the vein wall, shear veins at a low angle or parallel to the vein wall with opening of jogs

Bryne 1990). This could be due to partial filling of a crack related to different growth rate of crystals with different orientation; slowly growing crystals may not be able to fill a crack completely until the next opening cycle. Such interrupted inclusion bands, and the presence of wall rock fragments along inclusion bands are evidence for activity of the crack-seal mechanism (Fig. 6.12).

Aggregates of solid or fluid inclusions may also occur in surfaces oblique to the vein-wall contact, usually connecting jogs in the inclusion bands, and corresponding points in the vein-wall contact. These inclusions define *inclusion trails* (Fig. 6.10; Ramsay and Huber 1983; de Roo and Weber 1992; Köhn and Passchier 2000). While inclusion bands are parallel to the vein wall, inclusion trails are thought to track the opening trajectory of a vein (Ramsay and Huber 1983; Köhn and Passchier 2000). Isolated elongate fluid inclusions or fibres such as mica grains may also lie oblique to the edge of a vein and to inclusion bands or trails (Fig. 6.10). None of the inclusions is necessarily parallel to fibres or elongate grains that build up the bulk of the vein (Figs. 6.4, 6.9), since these

Fig. 6.9. Detail from an antitaxial shear vein of quartz and calcite (only quartz visible). The contact with the wall rock is parallel to the long axis of the photograph. Bands of solid inclusions trend from *top left* to *bottom right*; each band probably represents a separate crack-seal event. Jogs in the bands are thought to indicate the opening direction of the vein (*horizontal*). Elongate quartz crystals are oblique to both the solid inclusion bands and the opening direction of the vein, and trend from *bottom left* to *top right*. Orobic Alps, Italy. Width of view 5 mm. CPL.

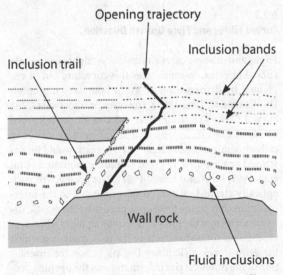

Fig. 6.10. Various types of solid and fluid inclusions in a fibrous vein. None of the inclusions is necessarily parallel to crystal boundaries in the vein. Surfaces of fluid inclusions are usually parallel to the vein wall, but solid inclusions may be both parallel to the opening direction of the vein (inclusion trails) or in planes parallel to the vein wall (inclusion bands)

may or may not track the opening direction of a vein as discussed in Sect. 6.2.1.

The nature of the interaction of fluids and growing fibres or elongate crystals in veins is still unclear (Bons 2000; Oliver and Bons 2001; Means and Li 2001; Hilgers and Urai 2002). In some cases, the crystals may grow from a fluid migrating through a central channel (syntaxial veins) or along the vein-wall contacts (antitaxial and unitaxial veins), but lack of gradients in composition of vein crystals along the length of the vein makes this unlikely in many cases, especially for fibres in antitaxial veins. Antitaxial vein fibres are usually strictly symmetric around the median line, something not to be expected in the case of repeated cracking and sealing of vein contacts (P. Bons, pers. comm.). Such veins also normally lack inclusion bands. Growth from a fluid introduced through the porous wall rock may therefore be responsible, although the exact mechanism of precipitation is unclear (Means and Li 2001). Such veins could develop without a through-going open crack along the vein-wall, the walls being pushed apart by the growing fibres. This process is known as *Taber growth* and the resulting veins as *Taber veins* (Taber 1916, 1918; Means and Li 2001).

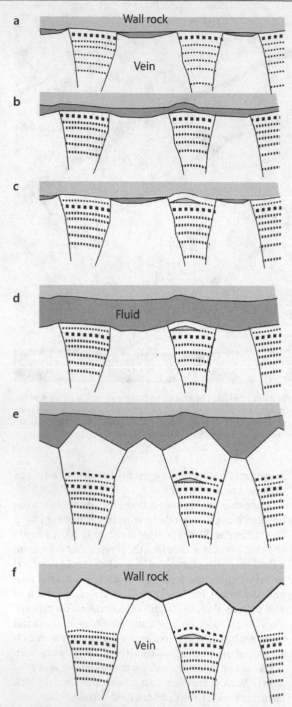

Fig. 6.11. Aspects of microstructural development in extension veins. **a–c** Development of discontinuous inclusion bands. If only the fastest growing crystals reach the boundary after crack opening, they will be the only ones that contain inclusion bands. **d, e** If opening rate is increased, no crystals can reach the boundary any more, euhedral crystal faces are formed, and no inclusion bands are trapped in the grains. **f** If the fluid pressure decreases sufficiently, the vein may collapse and the euhedral crystal faces can indent the wall rock

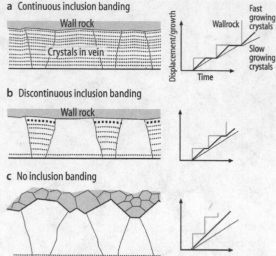

Fig. 6.12. Effect of relative growth rate of crystals and vein opening rate on the geometry of crystals in veins and on presence of inclusion banding. Crystals grow anisotropically in a unitaxial vein, from bottom to top. Their growth rate in the periodically opening crack at the top depends on their orientation. In this section, slow and fast growing crystals alternate. **a** Opening rate is slow, and both fast and slow crystals can fill an opened crack before the next opening increment. Each renewed cracking and growing episode produces a continuous inclusion band in all crystals. **b** Fast growing crystals can reach the contact after each cracking episode, but slow growing crystals cannot; the result is the formation of discontinuous inclusion bands. **c** Opening rate is so fast, that no crystals can reach the upper boundary any more and blocky, euhedral crystals form. Inclusion bands are completely lacking. (After Fisher and Brantley 1992)

6.2.3
Curved Fibres and Fibre Growth Direction

Fibres and in some cases elongate crystals are commonly curved in veins. In some cases this curvature can be explained by deformation or grain boundary migration (Williams and Urai 1989). However, in most cases, the curved fibres are strain-free and contain undeformed arrays of fluid inclusions or delicate banding visible in cathodoluminescence (Sect. 10.2.1), which would be destroyed or modified by grain boundary migration (Urai et al. 1991; Bons and Jessell 1997). The presence of such optically strain-free curved fibres has led to the interpretation that fibres grow in a particular kinematic direction, and that the growth direction changes when kinematic axes rotate relatively to the rock volume in which the veins develop (Ramsay 1980b); hence, the orientation of the fibres carries information on the opening vector of such dilating veins. For simple vein geometries, this direction was even suggested to be parallel to the extensional ISA of the flow (Figs. 6.5, 6.7; Durney and Ramsay 1973; Wickham 1973; Wilcox et al. 1973; Philip and Etchecopar 1978; Casey et al. 1983; Ramsay and Huber 1983; Ellis 1986). Such fibres or elongate crystals are said to be *displacement controlled*, or to be *tracking* the opening direction of the vein or the ISA. Figure 6.7 shows the geometry of fibres and elongate crystals as they would appear in veins by the four common types of displacement-controlled crystal growth in response to relative changes in movement direction of the wall rock (Sect. 6.2.1). Notice that syntaxial and antitaxial veins have mirror symmetry deflections due to their opposed growth directions. Composite veins show both senses. Non-localised fracturing in ataxial veins will usually give a 'mean' orientation without clear curvature. Unitaxial veins would only show a single curvature pattern.

In non-coaxial flow, veins and new-grown fibres or elongate grains will rotate as material lines with respect to ISA, and fibres that are growing in the (fixed) direction of the extensional ISA will therefore become curved. The curvature of these grains corresponds to certain clearly defined geometries treated in the next section.

The orientation of fibres or elongate grains is not in all cases associated with kinematic directions (Fig. 6.9). Comparison of quartz and calcite fibre and elongate grain orientation with that of trails of phyllosilicates in veins, or with off-set markers in the vein wall has shown that not all track the opening vector of the vein (Cox and Ethcridge 1983; Cox 1987; van der Pluijm 1984; Williams and Urai 1989; Bons 2001). In many cases, grains simply grow normal to the wall rock of the vein in which they nucleate, and fill the available void without change of growth direction. Williams and Urai (1989) have also shown that some curved, optically strain-free fibrous crystals are in fact recrystallised deformed fibres, which initially were straight and orthogonal to the vein wall.

Urai et al. (1991) presented a model for the bivalent growth behaviour of crystals in veins, largely confirmed by numerical experiments and observations (Fig. 6.13, ⊘Video 6.13; Bons 2001; Hilgers et al. 2001; Hilgers and Urai 2002; Nollet et al. 2005). According to the model, competition between grains that grow into an open space will normally lead to aggregates of equidimensional or slightly elongate crystals, but if crystals are forced to grow into a narrow crack and adapt to its shape, highly elongate crystals or fibres may develop. This will occur whenever growth rate of part of the crystals exceeds the mean local opening rate of the crack. If the growth surface of a vein is irregular in shape, e.g. because the original crack was not perfectly planar, grain boundaries tend to migrate rapidly towards asperities in the contact which point in the direction of the growing grains. The contacts will become fixed there during further growth of the grains into fibres (Figs. 6.13, 6.14, ⊘Video 6.13). In the case of Taber-growth from a fluid in the wall rock (Means and Li 2001), or if opening and growth occurs by small steps where a narrow fluid-filled void is filled between opening steps, the fibres will follow changes in displacement direction of the vein wall. This is the origin of displacement-controlled fibres (Figs. 6.13, 6.14, ⊘Video 6.13). However, if the vein opens more rapidly, or growth is slower, or if the irregularities on the growth

Fig. 6.13.
a If a crack with irregular shape opens and crystals grow isotropically and at equal rate to fill the crack, the grain boundary between them will be displaced normal to the crystal face till an asperity in the growth surface is reached that points towards the wall-rock and the growing grains; subsequently, this asperity will be followed by the grain boundary until the crack is filled. **b** This mechanism allows elongate grains and fibres to track the opening direction of the vein if growth and opening occur intermittently by small steps

a   Crystal growth in a fluid-filled crack

b   Elongate crystals tracking embayments

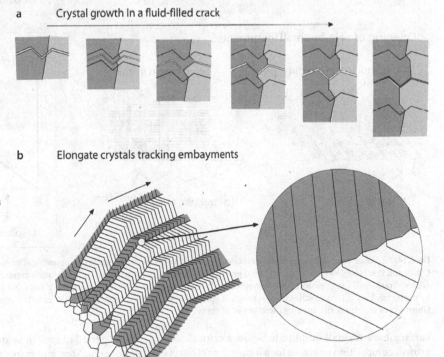

Fast opening of vein

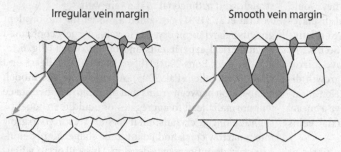

Irregular vein margin

Smooth vein margin

Blocky grains with euhedral crystal facets
non-tracking

Opening of vein by large increments

Irregular vein margin

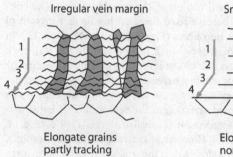

Smooth vein margin

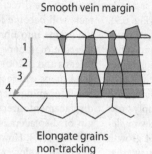

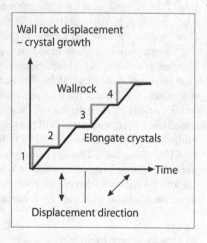

Elongate grains
partly tracking

Elongate grains
non-tracking

Opening of vein by small increments

Irregular vein margin

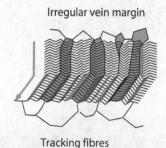

Smooth vein margin

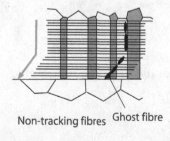

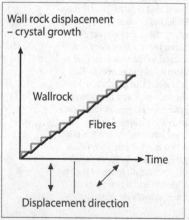

Tracking fibres

Non-tracking fibres   Ghost fibre

Fig. 6.14. Effects of vein-wall smoothness and opening rate of veins on crystal geometry in unitaxial veins, following Urai et al. (1991) and Oliver and Bons (2001) for anisotropic crystal growth. True fibres only form if opening is by small increments (*bottom left*); in this case, they are also tracking if the growth surface is rough. In case of larger opening rate, elongate crystals form due to growth competition, which can be partly tracking if the growth surface is rough. If opening rate exceeds growth rate, blocky euhedral crystals form that are non-tracking. Ghost fibres may track the opening direction in all cases

surface have a small amplitude below a critical value, growth competition proceeds for a longer time (Urai et al. 1991; Hilgers et al. 2001). In this case, crystals have a longer period of growth before reaching the wall rock in one opening step, and grains will expand by growth competition to few elongate grains (Bons 2001; Hilgers et al.

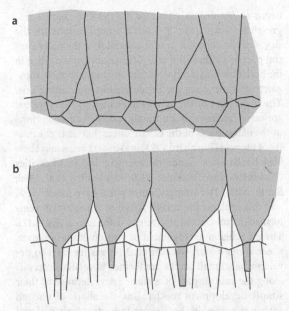

Fig. 6.15. Composition of the wall rock can have a strong influence on the size of elongate crystals in syntaxial veins. Elongate grains in a vein adjacent to a monocrystalline coarse grained wall rock **a** can be narrower than those in a fine-grained polymineralic wall rock **b**

2001; Hilgers and Urai 2002; Nollet et al. 2005; Fig. 6.15). No fibres form in this case. Since the number of grains decreases with growth, the mean grain width increases in the growth direction, and this can in many cases be used to reconstruct the growth direction in aggregates of elongate crystals. Elongate grains may become fixed to asperities in the contact, but may not track the opening direction as well as fibres (Fig. 6.14). If the growth surface is smooth and planar, e.g. because the original crack was straight, elongate crystals or fibres will grow at right angles to the growth surface irrespective of opening direction of the vein, and are said to be *face-controlled*. This may be the case when a crack opens along a planar crystal face such as that of a pyrite cube (Figs. 6.15, 6.21, ⊘Photo 6.21). Domains of face-controlled fibres can contain *ghost fibres* (Ramsay and Huber 1983), usually solid inclusions or fibres of a deviant mineral that cross the face-controlled fibres (Fig. 6.15). Ghost fibres are thought to track the opening direction of the vein. Veins can therefore be subdivided into end members with displacement- or face-controlled crystals, and mixed types that contain both displacement and face controlled crystals, or crystals that switch behaviour along their length.

If crystal growth of all grains is slower than the opening rate of the vein, crystals grow freely in a fluid filled space; in most cases, growth competition leads to formation of blocky, euhedral crystals with low-index crystal faces (Figs. 6.11, 6.12, 6.14).

Besides opening rate versus growth rate, and roughness of the growth surface, some other factors have been suggested to influence the shape of fibrous and elongate veins. Crystal growth can be *isotropic*, i.e. independent of crystallographic direction, producing approximately equidimensional grains in the case of free growth; or *anisotropic*, with more rapid growth in some crystallographic directions than in others; this would produce euhedral crystal faces and a crystallographic preferred orientation in the grains that survive growth competition (Figs. 6.11, 6.12; Bons 2001; Nollet et al. 2005). Although most crystal growth is thought to be anisotropic, the original model (Urai et al. (1991) was based on isotropic growth. Numerical experiments have shown that growth anisotropy can influence the shape of individual elongate grains in the case of growth competition where vein filling does not keep up with opening. However, it has little effect on fibre shape and development where the veins are filled completely upon opening (Bons 2001; Nollet et al. 2005). In the case of anisotropic crystal growth, it is also possible that fast growing crystals are tracking, while slowly growing ones are not.

Initial grain size of the wall rock can be of influence on the width of elongate grains or fibres in syntaxial veins. The grain size of monomineralic wall rocks will influence that in the veins if it exceeds the amplitude of antiformal asperities in the contact. In polymineralic rocks, however, the effect can be opposite; a schist with few small quartz grains has few nuclei for quartz crystals in a vein, so it can produce much coarser veins than a wall rock with more or bigger quartz crystals (Fig. 6.15; Fisher and Brantley 1992).

Another factor to consider is that the shape of the growth surface may change in the course of vein growth; fibrous "antithetic" calcite veins can nearly always be shown to be actually *composite veins* (Figs. 6.5, 6.6), with thin slowly growing quartz-chlorite selvages along the vein edge. The shape of the growth surface is therefore constantly changing in this case. Besides growth, fracturing of wall rock and fibres along the growth surface can also lead to considerable changes (Hilgers et al. 2001).

Summarising, in the model of Urai et al. (1991), displacement-controlled fibres or elongate crystals which track the *opening trajectory* of a vein can only form under special circumstances of an irregular growth surface, and small opening rate with respect to growth rate of crystals in the vein. The crystals are forced to grow into a shape that is unlike their low-index crystallographic directions. If grains cannot keep track of a moving boundary, non-tracking elongate grains will form with crystal facets in the case of anisotropic growth. The threshold between such "tracking" and "non-tracking" can be sharp (Hilgers et al. 2001) and may be a means to determine strain rate in rocks (Sect. 9.8; Fig. 6.14). In many geometrical situations, the 'free' surfaces on both sides of an opening crack will move apart in the direction of ISA, and in that case the vein will track the ISA direction. It will be clear, however, that this will not apply in the case of complex vein shapes. This means that curved fibres may track the opening direction of the vein (Fig. 6.7), but not necessarily the instantaneous extension direction in the bulk rock (Sect. 6.3). It is therefore necessary to be careful when using fibrous aggregates for kinematic analysis; the nature of the fibre growth process should first be established.

We limit the treatment of veins and strain shadows in this chapter mainly to pure shear and simple shear progressive deformation histories. Obviously, other flow regimes and more complex histories involving volume change are also possible. However, the end-member situations will in most cases allow assessment of at least shear sense, and permit establishment of the type of deviation from the ideal model.

6.2.4
Veins in Non-Coaxial Progressive Deformation

In brittle fault zones and some ductile shear zones, *tension gashes* develop in sets of veins that are arranged en-echelon ("Echelon tension gashes": Fig. 6.16a, ⊘Video 6.16; Olson and Pollard 1991; Peacock and Sanderson 1995). If such echelon tension gashes lie close together, the strips of wall rock separating them are known as *bridges* (Fig. 6.16b). Large tension gashes in shear zones usually have a characteristic, curved geometry (Fig. 6.16a). In both coaxial and non-coaxial flow, tension gashes open in a direction approximately parallel to the direction of maximum instantaneous extension. If the veins form during progressive deformation, they may both become wider and propagate laterally outward parallel to the instanta-

neous shortening axis (one of the ISA). In coaxial progressive deformation, this has no implications for the shape of the vein, but in non-coaxial flow the older central part of a vein will rotate just like any material line in the bulk flow (Fig. 5.39) while the tension gash still propagates outward along the instantaneous shortening axis. The older part of the vein is deformed in this process; antitaxial calcite veins commonly show deformation twins in the oldest parts of the calcite fibres that lie in the centre of the vein (Burkhard 1993) and quartz veins may show kink bands, shear bands, microcracks and small cavities (Nishikawa and Takeshita 2000; van Daalen et al. 1999; Smith 2005). The younger, outer parts have rotated less, and this explains the curved shape of tension gash veins formed in non-coaxial progressive deformation (Fig. 6.17). The 'en-echelon' arrangement of tension gashes in many shear zones occurs because tension gashes start to develop where small cracks are formed at regular intervals along the central part of a shear zone. Because of their simple development mechanism, the shape of tension gashes is a reliable shear-sense indicator; S-shaped and Z-shaped gashes indicate sinistral and dextral shear sense respectively (Fig. 6.16a, ⊘Video 6.16; Choukroune and Seguret 1968; Ramsay and Graham 1970; Beach 1975; Gamond 1983). In some cases, two generations of tension gashes are found, the younger one transecting the centre of the older one parallel to its tips (Fig. 6.16a). Such 'secondary' tension gashes are thought to form in response to rotation of the central segment of the older gash into the extensional field of the flow (Sect. 5.6.16; Fig. 5.39). Similarly, stylolites (Box 4.3) are sometimes found at a high angle to developing tension gashes. Notice that a foliation in a shear zone will form oblique to the gashes, and

Fig. 6.16.
a Schematic diagram of tension gashes in en-echelon arrangement in a shear zone. A second set of tension gashes develops in the centre of the older set. The sense of curvature for tension gashes and foliation in a shear zone is similar. Stylolites may develop cutting tension gashes or the foliation. **b** Tension gashes commonly pass into continuous veins laterally through feather veins. Feather veins probably form when a continuous vein grows over a series of en echelon tension gashes. If tension gashes are close together, the interfering strips of rock are known as bridges

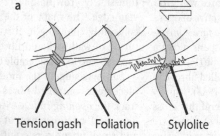

a

Tension gash   Foliation   Stylolite

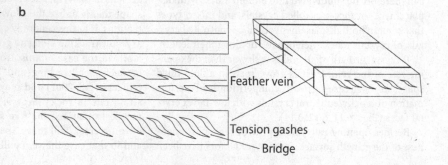

b

Feather vein

Tension gashes

Bridge

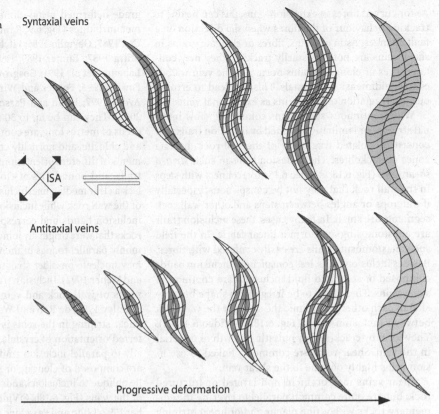

Fig. 6.17.
Schematic diagram of the growth of tension gashes (from *left* to *right*) in a dextral shear zone in the case of syntaxial and antitaxial fibre growth. Ornamentation indicates growth zones of the fibres in the gashes for each stage of deformation. At *extreme right*, growth zones have been omitted to show the geometry of fibres in the tension gashes

Syntaxial veins

ISA

Antitaxial veins

Progressive deformation

that foliation curvature due to a strain gradient has a similar geometry as the tension gashes, but with the low-strain (outer) parts of the foliation in the direction of the instantaneous extension axis (Fig. 6.16a).

In three dimensions, many tension gashes as described above consist of isolated lenses. In some cases however, echelon tension gashes are laterally grading into planar veins (Fig. 6.16b; Nicholson and Pollard 1985; Nicholson and Ejiofor 1987; Nicholson 1991; Tanner 1992b). In many cases it can be shown that echelon tension gashes lie at the tip of planar veins, especially where they cut a lithological contact. *Feather veins* may occur between the planar and en-echelon sections, and are interpreted to form by propagation of a straight vein trough tension gashes (Fig. 6.16b).

If flow in a shear zone is by simple shear, the instantaneous shortening direction will lie at 45° to the shear zone boundary (Fig. 2.6) and the tips of the tension gashes will lie in this direction (Fig. 6.16a, ⊘Video 6.16). If flow is between pure and simple shear ($0 < W_k < 1$; Sect. 2.5), the angle between the tips and the shear zone boundary will either be smaller or larger than 45°. Such structures are known as shortening- and stretching shear zones respectively (Sect. 5.6.3).

If fibres or elongate grains are present in curved tension gash veins, they may also have a complex shape that can be used to determine shear sense (Fig. 6.17). The pattern of curvature of the grains depends on the nature

of the growth process. Antitaxial grains have the same sense of curvature as the external shape of the tension gash; syntaxial grains have the opposite sense of curvature (Fig. 6.17). More complex internal structures are also possible (Smith 2005).

6.2.5
Shear Veins, Slickenfibres and Bedding Veins

If veins form by opening at a small angle to the vein wall, e.g. along a fault, their internal structure deviates in some aspects from that of tension gashes which open highly oblique to the vein wall. Such shear veins are easily recognisable if markers are present but in many cases displacement on them is significant, and markers are missing. In such cases, they can be recognised by the lensoid shape of the vein segments, commonly separated by faults filled with microbreccia or cataclasite. Many shear veins can be recognised because they contain fibres or elongate crystals that have grown at a small angle to the vein wall: The fibres may consist of carbonate (Labaume et al. 1991; Cosgrove 1993) quartz, mica or even minerals such as sillimanite (Argles and Platt 1999) or tourmaline. These structures are also known as *slickenfibres* in *slickenfibre veins*. (Figs. 6.8, 6.9; Sect. 5.2.2).

Slickenfibre veins commonly contain curved fibres parallel to the plane of the slickenside (Ramsay and Huber 1983).

As for curved fibres in extension veins, this can be due to tracking behaviour of the fibres when slip direction on a fault changes. As in all veins, fibres or elongate grains in shear veins are not necessarily tracking; they may contain fibres or elongate grains normal to the vein wall or even equidimensional crystals. This may lead to erroneous interpretation of such veins as extensional veins.

Wide continuous shear veins commonly show internal striping or lamination formed by inclusion trails that consist of isolated fragments of the wall rock, breccia zones or slickolites. The inclusion trails in such *striped shear veins* (Fig. 6.18, ⊘Video 6.18a,b,c) connect with steps in the wall rock that may not be conspicuous, especially if outcrops or angles between steps and other wall rock segments are small. In many cases, these inclusion trails are anastomosing and form a linear fabric. In the field such anastomosing trails are easily confused with fibres. Besides inclusion trails, less conspicuous inclusion bands composed of solid and fluid inclusions are common in shear veins. These tend to be irregular in shape but parallel to each other and mimic the shape of the wall rock between inclusion trails (Fig. 6.18b, ⊘Video 6.18b,c). They form by crack-seal or pulsating growth of material in the vein. Shear veins are commonly linked to extension veins highly oblique to the shear vein.

Shear veins may occur in non-layered or fractured rock, but are quite common parallel to layering in sedimentary rocks with a fine planar and/or linear striping parallel or subparallel to the edge of the vein. Such *bedding veins* (de Roo and Weber 1992) are common in low-grade deformed metasedimentary rocks, especially in metaturbidites (Fig. 6.9; Ramsay and Huber 1983, 1987; Cox 1987; Gaviglio 1986; Fitches et al. 1986; Mawer 1987; Mitra 1987; Tanner 1989, 1992a; Henderson et al. 1990; Labaume et al. 1991; Cosgrove 1993; Jessell et al. 1994; Fowler 1996; Fowler and Winsor 1997; Ohlmacher and Aydin 1997; Köhn and Passchier 2000; Passchier et al. 2002). They can be up to 30 cm thick and several hundreds of metres long, are commonly composed of quartz and chlorite, and normally contain three structural elements of different orientation: inclusion bands, inclusion trails, and boundaries of elongate crystals. Most veins lack a clear median line. Inclusion bands mimic the shape of the wall rock while inclusion trails connect jogs in the inclusion bands and corresponding points in the wall rocks that were originally joined. Inclusion trails are commonly parallel to jogs in the vein-wall rock contact that may nucleate on older structures such as folds (de Roo and Weber 1992). Inclusion trails are usually thin breccia zones of wall rock and vein material, or slickensides (Stanley 1990; de Roo and Weber 1992). The planar and linear striping in the veins is in some cases due to preferred orientation of crystals or fibres, but more generally to parallel inclusion trails. In such cases, the veins are composed of elongate or stubby quartz crystals that lie oblique to inclusion bands and trails, and to the edge of the veins (Fig. 6.18b, ⊘Video 6.18b,c; de Roo and Weber 1992; Köhn and Passchier 2000). Since elongate grains commonly nucleate on the vein boundary jogs, they tend to be normal to the inclusion bands.

Fig. 6.18.
a Development of a shear vein from a fault with jogs. After small displacement, isolated fibre packages lie along the fault, typical of slickenfibres. After larger displacement the fibre packages overlap and become separated by inclusion trails. This produces the characteristic banded nature of striped shear veins and bedding veins. **b** Detail of the internal structure of a typical striped shear vein. Inclusion bands, inclusion trails and boundaries of elongate crystals all have different orientation. Grain growth in this vein is unitaxial and occurs at *top right*, as shown by widening of a decreasing number of crystals in that direction. Inclusion bands change shape when wall rock fragments become detached. Grain boundaries commonly terminate or jog along inclusion trails

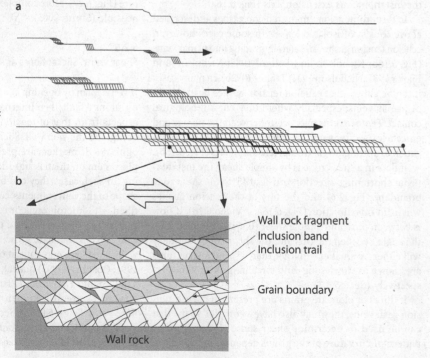

Wall rock fragment
Inclusion band
Inclusion trail

Grain boundary

Wall rock

Cox (1987) and Jessell et al. (1994) suggested an alternative mechanism for the formation of bedding veins, in which the different orientation of inclusion bands and trails is due to a changing opening direction of the vein, or even partly to dissolution in the vein. Mawer (1987) reported veins with two different types of inclusion bands, but lacking inclusion trails.

Bedding veins form by highly oblique opening of a vein and unitaxial, antitaxial, or even composite growth of quartz and mica. Crack-sealing may contribute to formation of the inclusion bands and trails. Vein opening at a relatively high rate normally precluded tracking behaviour of crystals in the vein, and led to development of elongate grains instead of fibres. These veins can be used as shear sense indicators from the arrangement of inclusion trains and elongate crystals. If opening of the vein is parallel to the inclusion trails (Cox and Etheridge 1983; Gaviglio 1986; Cox 1987; Köhn and Passchier 2000), shear sense is indicated by the angle between the inclusion band and the trails (Fig. 6.18, ⊛Video 6.18a,b,c). If grain shape and delicate deformation bands have been destroyed by static recrystallisation, the shear sense can still be determined from the angle between inclusion trails and the wall rock, or from the shape of jogs in the wall rock. If the opening direction is parallel to inclusion bands, however (Jessell et al. 1994), the inferred sense would be opposite to that discussed above.

Bedding veins can form by fault motion along bedding planes, as in thrust tectonics, but can also be associated with large scale folding; in many cases, the lineation in the veins is normal to fold axes. However, bedding veins are usually folded and commonly cut by stylolites associated with folds, and this suggests that they play a role at initial stages of fold development rather than during flexural slip in maturing folds (Jessell et al. 1994; Köhn and Passchier 2000).

6.3
Fringe Structures

6.3.1
Introduction

Rigid objects in a ductilely deforming rock cause local perturbations of the stress field and flow pattern. In the case of low temperature deformation and high fluid pressure, increased pressure solution may occur adjacent to the rigid object on the side of the shortening ISA, while extensional gashes may open on the contact of the object and the matrix on the side of the extensional ISA. New crystalline material may grow in these gashes and form strain shadows on both sides of the rigid *core object* (Figs. 6.1, 6.3). If the crystals in such strain shadows are elongate or fibrous, these are known as *strain fringes*. The combination of core object and fringes is called a *fringe structure* (Figs. 6.1, 6.3; Köhn et al. 2001a).

Because the presence of the rigid core object influences the geometry of the flow around a developing fringe structure, its shape and internal structure differ considerably from that of a vein. Fringe structures carry information on flow and deformation history in both their internal and external shape, and are therefore useful as kinematic indicators. Fringe structures can be syntaxial, antitaxial or complex. Notice that in fringe nomenclature, syntaxial fringes have the growth surface between the fringe and the wall rock, and antitaxial fringes between the fringe and the core object (Fig. 6.19). This can be confusing since it may seem to be contrary to the terminology used for veins.

Syntaxial fringes have been observed around crinoid stem fragments as core object in limestones (crinoid-type fringes; Ramsay and Huber 1983) but these are relatively rare. Complex fringes around pyrite crystals, with calcite and a quartz rim, have been reported, but most fringes are antitaxial (pyrite-type fringes; Ramsay and Huber 1983), probably because the core object is usually a mineral with contrasting properties from the matrix, such as pyrite in a pelitic, carbonatic or quartz-feldspar rock. The remaining part of this section deals with these most common antitaxial fringes.

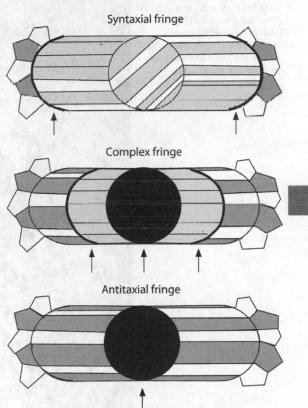

Syntaxial fringe

Complex fringe

Antitaxial fringe

Fig. 6.19. Schematic drawing of three types of strain fringes that occur in nature. The core object of the syntaxial fringes is a crinoid stem fragment with deformation twins. *Arrows* indicate the position of growth surfaces

6.3

Antitaxial fringe structures usually consist of quartz, calcite and locally also chlorite (Mügge 1928; Pabst 1931; Williams 1972b). In some cases, the fibres are deformed and tend to lose information in the distal, oldest and most strongly deformed parts. Rigid strain fringes have been studied for spherical, rectangular and irregular core objects (Durney and Ramsay 1973; Ramsay and Huber 1983; Ellis 1986; Etchecopar and Malavieille 1987; Aerden 1996; Köhn et al. 2001a,b, 2003). In general, the following behaviour is seen: If a rigid object in non-coaxial flow develops fringes at its side, fringes and core object will rotate with respect to kinematic reference axes at different and changing rates, while the fringes are pulled away from the core object. The external and internal geometry of fringes depends at least on the bulk shape of the core object, the roughness of the core object surface, its initial orientation with respect to kinematic axes, the flow regime in the surrounding matrix and growth conditions

for grains along the fringe-core object contact (Köhn et al. 2003). Previously, it was thought that fibre growth in fringes could be either entirely displacement-controlled (Figs. 6.20, 6.22; Choukroune 1971; Ramsay and Huber 1983; Etchecopar and Malavieille 1987) or face-controlled (Figs. 6.21, 6.22) and that displacement controlled grains grow in the direction of extensional ISA. However, fringe objects are much more complicated than veins, and therefore cannot be treated in such a simple way. The tracking or non-tracking behaviour of individual grains in fringes depends on a complex interplay of opening velocity of a crack between the growing fringe and the core object, and growth competition between grains in the fringe. Computer and analogue experiments have shown that just as in veins, grains can track the opening direction of a fringe but only if grain growth keeps pace with opening, and if the core-object surface is rough. If fibres are tracking, their growth direction

Fig. 6.20a.
Strain fringes of quartz adjacent to a spherical pyrite framboid in chert. The shape of the fringes and fibres resembles that modelled for simple shear progressive deformation (Fig. 6.24). Dextral shear sense. Yilgarn Craton, Western Australia. Width of view 17 mm. CPL (Photograph courtesy Dirk Wiersma)

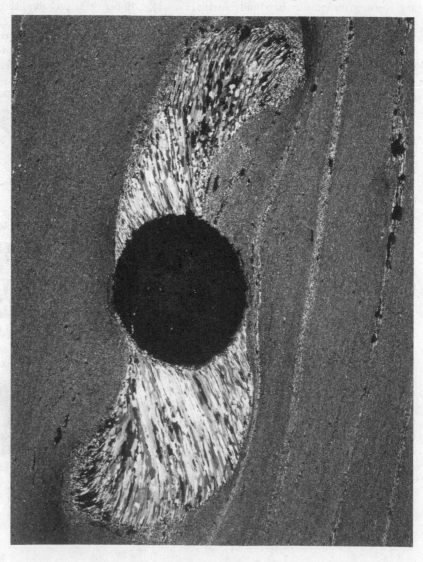

Fig. 6.20b. Detail of a fringe adjacent to another spherical pyrite framboid in the same outcrop. Most fibres are parallel and seem to be displacement controlled, but some small fibres radiate out from the central sphere and are face-controlled. Dextral shear sense. Yilgarn Craton, Western Australia. Width of view 12 mm. CPL (Photographs courtesy Dirk Wiersma)

Fig. 6.21. Quartz and calcite fibres in strain fringes around pyrite grains in carbonaceous slate. The object at *left* is shown enlarged in Fig. 6.27. Lourdes, Pyrenees, France. Sense of shear dextral. Width of view 20 mm. CPL. (Sample courtesy Henk Zwart)

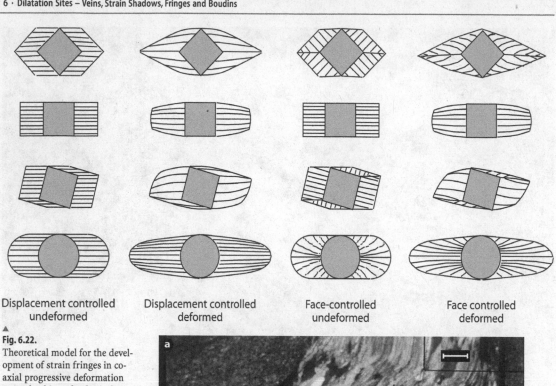

Displacement controlled undeformed

Displacement controlled deformed

Face-controlled undeformed

Face controlled deformed

Fig. 6.22.
Theoretical model for the development of strain fringes in coaxial progressive deformation around cubic and spherical core objects in the case of face-controlled and displacement controlled fibres. Rigid and deforming fringes are shown. In the case of cubic core objects, the shape of the fringe strongly depends on the original orientation of the core object with respect to the opening direction of the fringe

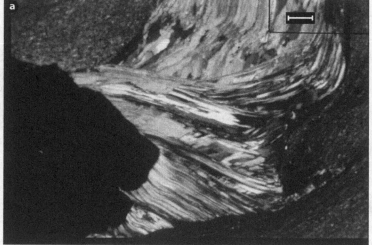

Fig. 6.23.
Natural strain fringe and its simulation after Köhn et al. (2001). **a** Natural strain fringe with displacement- and face-controlled fibres, and mixed fibres. **b** Simulation of a fringe that produces a similar pattern. Face-controlled parts of fibres (*grey*) mainly occur next to a suture line. Scale bar in **a** is 0.8 mm. (Courtesy Daniel Köhn)

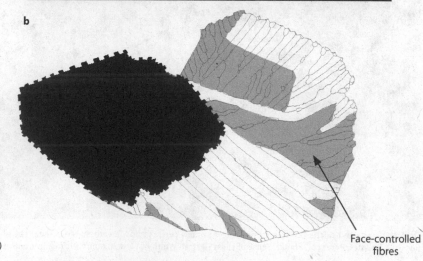

Face-controlled fibres

depends on the relative rotation of fringe an core object, which normally deviates from the orientation of the extensional ISA, even up to 90°. Fringes can therefore contain tracking, non-tracking and miscellaneous fibres, while tracking fibres are not generally opening in the direction of ISA (Fig. 6.23, ⊘Video 6.23).

The rotational behaviour of core-object and fringes is quite different from that of a single, isolated rigid object in a ductile flow, since all three objects influence each other (Köhn et al. 2003). Rotations of each element, and the fringe object as a whole, are slower than would be expected from their aspect ratio, and the rotation rate generally decreases with fringe growth (Box 5.4; Ghosh and Ramberg 1976; Passchier 1988a; Köhn et al. 2003). In fact, the rotation and growth of fringes is too complex to be approached analytically and can only be predicted by numerical or analogue modelling (e.g. Kanagawa 1996; Köhn et al. 2000, 2001a,b, 2003).

If a fringe is not rigid, it may deform internally. Since distal parts of antitaxial fringes are oldest, they are most strongly deformed. A non-deformed fringe can be recognised by an undeformed 'cast' of the core object at the distal part of the fringe (Fig. 6.20) and by fibres which lack undulose extinction and grain boundary migration structures. In some cases rigid strain fringes may start to act as nuclei for second-generation fringes at the distal ends of the fringe (Fig. 3 in Choukroune 1971).

6.3.2
Fringes on Spherical Core Objects

The simplest type of fringe develops on spherical core objects. Common examples are globular aggregates of pyrite with a raspberry-like external form known as *framboidal pyrites* (Fig. 6.20). Matrix material is pulled away from the rigid sphere by the flowing matrix and new fringe material is depositing in the gap. In coaxial progressive deformation displacement-controlled fibres form on rough objects and are simply parallel to the long dimension of the fringe. Face-controlled fibres form on smooth objects and can be predicted to show inward curvature of fibres that gradually decrease in width (Fig. 6.22; Köhn et al. 2000).

Modelling of the interaction of fringe growth, rotation and deformation predicts patterns as shown in Fig. 6.24 (Choukroune 1971; Malavieille et al. 1982; Etchecopar and Malavieille 1987; Köhn et al. 2000). In all cases, the fringes are curved into an S-shaped spiral for dextral shear sense (Fig. 6.24, ⊘Video 6.24a). If most fibres are tracking, they will show a similar curvature as the outline of the fringe, but be inclined at a steeper angle to the flow plane (Figs. 6.20, 6.24, ⊘Video 6.24a). If few fibres are tracking most will be face controlled and radiate outward. The curvature of the fibres will be mostly in the same sense as the curvature of the fringe, but fibres converge on the core object (Köhn et al. 2000). Curvature of the fibres is strongly dependent on the rotation rate of the core object with respect to the fringes (Fig. 6.25). This rotation rate, in turn, depends on *coupling* between the core object and the matrix (compare Sects. 5.6.7, 5.6.8).

6.3.3
Fringes on Angular Core Objects

Most fringes in nature develop on angular core objects with planar faces such as pyrite or magnetite crystals in fine-grained metapelite or carbonaceous slate at low metamorphic grade (Figs. 6.3, 6.21, 6.27). Because of the smooth crystal faces of such core objects, most elongate crystals or fibres tend to be non-tracking in adjacent fringes. However, one, two or more crystal faces can be in contact with a fringe at any time. As a result, a dividing line or *suture* may be present in the fringe separating populations of fibres or elongate crystals with different orientation. Such sutures are attached to the corner between two faces of the core object (Figs. 6.21, 6.23, 6.27, ⊘Photo 6.21, ⊘Video 6.24b, 6.27). In fact, sutures are trapped by corners of core objects just as individual fibres are on asperities in the contact (Köhn et al. 2000). Figures 6.22 and 6.24 show the development of strain fringes around an angular core object in coaxial and non-coaxial progressive deformation. In the case of displacement-controlled growth and non-coaxial progressive deformation, the fibres are strongly curved in the outer, oldest parts of the fringe and are more straight on the inside. This is a result of decreasing angular velocity of the fringe when its aspect ratio increases (Box 5.4). In the case of face-controlled growth, curvature of the fibres is complex and directed towards the suture lines, which are therefore more prominent than in displacement-controlled fibres. The shape of the sutures is usually less complex than that of the fibres and is more useful as a shear sense indicator; if the combined sutures of both fringes define an S shape, shear sense is dextral; if they define a Z shape, shear sense is sinistral (Fig. 6.24, ⊘Video 6.24b). The external shape of the fringes is usually curved or even "hooked" with a similar sense as the suture lines and as the external geometry of fringes around spherical core objects. However, the external geometry of fringes on angular core objects is less reliable as a shear sense indicator than those on spherical core objects; the orientation of the angular core object at the onset of fringe growth determines the final shape of the fringe. Fringes on elongate core-objects can therefore be subdivided into four geometry classes; nw, n, w and wn-type, depending on the orientation of the core object at the onset of fringe growth and its subsequent rotation (Fig. 6.26). An additional complication is that the aspect ratio of elongate core objects will also influence fringe shape (Passchier 1987b; Köhn et al. 2003; Sect. 5.6.7).

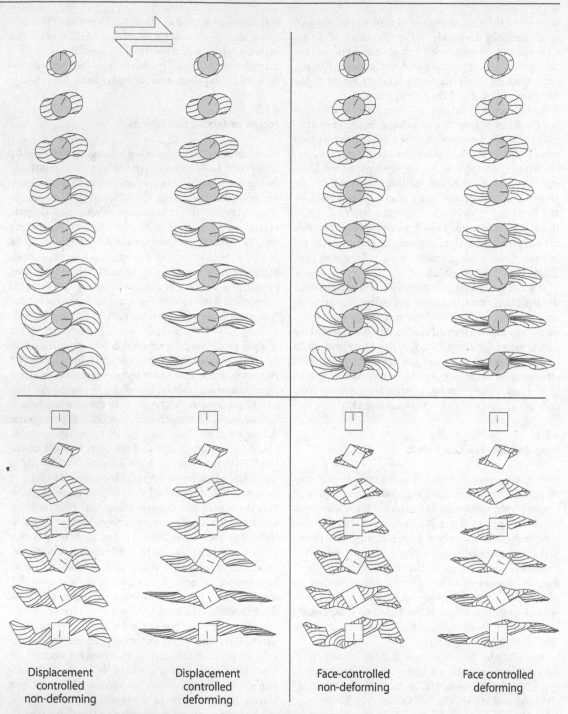

Fig. 6.24. Schematic diagram showing the geometry of fibrous fringes developing in simple shear matrix flow in the case of spherical and cubic core-objects, and for rigid and deformable fringes. The situation for cubic core objects is shown for only one initial orientation of the cube, and can be different for other initial orientations (see Fig. 6.22)

Aerden (1996) and Müller et al. (2000) interpreted hooked fringes as formed by polyphase deformation. Polyphase deformation can indeed produce such shapes, but they can equally be formed in a single phase of non-coaxial progressive deformation (Köhn et al. 2003). Detailed study of the internal structure of several fringes from a single sample can show which setting is more likely.

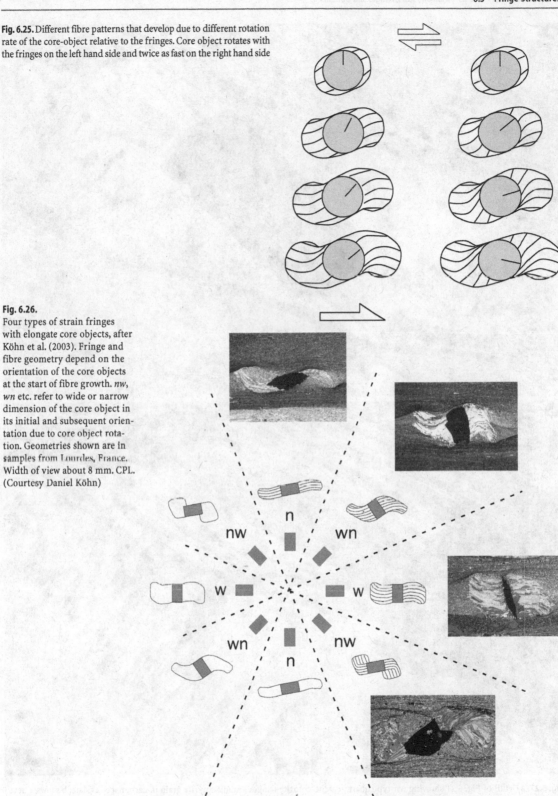

Fig. 6.25. Different fibre patterns that develop due to different rotation rate of the core-object relative to the fringes. Core object rotates with the fringes on the left hand side and twice as fast on the right hand side

Fig. 6.26.
Four types of strain fringes with elongate core objects, after Köhn et al. (2003). Fringe and fibre geometry depend on the orientation of the core objects at the start of fibre growth. *nw, wn* etc. refer to wide or narrow dimension of the core object in its initial and subsequent orientation due to core object rotation. Geometries shown are in samples from Lourdes, France. Width of view about 8 mm. CPL. (Courtesy Daniel Köhn)

Fig. 6.27. a Detail of Fig. 6.21 showing wn-type quartz-calcite-chlorite fringes around a pyrite grain in carbonaceous slate; **b** shows a detail of **a**. The assemblage resembles a face-controlled fringe with a suture formed in simple shear progressive deformation as shown in Fig. 6.24 but is slightly different, probably because this pyrite is elongate rather than square; the fibres have a sharp bend instead of a suture and may actually be displacement-controlled. Further explanation in text. Lourdes, Pyrenees, France. Width of view **a** 7 mm; **b** 2.5 mm. CPL

6.4
Fringes and the Deformation Path

Fringe structures are informative and may be used to estimate sense of shear and finite strain (Sect. 9.2; Durney and Ramsay 1973; Reks and Gray 1982, 1983; Ramsay and Huber 1983; Beutner and Diegel 1985; Gray and Willman 1991; Kirkwood et al. 1995; Kanagawa 1996; Köhn et al. 2003). If fibres in a fringe are straight, the long axis of the fringe is thought to represent the X-direction of finite strain. If the total length of both fringes and the rigid inclusion is divided by the length of the inclusion, a minimum value for the principal stretch in the X-direction is obtained. With this method, only extension (stretch > 1) can be measured, not shortening. It is therefore only possible to determine a strain value in the case of plane strain and known volume change, or if stretch in the Y- and Z-directions are known by other means. Framboidal pyrites are assumed to have grown during early diagenesis, and fringes around such pyrites can therefore monitor more than just the tectonic strain (Durney and Ramsay 1973; Ramsay and Huber 1983). Euhedral pyrite in mineralised zones may have grown after diagenesis. If both types of pyrite are present in a rock, it should be possible to separate dia-

genetic and tectonic strain to some extent. Even so, the obtained finite strain is a minimum value, since fringes may not have formed continuously during the entire deformation history.

Many fibrous veins and fringes have curving fibres with a geometry that is more complex than that in Figs. 6.17 and 6.24. Such fibres are interpreted to form by a deformation history with changing flow parameters. The irregular curves in fibres of Figs. 6.3 and 6.28, for example, can be explained as changes in the orientation of ISA with respect to the developing fringe and the foliation in the wall rock, provided the fibres are displacement-controlled. In fact, if fibres in fringes are tracking they trace the displacement path (Box 2.1) of a particle on the surface of the rigid core object with respect to the wall rock. The shape of fibres in a fringe therefore contains information on the deformation path in the wall rock (Sect. 2.2; Durney and Ramsay 1973; Elliott 1973; Wickham 1973; Gray and Durney 1979b; Casey et al. 1983; Ramsay and Huber 1983; Beutner and Diegel 1985; Ellis 1986; Gray and Willman 1991; Hedlund et al. 1994; Aerden 1996). Two models have been proposed to reconstruct the deformation path from fibre shape in fringes, both of which assume that fibres track the extensional ISA; the most commonly used model (Durney and Ramsay

Fig. 6.28. Quartz fringe adjacent to a spherical pyrite framboid in chert. All fibres show a sharp bend, which is probably due to a change in the instantaneous extension direction during progressive deformation. Leonora, Yilgarn Craton, Australia. Width of view 10 mm. CPL

Fig. 6.29.
Object centre path determined for a fringe structure from Lourdes, France. (After Köhn et al. 2001b; Courtesy Daniel Köhn)

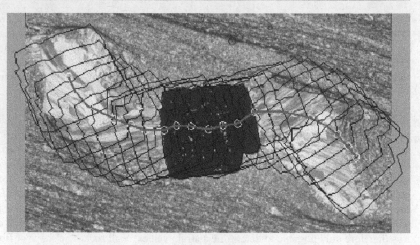

6.5

1973; Wickham 1973; Hedlund et al. 1994) assumes that fibres are rigid. The second model assumes that fringes deformed passively and homogeneously (Ramsay and Huber 1983; Ellis 1986). In both models, the fibres are subdivided into small segments that represent incremental deformation steps, and progressive deformation is restored by repeated multiplication of deformation tensors derived from the fibre segments (*fibre trajectory analysis*) The results of such deformation path analyses have been used in studies of folding (Wickham and Anthony 1977; Beutner and Diegel 1985; Beutner et al. 1988; Fisher and Anastasio 1994; Hedlund et al. 1994), foliation development (Fisher 1990) and tectonic evolution of deformed terrains (Fisher and Bryne 1990; Clark et al. 1993; Kirkwood et al. 1995). Unfortunately, these studies are not generally applicable because in most cases, individual fibres will not open parallel to ISA and the behaviour of single fibres is not representative for the entire fringe structure (Spencer 1991).

Aerden (1996) and Köhn et al. (2000) suggested a new method to analyse undeformed fringes where irregularities on the surface of the central object are fitted to certain fibres by sliding the central object along them in an analysis; this defines an *object-centre path* that the central object must have taken along the fringe to produce its present shape (Fig. 6.29). Rotation of the fringe relative to the core object can be separately determined. The object-centre path does not depend on behaviour of individual fibres, is unique for each fringe, and can potentially be used to determine finite strain and possibly the deformation path in the wall rock (Köhn et al. 2003).

A new type of development is the possibility to use quartz-fringes to determine strain rate in the host rock for dating purposes using Rb-Sr geochronology. Müller et al. (2000) report successful Rb-Sr micro-sampling and dating in fringes. Fringes clearly contain a lot of information on deformation conditions and history, and have the potential to become a powerful tool in the collection of geological data.

6.5
Non-Fibrous Strain Shadows and Strain Caps

Massive (non-fibrous) *strain shadows* are elongate domains on both sides of a core object in which the fabric is different from that in the rock matrix (Figs. 5.20, 6.30a, 6.31). The boundary between the strain shadow and the matrix may be sharp, as for a fringe, but is more commonly gradual (Figs. 6.1, 6.30a, 6.31, 7.35, 7.41, 7.42). Strain shadows are commonly enriched in soluble minerals such as quartz, carbonate and chlorite, whereas foliation-forming minerals such as micas or chain-silicates are under-represented. If a foliation is present in the matrix, it is usually more weakly developed or absent in the strain shadow (Figs. 6.31, 7.32, 7.41). The shape of massive strain shadows can be used as a tool to determine shear sense as in the case of fringe structures (Figs. 6.24, 7.35). Obviously, more care is needed than in the case of fringes, especially if the core object is angular.

In foliated rocks, strain shadows and fringe structures are commonly associated with *strain caps*, domains enriched in micas or insoluble minerals, where the main foliation is strongly developed (Figs. 6.30a, 6.31, 7.11, 7.24, 7.39). Strain caps occur at opposite sides of the core object, in the quarters oblique to the strain shadow or fringe. Some strain fringes may develop their own strain caps if deformation is advanced, when they may start to behave as independent core objects (Fig. 6.29).

The term strain shadow may suggest that strain in the 'shadow' of the core object is relatively low, but this is not generally correct. In fact, strain shadows and strain caps are structures representing complex partitioning of strain and volume change around a core object, the exact nature of which is usually unknown.

Massive strain shadows can be formed by non-fibrous infilling of a void at the surface of a core object as for strain fringes, by recrystallisation of a strain fringe, or by redistribution of mineral phases in response to inhomogeneous deformation around a core object. This may happen at low fluid pressure by pressure solution at the site of strain caps, solution transfer, and redeposition of material at grain boundaries in the developing strain shadow without opening of distinct voids. Figure 6.30b shows schematically some models for the development of different types of strain shadows.

Mantled porphyroclasts superficially resemble strain shadows (Sect. 5.6.5) but they have the same mineral composition as the core object (porphyroclast), form by different mechanisms and have a different kinematic significance as strain shadows (Sect. 5.6.7: Figs. 5.20, 5.21, 6.24). We therefore do not include mantled porphyroclasts in the category of strain shadows. It is preferable to reserve the terms strain shadow and strain fringe for domains of material that have a different composition from the core object. However, mantled porphyroclasts and strain shadows are end members of a range of possible combinations; some strain shadows may be difficult to distinguish from mantled porphyroclasts with wings that have undergone chemical or mineralogical changes (Robin 1979; Wintsch 1986).

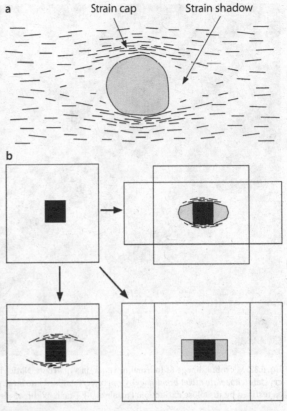

Fig. 6.30. a Geometry of strain shadows and strain caps around a rigid object in a foliated rock. **b** Schematic presentation of the development of three types of strain shadows, depending on strain and volume change as indicated by a *deformed square*. *Grey strain shadows* are formed by precipitation from solution. Notice that strain caps are not necessarily present

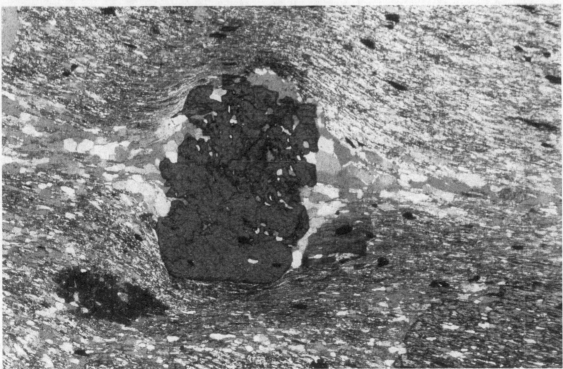

Fig. 6.31. Non-fibrous strain shadow around a garnet porphyroblast in micaschist. Orós, NE-Brazil. Notice the presence of strain caps and the gradual transition between the strain shadows and the matrix. Width of view 3.9 mm. Polars at 45°

Fig. 6.32. Microboudinage of tourmaline (*dark*) in a quartzite. Notice boudinage both perpendicular and parallel to the elongate tourmaline crystals (*chocolate tablet boudinage*). Quartz recrystallised into large grains overgrowing the necks of the boudins, and was deformed again as testified by undulose extinction. Fortaleza, NE Brazil. Width of view 4.5 mm. Polars at 45°. (Specimen courtesy Michel Arthaud)

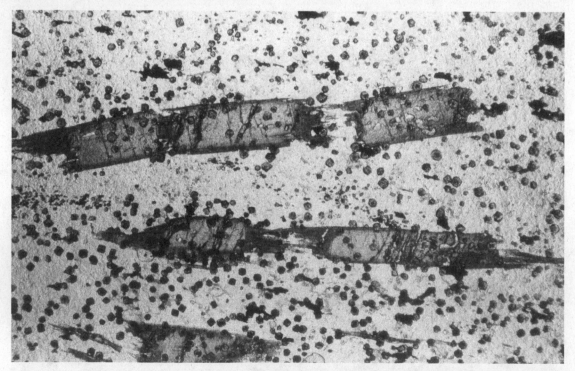

Fig. 6.33. Microboudinage of blue amphibole in a garnet bearing metachert. Notice how, after the boudinage of the light-coloured cores, darker amphibole overgrew the cores and tended to mend the separated boudins. Metamorphic conditions during boudinage can be estimated from the changing blue amphibole composition (cf. Fig. 6.34). Sesia Lanzo Zone, Italian Alps. Width of view 4 mm. PPL. (Sample courtesy Leo Minnigh)

6.6
Microboudinage

Boudinage affecting elongate mineral grains at thin section scale is often referred to as *microboudinage* (Misch 1969, 1970; Vernon 1976; Allison and LaTour 1977; Masuda and Kuriyama 1988; Masuda et al. 1989, 1990, 1995a,b, 2003, 2004; Ji and Zhao 1993; Ji 1998; Figs. 6.32, 6.33, ⊘Video 11.11c; see also Sect. 5.6.12). The structures created in the necks of these boudins may show many similarities with those in strain shadows and extensional veins.

Microboudins can help to establish metamorphic conditions during deformation, indicated by the mineral assemblage that grew in the necks of the boudins. In some cases, grain growth accompanies microboudinage to form zoned grains. Analysis of the zoning in these grains may also allow establishment of changing metamorphic circumstances during progressive deformation (Misch 1969, 1970; Figs. 6.33, 6.34, ⊘Photo 6.33).

Microboudinage can also be used as a strain gauge (Sect. 9.2; Ferguson 1981, 1987; Masuda et al. 1995a): the length of a boudinaged grain divided by the sum of the length of the boudins gives a minimum value for stretch along the long axis of the boudinaged grain. However, this axis does not necessarily coincide with the direction of maximum extension in the rock. It is also possible to use the length of boudin fragments to make estimates of differential stress (Lloyd et al. 1982; Masuda et al. 1990, 1995a,b, 2003; Ji and Zhao 1993) and aspects of the rheology of the matrix (Masuda and Kimura 2004).

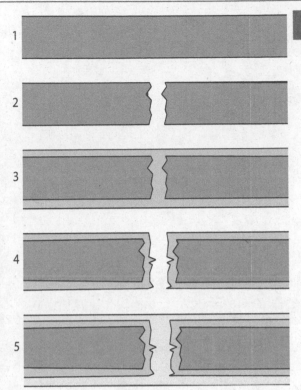

Fig. 6.34. Schematic deformation-growth sequence involved in microboudinage of zoned crystals, showing a single site of rupture. Although represented graphically as finite steps, both stretching and accretion are thought to proceed essentially simultaneously. (After Misch 1969, p 48)

Porphyroblasts and Reaction Rims

Information contained in the structure in and around porphyroblasts and in reaction rims is discussed in this chapter. Porphyroblasts often contain inclusion patterns that reflect the structure of the matrix during their growth. Once formed, a rigid porphyroblast protects these patterns from further modification, whereas the matrix may be subjected to later crenulation, recrystallisation, grain growth etc. For this reason comparison of porphyroblasts and their surrounding matrix gives valuable information on the sequence of deformation phases and the relative time interval of porphyroblast growth with respect to these phases. A distinction is made between pretectonic, intertectonic, syntectonic and post-tectonic porphyroblasts. Using this technique a relative sequence of metamorphic events and deformation phases can be established.

Special inclusion patterns, such as millipede structures, deflection folds and oppositely concave microfolds are explained and discussed. Attention is also given to possible pitfalls in the interpretation of porphyroblast microstructures, such as inclusion shapes and patterns controlled by crystallographic directions rather then representing passive inclusions reflecting a pre-existing matrix foliation.

The relative rotation or non-rotation of porphyroblasts with respect to their surrounding matrix or to geographic coordinates is still a hot topic in the literature and a revision is given of the current state of the art.

A new topic refers to amalgamated porphyroblasts, a possible source of misinterpretation that may produce in some cases apparently enigmatic inclusion patterns.

The final part of this chapter deals with reaction rims, an important tool to unravel the metamorphic history of the rock. Attention is given to coronas, moats, symplectites and pseudomorphs.

7.1
Introduction

A volume of rock involved in deformation and metamorphism will continuously undergo changes in structure and mineral content. This chapter treats mineral growth and replacement structures and the way in which their geometry can be used to reconstruct tectonic history. Two types of informative structures are treated: porphyroblasts and reaction rims.

Relatively large single crystals, which formed by metamorphic growth in a more fine-grained matrix, are known as *porphyroblasts* (Box 5.2). Porphyroblasts are a valuable source of information on local tectonic and metamorphic evolution. Inclusion patterns in porphyroblasts can mimic the structure in the rock at the time of their growth (Fig. 7.1) and allow a reconstruction of the relative timing of mineral growth, reflecting metamorphic conditions, and deformation. As such they can play a key role in the determination of pressure-temperature-deformation-time (*P-T-D-t*) paths experienced by metamorphic rocks. Apart from their inclusion patterns, porphyroblasts may also record the metamorphic evolution from core to rim; either a growth zoning may be present, or inclusions of certain minerals may show *P-T* conditions different from the matrix. Porphyroblasts can also be used in the study of the kinematic significance of their rotation or non-rotation with respect to specific reference frames, both with relation to folding mechanisms and to determine sense and minimum amount of shear in shear zones. A review of current trends in the study of porphyroblast microstructures is given in Johnson (1999b).

Porphyroblasts are not equally common in all rock types and under all metamorphic conditions. Most common and most informative are Al-silicate porphyroblasts like garnet, biotite, staurolite, chloritoid, andalusite etc. grown under upper greenschist to amphibolite facies conditions in metapelites. Garnet, plagioclase, epidote and hornblende may form interesting porphyroblasts in metabasites. Since porphyroblasts are such informative structures, it is usually advantageous to sample and study microstructures in available metapelites (and metabasites especially if they are garnet-bearing) in any area for large-scale tectonic studies. In order to evaluate the significance of porphyroblasts, we first describe why and how porphyroblasts grow and how they acquire inclusions. Important texts on the tectonic significance of porphyroblasts are Zwart (1962), Zwart and Calon (1977), Schoneveld (1979), Spry (1969), Vernon (1975, 1976, 1989), Bell and Rubenach (1983), Bell (1985), Bard (1986), Bell et al. (1986, 1992c), Yardley (1989), Yardley et al. (1990), Barker (1990, 1998), Shelley (1993), Passchier et al. (1992), Johnson (1993a,b, 1999b), Williams and Jiang (1999), Ilg and Karlstrom (2000) and Kraus and Williams (2001).

7.2
Porphyroblast Nucleation and Growth

The distribution and size of porphyroblasts in a metamorphic rock depend on the amount of nucleation sites and the rate at which the nuclei grow. The nucleation and initial growth stage of a new mineral in a metamorphic rock is hampered by the fact that small grains have a relatively high surface free energy and are therefore less sta-

Fig. 7.1.
a Diagram illustrating how an Al-silicate porphyroblast may grow in a mica-rich matrix by substitution reactions involving minor volume change. Opaque minerals (*right hand side*) and quartz are taken up as inclusions and their preferred orientation and distribution is mimicked by the inclusions. Later deformation may affect the matrix but will not change the included structure if the porphyroblast remains undeformed. The inclusion patterns may undergo rigid body rotation but will retain a record of the structure in the rock at the time of porphyroblast growth. **b** Commonly used terminology for porphyroblasts

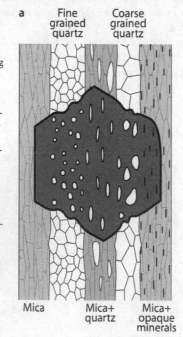

a   Fine grained quartz   Coarse grained quartz

Mica   Mica+ quartz   Mica+ opaque minerals

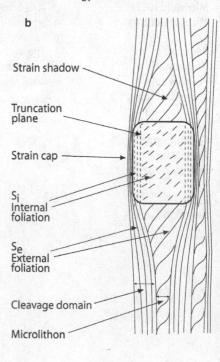

b

Strain shadow

Truncation plane

Strain cap

S$_i$ Internal foliation

S$_e$ External foliation

Cleavage domain

Microlithon

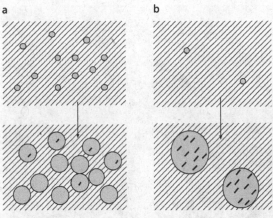

Fig. 7.2. a A large number of nuclei in a rock will lead to development of a large number of small porphyroblasts, each with few inclusions. **b** Few nuclei lead to few large porphyroblasts, which may contain a clear inclusion pattern and are therefore useful in fabric analysis

ble than large ones (Poirier 1985). This unstable stage can be overcome at specific sites controlled by small irregularities such as strongly deformed grains or microfractures (Yardley 1989). If many suitable sites are available, many small porphyroblasts may form; if few suitable nucleation sites are present, isolated large porphyroblasts develop (Fig. 7.2). Thus, nucleation rate and growth rate are competing processes. Matrix grain size also influences the size of porphyroblasts (Carlson and Gordon 2004).

Some minerals nucleate on the crystal lattice of another in a particular orientation, e.g. sillimanite on muscovite, amphibole on pyroxenes. This relationship is known as *epitaxy*. The special situation where the crystal lattices of both minerals are parallel is known as *syntaxy*.

Once a stable porphyroblast has formed, its radial growth rate is likely to decrease with time if its growth rate in terms of added mass is constant. However, in a study based on backscatter images and X-ray maps of garnets, it was found that multiple nuclei formed simultaneously and grew amoeba-like, to coalesce later to a single garnet with a constant radial growth rate, regardless of size (Spear and Daniel 1998). Although little is known about the absolute growth rate of porphyroblasts, theoretical considerations and available radiometric dating give conservative estimates for growth of a garnet 2 mm in diameter of less that 0.1 m.y. to 1 m.y. (Cashman and Ferry 1988; Christensen et al. 1989; Burton and O'Nions 1991; Paterson and Tobisch 1992; Barker 1994; Williams 1994).

7.3
Inclusions

The growth process of porphyroblasts is mainly controlled by diffusion, either in the solid state or through fluids present along grain boundaries. Elements necessary for growth that are not present have to be transported by diffusion to the surface of the porphyroblast. Minerals adjacent to the growing grain that do not or only partly participate in the mineral reaction have to be removed by dissolution and diffusion. In some cases, especially at high-grade metamorphic conditions, diffusion rates have been high enough to allow complete removal of reaction products and non-participating material, and clear, 'gem-quality' porphyroblasts result. In most cases, and especially at low to medium-grade metamorphism, minerals that do not participate in the reaction are not removed completely but are overgrown and enclosed by porphyroblasts as *passive inclusions*. If the rock adjacent to the growing porphyroblast had a compositional layering or a grain shape preferred orientation of grains, this fabric may be partly preserved when grains are included in the porphyroblast; an *inclusion pattern* results that mimics the original fabric (Figs. 7.1–7.3). In this way, straight foliation traces can be included, but also more complex patterns such as folds or even complete crenulation cleavages (Figs. 7.4, 7.5). Opaque minerals and quartz are most commonly included in this manner, but zircon, monazite, apatite, rutile, sphene and epidote-group minerals are also common. Mica inclusions are rare but do occur in some Al-silicate porphyroblasts; they may have been included as excess phases of reactants. However, care is needed, since mica overgrowths may resemble inclusions (see below).

Microstructural observation of inclusions in porphyroblasts by numerous workers has led to the conclusion that they are mostly included in a passive manner, without being significantly displaced by the growing porphyroblast (Zwart 1962; Spry 1969; Vernon 1975, 1976, 1989; Zwart and Calon 1977; Bell 1981, 1985; Bard 1986; Yardley 1989; Yardley et al. 1990; Barker 1990, 1998). Deflection of matrix foliation around porphyroblasts (Fig. 7.5a, ⊘Photo 7.5a,b) is therefore thought to form by deformation of the matrix around a rigid pre-existing porphyroblast (Zwart 1962; Vernon 1976; Barker 1990, 1998; Yardley 1989) and not by mechanical displacement of the matrix by the growing porphyroblast as earlier proposed by Spry (1969) and Misch (1971). However, growing porphyroblasts can displace graphite and white mica in rare cases, as explained in Sect. 7.7. Surfaces of aligned elongate inclusions within porphyroblasts are referred to as S_i (i for internal) whereas the foliation outside the porphyroblasts is called S_e (e for external; Fig. 7.1b). If deformation occurs after porphyroblast growth, S_i may have a different orientation from S_e.

The abundance of inclusions in Al-silicate porphyroblasts has been attributed to the limited mobility of Al-ions (Carmichael 1969). At greenschist and lower amphibolite facies conditions, Al-ions are far less mobile than Si, Fe, Mg, K or Ca-ions, unless the pH is extremely high or low, or if salinity is extremely high (Slack et al. 1993). This means that porphyroblasts of Al-silicates such as andalusite, cordierite, staurolite, chloritoid and gar-

7.3

Fig. 7.3. Example of the process visualised in Fig. 7.1. Post-tectonic porphyroblast of biotite (*centre*) and staurolite (*below*) grew over a layered structure in a fine-grained schist. The structure is mimicked within the porphyroblasts. South Africa. Width of view 5.5 mm. PPL

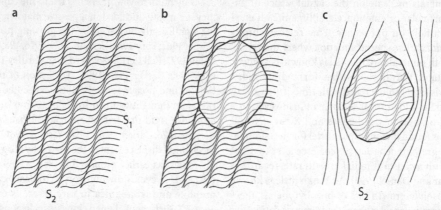

Fig. 7.4. Diagrammatic sketch of the evolution of complex porphyroblast structures as shown in Fig. 7.5a,b. **a** S_2 crenulation cleavage develops, overprinting an older foliation S_1 (Sect. 4.2.6). **b** A porphyroblast overgrows the structure and mimics it in its inclusion pattern. **c** Continued deformation and/or recrystallisation and grain growth (transposition: Box 4.9) destroys the folds in the matrix where a more or less continuous foliation (S_2) develops. Only the relict structure included in the relatively rigid porphyroblast records the structural evolution in the rock

net can be expected to grow at Al-rich sites such as mica layers and to have difficulty in replacing minerals that lack Al by an intact lattice: such minerals are included instead. This explains why Al-silicate porphyroblasts commonly contain numerous inclusions when they grow in quartz and graphite-bearing pelitic rocks. Such porphyroblasts may even obtain a *skeletal shape* (Fig. 7.6) when they grow into Al-poor domains, for instance

quartz-rich layers or strain shadows. They are unable to build a complete lattice at these sites, but rather follow grain boundaries. In many cases a compositional layering is included as inclusion-poor and inclusion-rich areas in porphyroblasts (Figs. 7.1, 7.3, 7.5b). This happens by overgrowth of mica-rich and quartz-rich domains (e.g. a differentiated crenulation cleavage). In extreme cases a porphyroblast grows only along layers of a specific com-

Fig. 7.5. a Micaschist with porphyroblasts of staurolite, biotite and andalusite. The large twinned crystal in the *centre* is staurolite with an inclusion pattern mimicking differentiated crenulation cleavage; a detail of the *right-hand rim* is shown in **b**. The *light-coloured* sub-horizontal bands in the crystal are inclusion-poor zones representing the differentiated mica-rich limbs of microfolds (cf. Fig. 4.13). The dominant foliation in the matrix is a fine-grained schistosity, strongly deflected around the staurolite porphyroblasts. This shows that since porphyroblast growth, crenulation cleavage in the matrix has been destroyed by a process as visualised in Figs. 7.4 and 4.18. (Example of syntectonic porphyroblasts: *F1* in Fig. 7.9). Rioumajou, Pyrenees, France. Width of view **a** 10 mm. Polars at 45°; **b** 2 mm. PPL

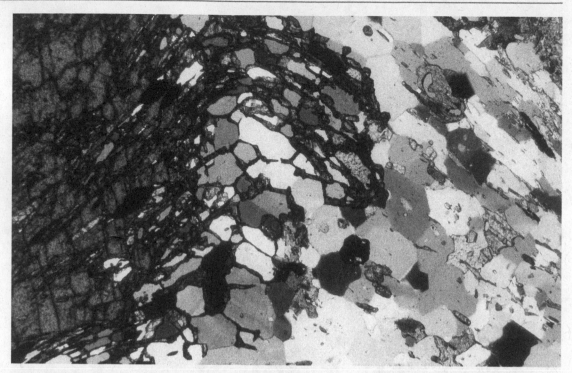

Fig. 7.6. Skeletal rim of a garnet porphyroblast (detail of Fig. 7.32) showing how quartz crystals of the strain shadow are incorporated into the growing crystal as slightly elongated inclusions. The curvature of S_i is caused by relative rotation between the porphyroblast and the matrix foliation (cf. Fig. 7.32). Baños, Ecuador. Width of view 2 mm. Polars at 45°

Fig. 7.7. Quartz-rich schist with elongate staurolite crystal (*dark crystal* in the *centre*) that grew along a mica-rich layer. Folding of the layering and layer-parallel foliation was probably later than the growth of the staurolite that contains a straight inclusion pattern. Pyrenees, Spain. Width of view 12 mm. Polars at 45°

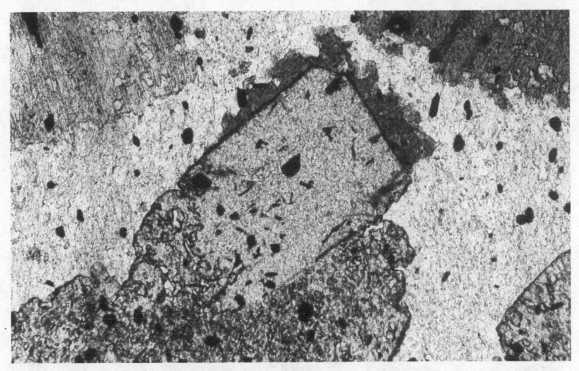

Fig. 7.8. Garnet biotite schist. The garnet crystal (*lower part* of the photograph) is very rich in inclusions except for an almost rectangular area in its upper part where only a few inclusions of opaque minerals are visible. The garnet apparently overgrew a biotite crystal in this area, the fringes of which are still visible at the outer contact. Since biotite has a similar Al concentration as garnet, the garnet crystal could substitute the biotite without incorporation of extra inclusions. Note also the more idiomorphic outline of the garnet crystal in this area. Once the biotite is completely overgrown its shape remains visible as a ghost structure of inclusion-poor garnet. Grenville Province, Canada. Width of view 2 mm. PPL

position and obtains an elongate shape, although its normal crystal habit may be more equidimensional. Figure 7.7 shows an example of elongate staurolite growing along pelitic layers in a metasediment (⊘Photo 7.7). The shape of pre-existing but now substituted minerals may also remain visible as inclusion-poor areas within some porphyroblasts. These structures are known as *ghost structures* (Fig. 7.8, ⊘Photo 7.8a,b).

Since the diffusion rate of ions is a function of temperature, inclusions become increasingly rare, more coarse-grained and less well defined at higher metamorphic grade (upper amphibolite and granulite facies), and are seldom frequent enough to define S_i surfaces.

7.4
Classification of Porphyroblast-Matrix Relations

7.4.1
Introduction

Porphyroblasts with inclusion patterns contain information on the nature of early deformation and metamorphic events, and on the relative age of mineral growth and deformation. Zwart (1960, 1962) elaborated a scheme with nine diagnostic relations based on the idea that crys-

---

Box 7.1	Terminology of porphyroblast-matrix relations

The terminology of porphyroblast-matrix relations as used in this book can be abbreviated by mathematical symbols, which can be useful in complex diagrams. These symbols are also illustrated in Fig. 7.9:

- Growth interval of mineral P
  post-tectonic with respect to D_n $\qquad D_n < P$
- Growth interval of mineral P
  pretectonic with respect to D_1 $\qquad P < D_1$
- Growth interval of mineral P
  syntectonic with respect to D_n $\qquad D_n \supset P$
- Growth interval of mineral P
  syn- to post-tectonic with respect to D_n $\quad D_n \leq P$
- Growth interval of mineral P
  intertectonic between D_n and D_{n+1} $\quad D_n < P < D_{n+1}$
- Growth interval of mineral P
  post D_n and pre to syn D_{n+1} $\qquad D_n < P \leq D_{n+1}$

---

tals may be older, younger or of the same age as specific deformation phases (Sect. 1.2). Zwart's scheme has long been used for basic reference, but, as pointed out by Vernon (1978), several criteria are ambiguous and relations have to be studied with great care. We use a modified version of the Zwart scheme (Fig. 7.9) in which porphyroblasts are classified as pre-, syn-, inter- and post-

7.4

tectonic (some texts use *kinematic* instead of tectonic), to describe the time relation between porphyroblast-growth and one or two specific phases of deformation, normally represented by a foliation or by folding in the matrix. As a shorthand notation, we also use symbols to show the time relationship of deformation and metamorphism as outlined in Box 7.1 and Fig. 7.10. Deformation phases (Sect. 1.2) are determined for individual thin sections; schemes of regional deformation phases have to be evaluated from schemes for individual thin sections (Fig. 7.10) combined with field observations (Sect. 7.9). For a correct interpretation of inclusion patterns it is desirable to determine their approximate three-dimensional shape by comparison of several parallel or orthogonal sections (Sect. 12.7). This applies specifically to inclusions with complicated patterns such as spiral-S_i garnets.

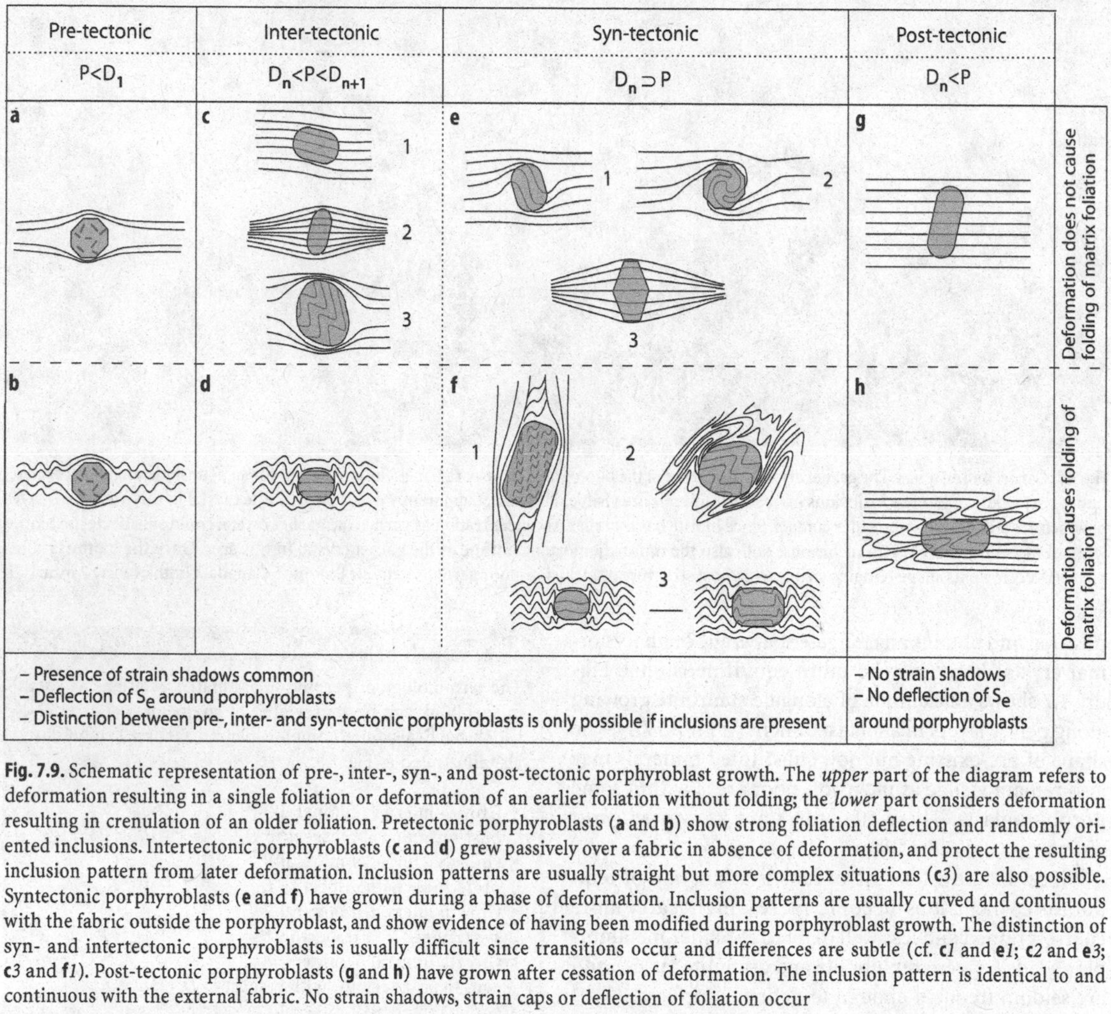

Pre-tectonic	Inter-tectonic	Syn-tectonic	Post-tectonic
$P<D_1$	$D_n<P<D_{n+1}$	$D_n \supset P$	$D_n<P$

Fig. 7.9. Schematic representation of pre-, inter-, syn-, and post-tectonic porphyroblast growth. The *upper* part of the diagram refers to deformation resulting in a single foliation or deformation of an earlier foliation without folding; the *lower* part considers deformation resulting in crenulation of an older foliation. Pretectonic porphyroblasts (**a** and **b**) show strong foliation deflection and randomly oriented inclusions. Intertectonic porphyroblasts (**c** and **d**) grew passively over a fabric in absence of deformation, and protect the resulting inclusion pattern from later deformation. Inclusion patterns are usually straight but more complex situations (**c3**) are also possible. Syntectonic porphyroblasts (**e** and **f**) have grown during a phase of deformation. Inclusion patterns are usually curved and continuous with the fabric outside the porphyroblast, and show evidence of having been modified during porphyroblast growth. The distinction of syn- and intertectonic porphyroblasts is usually difficult since transitions occur and differences are subtle (cf. **c1** and **e1**; **c2** and **e3**; **c3** and **f1**). Post-tectonic porphyroblasts (**g** and **h**) have grown after cessation of deformation. The inclusion pattern is identical to and continuous with the external fabric. No strain shadows, strain caps or deflection of foliation occur

Fig. 7.10.
Example of a relative age diagram for a single thin section of a micaschist. Chlorite is pre-D_1; garnet is intertectonic between D_1 and D_2. Staurolite is pre-syn D_2 and biotite has two growth phases, one syn-D_1 and the other syn-post-D_2. Albite grows post-D_2. In the diagram, abbreviations are given at *right*

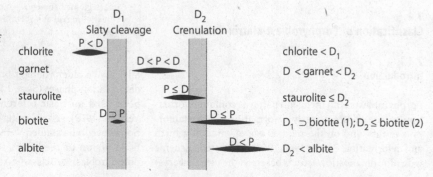

7.4.2
Pretectonic Porphyroblast Growth

Pretectonic porphyroblasts are rarely described (Zwart 1962; Fleming and Offler 1968; Vernon et al. 1993a,b) and seem to be uncommon in areas affected by regional metamorphism, except possibly in low-pressure/high-temperature metamorphism. Even in the case of contact metamorphism, some deformation may predate porphyroblast growth. If present, inclusions in pretectonic porphyroblasts are randomly oriented (a, b in Fig. 7.9; Fig. 7.11, ⊘Video 7.9a, ⊘Photo 7.11a,b) or show sector zoning (Sect. 7.7; Rice and Mitchell 1991). A primary compositional layering may survive as a ghost layering (Fleming and Offler 1968). It is incorrect, however, to interpret any crystal with random inclusions as pretectonic; in high-grade rocks, early foliations may be destroyed by grain growth, and subsequent porphyroblast growth may give rise to apparently pretectonic structures. Also, porphyroblasts with planar inclusion patterns may seem to contain randomly oriented inclusions in sections parallel to S_i (Fig. 10.2b). Pretectonic porphyroblasts may be surrounded by a matrix with polyphase deformation, as shown in case b in Fig. 7.9 (Vernon et al. 1993a).

Chlorite-mica stacks (Fig. 4.14), which are common in slates, may be one of the few examples of pretectonic (or early syntectonic) porphyroblasts. They occur as lens or barrel-shaped aggregates of chlorite and white mica in microlithons of slates, with (001) planes at a high angle to the foliation and commonly parallel to bedding (Beutner 1978; Weber 1981; Craig et al. 1982; van der Pluijm and Kaars-Sijpesteijn 1984; Woodland 1985; Gregg 1986; Li et al. 1994). Some chlorite-mica stacks may have formed as detrital grains (Beutner 1978), but evidence exists for their original growth as crystals parallel to bedding in a diagenetic foliation at very low to low-grade metamorphic conditions, prior to deformation (Craig et al. 1982; Woodland 1985; Gregg 1986). Partial dissolution and new growth parallel to a developing microlithon of secondary foliation causes their final shape (Talbot 1965; Weber 1981; Gregg 1986; Clark and Fisher 1995; Li et al. 1994).

7.4.3
Intertectonic Porphyroblast Growth

The term intertectonic is introduced here for porphyroblasts that have grown over a secondary foliation, and

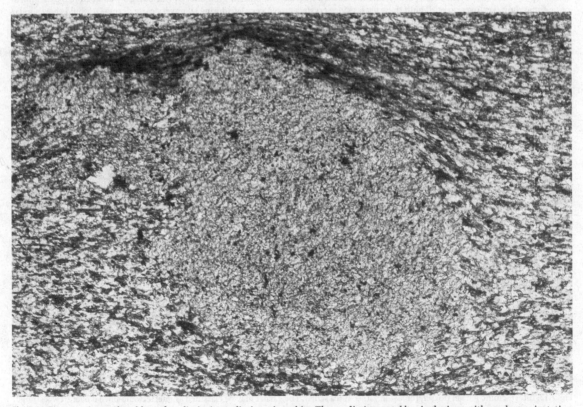

Fig. 7.11. Pretectonic porphyroblast of cordierite in cordierite-mica schist. The cordierite crystal has inclusions with random orientation showing that at the time of its growth the rock was a hornfels that lacked a directional fabric. Later deformation formed a foliation (S_1 *horizontal*), which is deflected around the cordierite crystal. Note the well-developed strain cap on top of the cordierite. Example of case a in Fig. 7.9. Leiden Collection. Width of view 2.5 mm. PPL

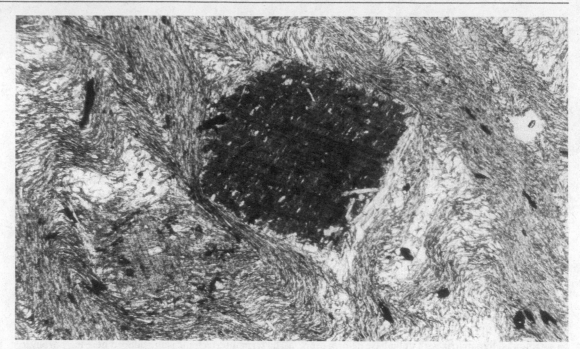

Fig. 7.12. Intertectonic biotite porphyroblast in biotite phyllite with approximately straight S_i (S_1) surrounded by a crenulation cleavage (S_2) that wraps around the crystal. This porphyroblast is labelled intertectonic since it grew after D_1 (including S_1) and before D_2. Its growth period may, however, overlap with both these phases since crystals grown very late during D_1 or very early during D_2 would show similar structures. Note mineral cleavage (001) within the biotite, almost perpendicular to S_i. Example of case d in Fig. 7.9. Concepción. Central-south Chile. Width of view 4 mm. PPL

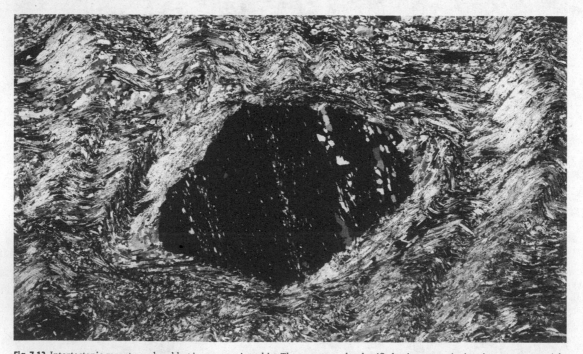

Fig. 7.13. Intertectonic garnet porphyroblast in garnet-mica schist. The garnet may be classified as intertectonic since it overgrew a straight secondary fabric ($S_i = S_n$) that has been deformed in the matrix by later deformation resulting in crenulation and relative rotation of about 90° of the garnet with respect to S_n. Since the continuity between S_i and S_e is almost completely destroyed by post-garnet deformation and grain growth in the matrix, an alternative explanation, considering S_i and S_n in the matrix as generated by entirely different phases, is also possible. Example of case d in Fig. 7.9. Tärnaby, Norrbotten, Sweden. Width of view 20 mm. Polars at 45°

Fig. 7.14. Hornblende crystal in quartz-albite-mica schist, with straight inclusions. S_i makes an angle with S_e of about 60°. If S_i and S_e are interpreted as resulting from the same deformation phase, the porphyroblast is syntectonic and must have grown relatively rapidly after a foliation had formed and before ongoing deformation produced the relative rotation of the crystal with respect to S_e. Alternatively the crystal may be intertectonic between D_n and D_{n+1} that caused the relative rotation. In this particular example the second interpretation is most likely because phyllites from the same area preserve a crenulation cleavage with intertectonic biotite porphyroblasts similar to those shown in Fig. 7.12. Example of case c1 in Fig. 7.9. Kittelfjäll, Västerbotten. Central Sweden. Width of view 4 mm. Polars at 45°

are surrounded by a matrix affected by a later deformation phase that did not leave any record in the porphyroblast (c and d in Fig. 7.9; Figs. 7.12, 7.13, ⊘Videos 7.9c, 9.7d). Porphyroblasts with a straight S_i that is oblique to S_e can also be named *oblique-S_i porphyroblasts* and are discussed in Sect. 7.6.8. (c1 in Fig. 7.9; Fig. 7.14, ⊘Photo 7.14).

7.4.4
Syntectonic Porphyroblast Growth

Syntectonic porphyroblasts have grown during a single phase of deformation D_n and are the most frequently encountered type of porphyroblasts in nature. This is probably due to the fact that deformation has a catalysing effect on mineral nucleation and diffusion rates (cf. Bell 1981; Bell and Hayward 1991). A large variety of microstructures can form in this group (Prior 1987; Figs. 7.9, 7.15–7.19, 7.32, ⊘Photos 7.15, 7.17, 7.18, 7.19a–c). The principal controlling variables are finite strain, the ratio of growth rate to strain rate, and the stage of progressive deformation during which the porphyro-

blasts grew (Barker 1994). The most characteristic syntectonic porphyroblasts form when growth and strain rates are of the same order of magnitude (Figs. 7.15, 7.32, ⊘Photo 7.15). Inclusion patterns are generally curved in syntectonic porphyroblasts, and random or straight in pre- and intertectonic porphyroblasts. S_i can be symmetrically arranged with respect to S_e, (e3, f3 in Fig. 7.9, ⊘Video 7.9f) or show oblique-S_i, sigmoidal or 'spiral' geometry (Sect. 7.6.8; e1 and e2 in Fig. 7.9). The latter is particularly common in garnet (e.g. Schoneveld 1979; Bell and Johnson 1989; Johnson 1993a,b; Williams and Jiang 1999; Fig. 7.39). Included folds in porphyroblasts (c3, f and h in Fig. 7.9) are known as *helicitic folds*. Porphyroblasts with oblique, sigmoidal and spiral-shaped patterns have sometimes been loosely referred to as *rotated porphyroblasts* (Sect. 7.6.8) but use of this term should be discouraged. The geometry may indicate *relative* rotation of porphyroblasts with respect to S_e but determination of movement of either fabric element in an external reference frame is more problematic (Ramsay 1962; Bell 1985; Johnson 1993a,b, 1999b; Sect. 7.6.8).

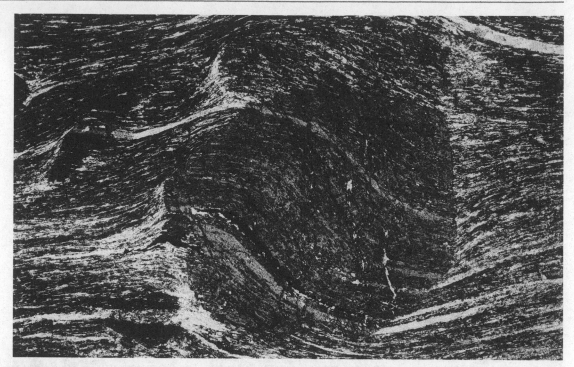

Fig. 7.15. Syntectonic garnet porphyroblast in garnet-kyanite schist showing about 45° of dextral rotation with respect to S_e during growth. Note the continuity between S_i and S_e, marked by opaque-poor bands that are continuous inside and outside the crystal. S_e is deflected around the porphyroblast and a poorly developed strain shadow is visible at its *lower left-hand side*. A late syntectonic porphyroblast of kyanite occurs at the *upper left side* of the garnet where it overgrew a strain cap. Example of case e1 in Fig. 7.9. Leiden Collection. Width of view 17 mm. PPL

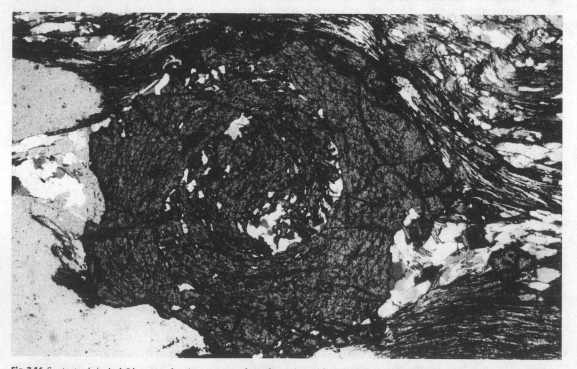

Fig. 7.16. Syntectonic 'spiral-S_i' garnet showing apparent dextral rotation with respect to S_e of approximately 300°, as deduced from the inclusion pattern of opaque minerals. Note the sharply curved trails of opaque inclusions in quartz-rich areas; these areas represent included strain shadows (cf. Fig. 7.37). Example of case e2 in Fig. 7.9. Rio Grande do Sul, Brazil. Width of view 5.5 mm. Polars at 45°

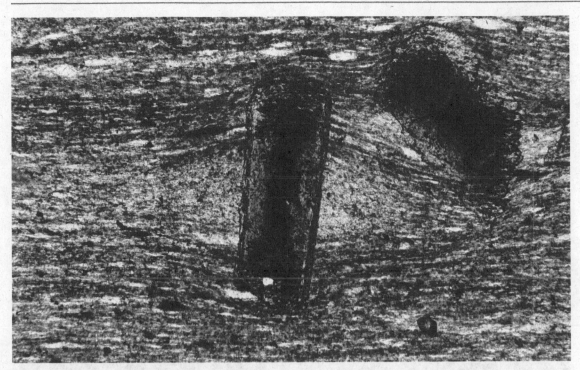

Fig. 7.17. Syntectonic chloritoid crystals (ottrelite variety) in slate. The central part of the main crystal does not contain any S_i, and the strain shadows next to this part seem to be free of cleavage, suggesting that the core of the crystal may be pretectonic. Only one phase of deformation can be recognised. Example of case e3 in Fig. 7.9. Curaglia, Switzerland. Width of view 1.8 mm. PPL

Fig. 7.18. Garnet porphyroblast in micaschist, syntectonic with respect to D_2. D_2 folds, traced by included opaque grains, are gentle in the garnet core becoming tighter towards the rim. S_2 shows clear deviation around the lower part of the crystal. No strain shadows have developed, probably because the garnet grew in a mica-rich band. The crenulation cleavage (S_2) and S_i (= S_1) are mainly defined by opaque minerals. Example of case f3 in Fig. 7.9. São Felix de Cavalcante, Goiás, Brazil. Width of view 3 mm. Polars at 45°

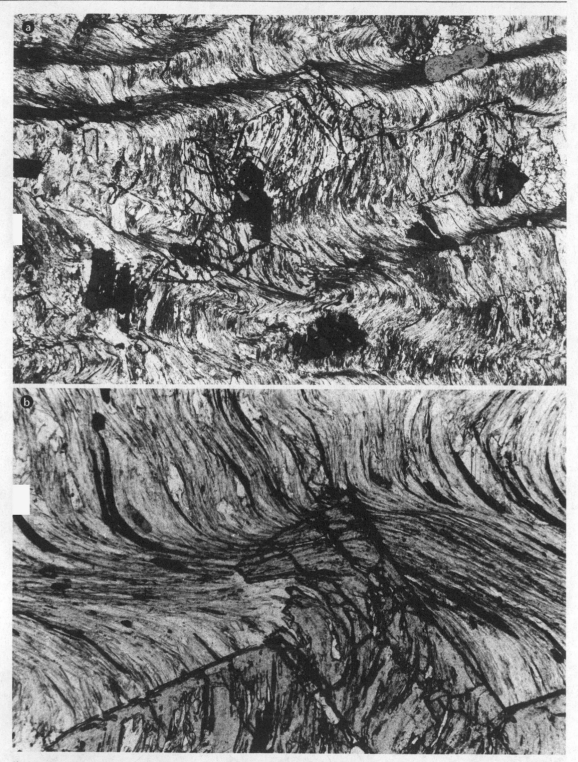

Fig. 7.19. Garnet kyanite staurolite schist with S_2 crenulation cleavage and staurolite porphyroblasts that are syntectonic with respect to D_2. **a** Inclusions in the central light-coloured staurolite crystal show gentle folding, becoming tighter towards the upper rim. Lukmanier Pass, Switzerland. Width of view 9.6 mm. Polars at 45°. **b** Detail of central upper part of photograph **a**, showing the abrupt curvature of S_1 in the outer rim of the staurolite crystal. S_2 is deflected around the crystal that partly overgrows the mica-rich domains of S_2. Example of case f3 in Fig. 7.9. Width of view 2.4 mm. PPL

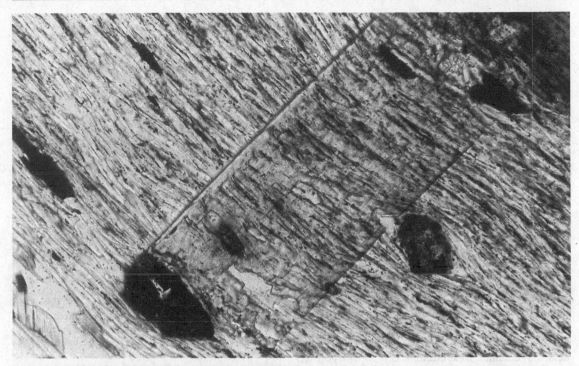

Fig. 7.20. Post-tectonic chlorite crystal overgrowing a slaty cleavage. The cleavage structure is perfectly continuous without any deflection or development of strain shadows. Older strain shadows around opaque grains are included without modification. Example of case g in Fig. 7.9. Araí, Goiás, Brazil. Width of view 0.6 mm. PPL

Fig. 7.21. Post-tectonic biotite porphyroblasts overgrowing a vertical crenulation cleavage (S_2). Although no inclusion patterns have developed, the lack of deflection of S_2 and the lack of strain shadows show that the biotite crystals grew after the crenulation cleavage development. The opaque crystals at *left* are probably of the same age. However, very late syntectonic growth is also possible, as discussed in the text. Example of cases c2 and d in Fig. 7.26. Ouro Preto, Minas Gerais, Brazil. Width of view 1.8 mm. PPL. (Thin section courtesy Hanna Jordt-Evangelista)

7.4.5
Post-Tectonic Porphyroblast Growth

This group is easy to define by the absence of deflection of S_e, strain shadows, undulose extinction or other evidence of deformation, which is common to pre- syn- and intertectonic porphyroblasts (g and h in Fig. 7.9). If inclusions are present, S_i is continuous with S_e (g in Fig. 7.9), even if folded (h in Fig. 7.9, ⊘Video 7.9h). Some care is needed with apparently post-tectonic porphyroblasts. It is not uncommon to find weak deformation effects, including strain shadows, in or around some crystals in a population of apparently post-tectonic porphyroblasts. In fact, there are no reliable criteria to distinguish between very late syntectonic porphyroblasts and post-tectonic ones (Figs. 7.3, 7.20, 7.21, ⊘Photos 7.20, 7.21).

7.4.6
Complex Porphyroblast Growth

A large number of combinations of the categories of porphyroblast-deformation relations mentioned above are possible, especially if a mineral has several growth phases. Relatively common are syntectonic crystals with post-tectonic rims, but combinations of porphyroblasts with pretectonic cores and syntectonic rims also occur (Fig. 7.17, ⊘Photo 7.17). These complex relations are easily overlooked but may be recognised by an unusual geometry

or abrupt changes in the geometry of the inclusion pattern, especially if this is associated with zoning in the porphyroblast (Sect. 7.6.8).

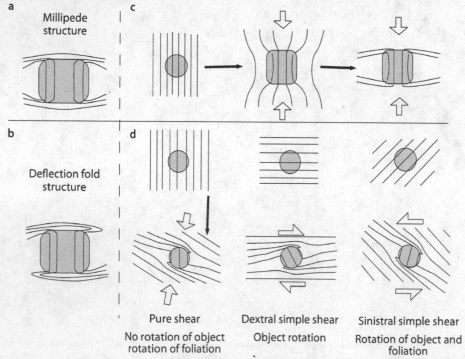

Fig. 7.22. a Millipede structure and **b** deflection fold structure for two inter- to syntectonic porphyroblasts. **c** Development of a millipede structure by coaxial flattening. **d** Three alternative ways to develop a deflection fold structure

7.5
Millipede, Deflection-Fold and Oppositely Concave Microfold (OCM) Microstructures

Bell and Rubenach (1980) and Bell (1981) have drawn attention to a microstructure in some inter- and syntectonic porphyroblasts, which they named *millipede microstructure* (Fig. 7.22a, ⊘Video 7.22). The term refers to syn- or intertectonic porphyroblasts around which S_e is deflected in opposite directions (Fig. 7.23, ⊘Photo 7.23). In a related type of structure, S_e is deflected through isoclinal folding at both sides of a porphyroblast in a way as shown in Fig. 7.22b. This structure has been called *deflection-fold structure* by Passchier and Speck (1994; Fig. 7.24). Johnson and Moore (1996) introduced the concept of *oppositely concave microfolds* (OCMs), referring to similar structures. These structures are an effect of foliation deflection adjacent to a rigid porphyroblast. Structures similar to deflection folds and millipede structures can be reproduced experimentally in homogeneous, non-partitioned flow around rigid objects (Ghosh and Ramberg 1976; Masuda and Ando 1988; Gray and Busa 1994). A possible alternative explanation for these structures is presented by Beaumont-Smith (2001) as the progressive overgrowth of conjugate crenulation cleavages. Bell (1981) claimed that deformation in rocks is generally partitioned into lenses ('pods') with little deformation or predominantly coaxial shortening, surrounded by an anastomosing network of shear zones (Fig. 7.25). He envisaged that porphyroblasts with millipede structures grow syntectonically in these pods until they impinge on the surrounding shear zones where they stop growing or dissolve. This does not seem to be generally valid, however; many millipede structures and deflection folds may be at least in part intertectonic, and may have grown over a straight foliation before a second foliation is developed; they should therefore be interpreted with care. The idea that porphyroblasts stop growing when they reach a shear zone or cleavage lamella (Bell 1981) seems also not universally applicable as illustrated by a staurolite porphyroblast that overgrew a cleavage lamella (Fig. 7.19b). In a review of five different types of OCMs Johnson and Bell (1996) conclude that these structures do in general not provide unequivocal evidence for non-coaxial or coaxial deformation histories. However, finite longitudinal strains can be measured in rocks that contain porphyroblasts with OCMs, by comparing the spacing between cleavage planes in and outside the porphyroblasts (Johnson and Williams 1998).

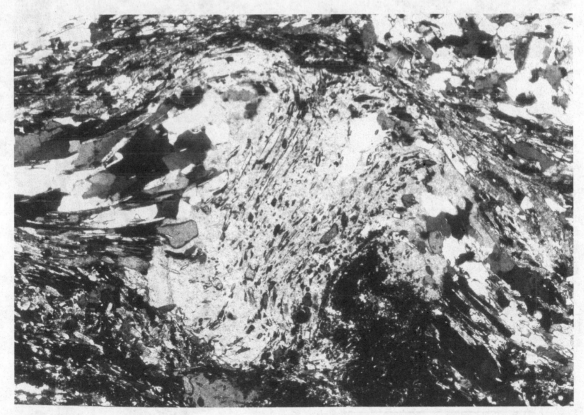

Fig. 7.23. Albite porphyroblast (*light grey*) with inclusion pattern characteristic of a millipede microstructure. S_i corresponds to S_1 and S_e to S_2. Garnet albite schist. Elephant Island, West Antarctica. Width of view 2.8 mm. CPL

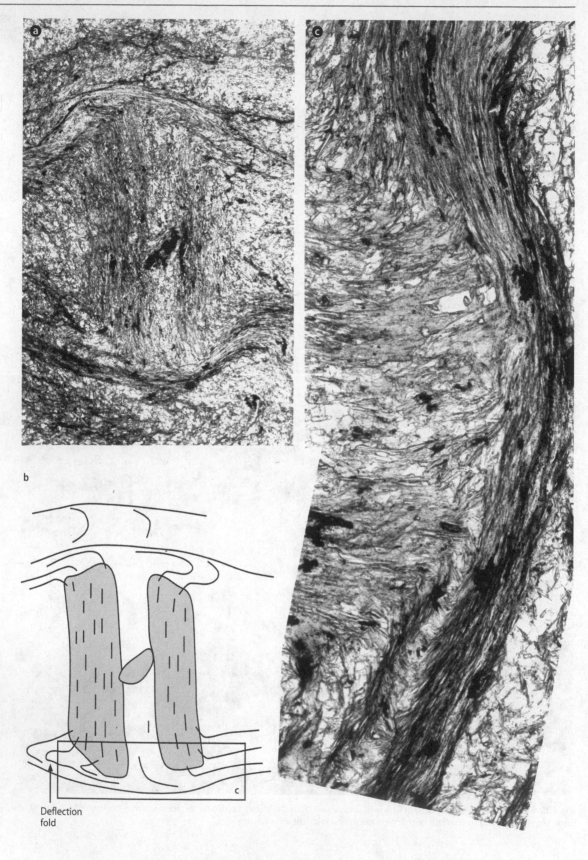

Deflection
fold

◀ **Fig. 7.24. a** Spool-shaped andalusite porphyroblast with symmetric distribution of deflection folds on both sides. **b** Schematic drawing of the structure in **a**. **c** Enlargement of the lower half of **a**, showing the deflection folds. Trois Seigneurs Massif, Pyrenees, France. Length of porphyroblast in **a** 2.9 mm. PPL

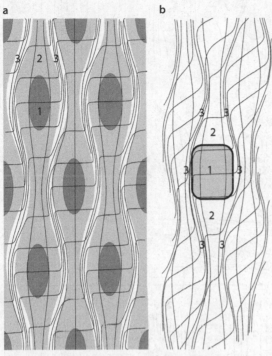

Fig. 7.25. a Deformation partitioning in bulk non-coaxial progressive deformation. **b** Deformation partitioning around a porphyroblast (*centre*) in bulk non-coaxial deformation. Millipede and deflection-fold structures are thought to develop by overgrowth of such partitioning patterns. *1* no strain; *2* coaxial progressive deformation; *3* non-coaxial progressive deformation. (After Bell 1985)

7.6
Problematic Porphyroblast Microstructures

7.6.1
Inclusion-Free Porphyroblasts

The presence or absence of inclusions in porphyroblasts cannot be used to date them with respect to other fabric elements (Fig. 7.26). If porphyroblasts do not have inclusions, it is difficult or even impossible to date their growth with respect to deformation. The relative age can in some cases be determined by the intensity of deflection of S_e or from the presence of strain shadows (Fig. 7.26). If there is no deflection of S_e, porphyroblasts may be post-tectonic (c in Fig. 7.26); if there is deflection of S_e or if strain shadows are present, porphyroblasts are pre-, inter or syntectonic (Figs. 7.26a and b, 7.46 at right, ✆Photo 7.46). Care has to be taken since deflection may be caused by later deformation phases and late

shortening normal to an earlier foliation (e.g. Bell 1986; Bell et al. 1986a). Strain shadows are not always accompanied by deflection of S_e.

Large, isolated elongate mineral grains such as micas and amphiboles which lie parallel to a foliation and which lack inclusion patterns can also be difficult to date relative to deformation (Fig. 7.26b and d). They may be inter-, syn-, or post-tectonic. Syntectonic porphyroblasts are, if elongate, usually well aligned with the foliation. Intertectonic porphyroblasts may have rotated towards the foliation plane and may be recognisable by slight but consistent obliqueness to the foliation and evidence of internal deformation or replacement along the edges (b in Fig. 7.26). Post-tectonic porphyroblasts are difficult to recognise if they have grown mimetically parallel to a pre-existing foliation. However, their post-tectonic nature can occasionally be recognised if some of the crystals overgrew the foliation obliquely (Figs. 7.21, 7.26d).

7.6.2
Shape and Size of Inclusions compared to Matrix Grains

In many cases where a porphyroblast overgrew a structure without later deformation, the size and shape of inclusions differs little from those in the matrix (Figs. 7.1, 7.3, 7.6, 7.8, 7.27a). This is especially notable in the case of opaque minerals; structures like grain shape preferred orientation, crenulation cleavage and compositional layering can be perfectly preserved in this way as inclusion patterns. Since many porphyroblasts behave as rigid bodies during later deformation, the inclusion patterns, once incorporated, can remain unaffected by later deformation or modification by grain growth, dynamic recrystallisation and transposition that may affect the matrix. In this way porphyroblasts often preserve stages in the tectonometamorphic evolution that would otherwise be lost (e.g. Figs. 7.4, 7.5). In some cases, inclusions in porphyroblasts may have another size and shape than matrix grains of the same mineral (Fig. 7.27b,c). Commonly, the inclusions are smaller and somewhat more rounded in shape than similar grains in the matrix (e.g. Fig. 7.6). This may be caused by partial diffusion, or by a reaction involving the included mineral. A decrease in size of included grains from core to rim in a porphyroblast may reflect a gradual change in diffusion rate or porphyroblast growth rate. More common is an increase in size from core to rim often accompanied by a sharp contrast with much larger matrix minerals. This can be explained by progressive coarsening of the matrix during and after porphyroblast growth, as a consequence of temperature increase and accompanying grain boundary area reduction (GBAR; Fig. 7.27b; Sect. 3.10). The presence of large inclusions compared to finer-grained matrix minerals, generally indicates grain size reduction due to deformation in the matrix (Fig. 7.27c). This is particularly common in mylonites (Sect. 5.3.4).

7.6

Fig. 7.26.
Schematic representation of pre-, syn-, and post-tectonic porphyroblast growth for porphyroblasts that lack passive inclusions. Only distinction of post-tectonic porphyroblasts is relatively easy. Distinction of pre-, inter- and syntectonic porphyroblasts is usually impossible, but they can commonly be distinguished from post-tectonic porphyroblasts by internal deformation features such as undulose extinction

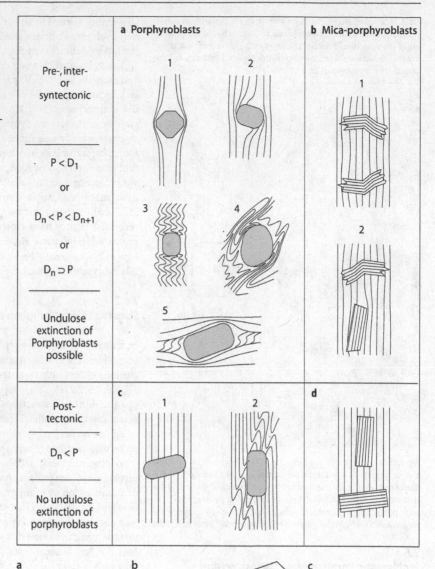

a Porphyroblasts

b Mica-porphyroblasts

Pre-, inter- or syntectonic

$P < D_1$

or

$D_n < P < D_{n+1}$

or

$D_n \supset P$

Undulose extinction of Porphyroblasts possible

Post-tectonic

$D_n < P$

No undulose extinction of porphyroblasts

Fig. 7.27.
a Growth of a porphyroblast over matrix material. Inclusions are more rounded but of similar size to those in the matrix. After porphyroblast growth, grains in the matrix may become coarser by static recrystallisation and grain growth (**b**), or smaller by dynamic recrystallisation due to deformation (**c**)

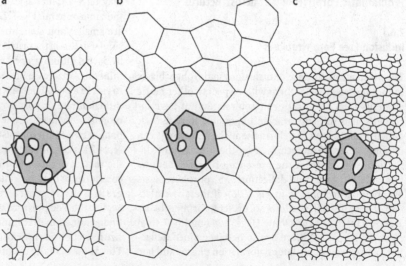

7.6.3
False Inclusion Patterns

Some structures in porphyroblasts may resemble patterns of passive inclusions, but form in another way and can be a source of error. *Rutile needles* in biotite or quartz could be misinterpreted as passive inclusions, but usually lack a clear preferred orientation. Alteration (e.g. of feldspar to sericite) or exsolution structures along crystallographically controlled planes may also be difficult to distinguish from inclusion patterns. Minerals that are solid solutions of two or more phases can show exsolution when metamorphic conditions change. This is especially common for minerals that crystallised at high temperature. During retrogression, small grains of the minor phase may form in the host crystal (Figs. 7.28, 7.29). The most common examples are found in feldspar (Fig. 7.28), amphiboles, pyroxenes (Fig. 7.29) and spinel.

In the case of feldspars, the new phase may occur as elongate grains or lamellae with a strong preferred orientation within the host crystal, parallel to crystallographic directions. In K-feldspar, this exsolution structure is known as *perthite* (Fig. 7.28); in plagioclase as *anti-perthite*. In K-feldspar, deformation of old crystals may cause local enhancement of exsolution, especially when high-grade feldspars are deformed at low grade (Sect. 3.12.4).

In general, exsolution structures may be distinguished from passive inclusions by their composition and their control by crystallographic planes.

7.6.4
Mimetic Growth

Mimetic growth (Sect. 4.2.7.6) can be a problem in porphyroblast analysis. In some cases porphyroblasts of a mineral A may be mimetically replaced by a mineral B that inherits the inclusion pattern of A and the deflection pattern of the foliation in the matrix around it. This is a potential source of error; however, in our experience this replacement is seldom complete. Generally, it can be detected by the presence of relicts of mineral A, or by other crystals of mineral B that show the correct microtectonic relation. This kind of (partial) replacement structures can give important information on the metamorphic evolution and is explained in Sect. 7.8.

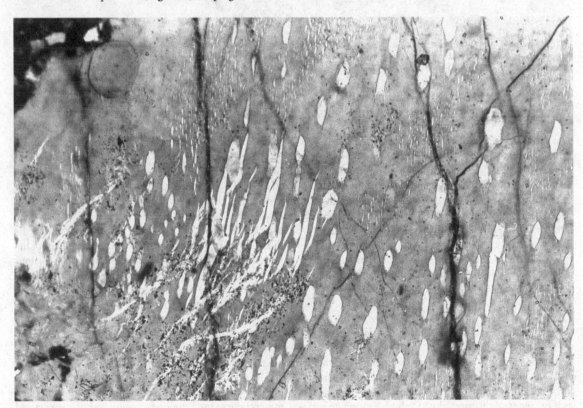

Fig. 7.28. Ellipsoidal and flame-shaped albite lamellae in perthitic K-feldspar from a granodiorite. Two sets of ellipsoidal lamellae of different size are present, possibly reflecting two stages of dissolution at different temperature (Sect. 9.9). The flame-shaped perthite-lamellae result from unmixing during greenschist facies deformation. The preferred orientation (N-S) is controlled by crystallographic directions. This microstructure should not be mistaken for passive inclusions. St. Barthélemy Massif, Pyrenees, France. Width of view 0.6 mm. CPL

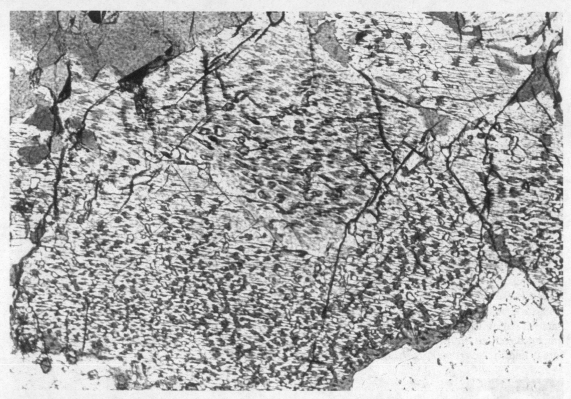

Fig. 7.29. Large clinopyroxene crystal showing substitution by hornblende in an exsolution-like structure that may be mistaken for passive inclusions. The crystallographic control of this type of false inclusions reveals their true nature. Retrograded granulite. Southern Minas Gerais State, Brazil. Width of view 1.8 mm. PPL

7.6.5
Deformed Porphyroblasts

The relations in Fig. 7.9 can also be recognised if pre-, inter- or syntectonic porphyroblasts show evidence of intracrystalline deformation such as microboudinage, undulose extinction and formation of subgrains. However, special attention is needed for crystals with helicitic folds; if an entire crystal is folded with the same wavelength and amplitude as the inclusion pattern, S_i may have been originally straight. Such crystals, recognisable by their undulose extinction, must not be mistaken for undeformed porphyroblasts containing true helicitic folds (Sect. 7.4.4).

7.6.6
Uncertain Age Relation of Host and Inclusions

In the previous sections, the recognition of passive inclusions, that is the inclusion of mineral grains that totally predate their host porphyroblast, has been presented as a rather straightforward procedure. If elongate inclusions of a certain mineral occur in groups in a porphyroblast and are not oriented parallel to a crystallographic direction, or even better, if they trace structures such as layering or folds in connection with similar structures in the matrix, they are most probably passive inclusions. Other inclusions that are probably included passively are quartz grains, which form part of overgrown strain shadows (Figs. 7.6, 7.39a) and inclusions with identical size, shape and distribution as grains of the same mineral in the matrix. Using these passive inclusions, the age relation of deformation and porphyroblast growth can be established (Sect. 7.4). However, in some cases it may be difficult to determine the precise *growth sequence* of the mineral species that form porphyroblasts and inclusions (cf. Flood and Vernon 1988; Barker 1994). If isolated grains of mineral A are included in grains of mineral B, they may either have grown in B (e.g. as sericite in feldspar), or they may have been included during growth of B. The latter situation, however, does not necessarily mean that mineral B formed *entirely* later than mineral A. It is possible to establish from such relations that minerals A and B were stable or growing together for a period of time, but it is very difficult or even impossible to decide which mineral started and stopped growing first (e.g. garnet and blue amphibole in Fig. 6.33). The only possible exceptions to this rule occur if all grains of A are enclosed in the centre of grains of B, or if single grains of A are rimmed by coronas of B (Sect. 7.8). Figure 7.30 shows an example where either staurolite or garnet start growing. In the final stages of both situations, the garnet includes a staurolite grain in its rim, giving the impression that garnet is the younger mineral (Fig. 7.30, ⊘Video 7.30a,b). The inclusion of one mineral by the other is therefore not always indicative of different age, but may result from different nucleation and growth rates of both minerals (Fig. 7.2). Minerals with low nucleation rate and rapid growth rate engulf those with high nucleation rate (Fig. 6.33).

As a conclusion, the relation between periods of mineral growth and deformation can be established in most cases, using the relations shown in Figs. 7.9 and 7.26; however, the relative age of growth periods of different mineral species is more difficult to establish.

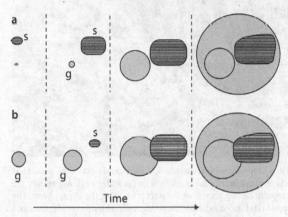

Fig. 7.30. Sequence of events to show that it is difficult to estimate the relative age of two minerals from inclusion relations. In **a** staurolite (*s*) starts to grow and becomes included in garnet (*g*) because the latter mineral has a higher growth rate; in **b** garnet starts to grow, including staurolite that starts to grow later, leading to the same inclusion relationship

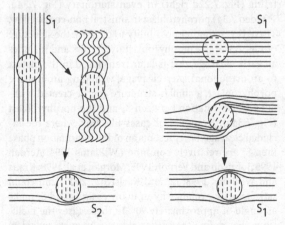

Fig. 7.31. If all connection is lost between S_i and S_e, they may either represent two different foliations S_1 and S_2 (*left*), or early and late stages of development of a single foliation (*right*)

7.6.7
Discontinuous S_i and S_e

If S_i is discontinuous with S_e (Figs. 7.5, 7.13, 7.31), deformation must have taken place after porphyroblast growth. However, it is not correct to attribute S_i in such cases always to the last preceding deformation phase, since it may also represent any earlier phase or an earlier stage in the progressive development of the external foliation, S_e (e.g. Figs. 7.5, 7.34e). In multiple deformed regions with discontinuous S_i and S_e no reliable relation can be established (Johnson and Vernon 1995a) by textural criteria only and several possible interpretations should be tested by other means, such as absolute age dating of minerals.

7.6.8
Rotation of Porphyroblasts

According to theory, rigid objects suspended in a homogeneously deforming matrix are expected under certain circumstances to rotate with respect to the ISA of flow (Box 5.4); this applies to elongate objects oriented oblique to shortening axes in coaxial progressive deformation, and to equidimensional and many elongate objects in non-coaxial progressive deformation (Fig. 5.23; Ghosh and Ramberg 1976; Masuda and Ando 1988). Equidimensional inter- and syntectonic porphyroblasts with oblique, sigmoidal and spiral-shaped S_i patterns (Figs. 7.14–7.16, 7.32, 7.33, ⊘Photo 7.32a,b) have therefore intrigued the imagination of many geologists as natural examples of objects that rotated with respect to ISA, capable of indicating sense of shear (e.g. Zwart 1960; Rosenfeld 1968, 1970, 1985; Cox 1969; Powell and Treagus 1969, 1970; Spry 1969; Wilson 1971; Trouw 1973; Schoneveld 1977, 1979; Powell and Vernon 1979; Olesen 1982; Lister et al. 1986; Vernon 1988; Miyake 1993; Johnson 1993a,b, 1999b; Williams and Jiang 1999). Two basic types can be distinguished: those with a straight or slightly curved S_i that makes an angle of up to about 90° with S_e (*oblique-S_i or sigmoidal-S_i porphyroblasts*; Sect. 7.4.3), and those with a spiral-shaped S_i showing apparent rotation angles in excess of 90° (*spiral-S_i porphyroblasts*). The sequence of events leading to the development of oblique-S_i porphyroblasts (Figs. 7.14, 7.34, Box 7.4; ⊘Video 7.34; Passchier and Speck 1994) can be as follows: (1) development of a foliation S_n; (2) growth of the porphyroblast, including S_n as S_i; (3) deformation that either causes relative rotation of the porphyroblast with respect to S_n or the development of a new foliation S_{n+1} without rotation. Such porphyroblasts are intertectonic, or syntectonic if the porphyroblast growth rate exceeded strain rate significantly. An intuitive explanation of oblique-S_i porphyroblasts (Fig. 7.34a, ⊘Video 7.34) is *dextral* rotation of the

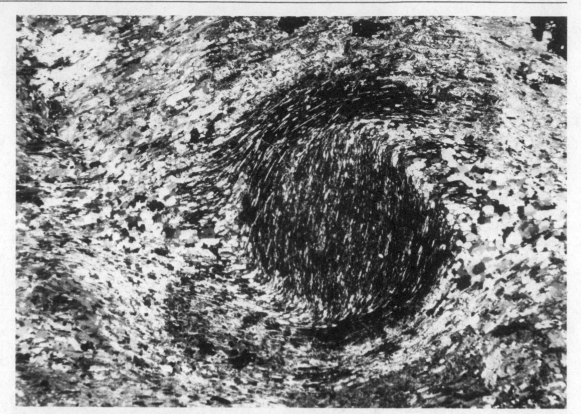

Fig. 7.32. Syntectonic garnet in micaschist showing almost 90° of sinistral rotation with respect to S$_e$. The central part of the garnet has approximately straight inclusions whereas close to the rim a deflection plane marks the transition to the outer part. Apparently the deformation responsible for the garnet rotation occurred relatively rapidly during growth of the part containing the abrupt bend. The outer rim grew over the matrix, apparently after the rotation had stopped (cf. Fig. 7.39a). Note the strong differentiation in the outer rim between strain shadows rich in quartz where the garnet grew in a skeletal way (enlarged in Fig. 7.6), and garnet rims at the upper and lower part where the garnet contains relatively few inclusions, corresponding to strain caps. Baños, Ecuador. Width of view 20 mm. CPL

Box 7.3   Rotation and reference frames

Many publications on microstructures describe the rotation of porphyroblasts or other structures. Unfortunately, in many statements, no mention is made of reference frames, e.g. 'the shape of the inclusion pattern shows that garnet rotated'. Without further explanation, it is not clear if the garnet rotated with respect to a foliation, with respect to bedding, or with respect to the earth's surface. This example illustrates that, without a proper reference frame, rotations cannot be defined (Sect. 2.4; Box 2.2; Fig. B.2.2). In all descriptions of rotation, a reference frame should therefore be given. Possible examples are the foliation in a rock, ISA, or geographical coordinates.

---

porphyroblast with respect to a stable S$_e$ and to flow ISA in a dextral non-coaxial flow (Figs. 7.22d centre, 7.34b). However, Ramsay (1962) pointed out that a pre-existing foliation at an angle to ISA would rotate with respect to the ISA in coaxial flow, while an equidimensional porphyroblasts in the same setting may remain stable. The structure in Fig. 7.34a can therefore also be explained by *sinistral* rotation of S$_e$ with respect to ISA and to a sta-

tionary porphyroblast in coaxial flow (Fig. 7.22d left, 7.34c, ⊘Video 7.34) or with respect to a more slowly rotating (Fig. 7.22d right) or even stationary (Fig. 7.34d, ⊘Video 7.34) porphyroblast in sinistral non-coaxial flow. There is yet another possibility to produce the structure of Fig. 7.34a: if a porphyroblast overgrows and includes an early cleavage, that is later transposed in the matrix to an orthogonal later cleavage, wrapping around the porphyroblast, a similar structure can be created without relative rotation between S$_e$ and the porphyroblast (Fig. 7.34e). Examples of cases like this, where the included cleavage, S$_i$, represents an older deformation phase then S$_e$ are relatively common (Williams 1994; Aerden 1995; Johnson and Vernon 1995; Morgan et al. 1998; Kraus and Williams 1998; Ilg and Karlstrom 2000) and could satisfactorily explain why in many cases S$_i$ and S$_e$ make an angle of approximately 90°. In these cases the inclusions that define the earlier cleavage may be considerably smaller then the matrix minerals that define S$_e$, due to progressive mineral growth in the matrix during the later cleavage development. These alternative interpreta-

Fig. 7.33. Syntectonic spiral-S_i garnet in micaschist, showing a double spiral of inclusions. One spiral consists of coarse quartz grains that nearly divide the garnet into two parts; the other spiral consists of fine graphite inclusions in massive garnet. Notice the tight folds in garnet and graphite at *lower centre* and *top left* which have been predicted in the model of Schoneveld (1977) as shown in Fig. 7.37. Aiuruoca, Minas Gerais, Brazil. Width of view 10 mm. PPL

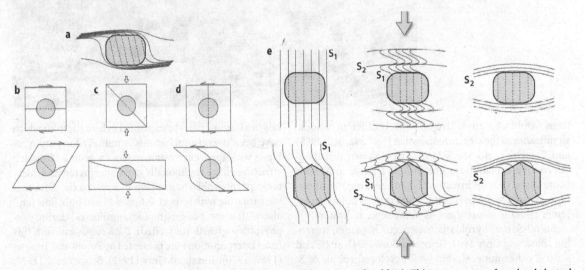

Fig. 7.34. a Intertectonic porphyroblast with straight S_i oblique to S_e (oblique-S_i porphyroblast). This structure can form by: **b** dextral rotation of the porphyroblast with respect to a less-rotating foliation in dextral non-coaxial flow, **c** rotation of the foliation around a stationary porphyroblast in coaxial flow and **d** sinistral rotation of the foliation with respect to a stationary porphyroblast in sinistral non-coaxial flow. The porphyroblast can be stationary in non-coaxial flow if it lies in, and is coupled with a non-deforming or coaxially deforming microlithon. **e** Progressive development of crenulation cleavage around a stationary porphyroblast; the crenulation cleavage may later be modified to a continuous foliation as illustrated in Figs. 4.18 and 4.19. The upper part of the figure refers to an intertectonic porphyroblast, whereas the lower part shows a porphyroblast that is early syntectonic with respect to D_2

Fig. 7.35.
a Syntectonic albite porphyroblasts with sigmoidal S_i pattern
and asymmetric strain shadows
of quartz. The asymmetry of the
strain shadows indicates dextral
shear sense, which implies that
the albite rotated in a clockwise
sense with respect to the kinematic frame during its growth.
b Detail of **a**. Sanbagawa Belt,
Japan. Width of view **a** 2.5 mm;
b 1 mm. PPL. (Photographs
courtesy S. Wallis)

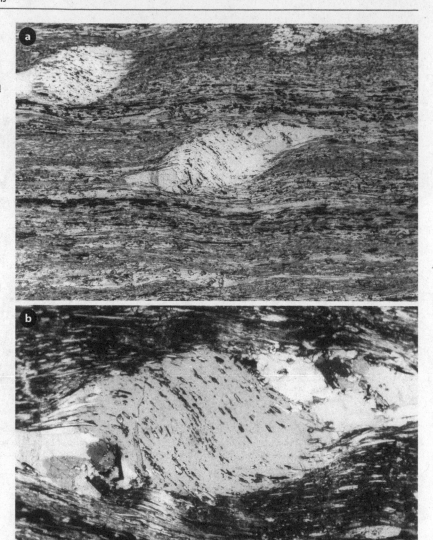

tions would, of course, give an entirely different tectonic
significance to the structure shown in Fig. 7.34a. Bell (1985)
and Passchier and Speck (1994) have shown that these
alternative explanations are indeed feasible in at least
some cases. Careful investigation of the geometry of ob-
lique-S_i porphyroblasts is therefore needed before an attempt is made to use them as shear sense indicators.

Spiral-S_i porphyroblasts may occur in garnet, staurolite (Busa and Gray 1992; Gray and Busa 1994), albite and
several other minerals, but well-developed spirals of S_i
with a relative rotation angle exceeding 180°, also known
as *snowball structures* (Figs. 7.16, 7.33), seem to be restricted to garnet (e.g. Rosenfeld 1970; Schoneveld 1977,
1979; Powell and Vernon 1979; Johnson 1993a,b, 1999b;
Moore and Johnson 2001). Porphyroblasts with open spi-

rals can in some cases be explained as helicitic folds which
have been overgrown, but this is unlikely for porphyroblasts with apparent rotation angles exceeding 180°; such
structures have traditionally been interpreted as syntectonic porphyroblasts, growing in non-coaxial flow, that
were rotating with respect to S_e and ISA of bulk flow (and
also with respect to geographic coordinates) during porphyroblast growth (refs. cited). Data consistent with this
latter interpretation were presented by Powell and Treagus
(1969, 1970), Busa and Gray (1992), Schoneveld (1977,
1979), Powell and Vernon (1979), Passchier et al. (1992),
Johnson (1993a,b), Williams and Jiang (1999).

Powell and Treagus (1969, 1970) have shown that inclusion patterns in spiral-S_i garnets are not cylindrical
in three dimensions, but commonly mirrored around a

Fig. 7.36.

a Schematic diagram of the geometry of three S_i inclusion surfaces (*1, 2* and *3*) in a spherical syntectonic spiral-S_i model porphyroblast. The porphyroblast is embedded in a matrix undergoing simple shear and grew during rotation over 90° with respect to ISA of bulk flow. Orthogonal reference axes represent the rotation axis of the porphyroblast (*R*), the pole to the flow plane (*S*) and the fabric attractor (*FA*) of bulk flow, respectively. **b** Serial sections through the model normal to the R-axis and S_e. The angle Θ, which defines the amplitude of the curve of S_i, decreases outward from the centre of the sphere and the trend of the inclusion trails changes systematically. **c** S_i patterns for various orientations of section planes through the centre of the model porphyroblast in **a**. (After Powell and Treagus 1970; Gray and Busa 1994)

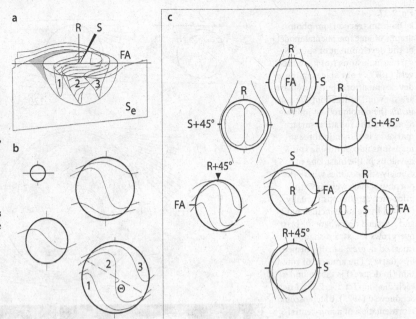

plane through the centre of the porphyroblast and normal to a twofold symmetry axis (Fig. 7.36a,b). They also demonstrated that central sections normal to the symmetry axes in such crystals show larger rotation angles than side sections (cf. Fig. 7.36c). Interestingly enough, some cross-sections through spiral-S_i porphyroblasts (Fig. 7.36c – R/FA section at right) strongly resemble millipede structures (Sect. 7.5; Gray and Busa 1994; Johnson and Moore 1996).

Busa and Gray (1992) studied oblong staurolite porphyroblasts that lie parallel to S_e with a variable orientation of the long axis within S_e. They found a three-dimensional geometry of S_i similar to that described by Powell and Treagus (1969, 1970), with the symmetry axis of the S_i-spiral parallel to the long axis of the porphyroblast, despite the variable orientation of the porphyroblasts in space. They also found a systematic relation between the 'rotation angle' of S_i and the orientation of long axes of the porphyroblasts. Such a three-dimensional geometry of the S_i pattern agrees well with theoretically predicted S_i patterns formed by syntectonic growth of a porphyroblast that was rotating with respect to S_e and ISA of bulk flow (Schoneveld 1979; Masuda and Mochizuki 1989; Gray and Busa 1994).

Schoneveld (1979) discussed garnet crystals in which two sets of spiral inclusion trails are present, one of quartz inclusions, and one of opaque inclusions (Fig. 7.33). He modelled these spiral S_i trails as resulting from the inclusion of quartz from strain shadows and opaque grains from strain-caps during rotation of the growing porphyroblast with respect to the ISA of bulk flow (Fig. 7.37, ⊘Video 7.37, 7.37a). Schoneveld (1977) has shown how the

ratio of growth rate versus rotation rate of porphyroblasts can determine the shape of spiral trails of quartz and opaque inclusions (Fig. 12 in Schoneveld 1977). This model, elaborated in three dimensions, has been tested in natural examples and seems to explain observed structures in a satisfactory manner (cf. Powell and Vernon 1979; Johnson 1993a,b; Gray and Busa 1994; Fig. 7.38). More recently, Samanta et al. (2002) presented a theoretical study, showing how complex inclusion patterns may form in rotational porphyroblasts with varying angle between the foliation and the shear plane and with varying ratio between pure and simple shear.

Wallis (1992b) reported a structure where the geometry of an S-shaped inclusion pattern in albite porphyroblasts and the asymmetry of flanking quartz strain shadows on these porphyroblasts can both be explained by dextral rotation of the porphyroblasts with respect to ISA of dextral non-coaxial flow in the matrix (Fig. 7.35).

Despite the microstructural and theoretical support for a 'rotational' origin of spiral-S_i porphyroblasts mentioned above, Bell (1985) and Bell et al. (1986a), first questioned the development of spiral-S_i garnets by porphyroblast rotation with respect to ISA of bulk flow, and advocated the theory that porphyroblasts do not rotate in a reference frame fixed to geographical coordinates. In many spiral-S_i garnets Bell and Johnson (1989) identified so-called *truncation planes* where the spiral inclusion pattern is interrupted (Figs. 7.1b, 7.39b). They interpreted S_i on both sides of a truncation plane as representing separate deformation phases, and reinterpreted the spiral fabrics as successively overgrown helicitic folds during up to eight subsequent deformation phases, in-

Fig. 7.37.

a Diagram traced from photographs of an experimental model of the development of spiral-S_i garnet inclusions from Schoneveld (1977). Four stages in the development of a porphyroblast are shown. The lines *outside* and *inside* the porphyroblast indicate S_e and S_i. The foliation surface marked *s* is shown in three dimensions in **b**. A double spiral develops in the blast, one of densely spaced lines where the porphyroblast overgrows the mica-rich strain caps (usually formed by opaque inclusions), and a second one where the blast overgrows the strain shadow (marked in *grey*; usually formed by quartz). The amount of rotation (in degrees) is shown under each diagram (cf. Fig. 7.33). After Schoneveld (1977). **b** Stereoscopic representation of a non-central S_i surface in a model garnet with a rotation angle of 400°, marked as *s* in lower part of **a**. (Schoneveld 1979)

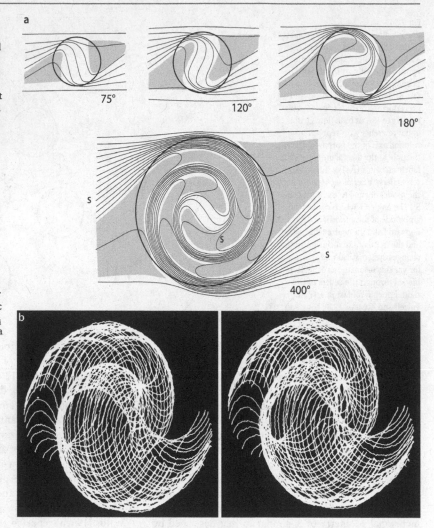

cluded without rotation of the porphyroblast with respect to geographical coordinates. However, the truncation planes can also be explained by overgrowth of strain caps by the porphyroblast (Fig. 7.39; Sect. 6.5; Passchier et al. 1992). Since strain caps in metapelites are mica-rich, Al-silicate porphyroblasts tend to overgrow their own strain caps (Fig. 7.32). Pulsating growth of a rotating porphyroblast, alternating with periods during which mica-rich strain caps develop can cause development of sharp *deflection planes* (Figs. 7.32, 7.39a). If porphyroblast growth is temporarily alternating with local dissolution, truncation planes as reported by Bell and Johnson (1989) can also develop during progressive deformation (Fig. 7.39b).

The models of Bell (1985, Fig. 7.25) and Bell and Johnson (1989) have led to extensive debate (e.g. Bell et al. 1992a–c; Busa and Gray 1992; Passchier et al. 1992; Visser and Mancktelow 1992; Wallis 1992; Lister 1993; Johnson 1993a,b, 1999b; Gray and Busa 1994; Williams 1994;

Williams and Jiang 1999; Kraus and Williams 1998, 2001; Ilg and Karlstrom 2000; Jiang 2001; Jiang and Williams 2004). The implications are important since inferred shear sense is opposite for the rotational and non-rotational models respectively and non-rotation of porphyroblasts would imply that these structures retain information on the orientation of foliations at early stages of deformation (Bell et al. 1997). Several studies (e.g. Williams 1994; Aerden 1995; Bell et al. 1997; Ilg and Karlstrom 2000) have demonstrated that porphyroblast inclusion patterns of oblique-S_i or sigmoidal porphyroblasts may be remarkably constant in their spatial orientation over large areas, irrespective of relatively heterogeneous post-porphyroblast deformation. A number of authors (Bell 1985, 1986; Bell and Johnson 1989, 1990, 1992; Steinhardt 1989; Hayward 1990, 1992; Johnson 1990; Bell et al. 1992a–c, 1997, 2003; Aerden 1994, 1995; Bell and Forde 1995; Bell and Hickey 1997; Stallard 1998; Hickey and Bell 1999; Bell and Mares 1999; Jung et al. 1999; Stallard and Hickey

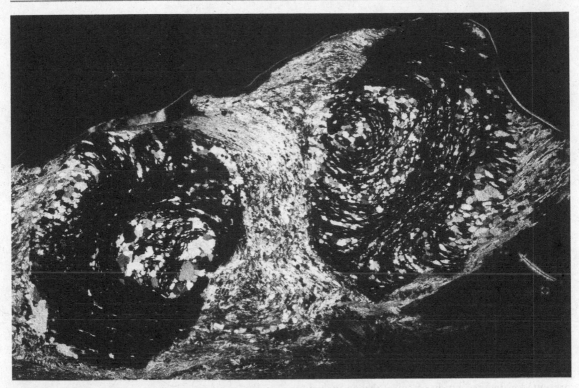

Fig. 7.38. Two garnet porphyroblasts in a garnet-mica schist containing inclusion patterns similar to the pattern at *right* in Fig. 7.36c (section normal to the S-axis of Fig. 7.36). In fact, the section is slightly inclined to the R(rotation)-axis (Fig. 7.36a). Leiden Collection. Width of view 45 mm. CPL

Fig. 7.39. Development of deflection planes and truncation planes around a syntectonic periodically growing porphyroblast in non-coaxial progressive deformation. The foliation and the porphyroblast rotate with respect to each other. **a** Strain caps and strain shadows develop around the porphyroblast during progressive deformation. After mica-rich strain caps are developed, they are overgrown by the porphyroblast, forming a deflection plane (e.g. Fig. 7.32). **b** If development of a syntectonic porphyroblast is temporarily interrupted by local dissolution in the strain caps, renewed growth over the strain caps causes development of truncation planes. The geometry of deflection and truncation planes is shown in the enlarged insets

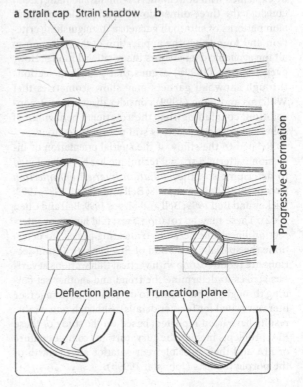

Box 7.4   Rotation or non-rotation of spiral-S$_i$ porphyroblasts

Discussions by several authors have shown that it is difficult to find reliable criteria to determine if spiral-S$_i$ or oblique-S$_i$ porphyroblasts rotated with respect to kinematic axes of bulk flow (as in Fig. 7.34b), or not (as in Fig. 7.34c,d,e) (e.g. Bell 1985; Bell and Johnson 1989; Bell et al. 1992; Passchier et al. 1992; Johnson 1993a,b, 1999). Below we list some criteria that might be of help.

A rotational interpretation is likely to be correct if

- the sense of rotation of the porphyroblast is confirmed by other sense of shear markers (Busa and Gray 1992; Fig. 7.35).
- S$_i$ spirals in three dimensions have gradually decreasing amplitude towards the rim of the crystal in both directions from the core of the crystal along the symmetry axis of the spiral (Figs. 7.36b, 7.37b; Powell and Treagus 1969, 1970; Busa and Gray 1992).
- included "folds" in S$_i$ have an axial surface trace that is strongly curved (Figs. 7.16, 7.33, 7.37).

- S$_i$ spirals that can be interpreted as included strain shadows are present (Figs. 7.33, 7.37a; Schoneveld 1977, 1979).
- the relative rotation angle of the porphyroblast exceeds 180°.
- in elongate crystals, the symmetry axis of the S$_i$-spiral in the crystals is parallel to the long axis, even in sets of porphyroblasts with variable orientation of the long axis (Busa and Gray 1992).

An interpretation of a spiral- or oblique-S$_i$ inclusion pattern as the result of matrix deformation around a stable porphyroblast is probable if

- evidence is present that the foliation in the matrix has passed trough folding and transposition (e.g. Fig. 7.34e).
- clear truncation planes are present in S$_i$.
- regional shear sense, determined by independent criteria, is not compatible with rotation of the porphyroblasts.

2001a,b, 2002; Bell and Chen 2002) have supported the non-rotational model. However, there are indications that the non-rotational model is not generally valid (Passchier et al. 1992; Visser and Mancktelow 1992; Wallis 1992; Gray and Busa 1994; Williams 1994; Chan and Crespi 1999; Williams and Jiang 1999; Kraus and Williams 1998, 2001; Jiang 2001; Jiang and Williams 2004). Johnson (1993a,b) presented a detailed study of spiral shaped inclusion trails in garnets in order to test rotational and non-rotational models and concluded that the geometries could be explained with both models. Williams and Jiang (1999) consider the three dimensional configuration of inclusion patterns of snowball garnets a distinguishing criterion and found that the application of this criterion to all the available data favours the rotational model. However, Johnson (1999b) argues that published sections through snowball garnets may show geometries that Williams and Jiang (1999) consider diagnostic for non-rotation, concluding that the rotational behaviour of spherical porphyroblasts is still poorly understood.

Related to the study of the spatial orientation of inclusion patterns a special technique has been developed to determine so called *foliation intersection/inflection axes* or 'FIA' (Bell et al. 1998; Bell and Hickey 1997, 1999; Hickey and Bell 1999; Bell and Mares 1999; Bell and Chen 2002). These may be rotation axes, fold hinges or intersection lineations between truncating planes. To determine the spatial orientation of FIA two sets of thin sections are required, one with vertical thin sections at various strikes to determine the trend and another set fanning through the horizontal to determine the plunge (see literature cited for further details). Although remarkable results have been reported, based on the study of these FIA, problems include the curvature and complex nature of FIA and their possible reorientation after growth of the porphyroblasts (Johnson 1999b).

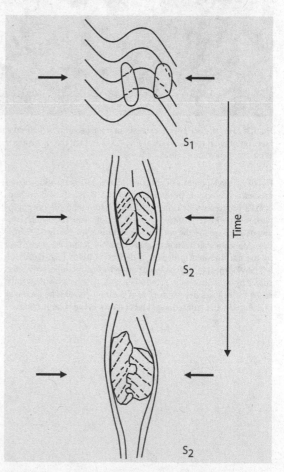

Fig. 7.40. Two porphyroblasts overgrow alternating limbs of a D$_2$ fold; continuing deformation brings the porphyroblasts in contact, producing a sharp angle between their inclusion patterns; finally the amalgamation and interpenetration by pressure solution may result in an apparently enigmatic inclusion pattern with orthogonal inclusions in intergrown parts. Compare Figs. 7.44, 7.45

Schoneveld (1979) has described a number of garnets that are rotational with relation to a progressively developing crenulation cleavage. This leads to fairly complicated inclusion patterns that, judging from the literature, are not very common in nature. In Box 7.4, a short list of criteria is given that may help to distinguish between porphyroblasts that are interpreted to have rotated with respect to an external foliation, and those that formed by overgrowth of helicitic folds.

7.6.9
Amalgamated Porphyroblasts

During progressive deformation porphyroblasts may approximate each other, get in touch and eventually grow together or amalgamate (Fig. 7.40). In most cases the anisotropic nature of the porphyroblasts will show readily the limit between individual crystals, but if this happens with garnet the amalgamated crystals may be easily mistaken for one big porphyroblast, possibly leading to an erroneous interpretation of the inclusion pattern.

An example of this type of structure occurs in garnet porphyroblasts from a chloritoid-garnet schist that crops out in the south of Minas Gerais State, Brazil. The schist contains garnet porphyroblasts with sigmoidal to spiral-S_i inclusion patterns (Fig. 7.41). The apparent rotation between the porphyroblasts and the matrix foliation varies between 60 and 230°. The garnets are interpreted as syntectonic with respect to the main deformation phase, D_1, responsible for the main foliation, S_1 that is continuous with S_i and deflects around the porphyroblasts, producing strain shadows (Fig. 7.41). A later deformation phase, post dating the garnet growth caused local crenulation of S_1. This crenulation often nucleates on garnet crystals where S_1 was already deflected and offered therefore an easy nucleation site for folding. Some

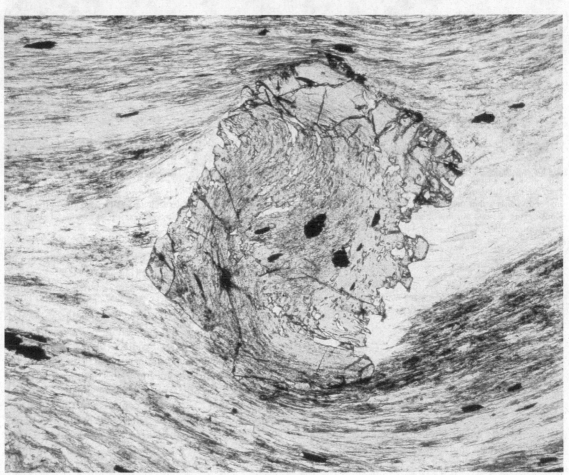

Fig. 7.41. Garnet porphyroblast in a fine-grained muscovite-chloritoid-quartz matrix; the horizontal foliation is a continuous cleavage, S_1, generated during D_1. The garnet is interpreted as syntectonic with respect to D_1 and shows a dextral rotation with respect to the foliation of about 180°, coherent with the regional nappe movement. Note the deflection of S_1 and the strain shadows around the garnet, consistent with syntectonic growth. Itumirim, southern Minas Gerais State, Brazil. Width of view 4 mm. PPL

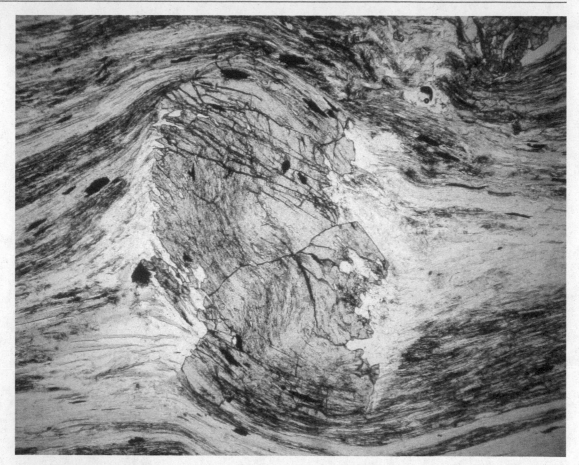

Fig. 7.42. Elongate garnet porphyroblast with an inclusion pattern that looks like helicitic folds with two complete wave-lengths; close inspection shows that the garnet is probably the result of amalgamation of two porphyroblasts along a black line close to the centre. Considering the two porphyroblasts separately they show similar patterns as in Fig. 7.41, with relative dextral rotation of about 90°. Itumirim, southern Minas Gerais State, Brazil. Width of view 6 mm. PPL

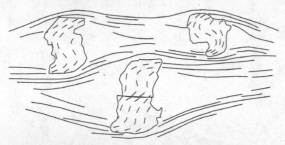

Fig. 7.43. Two rotational garnets that amalgamate may show an inclusion pattern of apparent helicitic folds with more than one wavelength

porphyroblasts with relatively elongated shape contain inclusion patterns that look like helicitic folds with two complete wavelengths (Fig. 7.42). At first sight this could be interpreted as growth during or after a crenulation phase that was erased from the matrix since it clearly does not mach the crenulation that postdates garnet growth described above. However, detailed observation reveals that the structure is in fact the result of amalga-

mation of two individual garnet crystals, both with an S-shaped inclusion pattern (Fig. 7.43).

Another example, in this case of amalgamated albite porphyroblasts, was observed in an albite-garnet schist from Elephant Island, South Shetland Islands, Antarctica. This schist contains numerous albite porphyroblasts that are frequently in mutual contact (Fig. 7.44). Along several of these contacts an abrupt change in angle of the inclusion pattern shows how the porphyroblasts approached each other in the deforming matrix until they touched. In a few cases the contacts were modified, apparently by dissolution, leading to the penetration of one porphyroblast into the other along interlobular contacts with stylolite-like appearance (Fig. 7.45). The inclusion patterns of the intergrown porphyroblasts are almost orthogonal showing the provenance of the porphyroblasts from parts of the matrix with different cleavage orientation. Locally this process can even produce isolated parts (in 2D) of a former porphyroblast within another (Figs. 7.40, 7.45).

Fig. 7.44. Two albite porphyroblasts with approximately straight inclusion patterns that make a sharp angle. The crystals apparently approached each other during progressive deformation. Compare with Fig. 7.40 second stage. Elephant Island, West Antarctica. Width of view 6 mm. CPL

Fig. 7.45. Intergrown albite porphyroblasts with orthogonal inclusion patterns; the contact has a stylolitic appearance typical for inter-penetration by pressure solution; compare Fig. 7.40, last stage. Elephant Island, West Antarctica. Width of view 5 mm. Polarizers at 30°

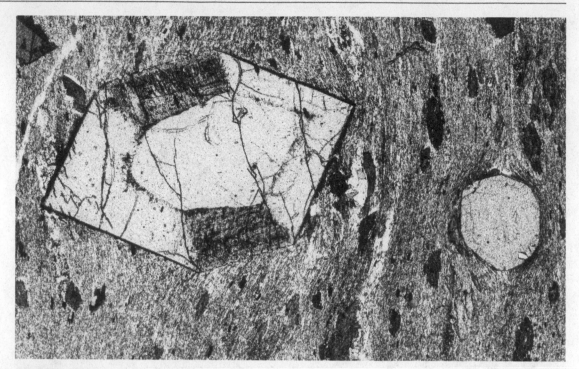

Fig. 7.46. Staurolite garnet biotite schist with a staurolite crystal (at *left*) that shows textural sector zoning and cleavage domes. S_i in the sectors mimics S_e rotated over about 30°. The foliation is strongly deflected around the garnet crystal without inclusions at the *right-hand side* of the photograph. Based on the deflection patterns, it can be concluded that garnet grew first, then staurolite and finally biotite (the *dark patches* in the matrix), all during the deformation phase that produced the foliation (syntectonic porphyroblast growth). Orós, NE Brazil. Width of view 10 mm. PPL

Fig. 7.47. Biotite schist with porphyroblasts of chiastolite showing re-entrant zones with feather-edge structures. Density of inclusions is controlled by crystal habit. Growth inclusions occur in the larger crystal. Laraquete, central Chile. Width of view 16 mm. Polars at 45°

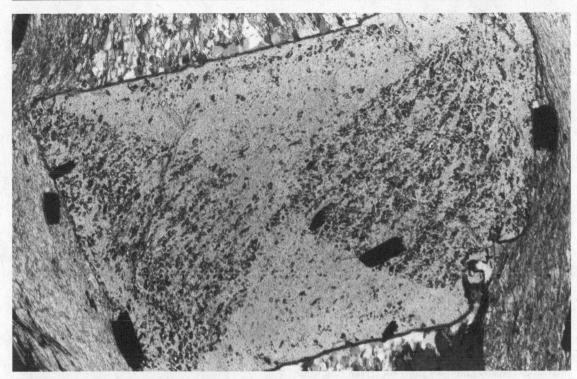

Fig. 7.48. Chloritoid crystal with textural sector zoning in hourglass shape, controlled by crystal habit. A shape orientation of S_i is visible in two sectors, suggesting a relative crystal rotation of 60° with respect to the matrix. Leiden Collection. Width of view 4 mm. PPL

Fig. 7.49. Layered chloritoid schist with almost euhedral chloritoid crystals that show both hourglass zoning and a rhythmic concentric layer-zoning of inclusion-rich and poor areas. A gentle crenulation has developed by D_2 that is also present in the crystals, indicating that they grew late syn-D_2. Açunguí, São Paulo, Brazil. Width of view 17 mm. PPL

7.7
Crystallographically Determined Inclusion Patterns

Passive inclusion as described in Sects. 7.3–7.5 is not the only process that controls the inclusion of foreign matter in a porphyroblast. In some porphyroblasts the distribution and shape of inclusion density pattern are associated with crystal habit (Figs. 7.46–7.49). Characteristic microstructures are *textural sector zoning* and *re-entrant zones* (Fig. 7.50, ⊘Video 7.50; Rice and Mitchell 1991). Textural sector zoning (Figs. 7.46, 7.48, 7.50a, ⊘Video 7.50, ⊘Photo 7.48) probably develops by differences in growth rate and diffusion rate at different crystal faces or from preferential adsorption of impurities at certain crystal faces (Frondel 1934; Barker 1990, 1998). If textural sector zoning develops on two pairs of faces only, it is known as *hourglass zoning* (Figs. 7.48, 7.49, ⊘Photo 7.48). Re-entrant zones are bands of inclusions, commonly with a preferred orientation, that follow the bisectrix of dominant crystal faces and are characteristic of chiastolite (Figs. 7.47, 7.50b). They commonly show *feather-edge structures*, also a growth feature (Figs. 7.47, 7.50, ⊘Video 7.50; Rice and Mitchell 1991). Re-entrant zones are thought to form if a crystal only incorporates inclusions (mostly graphite) along the crystal ribs, while the same material is dissolved or displaced along crystal faces (Spry 1969; Harvey and Ferguson 1973; Ferguson 1980; Ferguson and Lloyd 1980; Rice and Mitchell 1991). In the latter case, *cleavage domes* may form, accumulations of displaced material (usually graphite) along crystal faces (Figs. 7.3, 7.46, 7.50b, 7.51, ⊘Video 7.50). Observations by Rice and Mitchell (1991) suggest that an isotropic stress field (no or very small differential stress) is needed for inclusion displacement and accumulation of graphite. This would occur during contact metamorphism as in the thermal aureole of an intruding batholith in the absence of deformation.

Another controlling factor for inclusion orientation is crystal cleavage (Vernon 1976) that may influence both abundance and orientation of inclusions, especially in micas (Fig. 7.52).

Crystals with textural sector zoning or re-entrant zones commonly contain *growth inclusions* (Figs. 7.47, 7.50b, 7.51, ⊘Video 7.50). Unlike the passive inclusions treated above, growth inclusions form by new growth at the growing crystal face of the porphyroblast. They are usually recognisable by their strict parallel or orthogonal orientation with respect to porphyroblast crystal faces, and oblique orientation with respect to passive inclusions (Figs. 7.47, 7.50b, ⊘Video 7.50). In some cases, healed fractures may resemble passive inclusions and form a possible source of error.

Some porphyroblasts (e.g. garnet; Fig. 7.53) contain inclusion-rich cores surrounded by rims that are almost inclusion-free or vice versa. This may reflect changes in diffusion rate and crystal growth rate due to changing metamorphic conditions, possibly related to a growth interval. However, it may also be due to a change in the porphyroblast-forming reactions, if a new reaction becomes active which consumes the mineral that forms inclusions.

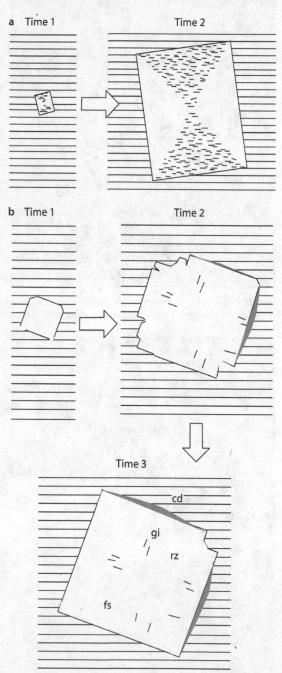

Fig. 7.50. a Development of porphyroblast with textural sector zoning with preferred orientation of passive inclusions (S$_i$). **b** Development of porphyroblast with re-entrant zones (*rz*) with feather-edge structures (*fs*) and growth inclusions (*gi*). Cleavage domes (*cd – dark grey*) occur on two sides

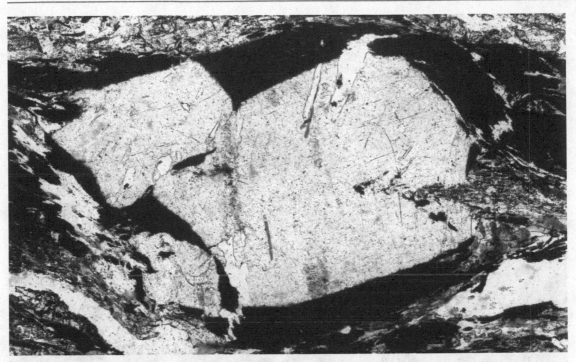

Fig. 7.51. Two intergrown andalusite porphyroblasts in micaschist, with cleavage domes of graphite. The porphyroblasts are almost free of inclusions, but graphite is a major constituent of the micaschist matrix. Some growth inclusions are visible in the porphyroblast. A weakly developed re-entrant zone is visible at *lower right*. Menderes Massif, Turkey. Width of view 2.5 mm. PPL. (Photograph courtesy R. Hetzel)

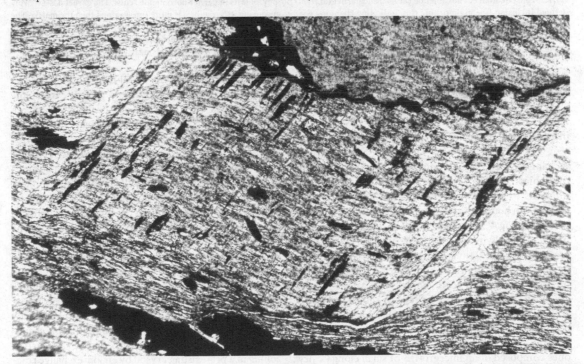

Fig. 7.52. Biotite porphyroblast in biotite schist with inclusions of opaque minerals in two almost orthogonal directions; the straight NE-SW ones are parallel to, and probably controlled by the crystal cleavage, whereas the curved NW-SE ones constitute S_i (passive inclusions). The centre of the crystal shows a clockwise rotation of about 25° with respect to S_e. Strain shadows are visible at the *left* and *right hand side* of the biotite crystal where the matrix with opaque grains was apparently detached from the crystal (Sect. 6.3). Rioumajou, Pyrenees, Spain. Width of view 4 mm. PPL

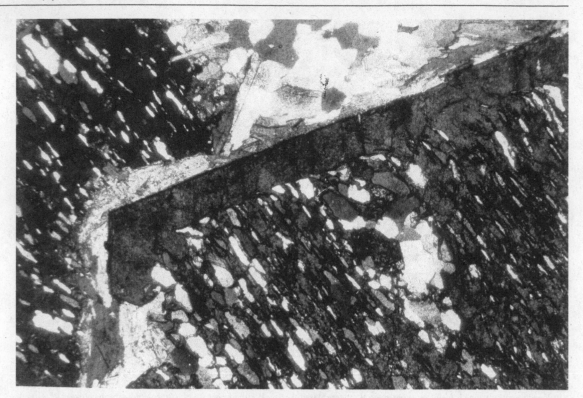

Fig. 7.53. Part of adjacent hornblende (at *left*) and garnet (at *right*) porphyroblasts in garnet-hornblende schist. The garnet has a sharply defined inclusion free idiomorphic rim, indicating changing conditions during its growth. S_i is more fine-grained then the matrix. The inclusions in the hornblende porphyroblast are similar in orientation and shape to those in the garnet, suggesting the contemporaneous growth of both minerals. Note the presence of secondary white mica along the contact of garnet and hornblende, suggesting a retrograde reaction consuming hornblende. Kittelfjäll. Västerbotten, Sweden. Width of view 4 mm. Polars nearly crossed

7.8
Reaction Rims

7.8.1
Introduction

A change in metamorphic conditions can give rise to porphyroblast growth, or to partial *replacement* of some minerals by others. Such replacement usually occurs along grain boundaries and causes development of *reaction rims* (Fig. 7.54, ⊘Video 7.54ad). Reaction rims are invaluable tools in the reconstruction of a sequence of metamorphic reactions. They can be monomineralic or polymineralic and can be divided into several geometric types. If they form closed rings around grains (shells in three dimensions), they are known as *coronas* (Fig. 7.54c–f, ⊘Videos 7.54ad, 7.54e,f). Monomineralic coronas are also known as *moats* (Figs. 7.54c, 7.55, ⊘Video 7.54ad, ⊘Photos 7.54a,b, 7.55a,d); polymineralic ones composed of an intergrowth of small elongate new grains are known as *symplectitic coronas* (Figs. 7.54d, 7.57–7.61, ⊘Video 7.54ad). The structure of lamellar or vermicular fine-grained intergrown material is known as *symplectite* (Figs. 7.54d, 7.57–7.61, ⊘Video 7.54ad).

Reaction rims form in several metamorphic settings in response to both retrograde and prograde metamorphic reactions. Substitution of andalusite by sillimanite is obviously prograde, while most coronas and symplectites are retrograde. Reaction rims form because progress of reactions generally depends on the presence of a fluid phase along grain boundaries for the transport of ions to and from the reaction site. Reaction rims are most common in high-grade rocks, eclogites and ultramafic rocks. This is probably due to the limited availability of aqueous fluids in these rocks, which inhibits reactions to reach equilibrium on the scale of a thin section.

A special type of reaction rim is formed by chemical *zoning* within a mineral. This may be visible by a gradual or abrupt change of the extinction angle or of the colour of pleochroic minerals; in other minerals it may only be detectable with a microprobe. Zoning can form in at least two different ways: (1) as *growth zoning* reflecting changing *P-T* conditions during growth, and (2) as *reaction zoning* in a pre-existing crystal by an ion-exchange reaction along the rim. Reaction zoning of an element A usually follows the outer rim of a zoned crystal and may advance further into the crystal, where it is in contact with another mineral rich in A, or along fractures. Growth

Fig. 7.54.
a–f Various types of reaction rims. Likely reactions causing the rims are given below each drawing. *A–F* are minerals involved in reactions, *P* and *Q* are not involved. The equations are open since phases other than the ones shown may be involved in the reactions as well. **g, h** Two types of pseudomorphic replacements. **i** and **j** show how reaction type can be established to some extent from the arrangement of inclusions. **k** and **l** show the significance of oriented inclusions in a reaction rim; the relative age of minerals *A*, *B* and *C* is given. In **k** the alignment of *C* is due to passive inclusion of aligned grains of *C* in *A*, and subsequent growth of *B*; in **l**, *C* grew with a preferred orientation in response to a high differential stress. Further explanation in text

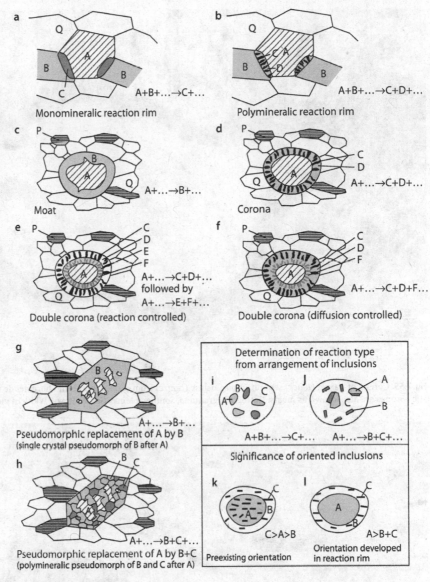

a Monomineralic reaction rim   A+B+…→C+…

b Polymineralic reaction rim   A+B+…→C+D+…

c Moat   A+…→B+…

d Corona   A+…→C+D+…

e Double corona (reaction controlled)   A+…→C+D+… followed by A+…→E+F+…

f Double corona (diffusion controlled)   A+…→C+D+F+…

g Pseudomorphic replacement of A by B (single crystal pseudomorph of B after A)   A+…→B+…

h Pseudomorphic replacement of A by B+C (polymineralic pseudomorph of B and C after A)   A+…→B+C+…

Determination of reaction type from arrangement of inclusions

i A+B+…→C+…   **j** A+…→B+C+…

Significance of oriented inclusions

k C>A>B   Preexisting orientation   **l** A>B+C   Orientation developed in reaction rim

zoning may also follow the outer rim or have a more complex shape due to growth-coalescence of smaller grains. Growth zoning can be truncated by grain boundaries if the crystal has been subject to erosion due to recrystallisation or pressure solution. In some cases, reaction zoning of an element A may even be superposed on growth zoning of an element B (Tucillo et al. 1990).

7.8.2
Coronas and Moats

If two dispersed minerals A and B in a rock react to form new minerals, these may form local isolated reaction rims along grain boundaries of A and B (Fig. 7.54a,b, ⊘Video 7.54ad). However, if a mineral A breaks down to other minerals along its outer rim, or if it reacts with a

mineral B that is abundant in the surrounding matrix, the newly formed minerals may form a ring-shaped corona around A (Figs. 7.54d, 7.55, ⊘Video 7.54ad, ⊘Photo 7.55a–d). In high-grade rocks, several coronas may be superimposed on each other as concentric shells as double or composite coronas (Figs. 7.54e,f, 7.57, 7.58, ⊘Video 7.54e,f). Careful monitoring and analysis of corona structures can allow detailed reconstruction of the pressure-temperature-time (*P-T-t*) path in a rock sample (Sect. 1.3). A special form of coronas are *atoll garnets*; ring-shaped garnets, usually filled with quartz and white mica. Atoll garnets possibly form as coronas on biotite or cordierite, where the core mineral is later replaced by an aggregate of other minerals. However, an origin by growth along grain boundaries in coarse-grained quartz-mica aggregates has also been proposed (Bard 1986).

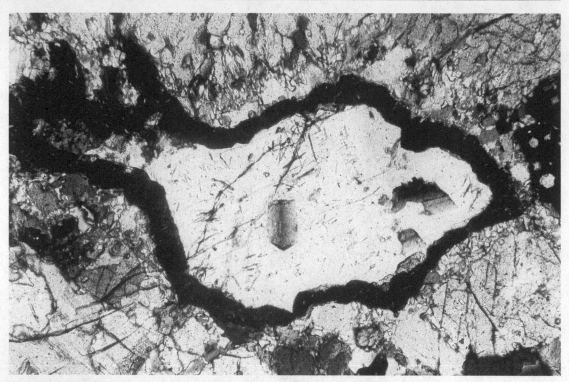

Fig. 7.55. Corona of garnet (*black*) separating plagioclase (*centre*) from clinopyroxene and hornblende in the matrix. Such monomineralic coronas are also known as moats. Retrograde granulite. Southern Minas Gerais, Brazil. Width of view 2 mm. PPL

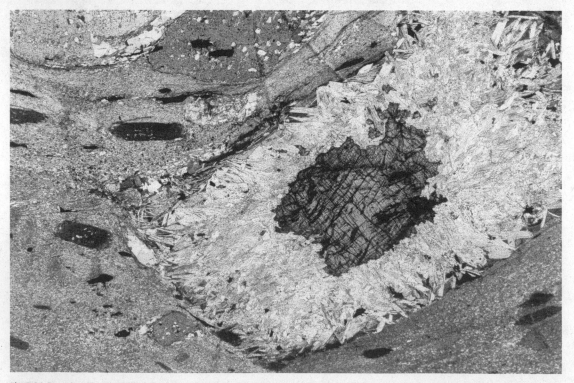

Fig. 7.56. Reaction rim composed mainly of muscovite around a staurolite crystal in staurolite biotite schist. The reaction rim forms a pseudomorph after staurolite. Pyrenees, Spain. Width of view 16 mm. CPL

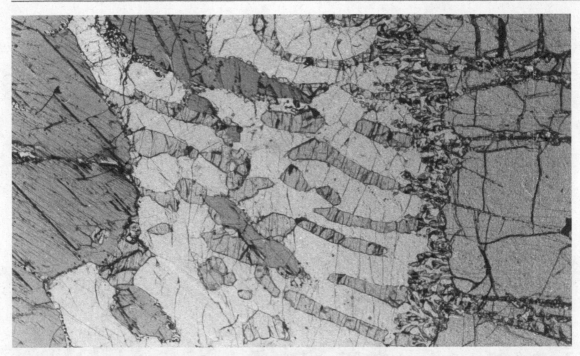

Fig. 7.57. Double symplectitic corona between hornblende (*left*) and garnet (*right*). The *left-hand*, coarse corona consists of plagioclase and orthopyroxene. The narrow corona at *right* consists of plagioclase, orthopyroxene and spinel. Notice that the narrow orthopyroxene-plagioclase-spinel corona occurs only along the garnet and along cracks in the garnet, but not near hornblende. Therefore, the coarse corona is thought to reflect a first reaction hornblende + garnet → orthopyroxene + plagioclase and the narrow corona represents a second reaction garnet → orthopyroxene + plagioclase + spinel. In this way, the sequence of reactions can be established in a thin section. Bolingen Island, Prydz Bay, Antarctica. Width of view 7 mm. PPL. (Photograph courtesy P. Dirks)

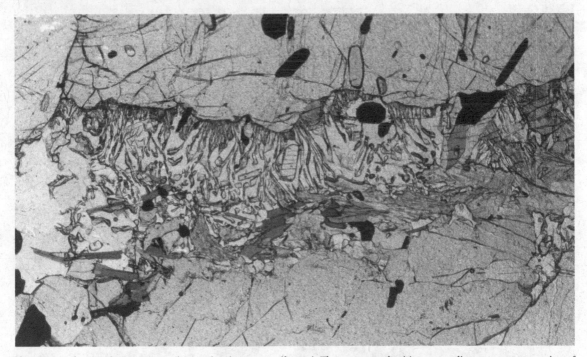

Fig. 7.58. Zoned corona between garnet (*top*) and orthopyroxene (*bottom*). The upper symplectitic corona adjacent to garnet consists of cordierite and orthopyroxene. The lower corona consists mainly of sapphirine. Central Sri Lanka. Width of view 7 mm. PPL. (Photograph courtesy P. Dirks. Thin section provided by L. Kriegsman)

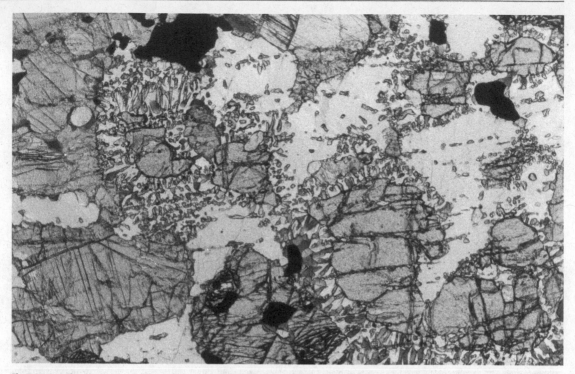

Fig. 7.59. Symplectitic coronas of plagioclase, orthopyroxene and hornblende between garnet and clinopyroxene grains, reflecting the reaction garnet + clinopyroxene + H_2O → plagioclase + orthopyroxene + hornblende. Note that some orthopyroxene grains are aligned in the coronas parallel to cracks in the garnet relicts; the coronas probably developed initially along cracks. The cracks therefore predate the reaction. Bolingen Island, Prydz Bay, Antarctica. Width of view 5 mm. PPL. (Photograph courtesy P. Dirks)

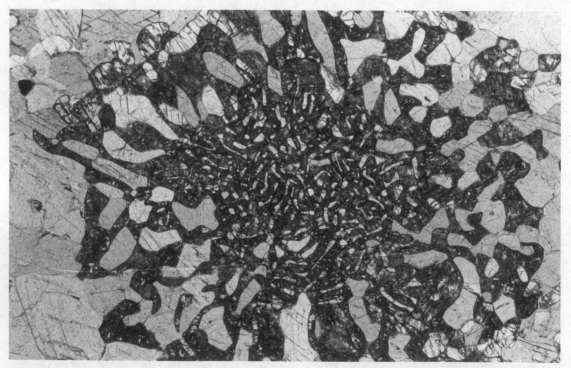

Fig. 7.60. Pseudomorph of plagioclase-orthopyroxene symplectite after garnet. The grain size of the symplectite decreases inwards. Wu Li Dong, N-Shanxi, China. Width of view 9 mm. PPL. (Photograph courtesy P. Dirks)

7.8.3
Symplectites

If two or more minerals are present in a corona, they may form a *symplectite* in which the minerals form an intergrowth of lamellae that may be straight, curved or vermicular (Figs. 7.57–7.61). Symplectites are thought to develop due to relatively rapidly proceeding reactions, or lack of a fluid phase to transport material towards and away from the reaction site. Nucleation sites of symplectite may be stress-controlled as well as *P-T* overstepping controlled (Simpson and Wintsch 1989).

Most symplectites are *reaction symplectites* that form by reactions of the type $A + B + ... \Rightarrow C + D + ...$ or $A + ... \Rightarrow C + D + ...$, where $C + D$ form the symplectite (Figs. 7.54b,d; 7.61a,c, ⊘Video 7.54ad). Discontinuous precipitation reactions of the type $A \Rightarrow A' + B$ constitute a special type that will be loosely referred to as *exsolution symplectites* (Fig. 7.61b,d). They develop when two grains of a supersaturated solid solution A with different orientation but identical chemistry are juxtaposed along a grain boundary. The grain boundary mi-

grates into one of the grains and leaves a symplectite of $A' + B$ behind (Fig. 7.61b). Exsolution symplectites can potentially be used as a temperature gauge (Sect. 9.9).

Symplectites may be affected by grain boundary area reduction (GBAR; Sect. 3.10) that causes the lamellae to neck and obtain a globular shape. Such symplectites have been named globular symplectites (Fig. 7.61c,d). They are particularly common in high-grade rocks; all examples in Figs. 7.57–7.60 can be classified as *globular symplectites*. Symplectites that are not or little affected by GBAR and in which lamellae are elongate with parallel boundaries are known as *lamellar symplectites* (Fig. 7.61a,b). They are commonly formed during metamorphic retrogression, especially in eclogites.

Kelyphite or *kelyphytic structure* is the name for a symplectitic corona structure around olivine, commonly in several concentric layers, which may contain orthopyroxene, clinopyroxene, amphibole and spinel or garnet. Spinel or garnet may be in symplectitic intergrowth with hornblende or orthopyroxene. Amphibole-plagioclase symplectites have also been called kelyphitic if they form clearly defined coronas.

A *myrmekite* is a bulbous symplectite of vermicular quartz in plagioclase (⊘Photo 7.54b). It is common in high-grade metamorphic and igneous rocks, mostly as breakdown product of K-feldspar during retrograde metamorphism (Smith 1974; Phillips 1974, 1980; Shelley 1993). Myrmekite may develop at stress-concentration sites during progressive deformation (Simpson and Wintsch 1989) and in that case can serve as a shear sense indicator (Sect. 5.6.9).

7.8.4
Establishing the Nature of Reactions

In coronas and symplectites it is important first to establish which components are likely to be new minerals, and which are old grains. An important principle is that silicates usually have slow diffusion through the crystal lattice, and that reactions therefore mainly occur at grain boundaries where grains are in contact with each other or with the scarce metamorphic fluid. It is therefore usually assumed that a corona or symplectite grows from the outside inwards, and that large grains with irregular shape, completely surrounded by another mineral, are older than this surrounding material (Figs. 7.54, 7.56, ⊘Video 7.54ad–l). In exsolution symplectites, growth is thought to proceed from the grain that is in crystallographic continuity with one of the phases of the symplectite, towards the second grain (Fig. 7.61b). Replacement commonly also takes place along cracks in old grains, and old grains therefore may be split into various fragments (Figs. 7.54g,h, 7.57, ⊘Video 7.54g). If replacement of the old grain is advanced, its remnants can be recognised by the fact that

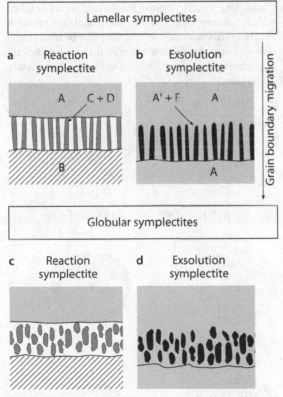

Fig. 7.61. Symplectites can be subdivided into reaction (**a, c**) and exsolution (**b, d**) symplectites, depending on the reactions that take place. Another possible subdivision is in lamellar (**a, b**) and globular (**c, d**) symplectites. Globular symplectites are thought to develop from lamellar ones through grain boundary area reduction, especially at high metamorphic grade

they all have the same lattice orientation (Fig. 7.54g,h), and by concave-inward boundaries with the new mineral (Fig. 7.54c,g, ⊘Video 7.54g). If old grains of two minerals are not in contact, they are likely to have reacted with each other to form the corona or symplectite (Fig. 7.54i). It is important to realise that contacts may be missed in thin sections with few observation sites (Fig. 12.2i). If old grains of two minerals are in contact, one of them is probably not involved in the reaction (Fig. 7.54j). New grains form either the corona or part of a symplectite, or are finely dispersed in the corona. New grains may be aligned, but will lack the identical lattice orientation of old grain fragments. Small grains in a corona with a weak preferred orientation may have grown as new grains in the corona when it was subject to a high differential stress (Sect. 4.2.7.7), but may also be relicts of inclusions in an old grain that has been replaced. This can usually be decided by the presence of relicts of the old grain (Fig. 7.54k,l, ⊘Video 7.54k,l). If *double coronas* are present (Fig. 7.54e,f), it can be more difficult to establish a relative age. It is sometimes assumed that the outer rim is youngest, but there are other possibilities. With changing metamorphic conditions, reactions may also occur between the included old grain and its corona (Fig. 7.62a), between the corona and phases outside (Fig. 7.62b), or between coronas. The old grain may also be replaced inward by a new corona while the first corona is unaffected (Fig. 7.54e), or the reaction may change while the corona

grows due to increasing length of the diffusion path (Figs. 7.54f, 7.62c). Especially if Al-rich minerals are involved, this possibility should be investigated. In all cases, interpretations based on geometrical arguments should be checked to see if proposed reactions are chemically and thermodynamically possible.

Reaction rims are interesting sites for geothermometry and geobarometry since they form under conditions other than the original mineral assemblage and therefore may define different stages along a *P-T-t* path (Sect. 1.3). However, there are some potential pitfalls; old grains and new grains may be strongly zoned, in which case it may be difficult to decide which compositions should be used for thermo-barometry; careful reconstruction of the growth of the reaction rim structures is useful in such cases. Moreover, the composition of old grains may have been modified during the reactions. This can usually be determined if zoning in old grains is carefully monitored by SEM observation or by a fine grid of measurement points over the grain in a microprobe. If zoning is parallel to the edges of the old grain, it is probably modified during the reaction. If the zoning is truncated by reaction rims, especially along cracks, it is probably original. Finally, care should be taken with the assumption that only old grains and reaction rims took part in a reaction; reactions may be rather complex and may also involve minerals that are not directly in contact with the reaction rims or that have completely disappeared.

Fig. 7.62.
Three mechanisms to form double coronas. *A–D* are minerals. The equations are open since phases other than the ones shown may be involved in the reactions as well. A second corona forms: **a** from reaction between the core mineral and the first corona; **b** by growth on the outside or reaction of the first corona with minerals in the matrix, or **c** by a change in the reaction that causes breakdown of the old grain

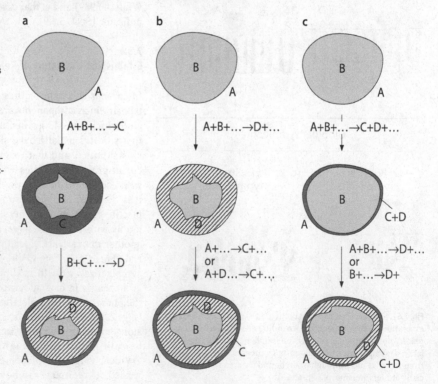

7.8.5
Pseudomorphs

If a crystal is largely or completely replaced by another mineral or an aggregate of minerals, the new minerals may preserve the shape of the original grain (Fig. 7.54g,h). Such an aggregate is known as a *pseudomorph*; for example, if chlorite replaces garnet we speak of a 'chlorite pseudomorph after garnet'. Pseudomorphs can be recognised if relics of the old mineral grain are still present (Fig. 7.54g,h), or if the old grain was euhedral with a characteristic crystal shape. It is sometimes difficult to decide in an aggregate of minerals which ones are involved in a reaction and in which direction the reactions were proceeding. Figure 7.56 shows an example of a partial muscovite pseudomorph after staurolite, and Fig. 7.60 a complete pseudomorph of plagioclase and orthopyroxene after garnet.

A distinction can be made between prograde and retrograde pseudomorphs according to the minerals involved. For example, a staurolite pseudomorph after chlorite is prograde, while a chlorite pseudomorph after garnet is retrograde. Since pseudomorphs show the subsequent stability of at least two metamorphic minerals, they are important for the reconstruction of P-T-t evolution. Guidotti and Johnson (2002) provide an example of how pseudomorphic replacement can be used in the study of superposed contact metamorphism on regional metamorphism.

7.8.6
Relation with Deformation

As mentioned above, some symplectites may form in response to deformation. In general, however, it may be difficult to decide whether pseudomorphs or coronas have replaced a mineral before, during or after a phase of deformation. In principle, the criteria mentioned in Sect. 7.4 can be used, but unfortunately reaction rims rarely contain passive inclusions. Intracrystalline deformation of the new minerals in the aggregates can be used to decide about relative age. For example, if undeformed chlorite replaces strongly deformed biotite grains, the chlorite is likely to be post-tectonic. Figures 7.57 and 7.59 show interesting examples where the relative age of intracrystalline fractures and symplectite can be established.

Box 7.5   Reconstruction of tectono-metamorphic evolution

Porphyroblast-matrix relationships and reaction rims may help to unravel the tectono-metamorphic evolution in an area (see also Chap. 1). Figure B.7.1 shows a schematic evolution in space and time to illustrate the different relationships that may be expected between mineral growth and deformation (⊘Video B.7.1). The following course of action could be undertaken in an area with a history as shown in Fig. B.7.1.

1. Define in each outcrop or tectonic domain (Fig. B.7.1α–ε, ⊘Video B.7.1) the sequence of deformation phases that can be recognised in thin sections from overprinting relations (Sect. 4.2.10.2). The phases have to be properly defined, for example as 'responsible for slaty cleavage, for crenulation cleavage, relative rotation of porphyroblasts, kink bands' etc. The phases may be labelled D_1, D_2, D_3 if field relations indicate that no earlier deformation phases were present. If this is not clear, labels like D_n, D_{n+1} etc. are more appropriate.
2. Plot the deformation phases on a horizontal relative time axis, leaving some space in between for possible intertectonic growth, unless there is evidence for continuity (Figs. 7.10, B.7.2a).
3. Determine the growth period for each mineral phase and plot it with horizontal bars or dashed lines where in doubt (Figs. 7.10, B.7.2a). 'Pillows' may be used to indicate main growth versus subsidiary growth (Fig. 7.10).
4. The terms pre-, syn- and post-tectonic have to be clearly defined with respect to a specific deformation phase or episode. For example, in Fig. B.7.2 mineral C in locality ε grew syn- to post D_2 and pre D_3.
5. Use field data and geometrical arguments to correlate deformation phases in different thin sections and from different sample locations (Fig. B.7.2).

At this stage it is clear that the reconstruction that can be reached with porphyroblast analysis alone is insufficient for a complete understanding of the tectonic evolution; the scheme of Fig. B.7.2 is only a coarse approximation of the true pattern in Fig. B.7.1b. Additional data related to the P-T-t evolution of each sample site, absolute age dating, nature and chemistry of igneous intrusions, and sedimentary environment of protoliths all play an equally important role in the final reconstruction. Particularly promising is the absolute age dating of mineral grains in thin section as outlined in Sect. 10.4.7 which may provide the approximate age of porphyroblast growth and of deformation phases (Montel et al. 1996; Williams and Jercinovic 2002).

Many examples could be cited of the use of porphyroblast-matrix relations for the better understanding of the tectonic evolution of certain areas (Johnson 1999b). The following are only a few recent ones amongst many others: Williams (1994), Aerden (1995), Johnson and Vernon (1995b), Kraus and Williams (1998), Solar and Brown (1999), Ilg and Karlstrom (2000).

An interesting example of how porphyroblast-matrix relationships can help to constrain the intrusion model of a pluton is given in Morgan et al. (1998). They demonstrate how a sill caused initial andalusite growth in the adjacent contact aureole, including the regional cleavage at a small angle to bedding as S_1 inclusion trails. Later inflation of the sill, transforming it to a laccolith-like body, produced vertical uplift and translation accompanied by strong attenuation of the country rocks, and rotation of bedding, including the andalusites. During this stage a second foliation, S_2, was created in the host rock, parallel to bedding and to the contact. Andalusite continued to grow over the curved transition of S_1 into S_2 in the rims of the crystals.

Box 7.5   Continued

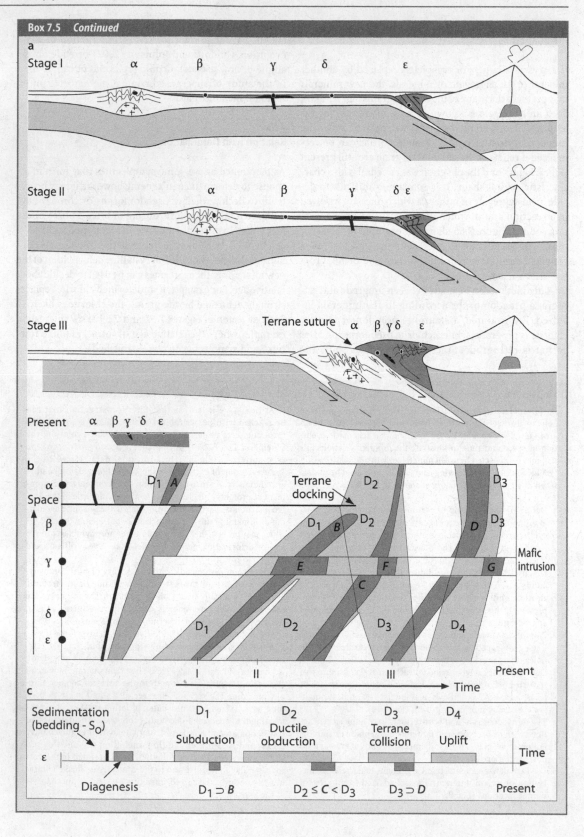

Box 7.5   Continued

◄ **Fig. B.7.1. a** Diagram showing three evolutionary stages of a subduction complex at an active continental margin, culminating in terrane collision. Five sample locations α to ε, are indicated. They are progressively involved in successive deformation phases and metamorphism. γ represents a mafic intrusion. **b** Diagram showing relationships between porphyroblast growth (*dark grey*, minerals A–G) and deformation (*light grey*, deformation phases D_1–D_4) in time and space for the region illustrated in **a**. The sequence of events starts with sedimentation (*bold line* at left), followed by diagenetic compaction (*medium grey*). Deformation phases are continuous in space, except along major fault zones such as at terrane boundaries; they can be diachronous and even joining in space (e.g. D_1 and D_2). D_1 at α represents early deformation related to the genesis of this terrane; D_1 at β to ε reflects the subduction movement and is therefore strongly diachronous; D_2 at δ and ε represents ductile obduction or backthrusting related to the shape of the continental margin. The arrival of the terrane causes renewed deformation, propagating outwards from the collision area, labelled D_2 at sites α to γ, and D_3 at δ and ε. Finally, D_3 at α to γ and D_4 at δ and ε reflect orogenic collapse and associated uplift. Metamorphic evolution is presented in a simplified way as growth of single minerals rather than mineral associations. Such growth can also be diachronous, but can be discontinuous both at major faults and at lithologic contacts. A is a high T/low P mineral related to arc activity, whereas B is a high P/low T mineral related to subduction. C may be a medium P/T mineral and D is a retrograde mineral. Minerals E, F and G reflect a similar history in the mafic intrusion, with somewhat different growth periods. **c** Shows the sequence of events at site ε as an example of the terminology used, and shows how structural and metamorphic analysis of individual samples can help to unravel the tectonic evolution of a region

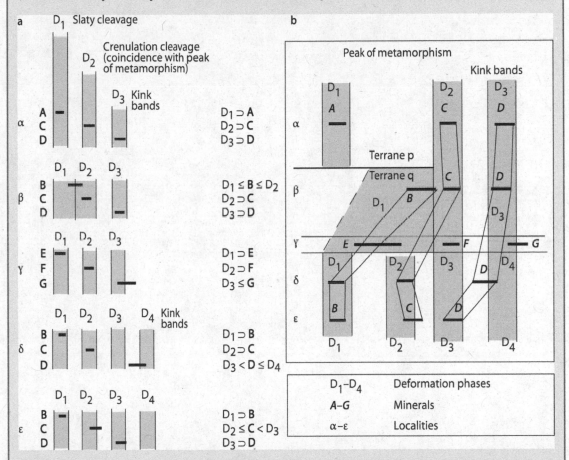

Fig. B.7.2. Diagram illustrating how data on relative age of minerals and deformation can be used in reconstructions of tectono-metamorphic history. The scheme is based on the imaginary evolution of Fig. B.7.1. **a** Relative age of mineral growth with respect to each phase of deformation is established for hand specimens α–ε and plotted in mineral-time diagrams as shown. Notice that generally space is left between phases for intertectonic growth. Only in β D_1 and D_2 are continuous as may be indicated by the inclusion pattern in mineral B. Labelling can be done fore minerals A–G as shown at right. If field-data are available, the schemes represented in **a** can be correlated into a diagram like **b**. Comparison of diagram **b** and Fig. B.7.1 shows that only a coarse reconstruction is possible, unless more samples are analysed and absolute age data are added to the diagram. Nevertheless, the basic tectono-metamorphic history of Fig. B.7.1 is contained in diagram **b**

Primary Structures

8.1
Introduction

8.2
**Primary Structures
in Rocks of Igneous Origin
or in Migmatites**

8.3
**Primary Structures
in Sedimentary Rocks**

In this chapter a number of microstructures are presented that originated in igneous or sedimentary rocks and that could be mistaken for microstructures from metamorphic rocks. This problem may appear in low grade metasediments and in igneous rocks that were subjected to solid state deformation after crystallisation.

In igneous rocks and in migmatites a distinction can be made between magmatic, submagmatic and solid state flow. Criteria to recognise these flow types in thin section are listed and discussed.

Many primary structures in sedimentary rocks, such as slump folds and sedimentary breccias may be mistaken for tectonic structures induced during the post-sedimentary evolution. Thin section study of these structures will rarely provide conclusive evidence, but in specific cases such as compaction structures around concretions, or pseudo boudinage, microscopic observation may help.

Some special structures, such as "loop bedding", resembling isoclinal folds, are also discussed in this chapter.

8.1
Introduction

The objective of this chapter is to review various microstructures from igneous and sedimentary rocks that are similar to structures in metamorphic rocks. Igneous rocks tend to crystallise gradually passing through transitional stages between the liquid and the solid state, that is with various percentages of solid crystals within a melt. If they experience deformation during this process a number of specific microstructures may form indicative of these stages; these will be revised below. Sedimentary rocks may contain a large variety of sedimentary structures (e.g. Collinson and Thompson 1982), some of which can be conveniently studied on the microscopic scale. It is not our objective to review these structures here; only a few structures that may be mistaken for tectonically induced structures in fully lithified rocks will be discussed.

8.2
Primary Structures in Rocks of Igneous Origin or in Migmatites

Many bodies of igneous rocks contain structures attributed to magmatic flow that are partially or completely overprinted by "solid state" deformational fabrics, often related to regional deformation. To unravel the igneous, metamorphic and structural evolution of these rocks it is fundamental to analyse microtectonic evidence for magmatic, submagmatic and solid state flow. This analysis is also essential for the understanding of magma formation and migration in anatectic migmatites (e.g. Sawyer 2001).

We will first shortly discuss the concepts of magmatic, submagmatic and solid state flow and then review the microstructural evidence for each of them.

8.2.1
Magmatic and Submagmatic Flow

Magmatic flow is defined as flow by displacement of melt, with consequent rigid-body rotation of crystals but without sufficient interference between crystals to cause crystal plastic deformation (Paterson et al. 1989; Smith 2002). Submagmatic flow can be defined as deformation involving flow of melt and crystals, assisted by crystal plastic deformation.

Experiments have indicated that the strength of a rock decreases by about an order of magnitude when the melt fraction increases from zero to only a few percent. Increasing the melt fraction to a value around 30% the viscosity of the rock system with melt (magma) decreases again abruptly by several orders of magnitude. This value is known as the *critical melt fraction* (CMF; Arzi 1978;

van der Molen and Paterson 1979) and although it is usually around 30%, it may be as high as 50% (Vernon et al. 1988) or as low as 10–20% for gabbroic rocks (Nicolas et al. 1988). The transition from submagmatic to magmatic flow probably roughly coincides with the critical melt fraction (e.g. Blenkinsop 2000, his Fig. 6.1), and corresponds to the transition from grain-supported flow to suspension flow.

8.2.2
Evidence for Magmatic Flow

Most of the criteria mentioned below are also discussed by Vernon (2000) and Blenkinsop (2000).

Grain Shape Preferred Orientation of Euhedral Crystals

The best criterion to recognise magmatic flow is a grain shape preferred orientation of inequant euhedral crystals that are not internally deformed (Figs. 8.1, B.5.3). The preferred orientation may be an S, L or, more commonly, an SL fabric and is usually defined by feldspar and/or mica in felsic rocks and by feldspar, pyroxene, amphibole or olivine in mafic rocks. Characteristic is the lack of rounded corners and lack of deformation in an isotropic matrix that can be very reduced. Some relatively weak undulose extinction in quartz is not uncommon. One has to realise that the preserved fabric relates probably to the later stages of magma solidification, since earlier stages are easily destroyed by the flowing magma. Magmatic foliation does not always reflect magma flow planes (Yuan and Paterson 1993; Paterson and Vernon 1995; Paterson et al. 1998) but in several studies that interpretation explains best the field and laboratory data (e.g. Cruden et al. 1999; McNulty et al. 2000).

Imbrication ("Tiling") of Elongate Euhedral Crystals that Are Not Internally Deformed

This structure also implies the presence of enough melt to enable the crystals to rotate without plastic deformation (e.g. Blumenfeld 1983; Shelley 1985; Blumenfeld and Bouchez 1988; Mulchrone et al. 2005). Blenkinsop (2000) considers this structure as non-diagnostic since it may also be formed in mylonites (compare Sect. 5.6.17), but if the crystals are euhedral and the matrix isotropic it seems a reliable microstructure.

Ophitic Fabric, Oscillatory Zoning and Growth Twins

These microstructures, especially the first two (Figs. 8.2, 8.3), are characteristic for magmatic rocks. Twinning is more difficult to interpret since many types of twins may also form in metamorphic rocks.

Fig. 8.1. Igneous foliation originated by magmatic flow. The large subhorizontal crystals are K-feldspar (some with Karlsbad twins) with some plagioclase and interstitial quartz along the contacts. Equidimensional relatively high relief grains are hornblende. Quartz syenite from southern Brazil. Width of view 2 mm. CPL

Fig. 8.2. Subophitic microstructure typical for magmatic crystallisation. Note the elongated plagioclase crystals penetrating pyroxene. Some plagioclase crystals are included within the pyroxenes, characterising ophitic microstructure locally. Gabbro from Angra dos Reis, SE Brazil. Width of view 5 mm. CPL

Fig. 8.3. Oscillatory zoning in plagioclase, typical for magmatic crystallisation. Quartz diorite from Tijuca, Rio de Janeiro, Brazil. Width of view 2 mm. CPL

Lack of a Deformation Fabric

This can be a strong indication for magmatic flow. However, it is not conclusive since static recrystallisation and grain growth may produce, especially in the granulite facies, isotropic fabrics, usually hard to distinguish from magmatic fabrics.

Distribution of Mineral Phases

In igneous fabrics the mineral phases are usually uniformly dispersed whereas in recrystallised metamorphic rocks they are often concentrated in clusters. This is because in the latter case they may be derived from few large parent grains by recrystallisation (Blenkinsop 2000).

Crystallographic Preferred Orientation

Since magmatic fabrics are produced by flow of euhedral phenocrysts these have, apart from a grain shape fabric, also a pronounced lattice preferred orientation. According to Benn and Allard (1989) olivines, pyroxenes and feldspars have their (010) crystal faces parallel to the magmatic foliation. Olivine and clinopyroxene have their

[001] direction parallel to the magmatic lineation, and feldspars in gabbros and tonalites have [100] parallel to the magmatic lineation (Benn and Allard 1989). In the case of olivine distinction can be made between magmatic and high grade solid state deformation since in the former [001] is parallel to the lineation and in the latter [100] (Benn and Allard 1989).

Angular Fragments in Flow Structures

Flow structures in some lavas may resemble mylonitic foliation (Fig. 8.4). However, solid particles in lava may retain their angular shape during magmatic flow (Fig. 8.4), whereas porphyroclasts in mylonites are usually rounded (e.g. Figs. 5.22–5.25).

Elongated Shape of Enclaves

Although this book is essentially concerned with microstructures it is often advantageous to consider mesoscopic evidence in conjunction with microstructures. For example the elongated shape of microgranitoid enclaves that do not show evidence of internal plastic deformation indicates that they flowed as magma globules (Vernon 2000).

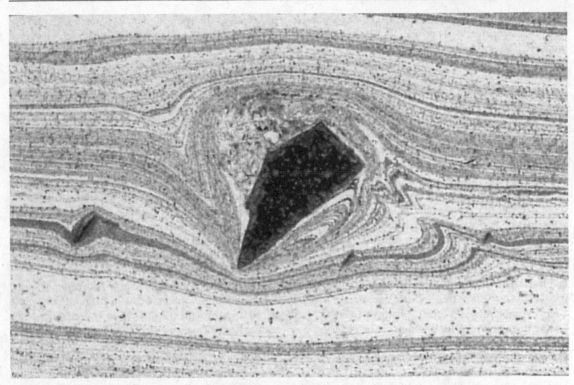

Fig. 8.4. Flow structure in lava with deviation and local folding around angular fragment. At first site this structure looks similar to a mylonite, but porphyroclasts in mylonites are usually rounded. Width of view 5 mm. PPL. (Sample courtesy Ron Vernon)

8.2.3
Evidence for Submagmatic Flow

Submagmatic flow can only be demonstrated if there is evidence for crystal deformation and contemporaneous presence of melt. The following processes play probably a role in submagmatic flow (Paterson et al. 1998): melt-assisted grain boundary sliding, contact-melting assisted grain boundary migration, intracrystalline plastic deformation, diffusive mass transfer, strain partitioning into melt-rich zones, transfer of melt to sites of low mean stress and melt-enhanced embrittlement. However, most of these processes are difficult to detect from microstructures.

The following criteria may be useful indicators of submagmatic flow (see also Blenkinsop 2000; Vernon 2000):

Grain Shape Fabrics

These may be formed both by rigid body rotation and by crystal deformation (Box 4.2). If there is evidence for the latter and for presence of melt, for instance by a matrix with igneous microstructures, then there is evidence for submagmatic flow. Leucosomes parallel to the foliation in anatectic migmatites may provide macroscopic evidence for submagmatic flow.

Intracrystalline Plasticity

It is quite common to find quartz with undulose extinction in granites that are not affected by regional deformation, showing crystal plastic deformation related to the magma crystallisation, probably by submagmatic flow. Deformation twins and bent twins in feldspar may also indicate submagmatic intracrystalline plasticity, but they do not prove the presence of melt.

Cataclasis

Some of the best submagmatic microstructures are fractures filled with melt (e.g. Hibbard 1987; Bouchez et al. 1992; Karlstrom et al. 1993; Blenkinsop 2000). Various experiments demonstrated that in the presence of up to about 15% of melt, cracks can be formed and filled with melt, especially in conditions of high melt pressure (e.g. Rutter and Neumann 1995; Dell'Angelo and Tullis 1988). Blenkinsop (2000) cites four criteria to establish whether the microfractures are filled by melt: the filling should be continuous and equal in composition to the igneous matrix; the composition of the filling should be compatible with the later stages of the igneous petrographic history of the rock; the fractures should be intragranular; early crystals like biotite or sphene may be trapped within the filling.

In metatexites it can often be observed, on the mesoscopic scale, that competent parts of the rock, like calc-silicate or amphibolite layers, are fractured or boudinaged with leucosomes (melt) concentrated in fractures or necks of boudins.

Presence of Late Magmatic Minerals in Strain Shadows

If it can be demonstrated that the infill of dilatational sites other than cracks or microfractures is of magmatic nature and not precipitated from an aqueous fluid, this is a strong argument for submagmatic deformation (e.g. Bouchez et al. 1992).

The problem in proving submagmatic flow is that it is often difficult to demonstrate that crystal deformation was contemporaneous with presence of melt. The natural evolution of most crystallising plutons in a stress field would imply submagmatic flow followed by solid state flow and in such cases it may be difficult to separate crystal plastic deformation features that were generated during submagmatic flow from similar structures generated later when the rock was fully crystallised. In some cases a clear angular overprint relation can be observed between an earlier magmatic foliation (e.g. as aligned elongated enclaves or rectangular phenocrysts) and a later solid-state foliation (e.g. Vernon 2000, his Fig. 19). In other cases the two superposed foliations may be sub-parallel and the later one may progressively erase the earlier one. A special case where C/S fabrics have the same orientation and sense of shear as inferred magmatic flow imbrication of the same minerals, may strongly suggest that the deformation was continuous from the magmatic to the solid state and that therefore at least part of it was by submagmatic flow.

8.2.4
Evidence for Solid State Deformation

It has been argued that "solid state" is not an appropriate term since there is usually a liquid present and that "non-magmatic" deformation might be a better choice (Blenkinsop 2000). However, we prefer to continue the usage of the term "solid state" since it is well established and also widely used for metamorphic processes where the same restriction holds.

In the context of distinguishing between magmatic, submagmatic and solid state deformation it is interesting to review briefly the principal lines of evidence for solid state deformation (Vernon 2000; Blenkinsop 2000):

1. internal deformation often visible as undulose extinction, formation of subgrains and recrystallisation into smaller grains (compare Chap. 3);

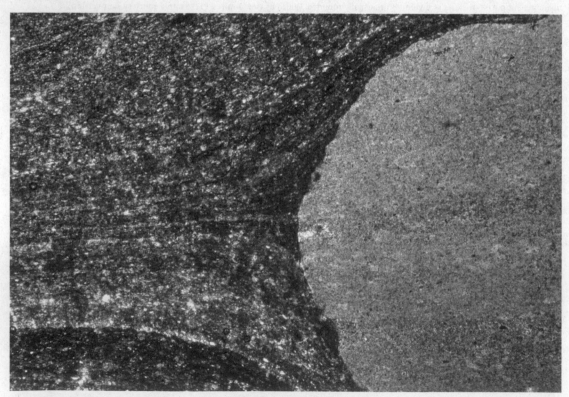

Fig. 8.5. Diagenetic compaction around carbonatic concretion in lithified lacustrine sediment of Taubaté Basin, SE Brazil. Width of view 7 mm. PPL. (Sample courtesy Margareth Guimarães)

2. recrystallised wings on deformed grains or porphyroclasts; these are particularly common in augen gneisses and mylonites (see Chap. 5);

3. elongation, usually in LS shape fabrics, of recrystallised aggregates (e.g. of quartz and mica) is very common in deformed granitic rocks;

4. prism <c> slip in quartz, recognised by quartz c-axes with a preferred orientation close to the stretching lineation (e.g. Law et al. 1992; Lagarde et al. 1994);

5. fine grained foliation(s) anastomosing around less deformed lenses or porphyroclasts, reflecting heterogeneous strain; they may constitute C/S fabrics (Sect. 5.6.3); not to be confused with magmatic flow foliation (Fig. 8.4);

6. boudinage of strong minerals such as feldspar, hornblende and tourmaline (Sect. 6.6);

7. myrmekite is considered by several authors (e.g. Vernon 2000) as evidence for solid-state deformation. However, it has been reported to form also by direct crystallisation from a melt (Paterson et al. 1989). Magmatic myrmekite may be distinguished from solid-state deformation related myrmekite (Hibbard 1987) since the former usually grows in dilatational sites around phenocrysts, whereas the latter forms at sites of stress concentration (e.g. Simpson and Wintsch 1989; Sect. 5.6.9).

8.3
Primary Structures in Sedimentary Rocks

Bedding, diagenetic foliation and tectonic foliation may be transitional structures and criteria to distinguish between them are given in Chap. 4. Slump folding and convolute bedding may be confused with tectonic structures of an early folding phase, but microstructural evidence is rarely conclusive in these cases. Compaction structures around concretions in sediments that are not tectonically deformed may resemble tectonic deformation (Fig. 8.5) and irregular lenses of coarser sediment in a finer grained matrix may produce structures very similar to boudinage (Fig. 8.6) also by compaction. Criteria to distinguish these structures from tectonically induced ones include the recognition of a tectonic foliation (Chap. 4), deformed pebbles in conglomerates and deformed fossils. Stylolites may form both by compaction and by tectonic deformation (Box 4.3). In the case of boudinage or apparent boudinage one always should consider the evidence that the fragments were once joined.

An interesting case is so-called "loop bedding", that consists of structures similar to isoclinal folds parallel to the bedding plane (Fig. 8.7), reported from lacustrine laminite sequences (e.g. Calvo et al. 1998; Rodriguez-Pascua et al. 2000) and interpreted to have been gener-

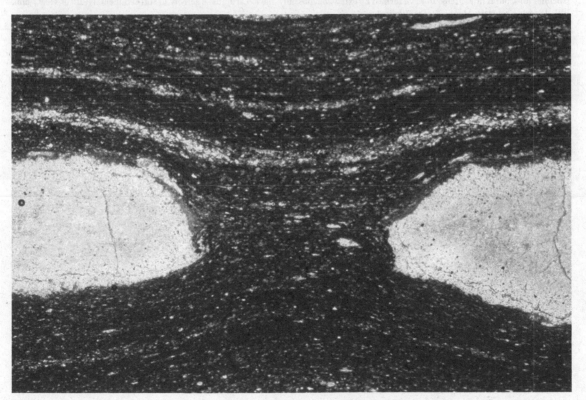

Fig. 8.6. Diagenetic compaction around silty lenses in mudstone, producing a structure similar to boudinage. Lithified lacustrine sediment of Taubaté Basin, SE Brazil. Width of view 6 mm. PPL. (Sample courtesy Margareth Guimarães)

Fig. 8.7. "Loop bedding" in fine grained lithified lacustrine sediment of Taubaté Basin, SE Brazil. The structure shows similarities with an isoclinal fold, but is interpreted to have formed by extension, possibly due to earthquake activity in soft sediment. Width of view 7 mm. PPL. (Sample courtesy Mauro Torres Ribeiro)

Fig. 8.8. Apparent thrust folding and faulting with divergent vergence in lacustrine sedimentary rocks of Taubaté Basin, SE Brazil. The structure is interpreted as due to earthquake activity. Width of view 4 mm. PPL. (Sample courtesy Mauro Torres Ribeiro)

Fig. 8.9. Pseudo chevron folds in carbonatic rock. The apparent folds were not generated by a process of folding but by differentiated growth of elongated carbonate crystals. Origin of sample unknown. Width of view 1.2 mm. CPL

ated in soft sediment by earthquakes. This structure can be misinterpreted as early phase isoclinal folding. An obvious difference is that no axial plane cleavage should be present, but diagenetic foliation may have formed, parallel to bedding and axial planes. In the same lacustrine tectonically undeformed sediments some structures occur with the appearance of early thrust folding (Fig. 8.8). These have also been interpreted as related to earthquakes.

Another case of apparent folds is the one illustrated in Fig. 8.9, where elongated carbonate crystals generated pseudo chevron folds, due to their differentiated growth (compare also Figs. 6.12, 11.5).

Rounded detrital grains in a sediment are commonly masked by overgrowth during diagenesis or low-grade metamorphism. However, with special methods such as cathodoluminescence or Raman mapping, the original contours of the grains can still be recovered (Figs. 5.1, 10.9).

Natural Microgauges

Microstructures have been used in the past to determine a sequence of deformation structures and shear sense. Many microstructures, however, seem to contain quantitative information as well. Obviously, the chemistry of minerals in rocks has long been used to obtain data on metamorphic conditions and isotope ratios have been used to determine the age of minerals and rocks, but such data can often only be obtained from powders or solutions, destroying the microstructure in the rock. This chapter presents a collection of microstructures which have been described in the literature as providing quantitative data from their geometry alone, observed by non-destructive methods. We have named this type of structures "microgauges". There are presently microgauges for strain, vorticity, stress orientation, differential stress, pressure, strain rate, temperature and rheology. The subject is in full development and new tools are added at a regular pace.

Some geologists may wonder if this type of geometric microgauges cannot simply be replaced by geochemical tools. Clearly, some types like those for stress orientation, differential stress and vorticity cannot, but even gauges for pressure and temperature are useful in rocks where the chemistry cannot give a clear answer, such as quartzites, or as a method to check results of geochemical techniques.

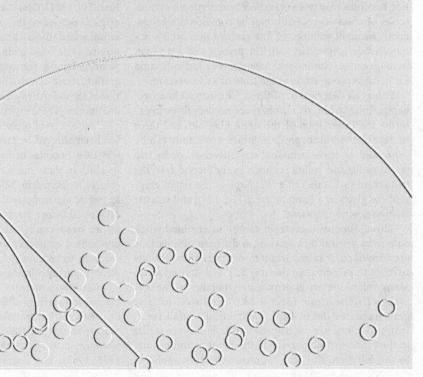

9.1
Introduction

Structural geologists have long used the macroscopic and microscopic geometry of the fabric of deformed rocks to determine a sequence of tectonic and metamorphic events, finite strain, or sense of shear. However, deformed rocks store a wealth of quantitative information that can be retrieved from the characteristic geometry of macro- and microstructures. Some microstructures can be used to determine parameters such as stress, temperature etc., and we therefore introduce the term *natural microgauges* for such features. This chapter gives some examples of presently available microgauges and possible future developments, and will hopefully stimulate readers into research on the subject. Microgauges can only be calibrated if the effects of all parameters that cause their geometric evolution are understood. At present, the study of microgauges is in its infancy and much theoretical and experimental work remains to be done. Finally, we try to indicate the limits and problems of the methods because the creation of numbers from rocks tends to give a (possibly) misplaced sense of confidence.

9.2
Strain Gauges

In many tectonic applications it is desirable to determine finite or tectonic strain. In order to express strain in simple values, it must be homogeneous (Sect. 2.5). In practice, natural strain will always be inhomogeneous at most scales of observation, but it may be considered homogeneous in small volumes of fine-grained rock with a homogeneous fabric (Fig. 2.4). The presence of a straight, homogeneous (continuous) foliation is a good indication for homogeneous strain at the scale of a thin section.

For a full description of three-dimensional homogeneous finite strain, six numbers are needed; three to describe the orientation of the strain ellipsoid, and three to describe strain magnitude. Strain magnitude can be expressed by three principal stretch values, or by two strain ratios and volume change. In the second case, the two strain ratios describe the *shape* of the strain ellipsoid (as given in a Flinn plot, e.g. Fig. 4.41), and volume change describes its *size*.

Three-dimensional strain can be determined when data from several thin sections in different orientations are combined. It is important to cut these thin sections parallel to principal strain axes X, Y and Z, if possible. Many foliations are approximately parallel to the XY-plane of finite strain (Sect. 4.2.9.2), and stretching or grain lineations (Sects. 4.1, 4.3) are generally parallel to X. If thin sections are cut normal to the lineation, parallel to the foliation, and parallel to the lineation but normal to the foliation, this will approximately correspond to

YZ-, XY-and XZ-sections. In thin section, two-dimensional finite strain can be completely described by three numbers representing the orientation of the strain ellipse and strain magnitude (Sect. 2.6). Strain magnitude can be expressed as two principal stretch values, or a strain ratio and area change. Methods to measure strain are mentioned in most structural geology textbooks. Below, we give an outline of some methods that can be applied in thin section to allow the reader to determine whether his material is suitable for strain analysis, and to make the best choice between different methods.

In some cases, strain ratio can be measured directly using objects with known original shape such as spherulites and oolites. However, it is important that the measured objects have the same rheology as the matrix in which they lie; for example, feldspar grains in a quartzite at low metamorphic grade cannot be used since they are stronger than quartz and therefore show only part of the tectonic strain. If objects were not initially spherical, the R_f / ϕ method can be used; the elliptic ratio of objects (R_f) is plotted against their orientation (ϕ) and the geometry of the resulting graphs is compared with standard patterns to determine strain (Ramsay and Huber 1983; Robin and Torrance 1987; Mulchrone et al. 2003; Meere and Mulchrone 2003).

If the exact original shape of objects is not known, but if the statistic mean is assumed to approach a spherical shape as for detrital grains in a quartzite, measurement of the dimensions of a large number of grains can be used to determine the strain geometry (Dayan 1981; Law et al. 1984). Care must be taken in this case that the original outline of the grains can still be seen, that no sedimentary shape fabric was present, and that only old, non-recrystallised grains are measured. In a passively deformed grain aggregate without grain boundary migration, strain can also be determined from the orientation of the deformed grain boundaries using the method of Panozzo (1984) explained in Sect. 10.6.2.

If the shape of objects such as grains in a sandstone has been affected by pressure solution, the Fry method (Fry 1979) or centre-to-centre method (Erslev 1988) should be used. In these methods, the distances between grain centres measured in different directions in a deformed aggregate are compared; if these distances were statistically equal before deformation (as in a well sorted sandstone), strain can be calculated (McNaught 2002). In recrystallised quartzite, however, it may be dangerous to use such techniques since the present centres of grains do not normally coincide with the original centres (Box 4.2).

Foliations which are thought to have developed by mechanical rotation of fabric elements can, in principle, be used to estimate strain ratios and the geometry of finite strain (Oertel 1983, 1985). This applies to feldspar phenocrysts in a granite and mica grains in a slate (Oertel 1983, 1985). The degree of preferred orientation of inequant grains is thought to reflect the intensity of deformation; the higher the strain, the more pronounced the preferred orientation becomes. This preferred orientation can be measured by goniometer (Sect. 10.3.5). Some care should be taken since mineral grains are not passive material lines; at low strain, the fabric in mica-rich aggregates may actually be stronger than that predicted by a model of passive rotation of material lines (Means et al. 1984). This is probably due to folding of micas normal to the shortening direction (Fig. 4.16(3)); material lines in this orientation will shorten but will hardly rotate (Fig. 4.16(1)). At high strain values, the fabric may be less intense than the theoretical prediction because micas do not stretch passively and rotate more slowly than passive lines because of their low aspect ratio (Fig. 4.16(1, 3); Means et al. 1984). Another point is that different minerals in a foliation may rotate at different velocities (Kanagawa 1991). Fabric intensity in slates can therefore be used only as an approximate measure of strain (Etheridge and Oertel 1979; Siddans 1977; Gapais and Brun 1981; Kanagawa 1991). The fabric of feldspar phenocrysts in granitoids may be more reliable (cf. Sect. 5.6.12).

Strain analysis is also possible using lattice preferred orientation patterns (LPO patterns). LPO patterns cannot be used to determine individual strain ratios or volume change, but carry information on the shape of the strain ellipsoid; the LPO pattern in a single thin section can give information on the full three-dimensional strain geometry (Fig. 4.41; Lister and Hobbs 1980; Wenk 1985). An LPO pattern is easily re-equilibrated, however, and may reflect only the last part of the deformation history (Law 1990). Comparison of the LPO pattern with another method of strain analysis in the same rock may help to decide if strain geometry changed during the deformation history.

One disadvantage of the determination of strain ratios as described above is that even if ratios in three perpendicular directions are known, the geometry of the strain ellipsoid can be determined, but not its size. This means that volume change cannot be determined in this way. Volume change is difficult to measure in rocks. Most techniques compare the chemistry of undeformed and deformed rock volumes, assuming that they were originally identical and that the undeformed volume did not change its composition (e.g. Mancktelow 1994). Unfortunately, these assumptions are not always valid (Sect. 4.2.9.3). Alternatively, volume change can be determined if stretches in several directions or a combination of stretches and strain ratios can be measured. Examples of structures that can be used to determine stretch values in thin section are deformed microveins (Sect. 6.2), strain fringes (Sect. 6.3) and microboudinage of crystals (Sect. 6.6). Sets of folded or boudinaged microveins can be used to determine stretch values provided that layer parallel shortening or extension in the veins is minimal. This is the case if little difference in layer thickness exists between limbs and hinges of folds as in some ptygmatic folds, or between the centre and edge of boudins. If many deformed veins in different orientations are present in a sample, they can be used to determine principal strain values, sense of shear (Sect. 5.6.12) and even the kinematic vorticity number (Sect. 9.3.2). Care has to be taken, however, that area change and volume change are not confused (Box 4.8) and that a three-dimensional reconstruction of the strain is made wherever possible.

An interesting method to determine principal stretch values and thereby volume change was proposed by Brandon et al. (1994) for sandstones which deformed by pressure solution and solution transfer, and where quartz grains with strain fringes show no effects of intracrystalline deformation (Ring and Brandon 1999; Ring et al. 2001). Thin sections are cut parallel to principal strain axes to show where fringes are developed (principal stretch > 1), and where grains have been dissolved (principal stretch < 1). The mean original diameter of grains can be determined in the direction of fringe growth since it did not change in that direction. Principal stretches > 1 can now be determined from the length of grains with their fringes, divided by their mean undeformed diameter; principal stretches < 1 are found from the mean diameter of dissolved grains in the direction of that principal strain axis, divided by the mean undeformed diameter.

A method to determine finite strain in shear zones was devised by Ramsay and Graham (1970). They showed that the curvature of a foliation into a shear zone reflects a gradient in finite strain from its peak in the core of a shear zone outwards; it develops because foliations rotate from a position between the instantaneous extension axis and the fabric attractor towards the latter with increasing strain (Sect. 5.5.3). The orientation of the foliation can be used as a strain gauge, since the angle between the foliation and the shear zone margin diminishes systematically with increasing strain. However, it is a function of W_k and A_k of flow as well (Sect. 2.5.2). Only if W_k and A_k can be estimated, e.g. if flow in a shear zone was by simple shear ($W_k = 1, A_k = 0$), is it possible to calculate principal stretch values at any site in the zone from the orientation of the foliation. The total displacement over the shear zone can also be determined by integration of the strain profile (Ramsay and Graham 1970; Ramsay and Huber 1983). In practice, this method is reliable only at relatively low strains. At high strain values the angles become very small and difficult to measure accurately.

Finally, localised intragranular deformation can be used to measure strain. Deformation bands in quartz grains can be used to determine 3D strain using sections in several orientations and a U-stage (Wu and Groshong 1991b). Mechanical twins in calcite and other minerals record small strains, and if twinning is the only mechanism of deformation, strain orientation and magnitude can be estimated in weakly deformed rocks with up to 15% strain

(Groshong 1972, 1974; Groshong et al. 1984a,b; Wiltschko et al. 1985; Kilsdonk and Wiltschko 1988; Ferrill and Groshong 1993; Evans and Groshong 1994; Harris and van der Pluijm 1998; González-Casado and García-Cuevas 1999, 2002; Craddock et al. 2000; González-Casado et al. 2003). Thin sections have to be prepared carefully in order to avoid twinning and fracturing during sample preparation. About 50 grains are usually sufficient to determine strain, but the method only works for coaxial progressive deformation (Evans and Groshong 1994; Burkhard 1993) and may be more suitable to find strain orientation than strain magnitude (González-Casado et al. 2003). There is also a linear relationship between the ratio of twinned to untwinned crystal as can be measured on a U-stage, and finite shear strain in a sample. This relationship is independent of temperature (Ferrill et al. 2004). Plots of mean twin intensity to mean twin width contain information on finite strain and temperature of deformation (Ferrill et al. 2004; Sect. 9.9). However, twinning only records the crystal-plastic component of the deformation, and brittle fracturing and pressure solution may also contribute considerably (Groshong et al. 1984a; Ferrill and Groshong 1993). Twinning is also used to determine the orientation and magnitude of stress during deformation (Sects. 9.5.1, 9.6.3).

9.3
Vorticity Gauges

9.3.1
Introduction

Several methods have been proposed to establish the kinematic vorticity number W_k of flow in rocks, i.e. the ratio of pure shear to simple shear (Sect. 2.5.2; Passchier 1988a; Means 1994). A useful tool in such reconstruction is the Mohr diagram for stretch (Means 1982; Passchier 1988b, 1990b). All applied methods are still in the phase of development and several microstuctures such as asymmetric boudins (Passchier and Druguet 2002), flanking structures (Passchier 2001; Graseman et al. 2003) and crystal tiling (Mulchrone et al. 2005) could be developed as new tools. In many cases, W_k of flow may change during progressive deformation, and therefore most methods determine a mean value of W_k over time, named W_m (Sect. 9.3.9; Passchier 1988a). A short outline of the presently applied methods to determine W_k or W_m in rocks is given below.

9.3.2
Deformed Sets of Veins

The stretch history of a particular material line in homogenous flow depends not only on its initial orientation with respect to ISA, but also on W_k, the kinematic dilatancy number A_k and finite strain. In simple deformation histories, lines undergo extension (e), shorten-

ing (s), or a transition from extension to shortening (e–s) and vice versa (s–e) (Figs. 2.8, 5.39). Finite strain affects the shape of the patterns, but not the *relative* size ratio of the (e–s) and (s–e) fields. If the relative size ratio can be established in a rock, they are a measure of W_k and A_k (Passchier 1991a) Veins which become folded or boudinaged can be used for such an analysis provided that a large range of orientations is present (Fig. 5.39). The method has been applied to rocks by Passchier (1986b), Passchier and Urai (1988) and Wallis (1992a).

A problem of the vein-set method is that boundaries between material line fields (Fig. 5.39) will not coincide with boundaries between fields of folded and boudinaged veins, and a calibration is necessary which is, at present, difficult. The method will only work for deformation histories with flow conditions that remained constant or underwent little change. The method is therefore at best semi-quantitative and has been mainly used to determine sense of shear (Fig. 5.39).

9.3.3
Lattice-Preferred Orientation

The geometry of LPO patterns such as those for quartz and calcite is a function of finite strain, strain geometry, active slip systems and W_k (Sect. 4.4.2). LPO patterns can therefore be used to distinguish between coaxial and non-coaxial progressive deformation (Law et al. 1986; Law 1987), and to determine W_k (Lister and Williams 1979; Wenk et al. 1987; Ratschbacher et al. 1991; Erskine et al. 1993; Xypolias and Koukouvelas 2001). However, measurements of W_k by this type of analysis tend to be rather inaccurate because they are slow to respond to changes in kinematics (Bestmann et al. 2000). Platt and Behrmann (1986) and Wallis (1992a) combined high quality strain data with degree of quartz LPO pattern asymmetry to find W_k; Vissers (1989) combined the rotation angle of garnet porphyroblasts with garnet aspect

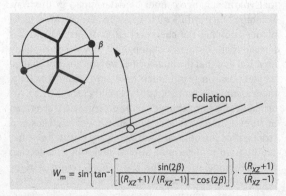

$$W_m = \sin\left\{\tan^{-1}\left[\frac{\sin(2\beta)}{[(R_{XZ}+1)/(R_{XZ}-1)] - \cos(2\beta)}\right]\right\} \cdot \frac{(R_{XZ}+1)}{(R_{XZ}-1)}$$

Fig. 9.1. Mean kinematic vorticity can be calculated from the angle between the foliation in a shear zone and the central axis of a C-axis LPO pattern of quartz from the same domain. Strain must be measured independently. (Based on Wallis 1992, 1995)

ratio and quartz LPO pattern asymmetry, and Ratsch-bacher et al. (1991) used the orientation of the c-axis maxima of calcite LPO patterns.

In flow with a monoclinic symmetry, the central axis of a quartz-c-axis pattern in a stereogram should be approximately at right angles to the flow plane of the deformation responsible for the fabric development. The orientation of a foliation formed in a single deformation event with respect to this flow plane is a function of W_k and finite strain. Therefore, if quartz fabric patterns are used to determine the orientation of the flow plane in a foliated rock, the angle between the foliation and this flow plane can be used to find W_m provided that finite strain can be determined independently (Wallis 1992a, 1995; Graseman et al. 1999; Law et al. 2004; Fig. 9.1).

9.3.4
Mantled Porphyroclasts, Fibrous Veins and Fringes

The rotational behaviour of rigid objects in non-coaxial flow depends, amongst other factors, on the shape of the objects and W_k of the bulk flow (Ghosh and Ramberg 1976; Freeman 1985; Passchier 1987b; Wallis et al. 1993; Masuda et al. 1995a; Box 5.4). In general flow types, between pure shear and simple shear, porphyroclasts with high aspect ratio can become blocked for further rotation, while those of lower aspect ratio are permanently rotated (Box 5.4). The geometry of the deformed mantle of porphyroclasts is influenced by this rotational behaviour (Sect. 5.6.7; Passchier 1987b, 1988a; Passchier and Sokoutis 1993). Consequently, the shape of porphyroclast mantles and the orientation distribution of the long axis of porphyroclasts, together with their aspect ratios can theoretically be used to determine W_m (Ghosh and Ramberg 1976; Passchier 1987b; Masuda et al. 1995c). Two methods based on this principle are presently used to determine mean vorticity W_m (Xypolias and Koukouvclas 2001; Law et al. 2004).

1. The *porphyroclast aspect ratio method* (Passchier 1987b). In this method the orientation, aspect ratio and mantle shape of porphyroclasts is used to determine which clasts were likely rotating, and which obtained a stable position; as seen in Fig. 9.2, for any W_k value there is a critical aspect ratio of clasts below which clasts can rotate freely, and will develop δ-type mantles (Sect. 5.6.7). Above this value, they will not rotate and develop σ-type mantles.
2. The *hyperbolic distribution method* (Simpson and dePaor 1993) is similar to the previous method but plots orientation and aspect ratio of porphyroclasts in polar coordinates, and fits the positions of σ-type mantled clasts with the highest aspect ratios to a hyperbolic curve on a hyperbolic net (Simpson and dePaor 1993). The opening angle of the hyperbolic net is a function of the mean vorticity (Fig. 9.3).

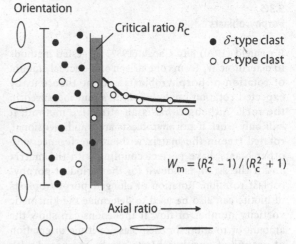

Fig. 9.2. Method to determine the mean kinematic vorticity number W_m from the critical object aspect ratio R_c, above which objects rotate permanently

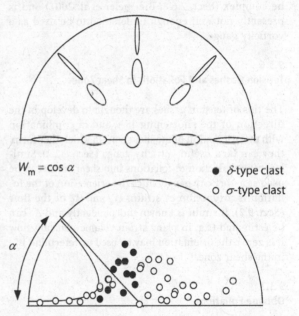

Fig. 9.3. Method to determine the kinematic vorticity number W_m from the aspect ratio and orientation of δ- and σ-type clasts in a mylonite sample, plotted in polar coordinates

The geometry of fibres in veins and fringes contains much information on W_k (Etchecopar and Malavieille 1987; Passchier and Urai 1988; Beam and Fischer 1999). Figures 6.17, 6.20, 6.21, 6.22 and 6.24 show that major differences exist between fibrous veins and fringes formed in pure shear and in simple shear. Since these geometries have not been calibrated they can presently only be used to distinguish end members of the flow range, but potentially these microstructures are powerful vorticity gauges. A pioneering attempt to use these structures was published by Passchier and Urai (1988).

9.3.5
Porphyroblasts

Rosenfeld (1970) and Ghosh (1987) suggested methods to determine W_k from comparison of the actual amount of rotation of porphyroblasts and the theoretically expected rotation angle for finite strain measured in the rock. Although this is an attractive method, it will only work if porphyroblasts are equidimensional, rotated free in the matrix without interference with other blasts, have perfect coupling with the matrix and if the strain is known for the period of porphyroblast rotation. Rotation of elongate porphyroblasts of biotite can also be used to determine the kinematic vorticity number of flow if inclusions can show the amount of rotation for each blast, or if the orientation of groups of porphyroblasts can be compared with some standard pattern (Holcombe and Little 2001). However, rotational behaviour of porphyroblasts can be complex (Sect. 7.6.8; Biermeier et al. 2001) and is presently not well enough understood to be used as a vorticity gauge.

9.3.6
Tension Gashes and Foliations in Shear Zones

The tips of tension gashes are thought to develop in the direction of the shortening ISA, and in combination with the orientation of the shear zone in which they form, they can be a useful vorticity gauge (Sect. 6.2.2). Similarly, the deflection of foliations into shear zones can be used as a vorticity gauge, since the orientation of the foliation is a function of strain, W_k and A_k of the flow (Sect. 9.2). If strain is known independently and A_k can be estimated (e.g. in plane strain volume-constant flow it is zero), the orientation may be used to determine W_m in the shear zone.

9.3.7
Oblique Foliations

Theoretically, the angle between an oblique foliation (Sect. 5.6.2; Box 4.2) and the fabric attractor is a function of W_k, strain rate and recrystallisation rate (and therefore probably of temperature). The angle can therefore be used to determine W_k and is expected to be maximal for simple shear. A problem with this method is that several other factors may influence the angle, and that finite strain should be sufficiently high to rotate fabric elements into parallelism with the fabric attractor (Sect. 2.9; Box 4.2). For example, some oblique foliations make an angle with the fabric attractor that exceeds 45°; such orientations are difficult to explain with the available theory.

9.3.8
Al-Cr Zoning in Spinel

Ozawa (1989) showed that many spinel grains in peridotite have an asymmetric sector zoning of Al and Cr, with Al concentrated in the direction of the stretching lineation, and Cr in the direction of the normal to the foliation in the peridotite. He suggests that Cr concentrated in the σ_1 and Al in the σ_3 direction due to unequal diffusivity of these ions when the spinel grains deformed by solid-state diffusion creep. In fact, the sector zoning is more likely to lie in the direction of ISA and may be useful to determine sense of shear and W_m if sector symmetry planes are oblique to the shape fabric in the peridotite.

9.3.9
W_k History and Accuracy

A problem with all determination of kinematic vorticity in naturally deformed rocks is the large number of assumptions in the methods used, and the uncertainties about progress of deformation. Most of the methods discussed above are two dimensional for practical reasons, using outcrop surfaces or thin sections, while in reality veins or porphyroclasts have a complex 3D shape. Other problems are assumptions of monoclinic flow, homogeneous deformation and invariable flow conditions during progressive deformation, all of which are unrealistic. The assumption of monoclinic flow can be checked by controlling the geometry of a large number of fabric elements for symmetry (e.g. Sawaguchi and Ishii 2003). Homogeneity of flow can be assured by using data from a small volume of rock. The possibility of variable flow conditions with time can be countered by using a mean value of W_k over time, W_m. This mean value W_m, established with the methods given above, is difficult to interpret but a possible solution to this problem is to measure W_m in a rock using several different gauges which re-equilibrate at different rate during the deformation history. For example, quartz fabrics are thought to reequilibrate relatively quickly, while deformed veins will record the whole deformation history if they predate deformation; if both methods give different values, this may indicate the trend of change in W_k during the deformation history (Passchier 1988a, 1990a; Wallis 1992a; Law et al. 2004). Reconstructions of deformation paths, including W_k history, may also be developed using fibrous veins and fringes (Ramsay and Huber 1983; Ellis 1986).

In conclusion, none of the methods to determine W_k are as yet very accurate and the best that can be hoped for in most settings is to show that flow was not simple shear, but contained a pure shear component. Great care should be taken that vorticity measurements are not over-interpreted.

9.4
The Concept of Palaeostress Gauges

The present state of stress in the lithosphere can be measured directly in the crust (Harper and Szymanski 1991) or assessed at greater depth using seismological or heat flow data (Molnar and England 1990). For states of stress in the geological past, microstructures can be used. Although several methods have been proposed to determine the orientation and magnitude of stress in a rock during deformation (also referred to as *palaeostress*), it is important to realise that stress is not preserved, only its effects. Moreover, stress is only defined at a point, and usually varies in magnitude and orientation from point to point in a rock, and also changes strongly with time. Stress leaves traces in a rock only when permanent deformation is realised and stress gauges are claimed to capture the properties of the stress field during this time. However, in inhomogeneous materials such as rocks on the grain scale, stress was probably strongly variable from crystal to crystal and even within individual crystals, and must have changed with time. Probably, stress gauges measure some kind of 'mean value' for stress and it is important to remain critical about the meaningfulness of such mean values. Microgauges to measure 'palaeostress' in deforming rocks are divided into three main types; 1. gauges for the orientation of the principal stress axes; 2. gauges for differential stress, and; 3. gauges for mean stress or pressure.

9.5
Gauges for the Orientation of Palaeostress Principal Axes

9.5.1
Twins in Calcite and Other Minerals

Calcite e-twins ($\{01\bar{1}2\}$) have been proposed as a tool for determination of the orientation of palaeostress principal axes (Turner 1953; Laurent et al. 1981; Dietrich and Song 1984; Borradaile and McArthur 1990; Shelley 1992; Burkhard 1993; Marrett and Peacock 1999; Nemcok et al. 1999; Laurent et al. 2000; Fry 2001). Since movement on twins can take place in only one direction, data on the presence of twins, the orientation of twins and crystallographic c-axes in a large number of grains can give an average value of the orientation of principal shortening and extension directions. Therefore, the quantity that is really measured is *strain* accommodated by the twins. Only in isotropic materials that undergo coaxial deformation will the orientation of stress and strain axes coincide. Hence, calcite twins can only be used to measure the orientation of principal axes of palaeostress in rocks that deform coaxially. Only straight twins can be used, which restricts use of the method to low temperature deformation (Sect. 9.9)

and low strain. In all cases, there will be a number of grains that do not fit in the determined directions of palaeostress axes (Groshong et al. 1984; Pfiffner and Burkhard 1987). The percentage of deviating grains can be taken as a measure of the reliability of the method in any particular case. Attribution of the deviating grains to other phases of deformation with a different orientation of the stress field is probably not realistic. Methods of determination of palaeostress orientation using twins have also been proposed for dolomite (Christie 1958), pyroxene (Raleigh and Talbot 1967; Trepmann and Stöckhert 2001), olivine (Carter and Raleigh 1969) and plagioclase (Lawrence 1970).

9.5.2
Fractures and Fluid Inclusion Planes

Tensional fractures that lack displacement probably form parallel to the local σ_1–σ_2-plane of stress (Sects. 2.11, 3.2, 6.1). Although this orientation of stress in a crystal can be influenced by crystallographic orientation, tensional fractures that cut several grains are probably close to the bulk stress orientation. Tensional fractures, however, can form late during the deformation history, mostly close to the surface or even during sampling or thin section preparation and are therefore difficult to date. Planes of fluid inclusions are more reliable in this sense, since they form at some depth and the density and composition of fluid inclusions can give some indication of *P-T* conditions of their development (Sect. 10.5). Also, the sequence of cross-cutting fluid inclusion planes can usually be determined. In principle, planes of fluid inclusions can be used to determine the changes in orientation of the σ_1–σ_2-plane in geological history and to link it with metamorphic conditions (e.g. Boullier et al. 1991).

9.5.3
Deformation Lamellae

Quartz deformation lamellae (Box 3.3) have been interpreted as planes of high resolved shear stress and can therefore be used to determine palaeostress directions in a way similar to calcite e-twins (Carter and Raleigh 1969; Law 1990; Twiss and Moores 1992). Since calcite e-twins can only accommodate movement in a single direction, they are probably more reliable as microgauges than quartz deformation lamellae. The same objections and restrictions to application of the method mentioned for calcite twins are valid for deformation lamellae.

9.5.4
Flame Perthite

Flame perthite occurs in K-feldspar and plagioclase bearing rocks deformed at greenschist facies conditions at high differential stress, especially in ductile shear zones (Fig. 7.28;

Debat et al. 1978; Passchier 1982a). It forms due to replacement of K-feldspar by albite with preservation of the crystallographic lattice orientation (Pryer and Robin 1995; Pryer et al. 1995). Pryer and Robin (1996) propose that flame perthite can be used as an indicator of principal stress axes orientation. If stress is less than that required for brittle fracturing of K-feldspar, flame perthite lamellae will grow if the ideal plane of flame growth is parallel or at a small angle to the maximal principal stress direction. This ideal plane is the *Murchison plane*, the plane of minimum lattice misfit between albite and K-feldspar (Smith 1974). Flame tips will lie parallel to the Murchison plane if only a single set of flame perthite is present (Pryer and Robin 1996). Analysis of flame perthite plane orientation in a large number of grains in a sample can therefore give the principal stress axis orientation.

9.6
Differential Stress Gauges (Palaeopiezometers)

9.6.1
Vein and Fracture Types

In brittle deformation, an indication of differential stress can be gained from the nature of developing veins or fractures (Sects. 3.2, 6.1; Etheridge 1983). If parallel veins oc-

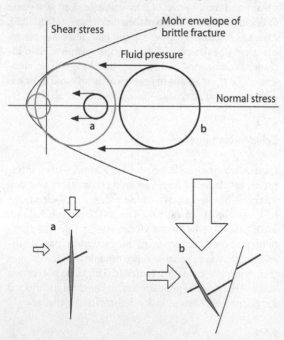

Fig. 9.4. Mohr diagram showing two states of stress; upon increased fluid pressure a Mohr circle (**a**) with small differential stress will touch the Mohr envelope for fracture in the tensional domain and tension veins will form without a shear displacement component. A large differential stress state (**b**) will mean that upon increase of fluid pressure the Mohr circle will touch the Mohr envelope in the shear domain and shear faults and veins will form

cupy tensional fractures without a shear component, differential stress must be small; at higher differential stress, veins will have a shear component or shear fractures develop which only have vein material in jogs (Fig. 9.4).

9.6.2
Dynamically Recrystallised Grain Size

The size of dynamically recrystallised grains in a deforming material (e.g. Figs. 3.29, 3.31) is a function of differential stress and has been proposed as a *palaeopiezometer*, i.e. a method to measure the magnitude of palaeostress (Twiss 1977, 1986; Mercier et al. 1977; Etheridge and Wilkie 1979, 1981; Christie and Ord 1980; Schmid et al. 1980; Ross et al. 1980; Karato et al. 1980; Koch 1983; Ord and Christie 1984; Ranalli 1984; Hacker et al. 1990, 1992; Michibayashi 1993; van der Wal et al. 1993; Rutter 1995; Post and Tullis 1999; Stipp and Tullis 2003). Dislocation density and subgrain size have also been proposed as palaeopiezometers and seem to give good results in metals. Until recently, these parameters could only be determined by labour-intensive work in a TEM, but fast development of SEM/EBSD techniques (Sect. 10.2.4) now allows easier data collection. However, the last two parameters seem to be highly sensitive to changes after deformation due to recovery. They are therefore less applied in geological samples (with complex, non-controllable histories) than in metals; dynamically recrystallised grain size is usually preferred.

For a particular differential stress during progressive deformation, each mineral has a particular mean size of recrystallised grains depending on the recrystallisation mechanism (Sect. 3.7; Drury and Urai 1990; Twiss and Moores 1992; Post and Tullis 1999; Stipp and Tullis 2003) and possibly on water content of grains(Bell and Etheridge 1976; Jung and Karato 2001a,b) and deformation temperature (de Bresser et al. 2001). The dependence on recrystallisation mechanism seems to be particularly sensitive for the transition from BLG to SGR recrystallisation, but less for the SGR-GBM transition. This probably reflects the fact that the former represents a more fundamental mechanical change in recrystallisation mechanism. The dependence on recrystallisation mechanism has been shown for quartz (Stipp and Tullis 2003) and may also apply to calcite (Schmid et al. 1980; Rutter 1995; Fig. 9.5) but not as clearly as in quartz (cf. ter Heege et al. 2002; Barnhoorn et al. 2004). In the case of quartz, temperature and the α/β transition have no influence on recrystallised grain size (Gleason and Tullis 1995; Stipp and Tullis 2003). It is important to note that we refer to mean grain size in this section; even amongst recrystallised grains, there can be a significant range in grain size. Grain size can be estimated with the mean linear intercept method (Smith and Guttman 1953; de Hoff and Rhines 1968) or, more accurately, with image analysis techniques (e.g. Panozzo 1984).

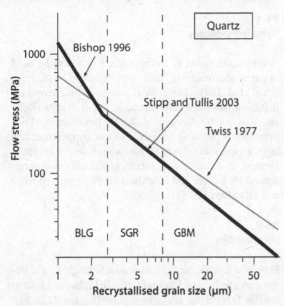

Fig. 9.5. Relationship between recrystallised grain size and differential stress, expressed as flow stress for quartz. BLG, SGR and GBM are mechanisms of dynamic recrystallisation (after Stipp and Tullis 2003). *Top part* of graph based on unpublished data from Bishop (1996) as quoted in Post and Tullis (1999). The older graph of Twiss (1977) is shown for reference

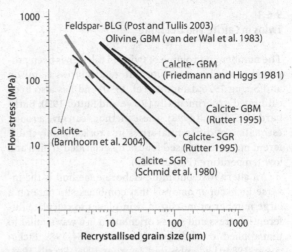

Fig. 9.6. Relationship between recrystallised grain size and differential stress, expressed as flow stress for feldspar, olivine and calcite. Data from indicated papers

Most available data on stable grain size and associated differential stress are related to quartz (Fig. 9.5; Twiss 1977; Post 1977; Ross et al. 1980; Michibayashi 1993; Stipp and Tullis 2003) but there are also data for other minerals (Fig. 9.6). Examples shown are for olivine (Karato 1984; van der Wal et al. 1993; Jung and Karato 2001a,b), calcite (Schmid et al. 1980; Rutter 1995; Barnhoorn et al. 2004) and feldspar (Post and Tullis 1999). In olivine, water content may influence the size of recrystallised grains by GBM (Jung and Karato 2001a,b). Estimates for differential stress in rocks based on grain-size palaeopiezometers range from a few MPa in high temperature deformation to 100–300 MPa in some low-temperature mylonite zones (Küster and Stöckhert 1999).

Grain size reduction in mylonites is due to the fact that differential stress in active ductile shear zones can be high, especially at low temperature; consequently the stable recrystallised grain size is small. However, extremely fine-grained rocks such as cherts may undergo grain growth during dynamic recrystallisation in a shear zone to reach the stable grain size (Masuda and Fujimura 1981). De Bresser et al. (2001) suggest that dynamic recrystallisation leads to a balance between grain size reduction and growth processes, set up in the neighbourhood of the boundary between the dislocation creep field and the (grain size sensitive) diffusion creep field. Figures 3.30 and 3.31 show examples of typical fabrics in quartz mylonites where differential stress can be estimated from dynamically recrystallised grain size. In some

cases, even stress differences in a single aggregate can be determined, such as near rigid porphyroclasts that can cause a local increase in differential stress.

Possible sources of error in the calculation of palaeostress are:

1. the presence of old grain relicts that may be smaller or (usually) larger than the recrystallised ones (Michibayashi 1993). Old grain relicts are recognisable by their irregular outline, deviant crystallographic orientation and well-developed intracrystalline deformation structures;
2. misinterpretation of the active recrystallisation mechanism (Post and Tullis 1999);
3. the presence of a second mineral that inhibits growth of the mineral to be measured, e.g. mica in quartzite (Krabbendam et al. 2003), or that causes local stress concentrations or strain shadows; therefore, only monomineralic aggregates should be used, or interpretation should not be extended beyond the boundaries of a monomineralic domain in a polymineralic rock;
4. static recrystallisation that may have affected grain size. Static recrystallisation can be recognised by the presence of straight, polygonal grain boundaries and 'strain-free' recrystallised grains (Box 3.10; compare Fig. 3.31 with Fig. 3.41), and evidence for grain boundary adjustment to euhedral shape in strongly anisotropic crystals such as micas; these will show evidence for static recrystallisation before quartz and feldspar do. Static recrystallisation is commonly associated with hydration and retrograde transformation of mineral assemblages;
5. the use of a piezometric relationship that has not fully been tested for possible effects of recrystallisation mechanisms, influence of water or role of temperature.

9.6.3
Twins in Calcite and Dolomite

The number and volume of twins in calcite has been proposed as a microgauge for differential stress (Jamison and Spang 1976; Laurent et al. 1990) and has also been calibrated experimentally (Rowe and Rutter 1990; Burkhard 1993; Ferrill 1998). These methods can give a rough estimation of differential stress in a rock, but only if different methods are used for calcite deformed at high and low temperature (Ferrill 1998).

An alternative but more elaborate method is the inverse Etchecopar analysis that combines slip data on a large number of measured twin planes to calculate differential stress and stress orientation in a way similar to that applied to brittle faults (Angelier 1984, 1989; Etchecopar 1984; Lacombe and Laurent 1996; Ferrill 1998; Laurent et al. 2000).

All methods are based on the fact, that twinning only takes place if the critical resolved shear stress (CRSS) on potential twin planes is exceeded. The CRSS for calcite is now estimated to vary between 5–15 MPa (Jamison and Spang 1976; Lacombe and Laurent 1996; Laurent et al. 2000) (Sect. 3.12.3); it may decrease slightly with increasing temperature and it may increase with increasing finite strain (Laurent et al. 2000). Grain size and porosity may also play a role (Rowe and Rutter 1990; Ferrill 1998). Small crystals seem to have fewer twins than large grains (Schmid 1982; Newman 1994). Differential stress is twice the maximum shear stress at any point, and for a differential stress of 10–30 MPa, twins can only form in the direction parallel to the plane of maximum resolved shear stress. If differential stress is larger, twins can also form in other orientations. In this way, the size and sharpness of the principal shortening direction maximum, determined for a population of twins, is a function of the differential stress (Jamison and Spang 1976; Laurent et al. 1990). The methods are also applicable to dolomite, where the critical resolved shear stress for twinning is expected to be higher than in calcite (Jamison and Spang 1976; Rowe and Rutter 1990; Newman 1994).

Unfortunately, the twin palaeopiezometer techniques rely heavily on the assumption of homogenous stress distribution in a sample, which is probably not realistic. Inhomogeneous stress distribution can have a similar effect on the spread in orientation of inferred principal stress directions as differential stress magnitude. Finally, since twin density increases with increasing strain, the twin palaeopiezometer is likely to give best results at very low strain values and in non-porous and relatively even-grained and coarse calcite and dolomite without an LPO (Burkhard 1993; Newman 1994; Ferrill 1998). It should be applied with care (Burkhard 1993) and not be used for strongly deformed rocks.

9.6.4
Twins in Pyroxenes

Deformation twins in clinopyroxene may also be used as palaeopiezometers (Tullis 1980; Kollé and Blacic 1982; Orzol et al. 2003). The critical resolved shear stress for twinning has been determined as 140–150 MPa (Kollé and Blacic 1982; Orzol et al. 2003). Twinning in clinopyroxene is probably restricted to low temperature and high strain rate (Godard and van Roermund 1995; Laurent et al. 2000). A practical application is demonstrated by Küster and Stöckhert (1999) and Trepmann and Stöckhert (2001).

9.6.5
Microboudins

Flow of a ductile medium around an elongate rigid object can produce an internal stress field that can lead to tensile fracturing and boudinage (Figs. 6.32, 6.33). Microboudins can be used to calculate differential stress based on the *fibre-loading principle* (Lloyd and Ferguson 1981; Lloyd et al. 1982; Ji and Zhao 1994). A rigid fibre parallel to the extension direction in ductile coaxial flow is though to experience a gradient in tensile stress that is maximum at the centre. Consequently, the fibre will break near the centre and this process will be repeated in the remaining fragments until some critical limit is reached that depends on mechanical properties of the fibre and on differential stress. The aspect ratio of microboudin fragments can therefore be used to estimate differential stress (Masuda and Kuriyama 1988; Masuda et al. 1989, 1990, 2003, 2004; Ji and Zhao 1993). The principle can be used in any rigid mineral, including feldspar at low-grade metamorphic conditions (White et al. 1980). The presence of tensional fractures (without a component of shear movement) is also useful because it restricts the possible magnitude of the differential stress to four times the tensile strength of the material under consideration (Etheridge 1983). The fibre-loading principle explains why some gaps between fragments are wider than others; the wider ones are thought to have formed earlier (Ferguson 1981, 1987; Masuda and Kuriyama 1988).

Although not strictly a boudinage mechanism, the extent of fracturing in rigid grains like garnet at low deformation temperature may also carry information on local differential stress.

9.6.6
Deformation Lamellae

Deformation lamellae have been proposed as palaeopiezometers (Drury and Humphreys 1987; Blenkinsop and Drury 1988; Trepmann and Stöckhert 2003). Subbasal deformation lamellae in quartz are thought to develop at differential stresses between 170–420 MPa. Spacing of deformation lamellae may also serve as a palaeopiezometer (Koch and Christie 1981).

9.7
Pressure Gauges

Lithostatic pressure is usually determined by classical petrological methods such as mineral composition of the rock and fluid inclusion density (Sect. 10.5). However, microstructural assemblages may help where older mineral assemblages or fluid inclusions have been destroyed. At low pressure, ductile and brittle calcite microstructures can be used as a gauge (Ferrill and Groshong 1993). The presence of gas bubbles or amygdales (gas bubbles filled with crystalline material) in pseudotachylyte indicates that the fault rock formed at shallow crustal levels, the exact magnitude of which depends on the composition of the rock. Another promising development is the recognition of deformed pseudotachylyte in many ductile shear zones. Pseudotachylyte is a brittle fault rock that mostly forms in the upper crust, and whose geometry is not easily destroyed by later deformation or metamorphism (Sect. 5.2.5). The presence of pseudotachylyte depends on several factors but its presence carries information on lithostatic pressure in the rock during its development (Sect. 5.2.5; Passchier et al. 1990a).

9.8
Strain Rate Gauges

Geological strain rates are estimated to lie usually between 10^{-13} and 10^{-15} s^{-1} (Pfiffner and Ramsay 1982; Carter and Tsenn 1987; Paterson and Tobisch 1992) and could theoretically be estimated in rocks if differential stress and temperature of deformation are known, using known flow laws derived from experimental data (Sect. 3.14). Differential stress values can be obtained using a palaeopiezometer as discussed above. This method to estimate strain rates has been applied to peridotite using flow laws for olivine (Karato et al. 1986; Suhr 1993) and for crustal rocks using flow laws for quartz (e.g. Stipp et al. 2002; Trepmann and Stöckhert 2003) and calcite (Ulrich et al. 2002). The discrepancies between results of different experiments give 'error bars' of one, or even two orders of magnitude. Sources of error are in the flow laws and their extrapolation to geological strain rates, in the estimate of differential stress if grain growth is inhibited or if recrystallisation is important, and in temperature estimates. For olivine, an error of 50 °C in the temperature results in an error of one order of magnitude in the strain rates. Another possible method is the use of LPO fabrics; a relation exists between the slip

systems that are active in a mineral and the strain rate (Lister et al. 1978); however, since other parameters (temperature, water activity) also influence the active slip systems and because it is not always possible to determine which slip systems were active in natural rocks, this method cannot (yet) be used. The irregularity of recrystallised quartz grain boundaries increases with increasing strain rate and decreasing temperature and may be calibrated, if temperature is known independently, as a strain rate gauge (Takahashi et al. 1998).

The shape of fibres and elongate crystals in veins and fringes is strongly dependent on the relation between crystal growth rate and opening rate of the vein or fringe; although the processes involved in the development of these structures are still incompletely understood, it is clear that they contain information on strain rate in the host rock (Sect. 6.2.3).

One of the most promising methods is direct isotopic dating of strain fringe increments in quartz fringes (Müller et al. 2000) and the growth and rotation rate of porphyroblasts (Christensen et al. 1989; Ridley 1986; Joesten and Fischer 1988; Paterson and Tobisch 1992). However, even cm-size porphyroblasts may grow in less than 1 Ma under some circumstances (Sect. 7.2; Burton and O'Nions 1991; Paterson and Tobisch 1992), and this is out of range for present direct dating methods.

δ- and Θ-type mantled porphyroclasts have been proposed as indicators of relatively high shear strain rate in ultramylonite since they indicate rotation with limited dynamic recrystallisation (Passchier and Simpson 1986; Whitmeyer and Simpson 2003). If dynamic recrystallisation rate is a function of differential stress, enhanced rotation may indicate a high shear strain rate; however, other factors may play a role as well so development of these structures should be further investigated.

In conclusion, estimates of strain rate are not yet accurate in practice, but promising progress is being made.

9.9
Temperature Gauges

Any experienced student of microtectonics is aware that there is a correlation between metamorphic grade during deformation and the presence and geometry of particular microstructures. Unfortunately, few of these structures have been calibrated to date; geometric temperature gauges could give independent data on temperature besides the classical petrological geothermometers and may be less easily modified by retrogression and later deformation than mineral composition.

One of the most promising temperature gauges is twin geometry in calcite. At temperatures below 400 °C, crystal-plastic deformation in calcite is mostly by mechanical e-twinning (Groshong 1988). The geometry of such deformation twins in calcite has been proposed as a temperature gauge (Fig. 9.7a–e; Jamison and Spang 1976; Mosar 1989; Ferrill 1991, 1998; Burkhard 1993; Ferrill et al. 2004). Narrow straight twins (less than 1 μm wide – Type I of Burkhard 1993; Figs. 9.7a,b) indicate temperatures below 200 °C and dominate below 170 °C. Wider twins which can be optically resolved (Type II > 1 μm) dominate above 200 °C up to 300 °C (Figs. 9.7a,c; Groshong et al. 1984; Rowe and Rutter 1990; Evans and Dunne 1991; Ferrill 1991; Ferrill et al. 2004). The reason is that increasing strain at temperatures below 170 °C leads to growth of new twins rather than widening of older ones (Fig. 9.8); above 200 °C, widening of existing twins dominates over the creation of new ones (Fig. 9.8; Ferrill 1991, 1998; Ferrill et al. 2004). It may be necessary to use a U-stage to distinguish narrow and wide twins, but all twins can be recognised by parallel colour fringes in sections oblique to the twin plane (Spang et al. 1974). At temperatures above 200 °C, Type III intersecting twins and bent twins are present (Figs. 9.7a,d). Bending of twins is thought to be due to activity of dislocation glide

on r- and f-planes (Burkhard 1993). At temperatures above 250 °C, twins obtain serrated boundaries due to twin boundary migration recrystallisation (Figs. 3.20, 9.7a,e; Sect. 3.7; Type IV twins of Vernon 1981; Burkhard 1993; Rutter 1995; Ferrill et al. 2004), which may sweep grains, and other types of dynamic recrystallisation may also occur. However, in large calcite crystals common in vein calcite, high strain rate at low temperature may lead to a high dislocation density, and recrystallisation may occur at temperatures below 250 °C (Kennedy and White 2001; Ferrill et al. 2004). Complete dynamic recrystallisation of calcite may occur above 300 °C (Evans and Dunne 1991; Weber et al. 2001).

The width of exsolution symplectites (Sect. 7.8.3) formed by isochemical reactions of the type (A ⇒ A' + B) can be used as a temperature gauge if properly calibrated (Joanny et al. 1991; van Roermund 1992). Such symplectites form by grain boundary diffusion mechanisms resulting in the nucleation of mineral B in the reacting interface as it migrates into the supersaturated adjacent parent mineral. Under such circumstances, grain bound-

Fig. 9.7.
a Schematic illustration of the influence of temperature on deformation by calcite twinning (after Burkhard 1993; Ferrill et al. 2004). **b–e** Photomicrograph examples of different twin types (all in crossed-polarized light). **b** Type I twins from the northern Subalpine Chain, France (sample 87–25d, Ferrill 1991). Width of view 0.68 mm. **c** Type II twins from the North Mountain thrust sheet (sample W91) in the Great Valley, Central Appalachian Valley and Ridge Province (Evans and Dunne 1991). Note that thin twins are locally developed within thick twins. Width of the photomicrograph is 0.22 mm. **d** Type III twins from the Ardon thrust slice of the Diablerets nappe (sample 691.1) in the Helvetic Alps (Burkhard 1990). Width of view 0.14 mm. **e** Type IV twins from the Doldenhorn nappe (sample 199.3) in the Helvetic Alps (Burkhard 1990). Width of view 0.14 mm. Photomicrographs **b** and **c** are from ultra thin sections (thickness of approximately 5 microns or less). (After Ferrill et al. 2004. Reproduced with permission of Elsevier)

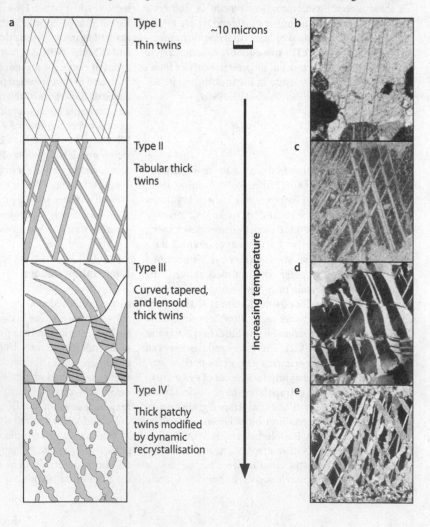

a

Type I
Thin twins

~10 microns

b

Type II
Tabular thick twins

c

Type III
Curved, tapered, and lensoid thick twins

Increasing temperature

d

Type IV
Thick patchy twins modified by dynamic recrystallisation

e

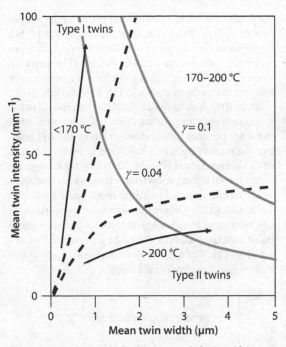

Fig. 9.8. Diagram showing the development trend of twins with increasing strain (*arrows*): at low temperature, more twins of Type I tend to be generated; at higher temperature, wider twins form which increase in width, but not in number with increasing strain. (After Ferrill et al. 2004)

ary diffusion rates are fully temperature dependent, which is directly reflected by the spacing of the symplectite (Shewmon 1969). Since the exsolved volume percentage of mineral B depends on the original composition of mineral A, it is the combined width of a pair of lamellae of A' and B that is critical, not their individual widths (Fig. 9.9; van Roermund 1992). Good examples of exsolution symplectites can be found in retrogressed eclogites in which the Na-bearing clinopyroxene becomes replaced by clinopyroxene-plagioclase symplectites (Boland and van Roermund 1983; van Roermund and Boland 1983).

Development of phase transformations such as exsolution structures can be shown in a *TTT diagram* (time-temperature-transformation) as pairs of curves, one for the time and temperature when a new phase is first detectable, the second for conditions when the reaction is completed (Putnis and McConnell 1980). At high temperature, approaching the critical temperature below which exsolution is possible, diffusion rates are high but nucleation rates are low; at low temperature, the reverse is true. Consequently, the curves in a TTT diagram show a minimum time for nucleation at some intermediate temperature.

Figure 9.9 shows a TTT diagram with a scheme of possible geometries of exsolved plagioclase lamellae in clinopyroxene. At high temperature, lamellae form ho-

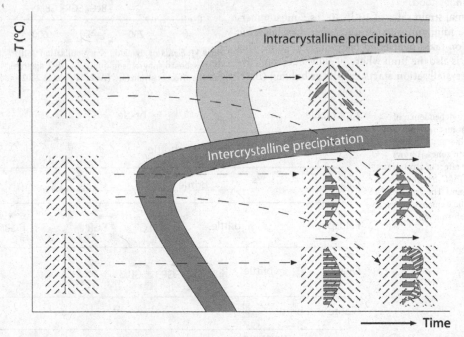

Fig. 9.9. Schematic TTT diagram showing the relative position of start and finish curves for inter- and intracrystalline exsolution of plagioclase from omphacite, after van Roermund (1992). *Straight dotted arrows* show isothermal trajectories of symplectite growth. *Insets* show schematic details of the developing microstructure along a grain boundary between two omphacite grains with identical chemistry but different orientation, indicated by *striping*. At high temperature, plagioclase nucleates inside the omphacite crystals. At lower temperature, exsolution symplectites are formed when the grain boundary migrates into one of the crystals. *Small arrows over insets* indicate movement direction of the grain boundary. The spacing of symplectite lamellae pairs decreases with decreasing temperature, as indicated. If temperature decreases significantly with time (*curved paths*), complex internal structures are formed. Such microstructures can be used to reconstruct cooling paths

mogeneously distributed inside the clinopyroxene crystals, but at lower temperature exsolution symplectites are formed at grain boundaries. Curved paths in temperature-time space can give rise to unique microstructures (Fig. 9.9), which can help to reconstruct *P-T-t* paths. Obviously, such temperature gauges are most powerful when properly calibrated. If such calibrations are absent, traditional geothermometry and barometry methods have to be used in order to quantify the results.

The geometry of deformed grains of some minerals can be used as temperature microgauges. For example, the presence of ductilely deformed ribbon feldspar in a rock indicates high-grade metamorphic conditions (Sect. 3.12.4) and the presence of ribbon garnet very high grade (Sect. 3.12.9; Ji and Martignole 1994). Greater accuracy can be attained using the geometry of biphase mineral aggregates (Sect. 3.13); if two minerals deform together, both usually deform at different rates at a particular temperature. This leads, for example, to the development of porphyroclasts of feldspar in a quartz-feldspar aggregate, and of orthopyroxene in peridotite. With changing temperature conditions, this difference in behaviour may diminish or reverse (Sect. 3.13.2; Fig. 3.42), and this change in fabric may be a potential microgauge for temperature. As for all microgauges, these structures can only be calibrated if the effects of other parameters are also understood.

At normal strain rates of 10^{-12}–10^{-14} s^{-1}, most minerals show a minimum temperature where crystalplastic deformation takes over from brittle deformation. This boundary is also the limit where, at sufficient strain, dynamic recrystallisation starts, mostly as bulging (BLG)

recrystallisation. The transition of brittle to ductile deformation depends also on other factors (Sect. 3.14) but mostly on strain rate; with increasing strain rate, the temperature at which crystalplastic deformation starts increases. Nevertheless, rough guidelines can be given for different minerals, as shown in Fig. 9.10 (Sect. 3.12).

In quartz, the change in dominant recrystallisation mechanism with temperature (Sect. 3.12.2) is potentially a useful temperature gauge, although strain rate will influence the results (Fig. 9.10; Okudaira et al. 1998; Stipp et al. 2002; Altenberger and Wilhelm 2000). Prismatic subgrains in quartz tend to form at low temperature, while chessboard subgrains are restricted to higher temperature conditions (Sects. 3.10, 3.12.2). The presence of prismatic or chessboard subgrains may be used as a geothermobarometer in quartz grains with the c-axes parallel to the plane of the thin section (Fig. 9.11; Kruhl 1996). However, the accuracy of this method is still unclear (Kruhl 1998).

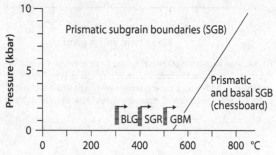

Fig. 9.11. Fields of dynamic recrystallisation and the appearance of chessboard subgrains in quartz in a *P-T* diagram. *Arrows* indicate the effect of strain rate. (After Stipp et al. 2002 and Kruhl 1998)

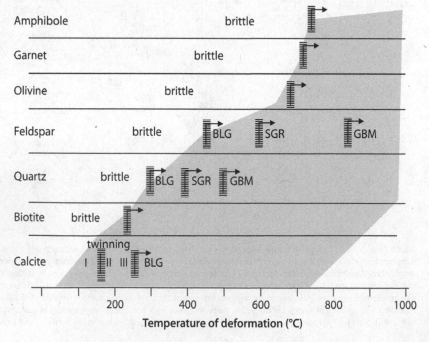

Fig. 9.10.
Temperature dependence of deformation mechanisms for different minerals. *Bars* indicate the transition zones. *Arrows* indicate the effect of strain rate. *BLG, SGR, GBM* – main types of recrystallisation. The *ornamented domain* is the domain of crystalplastic deformation

LPO patterns can in many cases be used to determine the active slip systems in crystals and indirectly temperature (Sects. 4.4.4, 4.4.5). The opening angle of quartz c-axis fabrics has been proposed as a temperature gauge with an accuracy of ±50 °C (Kruhl 1998; Law et al. 2004). In the case of plane strain deformation, the angle between girdles measured in the plane parallel to the lineation and normal to the foliation shows a linear dependence on temperature for most of its range (Fig. 9.12).

The ratio of the volume of clasts to melt-matrix in pseudotachylyte can be used as a geothermometer to estimate the initial temperature of the wall rock during pseudotachylyte generation (O'Hara 2001). Other potential temperature gauges are the presence of flame-perthite in K-feldspar and myrmekite in plagioclase (Sects. 3.12.4, 7.8.3; Simpson and Wintsch 1989; Pryer 1993) and the geometry of foliations (Sect. 4.2.9.4).

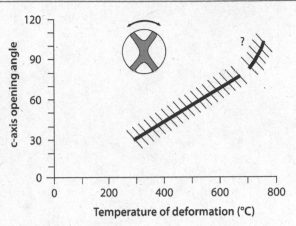

Fig. 9.12. Effect of temperature on the opening angle of small-circles of quartz-c-axes in an LPO diagram. (After Kruhl 1998 and Law et al. 2004)

9.10
Rheology Gauges

9.10.1
Introduction

Deformation experiments have taught us much about the rheology of rocks under a range of metamorphic conditions, but it would be useful to have more direct control from deformed rocks. There are indications that some fabric elements, which might be called rheology gauges, can be used to determine rheological constants (Sects. 2.12, 3.14). Rheology gauges are a promising new subject in geology, but few structures have been tested in practice yet. We restrict ourselves here to outline the potential of methods with published examples. Rheology gauges can only be further developed with much theoretical work and experimental deformation of rocks and rock analogues.

9.10.2
Folding and Boudinage

The wavelength of buckle folds in isolated competent layers in a less competent matrix depends on layer thick-ness and *viscosity contrast* between the layer and the matrix. The wavelength measured along the surface of a layer can therefore be used to determine absolute viscosity contrasts (e.g. Ramsay and Huber 1983). Boudins and mullions can be used to the same purpose (Kenis et al. 2004), but although buckle folding and boudinage in single layers both occur in response to a viscosity contrast, there is a difference; folding can occur in Newtonian and non-Newtonian materials, while boudinage is restricted to non-Newtonian materials (Sect. 2.12; Neurath and Smith 1982).

9.10.3
Fish and Mantled Objects

Fish and mantled porphyroclasts can have a complex shape, which depends on interaction with the deforming surrounding matrix (Sect. 5.6; Box 5.4). If metamorphic conditions of deformation are known, it should be possible to derive information on the rheology of the matrix and possibly on strain rate from such assemblages (Passchier et al. 1993; Pennacchioni et al. 2000; Bose and Marques 2004). However, much further research is needed before rigid objects can be used as rheology gauges.

9.10

Special Techniques

This book is mostly concerned with the study of microstructures in normal thin sections under the petrographic microscope. This is a fast, cheap method which gives a lot of information on rock deformation, but other techniques are also available for the study of small scale structures. In this chapter we give a short outline of all these techniques, some of which are relatively new. The aim is not to give a full introduction to each technique, or to allow the reader to use these techniques independently. Rather, we wish to introduce alternative techniques up to a level where the reader will be able to decide if a particular technique is useful and applicable to his specific type of samples and scientific problem. A choice of literature is provided for further reading on each subject. Information is also given on methods of sample preparation. Since this book deals with microstructures, we restrict ourselves to analytical laboratory techniques that use minerals in their original arrangement in the rock, similar to that in a thin section, excluding techniques that use powdered or dissolved rocks or minerals for analysis. The chapter treats the following themes; cathodoluminescence, Raman spectroscopy, SEM, TEM, X-ray tomography, U-stage, goniometers, electron microprobe, proton microprobe, mass spectrometry, in situ age dating techniques and fluid inclusion studies.

10.1
Introduction

The study of microstructures in thin section can give a lot of information for thematic and tectonic studies but has its limitations. In many cases, additional information has to be gathered by other techniques. In this chapter we wish to give the reader an outline of some other analytical laboratory techniques that use minerals in their original arrangement in the rock, similar to that in a thin section, excluding techniques that use powdered or dissolved rocks or minerals for analysis. Analysis may concern the fabric, the arrangement of minerals in the rock, the mineral chemistry and isotope composition, fluid inclusions, age of minerals and parts of minerals and lattice preferred orientation. We only discuss methods that study minerals in their original arrangement in the rock. We indicate what problems can be studied by these techniques, which type of materials can be studied and what type of sample is needed, and, finally, what the limitations of the different techniques are. The aim is to allow the reader to assess whether other techniques can help him to solve his problem and, if so, what material has to be prepared.

Finally a warning: it is not difficult to obtain data using unfamiliar techniques, but it is difficult to interpret the reliability of such data in any specific geological setting.

10.2
Techniques to Study Deformation Fabrics

10.2.1
Cathodoluminescence

Many minerals show luminescence when being excited with a beam of electrons. The impact of a primary electron beam on a sample causes processes like backscattering of electrons, energy transfer to the lattice resulting in local heating, and the generation of X-rays and secondary electrons. The latter are low enough in energy to transfer electrons of lattice ions to an excited energetic state. The return of ions from the excited to the basic state may cause a portion of the initial energy to be emitted as light photons in the visible range of the electromagnetic spectrum (Figs. 5.1b, 10.1, 10.9a). Because electron beams are commonly produced in cathodes, this type of electron beam-induced luminescence is referred to as *cathodoluminescence* (CL). One characteristic of CL is that the excited location and the location from which the CL is emitted may be several micrometres apart. As a result, CL images commonly have a somewhat blurred or "out-of-focus" appearance when compared with backscattered electron images (Sect. 10.2.4.3).

The CL signal emitted by an excited mineral is mostly complex in nature. The emission can be related to the mineral structure itself and this type of CL is mostly in the ultraviolet part of the spectrum. However, CL of many rock-forming minerals is mainly controlled by the presence of

Fig. 10.1.
Example of a SEM-cathodoluminescence image showing both sector and concentric zoning in hydrothermal vein quartz, possibly due to Al variation. The sample has been polished flat and is viewed looking down the (0001) axis. The sector zoning appears related to the positive and negative rhomb orientations and may therefore have potential for indicating dauphiné twinning. Note that the true trigonal symmetry at the very centre has been replaced by a pseudo-hexagonal symmetry throughout much of the grain. (Courtesy E. Condliffe, Leeds University)

100 µm

point defects. These may include ions of 3*d* elements (transition group metals, such as Mn^{2+}, Cr^{3+} and Fe^{3+}), ions of 4*f* elements (rare earth elements, such as Dy^{3+}), other ions [e.g., $(S_2)^-$ in sodalite and haüyne] and vacancies and related defect centres in the crystal lattice (Richter and Zinkernagel 1981; ten Have and Heynen 1985; Solomon 1989; Barbin and Schvoerer 1997; Götze 2000; Pagel et al. 2000; Nasdala et al. 2004a).

A large number of mineral species has been studied using CL, including carbonate minerals (Spötl 1991; El Ali et al. 1993; Yang et al. 1995; Habermann et al. 1996, 1998; Barbin and Schvoerer 1997; Gillhaus et al. 2000), diamond (Smirnova et al. 1999; Mineeva et al. 2000; Davies et al. 2003), zircon (Ohnenstetter et al. 1991; Cesbron et al. 1993; Hanchar and Miller 1993; Hanchar and Rudnick 1995; Nasdala et al. 2003), monazite (Seydoux-Guillaume et al. 2002), apatite (Barbarand and Pagel 2001; Kempe and Götze 2002), quartz and silica (Fig. 10.1; Matter and Ramseyer 1985; Shimamoto et al. 1991; Gorton et al. 1996; Götze et al. 2001), feldspar group minerals (Mora and Ramseyer 1992; Mariano and Roeder 1989; Finch and Klein 1999), and clay minerals (Götze et al. 2002). Cathodoluminescence is a well-established technique in sedimentology to study mineral growth processes and provenance of sediments (Marshall 1988; Augustson and Bahlburg 2003) but is also useful in microtectonic studies. The intensity distribution pattern and colour of CL may be independent of optically visible microstructures and can reveal mineral growth from solutions, e.g. in microcracks (Narahara and Wiltschko 1986; Laubach et al. 2004), in veins and along grain boundaries and details of the microstructure of cataclasites (Stel 1981; Blenkinsop and Rutter 1986). In veins, CL can reveal growth surfaces that are invisible in ordinary light (Dietrich and Grant 1985; Urai et al. 1991). Fracture patterns can be easily distinguished (Kanaori 1986). Overgrowth structures along grain boundaries, which are invisible in normal transmitted light, can be used for strain analysis (Onasch and Davis 1988). Dislocation networks can be made visible in some cases (Grant and White 1978).

Cathodoluminescence analysis can be carried out using either "hot cathode" or "cold cathode" systems attached to an optical microscope (then called OM-CL) and in a SEM or electron microprobe (called SEM-CL; (Fig. 10.9a, ⊘Photo 10.9; Yacobi and Holt 1990; Shimamoto et al. 1991; Lloyd 1994; Götze 2000; Nasdala et al. 2003). In both cases ordinary polished sections covered with a thin conductive layer (carbon or gold) are used. The CL technique involves the analysis of spectra or the generation of images. Cathodoluminescence spectroscopy provides information about the spectral composition of the emitted light, which is needed for the sound interpretation of causal CL-active defects etc. Cathodoluminescence imaging, in contrast, is predominantly used to reveal internal structures of minerals, for which it may not be absolutely necessary to know the causes for local variations of the CL. A combination of both types of analysis is monochromatic CL imaging. Here, distribution patterns are obtained separately for certain wavelengths. Most studies apply panchromatic (or integral) imaging, which uses the entire detected CL signal independent of its wavelength composition. One advantage of OM-CL imaging is that colour variations within minerals are easily observed, which provides a rough indication of the wavelength. A combination of trace element analysis and CL spectroscopy promises to become a powerful tool for mapping of trace element distribution in microstructures (Yang et al. 1995; Barbin and Schvoerer 1997).

10.2.2
Raman Spectroscopy

Raman spectroscopy is based on the interaction of the (vibrating) electric field vector of light with vibrations in the sample (molecule, crystal lattice). Light can be absorbed by a sample upon the excitation of a vibration, provided that its photon energy corresponds to the phonon quantum energy of an allowed vibration in the sample (vibrational quanta are called phonons). This is the case for light in the infrared range, and such absorption is used in infrared absorption analysis. In contrast, the phonon energy of visible light is too high to be transferred to the sample through the excitation of a vibration. However, it is possible that a small fraction of the photon energy of visible light is taken to excite the sample to vibrate. After such interaction, a scattered light photon will be lowered in energy and shifted in wavelength towards the red end of the electromagnetic spectrum, a process known as *Stokes-type Raman scattering*. Alternatively, it is possible that the vibrational state of the sample be lowered in which case the scattered light has gained energy and is blue-shifted. This is called *anti-Stokes Raman scattering*. In the case of *Rayleigh scattering* no energy shift of the light results. Raman analyses are mostly done in the Stokes region because of the higher intensity of Stokes-type Raman scattering compared with anti-Stokes scattering.

Raman scattering is a weak effect. Only a small fraction of the incident light interacts with vibrations in the sample as described above. The Raman shift is measured as the wave number difference between incident and scattered light. The wave number is defined as the reciprocal wavelength and is given in cm^{-1} (McMillan and Hofmeister 1988; Marfunin 1995; Smith and Carabatos-Nédelec 2001; Loudon 2001; Nasdala et al. 2004b).

Raman spectroscopy is nowadays generally done with a laser beam, which is directed at a sample using high-resolution objective lenses. Powerful Raman microprobe systems have a depth resolution as good as 2 µm when the beam is focused at the sample surface and a real lateral resolution of 1–1.5 µm, which may result in a real volume resolution better than 5 $µm^3$ (Markwort et al. 1995). The scattered light is analysed in a spectrometer. Apart from the analysis of single spectra, Raman is also

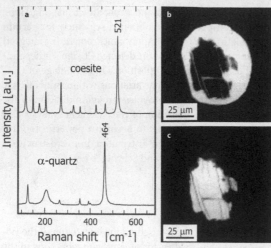

Fig. 10.2. Raman mapping applied to reveal the distribution of SiO$_2$ polymorphs within an inclusion in garnet. **a** Raman spectra of coesite and alpha quartz. **b** Distribution pattern obtained for the 464 cm^{-1} Raman band, showing the distribution of alpha quartz in white. **c** Analogous distribution pattern of the intensity of the 521 cm^{-1} coesite Raman band. The two images show that the central part of inclusion still consists of coesite whereas transformation to alpha quartz has started at the outer rim and at cracks. Ultrahigh-pressure gneiss from the Seidenbach reservoir, Saxonian Erzgebirge, Germany (Massonne 2001). (Photograph and diagram courtesy Lutz Nasdala and Hans-Joachim Massonne)

used to generate images of the internal structure of samples, using global illumination imaging and point-by-point mapping techniques (Lehnert 2000; Nasdala 2002).

Raman signals, usually referred to as "bands", reflect vibrations in the sample. These vibrations, in turn, are controlled by the size, valence and mass of atomic species in the sample, the bond forces between them, and their geometrical arrangement in the crystal lattice. As a consequence, each molecule and crystal structure is characterised by its own particular pattern of Raman bands. Spectral parameters of Raman bands, their variations (for instance, shifts, broadening, and relative intensity changes of bands) and other spectral peculiarities (for instance, band splitting and the observation of additional bands) may be used to identify a sample and characterise its structure. Particularly useful is the possibility to recognise polymorphs with identical chemical composition, for example quartz and coesite (Fig. 10.2). The main advantage of the technique is the possibility to analyse small samples non-destructively and without the need for sample preparation. Nearly all important minerals have been studied, and many institutions worldwide build-up reference databases. Apart from the analysis of solid minerals, Raman spectroscopy has a long history in the study of fluid inclusions (van den Kerkhof and Olsen 1990; Beny et al. 1982; Dubessy et al. 1989, 2001; Boiron et al. 1999; Giuliani et al. 2003). For a review see Nasdala et al. (2001, 2004b).

10.2.3
Electron Microscopy – Introduction

At magnifications exceeding 1 000 ×, objects seen by the optical microscope become fuzzy. This is a direct consequence of the use of light as a medium to transport information; the wavelength of visible light varies from 400 to 750 nm and no objects smaller than 100 nm can be observed. Depending on their velocity, electrons have much smaller wavelengths and can therefore be used to carry information about smaller objects. Two types of electron microscope are commonly used in geology; the scanning electron microscope (SEM), and the transmission electron microscope (TEM).

10.2.4
Scanning Electron Microscopy (SEM)

10.2.4.1
Introduction

Although this book deals with the interpretation of structures visible with the optical microscope, the scanning electron microscope (SEM) is now an important tool in the study of microstructures. Its accessibility in many laboratories and relatively easy use makes the SEM a powerful tool to complement and sometimes replace the optical microscope. In the SEM, a sample is placed in a specimen chamber under vacuum, and investigated by scanning a beam of electrons across the sample (Fig. 10.3). Only the surfaces of samples can be studied. The electrons are either scattered back from the surface of the sample (Lloyd 1987) or cause the sample to emit secondary electrons. Both types of electron signal are collected by detectors. The electrons that strike the detectors are used to build up an image of the sample on a monitor (Fig. 10.3). This technique allows magnifications significantly greater than those achieved via optical microscopy, depending on electron emission signal (e.g. >100 000 × for secondary electrons, or 100 times more than the optical microscope). This is an obvious advantage, but a disadvantage is that SEM images are always in grey tones. The recognition of minerals by their interference colour, as in the optical microscope, is therefore not possible. However, recognition of minerals with the SEM is usually not problematic. The amount of electrons that are emitted or back scattered in a volume of material struck by the electron beam is directly proportional to the atomic number (Z) of the constituent elements in the minerals, with heavier elements yielding more electrons. It is therefore possible to identify individual minerals in a sample by their grey tone on the screen (atomic number or Z contrast). For example, in most cases the following sequence can be observed

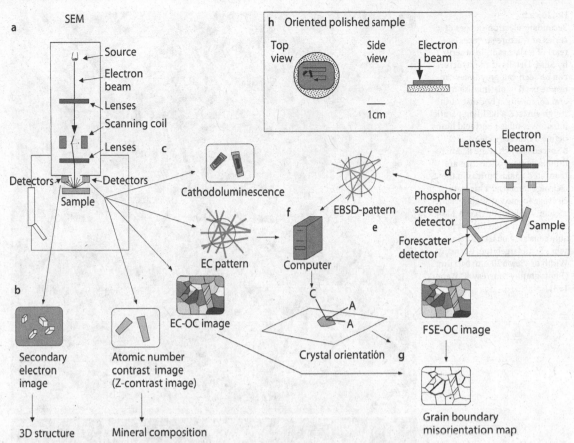

Fig. 10.3. Schematic presentation of a modern SEM with modes of operation. **a** Simplified cross-section through the SEM showing the operation of detectors in the top of the specimen chamber and a horizontal specimen close to the detectors to monitor secondary and backscattered electrons. Results are **b** secondary electron images and **c** cathodoluminescence-, atomic number contrast- and orientation contrast (OC) images produced by backscattered electrons. OC images produced from a horizontal sample high in the specimen chamber as shown in **a** are known as electron channelling orientation contrast (EC-OC) images. Electron channelling (EC) diffraction patterns can also be obtained. **d** operation of phosphor screen- and forescatter detectors at the side of the specimen chamber. For the use of these detectors, the sample is deeper in the specimen chamber and highly tilted. These detectors can be used to monitor **e** forescatter detector orientation contrast (FSE-OC) images and electron backscatter diffraction (EBSD) patterns. **f** Computer programmes can be used to index EBSD or EC patterns in order to obtain crystal orientations. **g** These in turn, combined with FSE-OC images, can be used to make grain boundary misorientation maps. **h** Typical oriented polished SEM sample and scanning motion of the electron beam over the sample

from dark to light; serpentine – quartz – feldspar – biotite – rutile – ilmenite – zircon. A more detailed list is given in Hall and Lloyd (1981).

Most SEMs have linked energy dispersive X-ray analysis (*EDAX*) facilities available (Goldstein et al. 1992). This tool measures the energy of X-rays emitted by the irradiated sample, and can identify the elements that occur in the sample. With this tool, minerals cannot only be easily identified, but their chemical compositions can be determined, either qualitatively or fully quantitatively. If a few grains in a sample have been analysed by EDAX, the rest can be recognised by their respective grey tones. Notice that EDAX cannot be used to analyse grains smaller than 5 µm as the X-rays are produced in a pear-shaped volume in the sample where the electron beam hits the surface. The diameter of this volume is much larger than the diameter of the electron beam.

10.2.4.2
Secondary Electron Mode

In secondary electron emission mode (Figs. 5.35, 10.4, 10.5), electrons, which are detached from the sample, are caught by a detector and used to build up a picture. This mode is useful to study topography in samples, and can also be used to get some information on the composition of minerals using grey-tones. This method has been much used to study fossils, the 3D geometry of small crystals, fluid inclusions, stylolite surfaces and other structures where 3D information is essential.

Fig. 10.4a,b.
Secondary electron images of a
series of fracture surfaces (see
text) of a slate sample observed
by SEM. Details of the structure
can be seen, but grey tones cannot be used to distinguish mineral composition because of the
rough surface. **a** Bedding parallel
foliation is folded and fold limbs
define a first slaty cleavage S_1.
b Spaced slaty cleavage S_1. Microlithons contain micas that are
truncated against micas in the
cleavage lamellae. This slaty
cleavage formed at slightly higher
metamorphic grade than **a**. Massive grains in microlithons probably consist of quartz. Rheinisches Schiefergebirge, Germany.
Width of view **a** 20 µm; **b** 95 µm.
(Photographs courtesy K. Weber
1981)

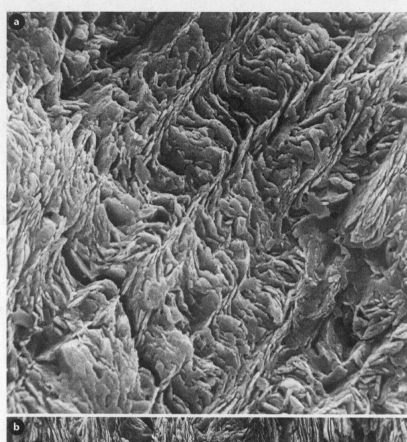

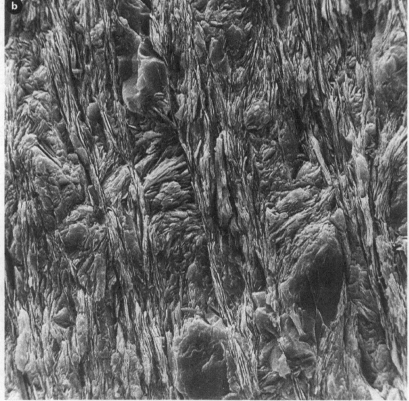

Fig. 10.4c.
Secondary electron image of a fracture surface (see text) of a slate sample observed by SEM. S_2 crenulation cleavage in slate overprinting an S_1 slaty cleavage. Rheinisches Schiefergebirge, Germany. Width of view 95 μm. (Photograph courtesy K. Weber 1976)

10.2.4.3
Backscatter Electron Mode

Backscatter electron mode (BSE) uses electrons from the primary beam, which penetrate the sample and are reflected back from crystals at the surface. The characteristics of the backscattered electrons depend on both the elements and crystal structure of the minerals in the sample. Backscattered electrons therefore carry information on the composition of the sample and crystal orientation (Randle 1992; Dingley and Field 1997; Prior et al. 1999; Wheeler et al. 2003; Trimby et al. 2000). The backscatter electron mode of SEM operation is most useful if fine details of mineral composition, texture or LPO are the subject of study (Humphreys et al. 1999; Lloyd et al. 1997). However, SEM backscatter electron mode operation requires a number of different electron detectors and configurations depending on the specific backscatter electron signal. These are (Fig. 10.3):

1. *Atomic number* or *Z contrast*, where different grey tones represent different mean atomic weights of elements in the mineral phases (Figs. 10.5, 10.6). In general, the mineral will appear darker when lighter elements are present. In this mode, only parts of the sample with different composition will show different grey tones.

2. *Orientation contrast* (OC). Here, parts of a mineral with different crystal orientation will show different grey tones (Fig. 10.7). Unfortunately, the intensity of the grey tone is not dependent on degree of misorientation, which means that subgrains and grains will show the same aspect. Nevertheless, OC patterns show a lot more detail than optical microscope images of the same domain. OC images can be obtained in two ways. *Electron channelling orientation contrast* (EC-OC) (Fig. 10.9b, ⊘Photo 10.9) involves horizontal (i.e. beam normal) samples positioned very close to the objective lens and backscatter electron detector such that the incident electron beam maintains a large angle during the scanning process that satisfies the Bragg conditions for diffraction. *'Forescattered' electron* OC images (FSE-OC; Fig. 10.7) involve highly tilted (typically 60–75°) samples relative to the incident electron beam and only electrons that are reflected or scattered forward are caught on a fluorescent screen and a forescatter detector (Fig. 10.3; Prior et al. 1996, 1999; Fliervoet et al. 1997). In this configuration, the angle of incidence of the electron beam and the tilted sample also satisfy the conditions for diffraction.

3. *Electron diffraction patterns.* These patterns appear as patterns of crosscutting lines (Fig. 10.8). They are in fact parts of very flat cones of diffracted electrons,

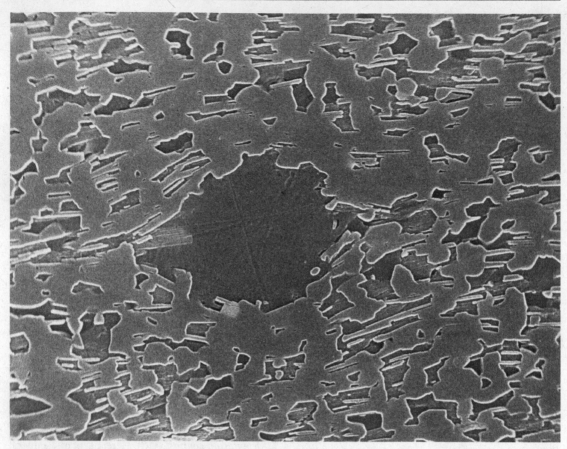

Fig. 10.5. Polished sample of ultramylonite, etched with HF vapour. Secondary electron mode. Gold cover. Fine detail is visible through the presence of *etch pits* and different *grey-tones* for the three minerals present. Quartz has not been etched much and forms the matrix (*light grey*). Feldspar and biotite have been etched. Feldspar is *dark grey* and forms the central porphyroclast. The etched *medium grey* elongate grains with cleavage are biotite. Biotite seems to replace feldspar at the left-hand side of the porphyroclast. St. Barthélemy Massif, Pyrenees, France. Width of view 40 μm

Fig. 10.6.
Atomic number or Z contrast images are based on the relationship between backscattered electron emission characteristics and sample composition (e.g. Lloyd 1987). The example shows a typical Z contrast image obtained from low-grade metasediments of the Clew Bay Complex, S. Achill Island, Mayo, Ireland (Yardley et al. 1996). Polyphase garnets have a detrital pyrope-almandine (*py-al*) core surrounded by spessartine-almandine (*sp-al*) and then grossular-almandine (*gr-al*), which is unmixing to a grossular-rich and an almandine-rich component. They sit in a matrix of quartz (*qtz*), chlorite (*chl*) and white mica (*mus*), with the latter zoned for Mg+Fe as revealed by the image contrast variations (Fe-rich compositions appear brighter). (Photograph courtesy Eric Condliffe and Geoffrey Lloyd)

which scatter at a small angle from lattice planes in a crystal. Such a diffraction pattern carries all information necessary to determine the exact orientation of a crystal with an error of less than 1°, provided the type and composition of the mineral is known (Krieger-Lassen 1996; Prior 1999). In order to find the crystal orientation from the pattern, the lines in the pattern must be *indexed*, i.e. the corresponding lattice planes should be identified. There are two different methods to obtain electron diffraction patterns:

3a. with the sample in the position for EC-OC (i.e. horizontal), diffraction patterns can be obtained if the incident electron beam is focused on a fixed position on the sample and rocked back- and forward (Figs. 10.3, 10.8a). The diffraction patterns produced

in this way are known as electron channelling (EC) patterns and have a spatial resolution of >1–10 μm but an angular resolution of <1°. The angular spread of EC patterns is typically <20°, which makes identification relatively difficult (Fig. 10.8a).

3b. with the sample in the position for FSE-OC (i.e. highly inclined), diffraction patterns can be obtained simply by focusing the incident electron beam on a fixed position on the sample. The diffraction patterns produced in this way are known as electron backscattered diffraction (EBSD) patterns and have a spatial resolution of 0.1–1.5 μm with an angular resolution of ~1°. The angular spread of EBSD patterns is typically >50°, which makes identification relatively easy (Fig. 10.8b).

Fig. 10.7.
Examples of 'forescattered' electron orientation contrast (FSE-OC) images from an uncoated lherzolite mantle nodule, Lhesoto, South Africa. *Top*, olivine (*ol*), orthopyroxene (*opx*) and clinopyroxene (*cpx*) crystallographic microstructure in lherzolite. *Bottom*, detail of dynamic recrystallisation in orthopyroxene achieved by the progressive development of deformation lamellae (*def. lam.*), subgrains (*sgrs*) and new grains. (Photographs courtesy Geoffrey Lloyd)

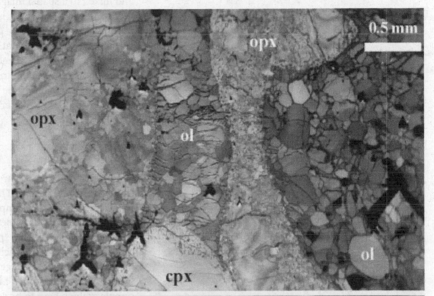

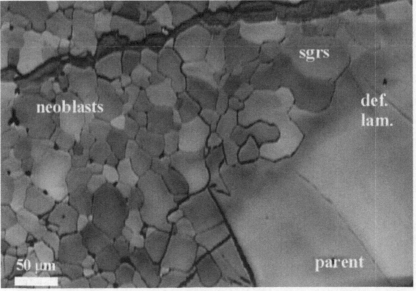

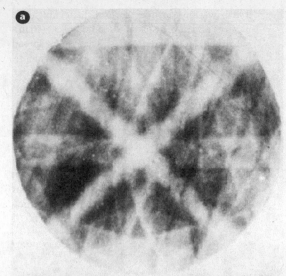

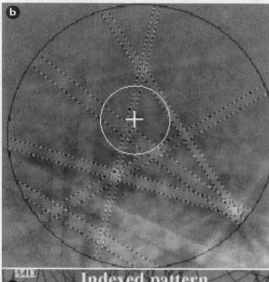

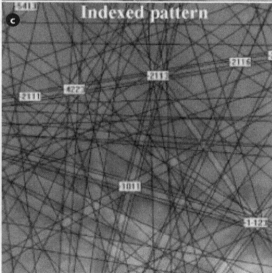

EBSD is a very powerful method to investigate microstructure in rocks, and presently more popular than EC patterns since resolution is higher and patterns are easier to determine (⊘Photo 10.3a–e). The interpretation of the diffraction patterns through indexing of lines in the diffraction patterns is now highly automatised and can be carried out by a number of computer programs (Fig. 10.8c, ⊘Photo 10.3d,e).

Combinations of atomic number contrast, OC and EBSD-pattern information for a sample provide a powerful tool in microtectonics. It is standard practice to make atomic number contrast and OC images of a microstructure, and then select a large number of points where EBSD patterns are determined and translated into orientation data. In this way, the complete crystallographic preferred orientation of a microstructure can be obtained quickly, and linked to a spatial image (⊘Photo 10.3a–e). A useful way to present such data is through a *grain boundary misorientation map*, where grain boundaries are marked by the angle of misorientation separating neighbouring grains and subgrains (Trimby and Prior 1999). Attempts are being made to automatise the process completely, to make complete maps of the atomic number contrast and the orientation of the material over a fine, regular grid.

10.2.4.4
Sample Preparation

Samples used in the SEM are studied under vacuum, and should therefore be dry. This is no problem with most rock samples, but clay samples may have to be dried before use. One specific problem of the SEM is that the electron beam, which hits the specimen, tends to cause local electrostatic charging which can cause beam deflection, thereby distorting the image. It is therefore necessary that samples are conductive. This is a problem in most rock samples, which have to be coated with a thin conductive layer of a metal such as gold (secondary electron-mode images) or carbon (back-scatter electron-mode images).

Fig. 10.8. a Electron channelling (EC) pattern from quartz; the *white cross* defines the pole to a positive rhomb (10Ī1) plane (Lloyd 1987). The diameter of the pattern covers ~20°. **b** Electron backscattered diffraction (EBSD) pattern from quartz. The pattern extends over a significantly larger angular area (~80° square in this case) than the EC pattern, indicated by a *white circle*, and is therefore much easier to index. **c** EBSD pattern indexing (e.g. quartz). A number of commercial computer-based systems are available to index EBSD patterns. They are all based on a comparison between the configuration of Bragg lines/bands in the observed pattern (**b**) with those predicted by theory (**c**) and stored in a database of crystal diffraction characteristics for each mineral phase (see Prior et al. 1999). Indexing can be either manual or automatic and can be used to define the complete crystal orientation (via three spherical Euler angles), provide mineral identification (i.e. quartz), the spatial coordinates of the pattern within the sample, a numerical indication of pattern quality (i.e. 'band contrast') and the mean angular deviation ('goodness-of-fit') between observed and predicted patterns. (Images courtesy Geoffrey Lloyd)

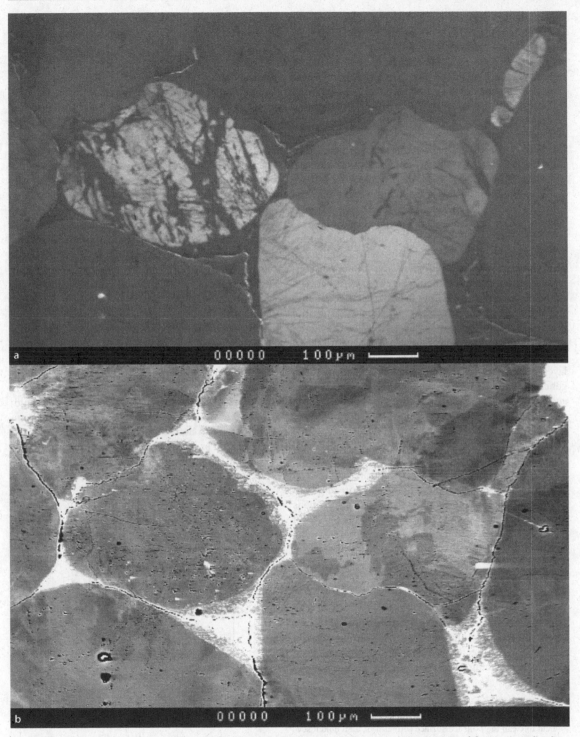

Fig. 10.9. a SEM cathodoluminescence (CL) image and **b** electron channelling orientation contrast (EC-OC) image of the same sample, where both images show different but complementary information. The images are of quartz grains in Cambrian quartzite, Assynt, NW Scotland (Lloyd 2000). The CL image clearly shows the different grains with different CL characteristics; one grain contains many microfractures healed by authigenic (*black*) quartz. The OC image shows the crystallographic related microstructures in the grains; the *light* and *dark areas* in the *right central* grain are dauphiné twin microstructures, whilst other grains show deformation lamellae and subgrains. All microstructures are due to mutual indentation of grains, a precursor to the development of cataclastic microstructures on through-going fractures. (Photographs courtesy Geoffrey Lloyd)

There are several methods to prepare samples for the SEM; as only the surface of the sample is studied, it is possible to use either polished thin sections or small chips of rock (Fig. 10.3). Small chips can be polished or broken. The advantage of a broken surface is that it is free of contaminations and usually breaks along grain boundaries; it therefore shows a lot of information on grain shape, grain size and internal structure of grains (Fig. 10.4; Weber 1981). A disadvantage is that the surface is irregular and difficult to interpret by scientists familiar with the interpretation of planar sections through a material. This problem can be overcome by using a polished thin section or polished chip of rock (Hall and Lloyd 1981); this will give familiar geometries as in thin section, which is an advantage if one wishes to compare optical and SEM microstructures. A disadvantage of this kind of sample preparation is that polished samples will only show structures if sufficient atomic number contrast or difference in orientation contrast exist between grains. Polishing also destroys the crystal lattice in a thin surface layer of the specimen, which may result in a blurred image of the grains. These problems can sometimes be solved by etching the specimen briefly in hydrofluoric acid (HF)[1]; this will remove the damaged top layer and may also etch some minerals more than others, generating a topography which outlines the fabric in the specimen (Fig. 10.5). Such a topography, however, is not wanted if backscatter electron mode is used. A more advanced technique, necessary for EBSD work, is therefore a second round of polishing using a silicone fluid, which removes the damaged top part of the sample without creating a relief (Sect. 10.3.6). The surface that is going to be studied should not be touched; this may spread grease over the sample and can decrease visible detail. Specimens are mounted on a special holder to fix it in the SEM with a special conductive glue. After surface preparation, the specimen is coated with a thin conductive layer.

10.2.4.5
SEM and Optical Microscopy

This book deals almost exclusively with optical microscopy, but it seems unavoidable that scanning electron microscopy will become an irreplacable tool alongside and partly replacing optical microscopy. This is especially

the case where use of polished thin sections can provide an ideal combination of optical and SEM observations (Trimby and Prior 1999).

The SEM provides higher resolution than optical microscopy, more complete information on twin-, grain- and subgrain boundary structures and on chemistry of minerals. Moreover, it provides easy and complete texture analysis, in contrast to the incomplete and more time-consuming optical techniques. Optical microscopy, however, also has a number of advantages, which will not make it obsolete in the near future. Large-scale structures and coarse-grained rocks are difficult to study by SEM, and are better served by optical microscopy. In an optical microscope, subgrains and grains are easier to distinguish than in an OC image. The thickness of a thin section provides a certain amount of 3D information, like the presence of fluid inclusions and tilt of grain boundaries, while the SEM is a purely two-dimensional tool. Organic substances used in analogue experiments and ice, are difficult although not impossible to study by SEM.

Although the large magnifications obtainable by SEM are an obvious advantage, it also means that it can be difficult to find a specific feature in a sample that was easily recognisable under the optical microscope. Turning the SEM to low magnification usually does not help, since very little detail will be visible. Therefore, if a specific feature in a sample such as a porphyroblast or a fault is to be studied, it is advisable to mark it under a reflected light microscope with an electrically conducting marker, or to use a photograph or sketch of the sample; once the sample is in the SEM, it may be very difficult to find the object.

Despite its restrictions, optical microscopy has the advantage that preparation is fast and easy and can be carried out at low cost. SEM is more versatile, but much more expensive and complex to acquire and maintain.

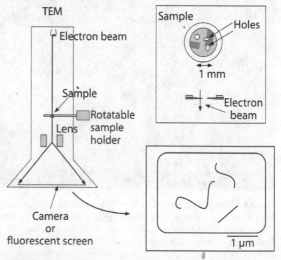

Fig. 10.10. Simplified cross-section through a TEM showing the path of the electron beam and the nature of the samples used

---

[1] WARNING: Take great care when etching in HF; it is usually sufficient to clean the polished surface with alcohol, then hold it several seconds over the vapour of a small reservoir of HF using a pair of tweezers; use suitable gloves and only work in a fume cupboard! HF is a very dangerous acid and should never be touched. It is useful to study the etched surface under a reflected light microscope; after a roughness becomes visible under the microscope, the sample is sufficiently etched. It is rarely necessary to submerge the specimen in the HF.

10.2.5
Transmission Electron Microscopy (TEM)

10.2.5.1
Introduction

Transmission electron microscopy (TEM) allows observation of structures as small as a few nanometres at magnifications of up to 350 000×. Dislocations, deformation lamellae, subgrain and grain boundaries and twins can be studied with the TEM, since it allows interpretation of their nature and their orientation with respect to the crystal lattice (Figs. 3.13, 3.14, 3.24, 10.11). Although TEM analysis of intracrystalline structures is very powerful, considerable experience is needed before it can be applied, and a description of TEM techniques is outside the range of this text. However, a short outline is given to allow readers to assess whether TEM work can help to solve their problems. Reviews can be found in McLaren (1991), Buseck (1992) and Champness (2001).

In a TEM a beam of high velocity electrons passes through a thin foil of sample material suspended in a high vacuum (Fig. 10.10). The electron beam is deflected by electronic lenses, which enlarge the image, and the result is focused onto a fluorescent screen, photographic plate or CCD camera. An intact crystal lattice is transparent for the electron beam, but a crystal defect such as a dislocation, either isolated or in a (sub) grain boundary is visible as a shadow on the image (Figs. 3.13, 3.14, 3.24, 10.11). A TEM has sample-holders in which a sample can be rotated in all directions (Fig. 10.10). In certain orientation and observation modes, a crystal will show diffraction patterns that can be interpreted in terms of crystal structure and orientation (Fig. 10.11, insets). Dislocations will

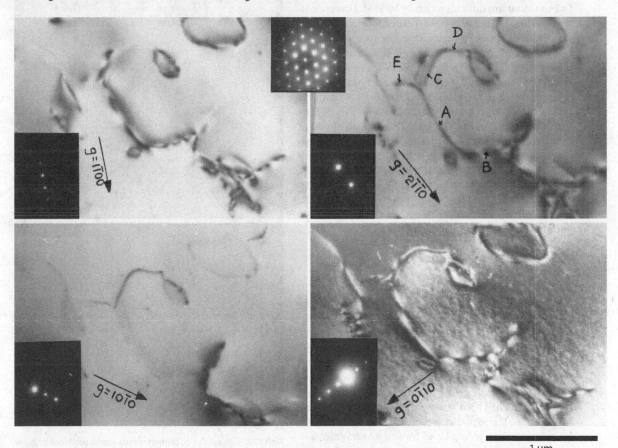

Fig. 10.11. TEM images of the same area in a quartz grain showing free dislocations. The four photographs were taken at four different orientations of the sample in the specimen holder. For each orientation, a diffraction pattern was obtained for the quartz grain under different diffraction conditions (insets). In each case, the diffraction vector is labelled as g on the photograph. The diffraction pattern at the *top centre* is the pattern taken with the [0001] direction parallel to the electron beam. The other images are taken by tilting away from this orientation so that only one set of lattice planes are diffracting. The Burgers vector of dislocation segments can be established from images where the segment produces no contrast in the image, such as at lower left for dislocation segment A–B. In this case the dislocations produce no contrast when the crystal is diffracting on {10$\bar{1}$0} planes. This means that the Burgers vectors are perpendicular to the diffraction vector (g), with b = <1$\bar{2}$10> parallel to the <a> directions in quartz. Image of free dislocations in quartz mylonite from St. Barthélemy massif, France. (Photographs courtesy Martyn Drury)

be visible in the TEM only if their Burgers vector (Sect. 3.4) has a particular orientation with respect to the electron beam; rotation of a crystal until a dislocation becomes invisible, together with determination of the diffraction pattern of the crystal in that orientation allows determination of the Burgers vector for that dislocation (Fig. 10.11). Similarly, the orientation of other intra- and intercrystalline structures can be determined with great accuracy using diffraction or Kikuchi patterns (Gapais and White 1982; Fitz Gerald et al. 1983). TEMs usually also have equipment for energy dispersive X-ray analysis and scanning (Sect. 10.2.4.1); therefore, the chemical composition of grains can also be determined, although it is time consuming to conduct quantitative analyses.

The TEM is most useful to identify and study microstructures on the grain scale, or to determine the mineral content and microstructure of very fine-grained rocks ($<1\,\mu m$) that are difficult to study by SEM. However, although the TEM is a very powerful devise for microstructural studies, it has its limitations. The volume of material which can be studied in a single session on the TEM is very small. Although it is possible to determine the orientation of single crystals, it is a rather tedious and time-consuming job to use the TEM to measure LPO patterns unless the microscope is equipped with on-line diffraction pattern indexing software. EBSD with the SEM (Sect. 10.2.4) is usually a better technique for LPO measurement unless the grain size is less than one micron.

Not all minerals can be studied by TEM; in general, minerals with a large water content such as clay minerals and muscovite are less suitable since they disintegrate in the electron beam; water-free minerals such as olivine and pyroxenes are most stable in the TEM.

10.2.5.2
Sample Preparation

TEM samples are small fragments of the rock slice of a thin section, attached to metal grids with a diameter of a few mm (Fig. 10.10). The thin section should be prepared with a resin or glue such as 'lakeside' that is dissolvable in alcohol; metal grids are glued with araldite onto interesting parts of the sample as determined by optical microscope; a single grid is insufficient since many samples fracture, or are otherwise damaged during further preparation. The rock slice is then removed from the glass by submersion in alcohol, and broken into fragments to remove the metal grids and small attached rock chips. The chips must be thinned to a few nm before they are transparent for the electron beam, and since the sample should not be mechanically damaged during the process of thinning, the superfluous material is evaporated by bombardment with Ar-ions in a vacuum. This process of *ion thinning* continues till small holes are visible in the sample; observation in the TEM is only possible adjacent to these holes, where the sample is sufficiently thin (Fig. 10.10). After thinning, coating of the sample with carbon is necessary to avoid charging. Because of this sample preparation method and thinning procedure, it is difficult to select a particular small grain or object in a thin section for observation by TEM; the chance that it will be located exactly on the edge of one of the 'observation holes' is rather small. SEM work is probably more effective if a particular isolated structure is to be stud-

Fig. 10.12. High resolution computed micrograph produced by X-ray microtomography of a garnet-bearing schist from the Main Central Thrust zone, Nepal Himalaya. *Light-coloured* areas are garnet porphyroblasts with spiral inclusion trails, which are wrapped by the principal foliation in the thrust zone. Image represents a slice through a rock cylinder 20 mm in length and 10 mm in diameter. The data were collected at the Consortium for Advanced Radiation Sources, University of Chicago, USA. Resolution is 17.2 microns per pixel. (Image courtesy of Scott Johnson and Richard Spiess)

ied. However, a new technique of preparing TEM samples using a focused ion beam (FIB) within a SEM allows precise selection of areas for TEM study.

10.2.6
Tomography

It is presently possible to make direct 3D observations of microstructures *inside* a small rock sample using tomography techniques (Fig. 10.12, ⊘Video 10.12a–c; Carlson et al. 2003). Because of the high density and presence of heavy elements in rocks, and the demand on calculation capacity for 3D analysis, only small samples can be investigated by this method at present. Available equipment ranges from small desktop X-ray tomographs for small samples (up to 2 cm³) and low resolution (down to 5 µm), to large-scale machines attached to a particle accelerator for larger samples and high-resolution.

A great advantage of tomography is that the method is non-destructive. Any small rock sample can be directly analysed without special preparation, although samples should preferably be circular or square in cross-section. Best results are obtained for high density contrast or high atomic number contrast. This means that pores in a sandstone, or e.g. galenite grains in carbonate are easy to resolve, but feldspar grains in quartz mylonite, or leucosomes in a gneiss, are difficult to see.

10.3
Methods to Measure Lattice-Preferred Orientation

10.3.1
Introduction

Several methods are used to measure lattice-preferred orientation in rocks (Leiss et al. 2000; Sects. 4.4.3–4.4.5). The classical method is the use of a *U-stage*, which allows rotation of a thin section or sample in three directions to determine the orientation of fabric elements. Other methods are optical bulk analysis, texture goniometry, and electron microscope techniques (Ullemeyer et al. 2000). For goniometry, the term texture is commonly used instead of LPO in the literature and therefore maintained here (Box 1.1). All methods require rather expensive equipment and each has its advantages and disadvantages.

10.3.2
U-Stage Measurements

In a universal stage or *U-stage* a thin section is mounted between two glass hemispheres (Fig. 10.13). The complete setting of hemispheres and thin section can be turned around several axes in such a way that a single crystal in the thin section at the centre of the half-spheres

Fig. 10.13.
Schematic top view and cross-section of a U-stage, showing the various parts. The thin section is wedged between two halfspheres, and the internal part can be tilted and rotated until specific features have a vertical orientation, after which the orientation can be read from the arcs and rings of the stage. *Detailed explanation on the CD*

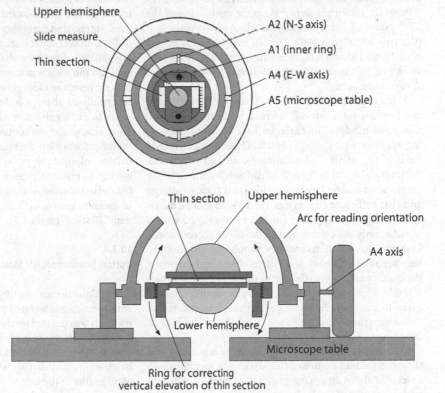

Upper hemisphere
Slide measure
Thin section

A2 (N-S axis)
A1 (inner ring)
A4 (E-W axis)
A5 (microscope table)

Thin section   Upper hemisphere
Arc for reading orientation
A4 axis
Lower hemisphere
Microscope table
Ring for correcting
vertical elevation of thin section

can be focussed in orientations up to 40° from the horizontal. In this way, crystals or microstructures such as twins, lamellae and inclusions can be rotated until they lie in a particular orientation with respect to the axes of the microscope. If the orientation of the U-stage axes is now read, these data can be used to calculate the orientation of the structure with respect to thin section axes, and eventually to geographic orientation. Modern versions of the U-stage have been modified to allow automatic storage of measurements in a computer.

Although the U-stage has a long history, it is still an important instrument to measure LPO patterns, especially for quartz. For those who do not have access to goniometers and electron microscopes it is the only means to assemble data on lattice-preferred orientation of minerals. A U-stage is also quite useful for study of microstructures such as twins, inclusions and cleavage planes which have been obliquely cut by a thin section. These will appear unsharp in a normal thin section, but in the U-stage they can be observed in a more suitable orientation, and their orientation can be measured with reasonable accuracy.

The U-stage can also be a useful tool to establish the shape of grain boundaries (Kruhl 2001; Kruhl and Peternell 2002) and the relative orientation of subgrain boundaries and the crystal lattice of quartz on both sides. This method has been used to determine the orientation of slip systems in minerals like quartz; the orientation of a subgrain boundary depends on the orientation of the slip system of the dislocations, which accumulate in it (Christie and Green 1964; Trepied et al. 1980; Mainprice et al. 1986). However, there is some doubt about the reliability of this method, and additional TEM analysis is always necessary.

U-stage measurements of LPO are inexpensive and can be done on a normal microscope. Notice, however, that some modern microscopes have insufficient working space to fit a U-stage. Full LPO can be determined for olivine, calcite and orthopyroxene, and with some difficulty also for minerals with lower symmetry like amphiboles and feldspars, provided that grains are large and that sufficient cleavage or twin planes can be measured. Full LPO of quartz cannot be measured with a U-stage, only the orientation of the c-axis. Nevertheless, U-stage measurements are very popular for quartz since they are relatively easy and fast to obtain, and normally the c-axis pattern is sufficient to determine shear sense. If fluid inclusions are present, it may be possible in some cases to use their negative crystal shape to determine the orientation of a-axes as well (Fig. 10.17a; Rosin et al. 2004).

Disadvantages of the U-stage are the laborious and time-consuming nature of the work and a minimum diameter of about 20 μm for grains to be measured. Strong undulose extinction also hampers measurement. U-stage measurement of LPO patterns does not allow a statistically unbiased analysis of the fabric; the volume of minerals with a particular orientation is difficult to determine since normally only one measurement per grain is taken, and the relative size of the grains is not taken into account. AVA diagrams (Sect. 4.4.3) have been constructed based on U-stage measurements but these tend to be inaccurate.

10.3.3
U-Stage Sample Selection

The presence of an LPO in an aggregate can be checked for minerals with low birefringence by inserting a gypsum plate with crossed polars; if a preferential colour occurs in some orientations when the microscope table is turned, an LPO is probably present. It is important that a thin section used for LPO analysis does not contain too many minerals other than the one to be measured, since such minerals will have influenced the LPO and inhibit comparison with standard patterns measured for monomineralic rocks. Grain size should not be less than 30 μm, otherwise it is better to use a texture goniometer. The thin section should have a basal glass plate of less than 1 mm thick; presence of a thick basal glass plate makes focussing on single grains difficult. Make sure that the thin section is properly oriented. The range of measurements is usually restricted to 30% of the thin section in the centre; grains in the periphery cannot be measured. Interesting parts should therefore lie in the centre of the thin section. It is important to realise that several populations of grains with different LPO may be present even in a monomineralic rock. For example, course grains may represent an older generation, and fine grains a new, recrystallised phase. It is therefore usually advantageous to try to distinguish several groups of grains on the basis of size, shape and content of inclusions or intracrystalline deformation structures. In many cases it is useful to have enlarged photographs of a thin section before an attempt is made to measure the fabric. This will also be helpful in the construction of AVA diagrams. The method to measure quartz-LPO using a U-stage is described in item "U-stage" on the CD.

10.3.4
Optical Semiautomatic Methods

A technique to measure LPO in an ordinary thin section with computer-integrated polarisation microscopy (CIP method) was devised by Heilbronner-Panozzo and Pauli (1993) and described in Heilbronner (2000). An ordinary microscope is used with added tilt stage and rotating polarising filters. A digital camera records the brightness of all grains in the field of view while the polarising filters are rotated and the thin section is stationary. Images for each rotation increment are stored in a computer and used to calculate the LPO orientation of each pixel in the image. A colour code (colour look-up table) is devised to colour-code the c-axis orientation of each pixel, thus creating an orientation image where each coloured pixel indicates a particular orientation. Such an image is a high resolution AVA diagram which can be used for accurate analysis of the grain scale fabric with image analysis techniques (Fig. 10.14 (in colour on CD); ⊘Photo 10.14).

Fueten (1997) devised a similar automatic rotating stage where a control box is connected to two rotating polarisers, one of which is fixed below the thin section. Rotation of the polarizer, acquisition with a digital camera and processing of the images is guided by a PC (Goodchild and Fueten 1998). This method can determine the trend and plunge angle, though not plunge direction for the c-axis of quartz grains.

Another tool for this type of analysis was developed by Wilson (Wilson et al. 2003). Here, a specially adapted microscope and stage is required, fitted with three digital cameras that can be linked to a laptop computer. The device slowly moves the thin section with respect to the microscope lens and thus scans the section mechanically with a resolution of up to one μm; it can measure the orientation of c-axes of each pixel. This device gives an accurate and high-resolution image of c-axis orientations of quartz or other uniaxial minerals.

Besides AVA, the three methods discussed above can also be used to produce other types of orientation images such as misorientation images, relating pixel orientation to some reference direction, or to that of its direct neighbours (e.g. Wilson et al. 2003). This marks grain and subgrain boundaries. Volume weighted pole figures can be derived and therefore statistically balanced analysis of the fabric can be carried out on the orientation, as in goniometer data. Some techniques allow automatic recognition and mapping of different minerals. The advantages of these methods are relatively low costs, the possibility to use normal thin sections, and high resolution. A disadvantage is that the methods only apply to uniaxial minerals such as quartz, calcite, dolomite, ice, and norcamphor and can only determine and plot the orientation of c-axes.

10.3.5
Texture Goniometers

Two types of goniometers are commonly used to study fabrics in rocks; the X-ray and neutron texture goniometers. The X-ray goniometer is most accessible and is most easily operated on monomineralic rocks, although bimineralic rocks can be studied in some cases (Braun 1994). Samples should be relatively fine-grained (<200 μm) and homogeneous, preferably with a high crystal and lattice symmetry. Neutron texture goniometry can only be carried out with a

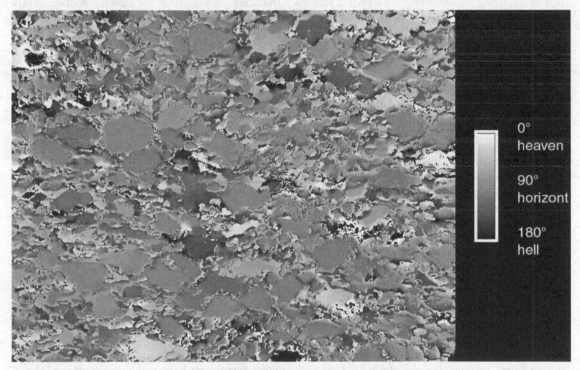

Fig. 10.14. AVA diagram produced by the CIP method. Only the plunge of C-axes is shown in this grey tone image. Experimentally deformed Black Hills quartzite. Deformation conditions 900 °C, 15 kbar and 10^{-5} s^{-1} (Regime 3 – Hirth and Tullis 1992). (Image courtesy Renee Heilbronner). Complete colour version of this AVA as Fig. 10.14 on the CD (see also ⊘Photo 10.14)

neutron diffractometer and suitable neutron source, few of which are presently available, and is rather time-consuming (Brokmeier 1994). The method has great advantages, however, since large and therefore coarse-grained samples can be used (Brokmeier 1994); samples can be up to 4 cm in diameter and need not be polished, although spherical or cylindrical samples are preferred. Neutron goniometry is suitable for polymineralic aggregates containing minerals with low symmetry such as amphiboles and plagioclase, and for measurement of weak fabrics (Wenk et al. 1986b; Brokmeier 1994; Siegesmund et al. 1994; Ullemeyer et al. 1994, 2000). Both types of goniometers can give full LPO patterns, which are statistically balanced for the volume of each of the present directions. Goniometers have been used to study the LPO of micas in foliated rocks for finite strain analysis (Sect. 9.2; Tullis and Wood 1975; Wood and Oertel 1980; van der Pluijm et al. 1994; Ho et al. 2001) and even sense of shear from the skewness of the fabric (O'Brien et al. 1987). For minerals like quartz and calcite goniometry has been used to determine full ODFs (Sect. 4.4.3; Casey 1981; Schmid and Casey 1986). A disadvantage of goniometers is that they cannot show how the different orientation peaks are distributed over the sample. This implies that no AVA diagram can be constructed and that no link can be made to microstructures, which could give information on the mechanisms by which the LPO originated. For this reason, additional U-stage observations are commonly made on thin sections prepared from a sample that was analysed by goniometry.

10.3.6
SEM-Techniques

EC and EBSD pattern observation modes in the SEM are possibly the most promising tool for determination of preferred orientation of minerals in thin sections (Fig. 10.8, ⊛Photo 10.3d,e). Although SEMs with equipment for EC and EBSD are still scarce, the method has many advantages; full LPO patterns for many minerals can be measured automatically in a complete sample, and maps can be made of the distribution of these LPO over the sample. Details are described in Sect. 10.2.4. (Lloyd and Hall 1981; Lloyd et al. 1981; Joy et al. 1982; Lloyd 1987; Lloyd and Freeman 1991, 1994; Mainprice et al. 1993).

Since the orientation of the fabric at the surface of the crystal is analysed, "superpolished" specimens are used. Normal polishing with diamond paste or similar abrasives damages the surface layer of a specimen, and this means that no good channelling patterns can be made. Therefore, ordinary polished rock chips, mounted in epoxy, are repolished using a silica solution in a special polishing rig to remove the damaged surface layer. Such "superpolished" sections are also useful to do careful work on small-scale structures in a normal SEM, since they give much better resolution than ordinary polished sections.

10.4
Chemical and Isotope Analysis

10.4.1
Electron Microprobe

The electron microprobe has been a standard analysing tool in most laboratories for 40 years to determine main-element composition of mineral grains in a polished thin section. A high-energy focused beam of electrons is used to generate X-rays characteristic of the elements in a sample. The resulting X-rays are diffracted by analyzing crystals and counted in detectors. The electron current in an electron microprobe is up to 1 000 times higher as in a SEM, because the purpose of the machine is to generate X-rays, rather than study topography or texture (Reed 1996). Chemical composition is determined by comparing the intensity of X-rays from standards (known composition) to those from materials in investigated samples, normally polished thin sections.

Polished sections are coated with carbon to reduce charging by the electron beam, and introduced in a specimen chamber that is operated in vacuum. Spot analyses and profiles of series of spot analyses give accurate main element compositions in each part of the sample with a resolution of a few micrometres. Normally, sites to be analysed are found and coordinates marked before introducing the samples in the machine in order to save time during the analysis steps. Parts of a sample as small as 3 μm can be analysed. Elements from fluorine ($Z = 9$) to uranium ($Z = 92$) can be analysed routinely down to 100 ppm. Both electron-generated images, similar to those in the SEM and optical microscope images of the sample can be seen during the analysis. The method is virtually non-destructive, fast, relatively cheap and uses normal polished thin sections.

Modern microprobes permit *high-resolution compositional mapping* of mineral phases often allowing the distinction between newly grown rims around older cores. This is a powerful method to study metamorphic reactions and mass transfer e.g. during cleavage formation (Williams et al. 2001).

10.4.2
Proton Microprobe

The proton microprobe, also known as the *nuclear microprobe* is a powerful tool for the analysis of trace elements and their distribution in the microstructure of a rock (Ryan and Griffin 1993; Ryan 1995; Ryan et al. 2001). Trace elements can be analysed with very low detection limits, down to sub-ppm, and with a high horizontal resolution. Samples can be up to thin-section size and should have a smooth surface. The sample is irradiated by a beam of high energy (several MeV) protons from a particle accelerator, leading to emission of protons, ions, X- and gamma rays, which can all be analysed. Several analysing techniques are presently available with the proton microprobe, the most important of which are mentioned below, notably Proton Induced X-ray Emission (*PIXE*) (Fig. 10.15).

Fast protons induce X-rays by dislodging electrons from the inner electron shells of atoms, which are replaced by outer-shell electrons with the emission of X-rays. The PIXE spectrum can be used to determine the sample composition down to the trace elements at ppm level and below. This is similar to EDAX in the SEM or electron microprobe, but much more accurate while standards are not needed. However, PIXE is generally only used for elements heavier than Al ($Z > 13$). Special approaches can be used down to Na ($Z = 11$), but commonly filters are used to absorb intense major element lines to improve trace element detection typically for elements above Ca-Fe ($Z = 20–26$).

Energetic protons approach atomic nuclei closely and cause nuclear reactions in light elements and emis-sion of particles and gamma rays. Proton Induced Gamma-ray Emission (*PIGE*) is suitable for the analysis of light elements such as Li, Be, B, F, Na, and Al. This fills part of the gap left by PIXE. Analysis of the secondary reaction products also gives information on light element concentrations. This is known as Nuclear Reaction Analysis (*NRA*). NRA with reaction product particle detection is particularly useful for lighter elements that cannot be analysed with PIXE, especially for analysis of C, N and O.

The energy loss of the backscattered protons gives direct information about the three dimensional distribution of target nuclei. This is known as Rutherford Backscattering Spectrometry (*RBS*).

Fast protons can also produce secondary electrons, which can be used to determine the topography of a sample, as in the SEM. Visible or infrared emissions from the sample, known as Ionoluminescence (IL) can be used on the same structures and give similar information as cathodoluminescence.

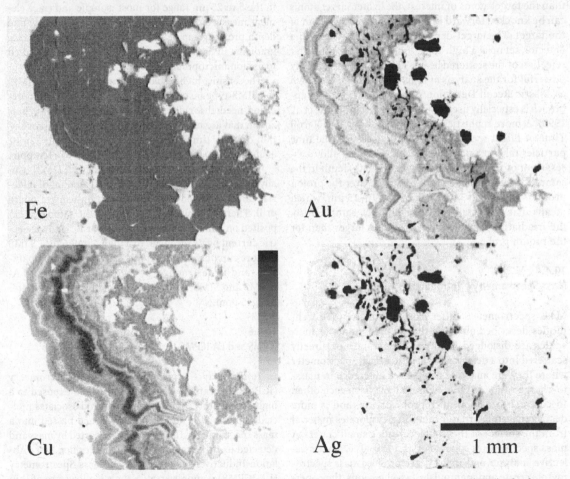

Fig. 10.15. Trace element zonation in a pyrite vein in quartz as observed using proton induced X-ray emission (PIXE). The data were collected using a 6 nA beam of 3 MeV protons focused into a 1.6 μm beam spot in the CSIRO nuclear microprobe in Clayton, Australia. Sample from the Emperor mine in Vatukoula, Fiji. (Photos courtesy Chris Ryan at CSIRO)

A proton microprobe is particularly suited for the study of *fluid inclusions*, since the high-energy protons can penetrate down to more than 40 μm into a sample.

Advantages of the proton microprobe are the high resolution and accuracy, the fact that the proton beam does very little damage to the sample, the high penetration depth, and the fact that low-vacuum can be used when needed for delicate samples. Disadvantages are that the high penetration depth precludes the analysis of small grains, and the high costs and limited availability of the equipment.

10.4.3
ERDA and RToF Heavy Ion Microprobes

Several types of ion microprobes are currently available, using ions heavier than protons (Breeze et al. 1996). In principle, the same analytical methods can be used as in the proton microprobe but the lighter elements can be more easily investigated. By using a beam particle heavier than the target atoms of interest, the lighter target atoms can be knocked forward in the collision and recoil out of the target. The target-derived particles are caught in a detector, set up at a high angle to the incident beam after rejection of the scattered beam. The method is especially powerful for the analysis of light elements and is known as Elastic Recoil Detection Analysis (*ERDA*). The approach is especially useful for hydrogen analysis (Sie et al. 1995). A more sophisticated technique, known as Recoil Time-of-Flight spectrometry (RToF) measures the time particles take to reach the detector. This also allows assessment of the distribution of phases with depth in the sample. A disadvantage of the ion probe over the proton probe is the larger size of the spot, 10–15 μm, which means that less detail can be seen in the sample. Also, the irradiation damage to the sample is larger than for the proton probe.

10.4.4
Mass Spectrometry – Introduction

Mass spectrometers differ from ion microprobe techniques described above in that ionized atoms or molecules are dislodged from the sample and not directly scattered into a detector, but sent to a mass spectrometer where they are analysed using their difference in mass-to-charge ratio (m/e) to separate them from each other. In contrast to the ion microprobe, the method is more destructive and, for laser sampling, evaporates material from the surface of the sample, creating small craters. A mass spectrometer consists of an ion source, a mass-selective analyzer, and an ion detector. Since mass spectrometers create and manipulate gas-phase ions, they operate in a high-vacuum system.

10.4.5
SIMS, TIMS and SHRIMP

In Secondary Ionization Mass Spectrometry (SIMS) analysis, a focused beam of "primary" ions bombards the surface of a polished sample. The bombardment removes atoms from the surface of the section, some of which are ionized and can be accelerated into a mass spectrometer. The method is useful for in situ quantification of many elements at sub-ppm concentrations. Like the ion microprobes, the method is an extension of the information that can be obtained from the electron microprobe. The method is used for measurements of isotope ratios for major and minor elements and in situ mapping of isotopic and trace element abundances in thin sections or individual grains. SIMS are in routine use for U-Pb zircon dating, measurements of trace element abundances, and stable isotope measurements including those of O, C and S. While the lateral spatial resolution can be sub-micron for major element mapping, it is more typically in the 5 to 25 μm range for most isotopic and trace element measurements. However, craters are shallow and depth profiling can be used to investigate micron-scale gradients in isotope composition. SIMS is also referred to as "ion microprobe".

The sensitive high-resolution ion microprobe (SHRIMP) is a SIMS-type instrument that can reach very high precision needed for dating portions of minerals such as zircon that are used for dating. Presently, spatial resolution of a few μm is reached and most elements can be analysed down to 5 ppb and isotopic ratios to a few ppm.

Thermal Ionization Mass Spectrometry (TIMS) usually uses mineral separates, but a small amount of material can be isolated from a thin section using a micro drill. Thermal ionization means that the sample is deposited on a metal ribbon, such as Pt or Re, and an electric current heats the metal to a high temperature. TIMS is more cumbersome in sample extraction and preparation, and is destructive, but more accurate than LA-ICPMS and SIMS; isotope ratios can be measured better than 15 ppm.

10.4.6
ICPMS and LA-ICPMS

In *Inductively Coupled Plasma Mass Spectrometry* (ICPMS) the sample is dissolved and then exposed to a high temperature argon plasma, which dissociates molecules and ionizes atoms, which are then passed into a mass spectrometer. Here, the ions are sorted by mass and detected using a scanning electron multiplier. Laser Ablation Inductively Coupled Plasma Mass Spectrometry (LA-ICPMS) is more versatile since it allows use of thin sections, albeit with a thickness of 80–100 μm. This technique combines laser ablation and ICPMS techniques and can be used to measure trace elements; especially the rare earth elements of minerals in a thin section can be measured directly. The laser ablates a small amount of material from the surface of the section, producing ablation-pits of 10–200 μm. Detection limit is sub-ppm. The method is useful to measure profiles over large mineral grains.

The advantage of a laser to ablate material from a sample is that the sample does not charge, and both conducting and non-conducting materials can be used, without need for a conductive coat and/or other charge balancing techniques, as in SIMS and the electron microprobe. The sample must be in an airtight chamber, but is not exposed to a vacuum. The ionization step is separate from the sampling step by laser ablation, and the ablated material is transported in Argon to the ICPMS.

Although LA-ICPMS is a very versatile method, it is still less accurate than SIMS or TIMS for determination of isotopic ratios analyses, but this is not a serious problem for the heavier isotopes. However, multidetector ICP-MS can measure isotopes more accurately.

10.4.7
In-Situ Age Determinations

Most age dating techniques are based on isotopic ratios of radiogenic elements, which can be measured with one of the mass spectrometer types described above. Some of these techniques allow dating of minerals or even parts of minerals in their original fabric, in a thin section (Johnson 1999b). Most important and allowing highest resolution is the SHRIMP for dating zircon and monazite in thin section, although LA-ICPMS, SIMS and similar methods can also give good results. Examples are monazite inclusions in garnet dated by SIMS (Zhu et al. 1997) and monazite and K-feldspar using LA-ICPMS, (Möller et al. 1998; Christensen et al. 1995).

Particularly useful is Ar-Ar dating in thin section of micas, amphiboles or feldspars where the entire thin section is irradiated before analysis (Kelley et al. 1994). A laser is used to evaporate some material from the surface of the crystals, after which it is sent to a mass spectrometer for analysis.

An alternative but not generally accepted method is *chemical age dating*, where concentrations of Y, U, Th and Pb in monazite (Williams et al. 1999; Williams and Jercinovic 2002), xenotime, uraninite, zircon and staurolite (Lanzirotti and Hanson 1997) measured by electron microprobe are used for dating. The method is especially promising for monazite since this contains virtually no original lead, so all lead can be assumed to be radiogenic, while lattice diffusion is apparently slow, preserving delicate chemical zoning patterns. The method can be used

for crystals older than about 200 Ma. The method uses a type of least-squares regression or plane fitting through a data cloud and needs a considerable spread in order to give workable results. Errors are in the range of 20–50 Ma. Advantages are high spatial resolution (down to 2–3 μm), the possibility to analyse parts of complexly zoned crystals and the non-destructive analysis compared with mass spectrometers which need to evaporate material. Analyses can be done in a relatively short time at low expense (Suzuki et al. 1991; Rhede et al. 1996; Montel et al. 1994; Terry et al. 2000). A particular advantage is the possibility to date monazite crystals in their relationship with porphyroblasts and foliations, and thus to date parts of the microstructure (Williams and Jercinovic 2002). Problems are Pb-contamination during sample preparation, analytical problems, uncertainties of the original composition and zoning of monazite growth and the possibility of lead loss (Scherrer et al. 2000).

Chemical dating cannot replace isotope dating, but is presently at least useful for first-approach dating and has the advantage that it can date zoning in crystals and inclusions in porphyroblasts. In the SHRIMP, resolution is a few μm and crater depth 1 μm. In the electron microprobe less than 1 μm and little damage is done to the specimen. In situ dating promises to aid precise dating of deformation phases, tectonic processes, mineral growth rates and to refine general *P-T-t* paths for metamorphic rocks.

10.5
Fluid Inclusion Studies

10.5.1
Introduction

Fluid inclusions are small pockets of fluid inside crystals. They are assumed to be samples of the fluid that was present in the rock during deformation and metamorphism, although the original composition and density may have been modified by leaking and preferential removal or addition of certain components (Roedder 1984; Shepherd et al. 1985; Crawford and Hollister 1985; de Vivo and Frezzotti 1994; Goldstein and Reynolds 1994; Boullier et al. 1997). Fluid inclusions of CO_2, H_2O, N_2 and CH_4 are relatively common (Figs. 10.16, 10.17a). Usually, inclusions contain more then one phase (Fig. 10.17b); many contain a gas bubble and some contain also solid phases such as small salt or graphite crystals (Fig. 10.17d).

It is possible to determine the composition and density of the fluid in inclusions without destroying the sample. To a first approximation, H_2O rich inclusions in quartz can be recognised by the presence of a small salt crystal and gas bubble, and by a small contrast in refractive index with quartz (Figs. 10.16, 10.17c,d). CO_2-rich

10.5

inclusions may have a gas bubble but lack salt crystals, and have a stronger contrast in refractive index with quartz, leading to thick 'shades' along the contacts (Figs. 10.16, 10.17b,c).

Fluid inclusions either occur isolated or as swarms that commonly lie along a plane. Isolated inclusions, many with a negative crystal shape (Fig. 10.16), may represent inclusions that were captured when the host grain grew from a solution; they are therefore also known as *primary fluid inclusions* (Figs. 10.17a,b). Planes of inclusions commonly contain inclusions of similar composition and density, and are thought to represent healed fractures on which fluid has been present (Figs. 3.18, 10.16, 10.17c,e); they are also known as *secondary fluid inclusions*. Such fractures can normally be followed over long distances in a section, and may transect grain boundaries or even mineral boundaries (Figs. 10.16, 10.17f). The ratio between the diameter of fluid inclusion and the length of fluid inclusion planes as measured in thick sections is around 10^{-2} when the average aspect ratio of microcracks in rocks is 10^{-3} after Kranz (1983). The lower aspect ratio (10^{-2} versus 10^{-3}) may be explained by dissolution crystallization processes during healing of mode I microfractures. Thus, the relative age of inclusions in the healed crack and structures or minerals in the rocks can be established; the composition and density of the inclusions can be related to deformation phases or metamorphic conditions. Where healed cracks transect grain boundaries, it is possible to determine whether they were tension fractures (without a shear component) or shear fractures. Tension fractures are most common. If their orientation is known, they can then be related to a stress tensor of a particular orientation and, in the case of tension fractures, an upper limit can be placed on the magnitude of the differential stress (Sect. 9.6). In many cases, several sets of healed cracks occur in a rock (Fig. 10.17e), and their relative age can be established at the intersection point, since inclusions of the older phase will have been destroyed by the younger fracture (Fig. 10.16). Thus, sequences of fracturing and associated fluid compositions and densities can be linked to phases of deformation and, in some cases, metamorphic conditions. Since fluids play a major role in deformation of rocks, such information is extremely valuable. However, there are also some disadvantages in the use of fluid inclusions:

1. inclusions are usually relatively young; the older inclusions are destroyed by deformation;
2. inclusions may have leaked or changed shape, composition and density, and this is difficult to check; especially inclusions with strongly irregular shape, or with surrounding swarms of small inclusions may have leaked, decrepitated or imploded.

In healed cracks the situation is probably relatively safe, since any inclusions with strongly deviant composition and density can be regarded as suspect. For these reasons, any work on fluid inclusions in metamorphic rocks should take into account their topology: shape, fluid composition, density and mutual relationships. For example,

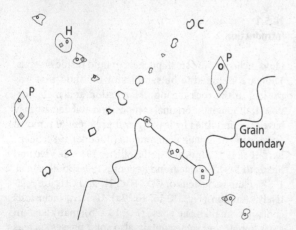

Fig. 10.16. Schematic drawing of fluid inclusions as they may appear in a thin section; large isolated inclusions with a negative crystal shape can be primary (*P*). Secondary inclusions occur in arrays that represent healed fluid-filled cracks, and have similar composition within one array; two types are shown, H_2O-rich inclusions (*H*) with gas bubbles and small salt cubes, and CO_2-rich inclusions (*C*), with thick shades along the edge due to a difference in refraction with surrounding quartz. The H_2O inclusion array cuts a grain boundary, and the CO_2-rich array cuts the H_2O rich array, thus establishing the relative age of the structures

Fig. 10.17. a Typical primary CO_2-H_2O fluid inclusions containing ▶ three phases. Camperio, Swiss Alps. **b** Primary CO_2-H_2O fluid inclusion containing three phases: from *centre* to *margin*: a bubble of CO_2 gas and CO_2 liquid (*centre*), and H_2O liquid (*lighter part* at the *right-hand side*). The well-developed negative crystal shape is typical of primary fluid inclusions. Camperio, Swiss Alps. **c** Two planes of secondary fluid inclusions in quartz. The *top* one of dark inclusions contains CO_2. The *lower* one, of low-relief inclusions contains H_2O. High-grade gneiss from the Grenville Province, Canada. **d** H_2O-NaCl fluid inclusion in a quartz crystal. A gas bubble and salt cube are present in the inclusion; the salt cube contains itself a small inclusion. The fluid inclusion formed late and indicates the percolation of seawater through the rock. Zabargad Island. (Red Sea, Egypt). **e** Quartz crystal with several sets of planes of secondary fluid inclusions. Bands of variable width composed of small inclusions are thin inclusion planes; width of the band increases with decreasing angle between the inclusion plane and the thin section. All are CO_2-H_2O inclusions. Some of the larger inclusions have an annular shape due to late differential stress increase and implosion of the inclusions. The planes of small fluid inclusions are probably formed from imploded large (now annular) inclusions. The two large irregular dark inclusions at *bottom left* are decrepitated, empty inclusions. Main Central Thrust, Himalaya. **f** Fluid inclusion planes transecting grain boundaries in quartzite. All are CO_2-H_2O inclusions. Grain boundaries also contain fluid inclusions, but these are now mainly empty. Main Central Thrust, Himalaya. 100 µm thick doubly polished section. PPL. (Photographs courtesy A-M. Boullier)

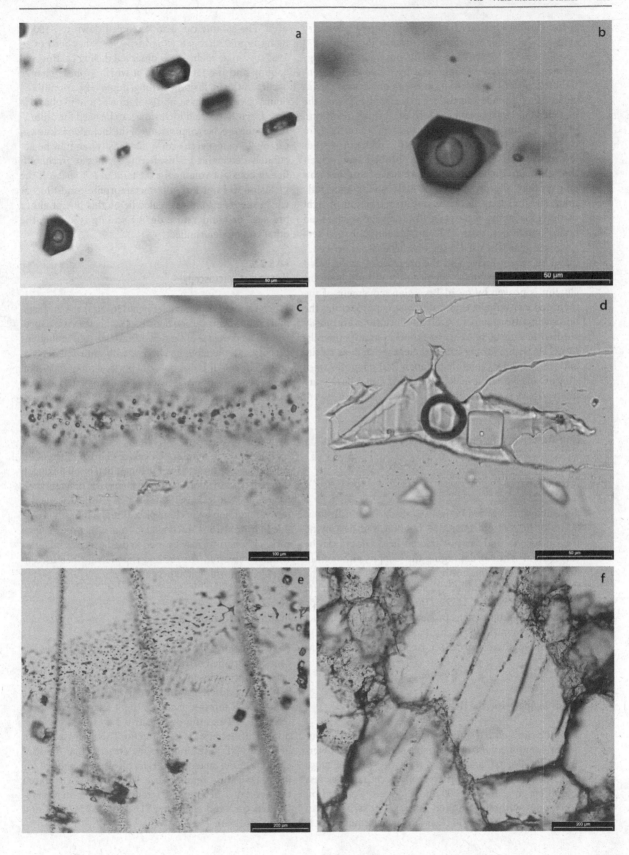

some fluid inclusions have a ring-shaped geometry (Fig. 10.17e). They are known as *annular fluid inclusions* (Boullier et al. 1991). These inclusions form from large primary or secondary inclusions by increased differential stress after their formation event; this causes the inclusions to implode and reduce to a ring shape while the expelled fluid forms planes of secondary small inclusions.

The relationship between healed cracks of fluid inclusions and other microstructures in a rock can be studied in ordinary thin sections or polished sections, although many inclusions will have leaked and are now empty; they appear as dark inclusions with unusually thick shades along the rims due to refraction of light through the empty space (Fig. 10.17e,f). For precise fluid inclusion studies, thick (100 µm) doubly polished sections are generally used. Intact inclusions can be recognised by a clear image and the presence of an included bubble or small crystals. The presence of fluid in an inclusion is proven by rapid Brownian movement of the included gas bubble, especially when approaching the homogenisation temperature. Fluid inclusions are mostly studied in quartz, but can be found in many minerals. Minerals with a strong mineral cleavage such as calcite and micas are less suitable for analytical purposes since they are likely to have leaked; similarly, minerals that are likely to have reacted with water or CO_2 are to be avoided.

10.5.2
Non-Destructive Techniques to Study Fluid Inclusions

10.5.2.1
Introduction

Several techniques can be used to study fluid inclusions. Some of these are destructive, opening and destroying the inclusions, and are not treated here. Others can be carried out without damaging the sample, although in many cases special thick polished sections are needed. The most important techniques to study fluid inclusions are thermal analysis (Sect. 10.5.2.2) and Raman spectroscopy (Sect. 10.2.2), proton microprobe (Sect. 10.4.2), ultraviolet microscopy for inclusions, which contain oil (Sect. 10.5.2.3), and infrared microscopy (Sect. 10.5.2.4) for opaque minerals.

10.5.2.2
Thermal Analysis – the Heating-Freezing Stage

For inclusions with a simple composition (less than four major components) composition and density are determined by thermal analysis on a *heating-freezing stage* that fits under a normal microscope. A chip of rock less than 1 cm in diameter, 100 µm thick, and polished on both sides without attached glass plates is heated to 300 °C, or even up to 600 °C if additional equipment is used. The sample can also be cooled down to −160 °C using liquid nitrogen (Roedder 1984; Shepherd et al. 1985). A single inclusion is observed during heating and cooling and the temperature at which a phase change occurs in the inclusion (melting, homogenisation to the liquid or vapour phase, or dissolution of a solid phase) is noted. Synthetic fluid inclusions can be used for calibration. If the precise composition of the inclusion is known, e.g. by Raman spectroscopy, the density can now be accurately determined. This density defines an *isochore*, a line of constant volume for the fluid in *P-T* space. If the inclusion did not leak, the metamorphic conditions at which the inclusion was sealed lie on this line; if either pressure or temperature during sealing is known, the other can be determined.

10.5.2.3
Ultraviolet Microscopy

When exposed to ultraviolet light (365 nm) the aromatic portion of a hydrocarbon fluoresces within the visible spectrum. The presence of fluid hydrocarbons (oil) in fluid inclusions can therefore be checked with ultraviolet microscopy. The colour is diagnostic of physical properties of oil such as composition and maturity or density, the latter expressed as degrees *API gravity* (Burruss 1987, 1991; Burruss et al. 1985; George et al. 1997, 2001). Fluorescence colour tends to change from yellow to blue as maturity or API gravity increases.

Differences in fluorescence colour of oil in different inclusions can be used to infer oil migration or oil mixing. If different types of fluorescence are observed in different trails or groups of inclusions, it is likely that several discrete water/oil migration events took place through the sample, e.g. due to hydrocarbon fractionation from a maturing source rock. This, in turn, may be used to predict gas or oil ratios within a reservoir. The method is powerful when used in combination with microthermometry.

10.5.2.4
Infrared Microscopy

Fluid inclusions also occur in opaque mineral phases but cannot be studied in transmitted light when inside a crystal. Fluid inclusions in some opaque minerals can be studied in transmitted near infrared light using microscopes with special infrared facilities (light source, lenses, objectives) and an infrared sensitive TV camera to transmit the infrared image to a monitor. This microscope can be fitted with a heating-freezing stage so that fluid inclusions in the opaque minerals can also be analysed for homogenisation temperature. The quality of infrared transmittance of a mineral strongly depends on its crystal structure and chemistry (Lüders 1996; Lüders and Ziemann 1999).

Inclusions, which are placed on the infrared path, partially absorb the light beam that is collected by an MCT detector. An infrared spectrum in absorbance units is obtained and can be Fourier transformed to produce the absorbed frequencies, and therefore the masses of the ions present. This is known as *Fourier Transform Infrared Microspectroscopy*.

10.6
Image Analysis

10.6.1
Introduction

Several powerful computer programs have become available over the last few years that allow 'automatic' analysis of certain microstructures (e.g. Masuda et al. 1991; Heilbronner-Panozzo and Pauli 1993; Heilbronner and Herwegh 1997; Heilbronner-Panozzo and Pauli 1994; Bartozzi et al. 2000). Microfabrics can be monitored with a video camera and the digitised image can be fed into a computer. The image can then be analysed in terms of the intensity and distribution of different colours. An image analysis program allows the calculation of percentages of particular colours visible in an image, and recognises sharp transitions in colour. Such transitions are commonly grain boundaries, and it is possible in some samples to identify individual minerals and mineral grains. With the use of filters, a single mineral may be selected, and the total percentage of this mineral in the image can be determined. In addition, the size and shape of the grains can be determined and plotted. Such techniques are obviously useful if large numbers of grain shapes, grain sizes and volumetric fractions of minerals have to be determined, e.g. the percentage, shape, size and orientation of plagioclase grains in a basalt. Unfortunately, even the best presently available programs have difficulties with samples in which the contrast between minerals is small, or where grain boundaries are of variable sharpness and lattice misfit angle, as in dynamically recrystallised rocks. Best results are obtained in rocks that consist of few, clearly distinct phases, e.g. pyrite grains in quartz, or quartzite with pores. Another disadvantage of image analysis is that most commercially available software is expensive, sometimes difficult to operate, and not adapted for geological applications[2]. However, *image analysis* programs are continuously being improved and are an important tool in microtectonic analysis. A single method is described below because it can be applied without access to expensive programs.

10.6.2
SURFOR and PAROR

A simple method to determine grain shape and orientation in a grain aggregate can be used without expensive software on a desktop computer (Panozzo 1983, 1984; Schmid et al. 1987). Drawings of grain boundaries are made from photographs or directly on a computer screen, and subdivided into segments of a fixed length, e.g. 2 mm on photographs. A projection line is then defined, all line segments are projected onto this projection line, and the total length of projections is added. This process is repeated after rotating the projection line over a small angle, e.g. 1°. In this way, 180 total projection lengths $A_{(\alpha)}$ are obtained for 180 angles of projection (α). This process can easily be automated with simple programming (program SURFOR: Panozzo 1983, 1984; Schmid et al. 1987). When $A_{(\alpha)}$ is plotted against (α) in a diagram, the symmetry of the distribution can be assessed. If grains are elongate and define a simple shape preferred orientation, the ratio of $A_{(\alpha)max} / A_{(\alpha)min}$ defines the elliptical ratio of the aggregate, which may be a measure of strain if grain boundaries were immobile (Sect. 9.2). The orientation (α) of $A_{(\alpha)max}$ is the orientation of the foliation defined by the longest axis of the grains (program PAROR: e.g. Heilbronner and Bruhn 1998; Molli and Heilbronner 1999).

---

[2] For Macintosh users, there is a public domain image analysis program called "NIH Image", developed by the Research Services Branch (RSB) of the National Institute of Mental Health (NIMH), part of the National Institutes of Health (NIH). For PC and LINUX users there is now a similar Java program called "ImageJ".

Experimental Modelling Techniques

All previous chapters deal with observations of microstructures in rocks, and conclusions that can be drawn from those observations. However, no science is complete without some degree of experimental work to test ideas and theories. Unfortunately, geology is a difficult subject to study by experimental techniques; most geological deformation processes are too slow to model in laboratory experiments; they occur at extreme pressure and temperature, and at a scale too large for experimental reproduction. To overcome this problem, two types of experimental work have been set up to date.

Analogue experiments use materials that are easier and faster to deform than real rocks, but which give results that can be compared with microstructures in rocks. Such experiments are carried out in small presses, usually at atmospheric pressure and low temperature, and even under the microscope. They can reveal the sequence of development of microstructures, and even movies can be made.

Numerical experiments are done on the development of all kinds of microstructures, on the grain scale and beyond. Several examples of such studies are discussed.

In this chapter, we restrict ourselves to experiments on actual microstructures, and do not treat the extensive work on large scale structures, thrust belts, orogenesis, granite intrusion and other crustal and mantle processes.

11.1

11.1
Introduction

As we discussed in previous chapters, microtectonics is mostly concerned with the interpretation of geometries in thin section formed by solid-state deformation. Our understanding of the way in which microstructures develop is mainly based on careful observation of fabrics from simple tectonic settings in various stages of development. The only way to test such interpretations unequivocally would be to carry out experiments identical to the deformation processes in the rock. Obviously, this is generally impossible because of the pressure and temperature conditions required, and especially because of the very slow rate of natural deformation processes in rocks. There are, however, ways to get around this problem. At least three different approaches are possible: (1) experimental deformation of rocks at high pressure and temperature, and at high strain rate; (2) experimental deformation of rock analogues at low pressure and temperature and at high strain rate; (3) numerical experiments.

Deformation experiments on rocks and minerals are possible at pressures equivalent to those in the crust for samples on mm–cm scale in special high-pressure deformation rigs (e.g. Rowe and Rutter 1990; Rutter and Neumann 1995; Kohlstedt et al. 1995; Gleason and Tullis 1995; Post and Tullis 1999; Brodie and Rutter 2000; Laurent et al. 2000; Mauler et al. 2000; Pieri et al. 2001; Heilbronner and Tullis 2002; Orzol et al. 2003). In these rigs, brittle and even ductile deformation structures can be produced by implementation of a high confining pressure and high temperature. Since samples cannot be deformed at natural strain rates, deformation is done at excessively high temperature and high strain rate to activate similar deformation mechanisms as in the geological environment under consideration (Chap. 3). Deformation of cylindrical specimens can be implemented to low strain by axial compression or to high strain in torsion. Samples used are either real rocks, or artificial rocks produced through sintering of a mineral powder.

The advantage of this type of experiments is that they give information on flow parameters of natural rocks, (Sect. 3.14) and on the generation of certain small-scale structures. A disadvantage is that only the initial and deformed state of the fabric can be studied, and that the intermediate stages and the actual deformation process cannot be observed during the experiment. In this chapter, we mainly concentrate on analogue and numerical experiments, which are better suited to study the development of microstructures.

11.2

11.2
Experimental Deformation of Analogue Materials

Since the 19th century, simple experiments have been set up with materials other than rocks with the aim to increase understanding of rock deformation through the deformation of analogue materials such as clay, sand, bee's wax and paraffin wax. Analogue materials have a different rheology than that of real rocks, but they deform at low pressure and temperature and at high enough strain rate to follow development of structures *in situ*. Although analogue materials deform much faster than rocks, they can form structures that are similar to those formed in rocks such as thrust fault systems, tension gashes, diapirs, but also microstructures such as shear zones, folds, mantled porphyroclasts and mineral fish. This means that analogue experiments are suitable as pilot experiments to create ideas for the interpretation of structures in rocks. It is also possible in some cases to use such experiments as an equivalent of a real rock experiment, provided the model and the original can be scaled in a proper way (Twiss and Moores 1992, p 437). If materials can be found such that length scales, velocities, forces and accelerations in the material can be scaled to that of the real rock, direct comparisons can be made. The main difficulties involved in such scaling are, however, that the rheological properties of real rocks are incompletely understood, that only a limited number of analogue materials and rheologies are available, and that gravity is equal in the real world and in the model (Twiss and Moores 1992).

11.3
Large-Scale Analogue Modelling

In large-scale modelling, cm–m scale machines are used to deform volumes of analogue material. Such studies are mostly directed to m–100 km scale geological features, but here we only discuss some of the types of experiments that have been used to model microstructures. The simplest type of device is similar to that shown schematically in Fig. 11.2 in which simple shear, pure shear and general flow in plain strain can be realised (Figs. 2.1, 11.1, ⊘ Video 11.2a–d; e.g. Ramberg 1955; Mancktelow 1988a; Sokoutis 1990; Piazolo et al. 2001; Zulauf et al. 2003). A disadvantage of this type of machine is that only a limited finite strain can be reached; once the walls of the machine approach each other, the experiment must be stopped (Piazolo et al. 2001; ⊘ Video 11.2a–d). High strain as in mylonites cannot be produced. To overcome this problem, another type of deformation rig called a circular or torsion rig can be used. In a torsion rig, a ring-shaped sample is deformed and very high strains can be reached. A disadvantage of the circular rig is that sample preparation is more difficult, and that flow cannot be homogeneous simple or pure shear as in a square shear box; material deforms along curved flow lines and with a strain rate and stress gradient from the inside to the outside of the ring (Fig. 11.1b). Both types of deformation rigs can be filled with clay, salt-mica aggregates (Williams and Price 1990; Williams and Jiang 2001), paraffin wax (Mancktelow 1988b), or with polymers and derived complex materials (McClay 1976; Zulauf and Zulauf 2004; Hailemariam and Mulugeta 1998; Weijermars 1986;

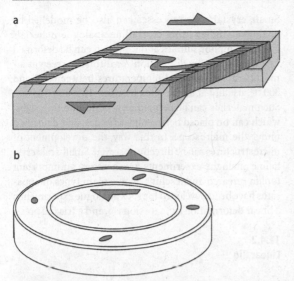

Fig. 11.1. a Simplified drawing of an analogue deformation rig for simple shear deformation. **b** Annular deformation rig for high strain experiments

ten Grotenhuis et al. 2002). Complex initial geometries can be set up such as rigid objects with mantles, multilayers and shear bands. Experiments are usually done at atmospheric pressure and room temperature and at strain rates between 10^{-1} and 10^{-6} s^{-1}. Observations are made at the surface of the sample if the material is not transparent, or by serial sectioning after an experiment is done. Three dimensional studies can be made by using transparent materials such as honey or polydimethysiloxane (PDMS; Weijermars 1986), or by using a rig that can be fitted in an X-ray tomograph (Coletta et al. 1991; Guillier et al. 1995).

Large-scale analogue modelling has been used for microtectonic studies in the following fields; mantled porphyroclasts (Passchier 1994; Piazolo and Paschier 2002b; Passchier and Sokoutis 1993; Passchier et al. 1993; ten Brink and Passchier 1995; ten Brink 1996), foliation refraction (Treagus and Sokoutis 1992); folding and boudinage of foliations (Williams and Price 1990), mineral fish (Pennacchioni et al. 2000; Ceriani et al. 2003), preferred orientation of phenocrysts (Ghosh and Ramberg 1976; Fernandez 1987, 1988; Arbaret et al. 1996, 2001; Ildefonse et al. 1992a,b;

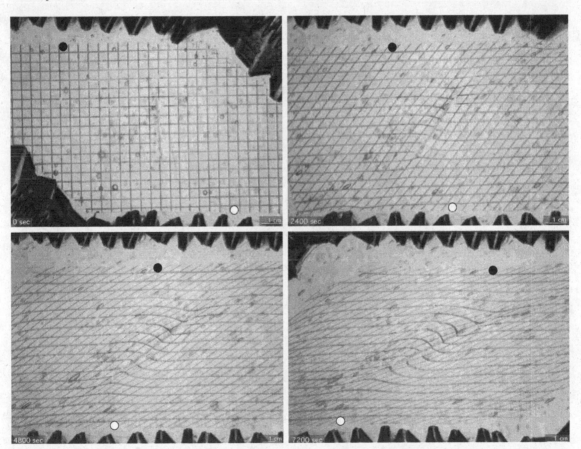

Fig. 11.2. Example of a shear box experiment using a transparent putty, polydimethylsiloxane, and a vertical fault filled with a lubricant in the centre of the box. Dextral shear in the shearbox leads to ductile deformation of the putty and deformation of the grid. The fault is active during this ductile flow, and this leads to development of flanking folds (see also ⊘Video 11.2a–d). The experiment is shown at 0, 2 400, 4 800 and 7 200 seconds. *Shaded spheres* indicate marker points on the grid. (Experiment and photographs by Vincent Heesakkers, Mainz)

Ildefonse and Mancktelow 1993; Piazolo et al. 2002; Marques and Coelho 2001, 2002, 2003), folding and boudinage (Kidan and Cosgrove 1996; Kobberger and Zulauf 1995). All these experiments have been designed to model structures at a scale larger than the real objects, and have provided many new ideas on development of microstructures. Disadvantages are that real scaling is difficult because of the limited number of experimental materials available and technical problems of sample preparation. Advantages are the fast, cheap way in which true 3D-experiments can be realised and the possibility to study complex systems.

11.4
Micro-Analogue Modelling

11.4.1
Introduction

Large-scale analogue modelling is suitable for the study of larger, polycrystalline microstructures since they treat rocks as fluids and ignore processes on the crystal scale.

Small, crystal scale processes can also be modelled by analogue studies using crystalline analogue materials such as camphor, nitrates and ice which can be deformed in a ductile way by dislocation- or diffusion creep at atmospheric pressure and temperatures between –10 and 300 °C, usually at strain rates between 10^{-2} and 10^{-6} s^{-1}. Such materials can be pressed or ground into thin slabs, which can be placed between glass slides, and deformed under the microscope. In this way, the development of microstructures can be directly observed. Small-scale crystalline analogue experiments are therefore an important tool in research and teaching. Three main types of apparatus have been developed for rock analogue experiments; a linear deformation rig, a torsion rig, and a triaxial press.

11.4.2
Linear Rig

In the linear rig (Fig. 11.3a), a thin sample of analogue material is mounted between two glass slides, usually similar slides as used for thin section preparation (Fig. 11.3b–d,

Fig. 11.3.
a Schematic drawing of a linear deformation rig used to deform analogue material. **b–d** Three setups of specimen chambers for the rig. The analogue material is deformed in experiments by movement of a piston **b** or of one of the two glass plates (**c, d**); arrangement of the pistons determines how the material is deformed, either in bulk pure shear or simple shear flow

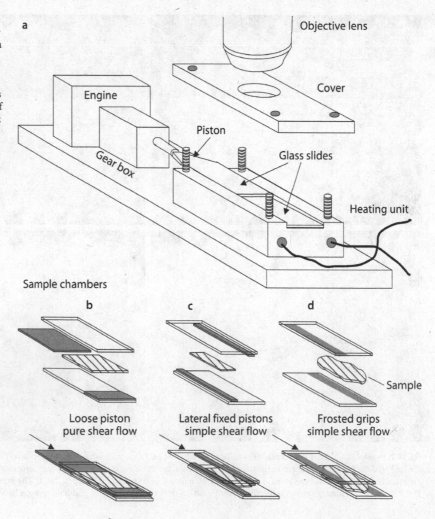

⊘Video 11.3; Means and Xia 1981; Wilson 1984; Burg et al. 1986; Means 1989). The sample can be deformed in co-axial or non-coaxial flow by the geometric arrangement of pistons. Pure shear deformation can be initiated by pressing a thin metal piston (at least 70 μm thick, otherwise the piston will buckle) between the glass slides against the sample (Fig. 11.3a), or by 'frosting' both glass slides at front or back with silicon carbide powder; the sample will remain attached to the frosted part of the glass, and will slide past the untreated glass, especially if the latter is lubricated with silicone grease (Fig. 11.3d). With this setup, samples as thin as 30 μm can be deformed. Simple shear can be initiated by the use of indented piston grips glued onto the glass plates (Fig. 11.3c), or frosted grips (Fig. 11.3d). General non-coaxial flow can be generated with obliquely mounted grips or pistons. The choice of thickness for the sample depends on the amount of detail that is required, and the birefringence of the mineral; minerals like ice and camphor have low birefringence and can be studied in relatively thick samples that are easy to prepare. In such thick samples, it may be difficult to resolve small grains, grain boundaries and other small details. Materials with higher birefringence can be studied in thin sections with frosted grips; this has the advantage that a similar view as a thin

section is obtained and that small details can be studied. Disadvantages are difficult sample preparation and the possibility that friction of the sample against the glass plates causes deviations or distortions. The sample and glass slides are contained in a metal frame, which can be heated and kept within a certain temperature range with a thermocouple and a temperature control unit. The sample is deformed by displacement of a piston or one of the glass slides at a constant rate by a small motor (Fig. 11.3a); different strain rates are realised by the use of a range of gearboxes. The deforming sample is observed through a microscope and can be monitored with a camera or video equipment. It is even possible to carry out experiments under conditions of controlled confining pressure by sealing the entire deformation rig inside a small pressure vessel that fits under a microscope (Bauer et al. 2000b).

11.4.3
Torsion Rig

The torsion rig (Fig. 11.4a) deforms a sample between two glass slides (Fig. 11.4b, ⊘Video 11.4). Each glass slide has a ring-shaped frosted grip, one of small and one of large diameter. When one of the glass slides rotates, a ring-shaped part of the sample is deformed between the

Fig. 11.4.
a Schematic drawing of a torsion deformation rig used to deform analogue material. **b** Setup of the specimen chamber for this rig; analogue material is deformed in torsion between two glass plates that rotate with respect to each other. Presence of frosted grips on the glass causes a small ring-shaped part of the sample to deform in non-coaxial flow. **c** Detail of the ring-shaped deforming part of the sample. (**b** and **c** courtesy C. ten Brink)

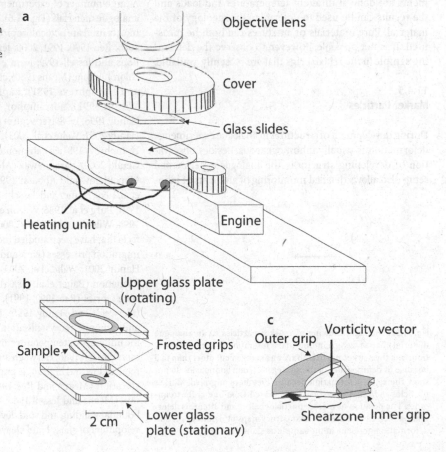

two frosted grips (Fig. 11.4b,c). The rotating glass slide is fixed in part of the metal frame of the rig, and both are driven by an engine as described above (Fig. 11.4).

The advantage of the torsion rig is that very high finite shear strain (exceeding 100) can be reached; in the linear rig, shear strains are limited to values of 5–10. In mylonites, shear strain values exceeding 10 are common, and may be required for the development of characteristic structures. A disadvantage of the torsion rig is that the deforming sample is circular and therefore does not deform in simple shear; in fact, there is a strain rate and stress gradient from inside to outside in the ring-shaped part of the deforming sample (Passchier and Sokoutis 1993; Masuda et al. 1995b). However, the gradient is accurately known and such gradients may also occur in natural shear zones.

11.4.4
Triaxial Rig

A triaxial deformation rig (Bons 1993) can be used to determine the rheology of analogue materials. Small columns of analogue material are compressed by a weight and the shortening rate is monitored. Normally, a confining pressure is applied using compressed air. The data are used to determine stress-strain rate curves. Such experiments are done at different temperatures and loads and the results can be used to determine the rheology of the material. Pure materials or mixtures can both be measured. It is not possible, however, to observe the deforming sample in the triaxial rigs that are presently available.

11.4.5
Marker Particles

During development of structures in analogue experiments, deformation is usually inhomogeneous. Besides observation of developing structures, the analogue experimental setup also allows detailed monitoring of gradients of inho-

mogeneous deformation over the sample. This is realised by insertion of fine silicon carbide powder in the sample (1 000-grid powder, normally used for preparation of thin sections). This powder does not react with the sample material and does not seem to interfere with the deformation process on the scale of observation. The particles act therefore as passive material points (Sect. 2.2; Figs. 2.1, 2.2); structures like grain boundaries cannot be used as markers since they may migrate through the material (Fig. 11.5). If photographs or video images of a deforming sample at various stages of development are combined, the positions of individual particles can be compared and used to reconstruct local deformation (Figs. 11.5, 11.6, ⊘Video 11.6b, 11.7b). If the position of at least three particles is known in two subsequent stages of the experiment, finite strain, vorticity and area change can all be determined (Figs. 11.7, 11.8, ⊘Video 11.7; Sect. 2.6; Box 4.8). The position of particles must be digitised and can be processed with a computer. Bons et al. (1993) presented a computer program that can process data on positions of particles into contours of strain, vorticity and area change.

11.4.6
Examples of Analogue Experiments

A large number of experiments has been carried out with analogue materials (Fig. 11.6). The materials that have been mostly used are octochloropropane (OCP – Fig. 11.6; Jessell 1986; Ree 1990, 1991, 2000; ten Brink and Passchier 1995; Bons and Jessell 1999; Nam et al. 1999), biphenyl, paradichlorobenzene (Means 1980), camphor (Urai et al. 1980; Urai and Humphreys 1981), naphthalene (Blumenfeld and Wilson 1991), norcamphor, benzamide (Herwegh and Handy 1996, 1998; Herwegh et al. 1997, 2000; Rosenberg and Handy 2001; Walte et al. 2003), sodium nitrate (Tungatt and Humphreys 1981a,b), ammonium nitrate (McLaren and Fitz Gerald 2000), paraffin wax (Abbassi and Mancktelow 1992; Mancktelow and Abbassi 1992; Chatterjee 1994; Rossetti et al. 1999; Hafner and Passchier 2000) and ice (Wilson 1982, 1994; Burg et al. 1986; Wilson et al. 1986; Wilson and Zhang 1994; Wilson and Marmo 2000; Wilson and Sim 2002). Effects that have been studied include melt-forming and melt migration processes (Park and Means 1996; Rosenberg and Handy 2001; Walte et al. 2003), fluid migration during deformation (Bauer et al. 2000a), the development of steady state fabrics (Ree 1990, 1991), shape preferred orientation (Herwegh and Handy 1998), lattice-preferred orientation and dynamic recrystallisation mechanisms (Jessell 1986), the influence of temperature and finite strain on these processes (Jessell 1986; Herwegh and Handy 1996), the development of low and high angle grain boundaries and their classification (Means and Ree 1988), grain boundary migration (Means and Jessell 1986; Ngoc Nam et al. 1999), grain boundary sliding and void development (Ree 1994), the development of shear band cleavage (Passier 1991; Chatterjee

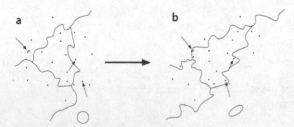

Fig. 11.5. Schematic drawing of marker particles in an analogue material that shows grain boundary migration during deformation. In **b** the marker particles have an arrangement other than in **a** because of deformation of the aggregate. Grain boundaries do not show the same deformation because they have migrated. Marker particles indicated by *arrows* have passed from one grain to another. The position of marker particles before and after this deformation step can be used to reconstruct maps of deformation and deformation gradient in the sample as shown in Fig. 11.8

Fig. 11.6.
Examples of a deformation experiment using a sample of octachloropropane (OCP), which is deforming in simple shear in a linear deformation rig (dextral sense of shear). The sample has already undergone a shear strain of 1.7 at stage **a**. Between **a** and **b** a shear strain of approximately 0.1 has accumulated, as shown by the displacement of two marker points indicated by *circles* at *top* and *bottom right*. Due to this extra shear stain, the shape of the grains has changed and grain boundaries have migrated in **b**. After step **b** the experiment was stopped. **c** Shows the same area as **b**, but after 16 h of static recrystallisation. A polygonal fabric has developed in response to grain boundary area reduction (GBAR); grain boundaries have become relatively straight and there is evidence for grain growth. Subgrain boundaries developed in some grains. Unpublished experiment by J.-H. Ree. Part of this experiment was published in Means (1989). Width of view 2.1 mm. CPL. (Photographs courtesy J.-H. Ree)

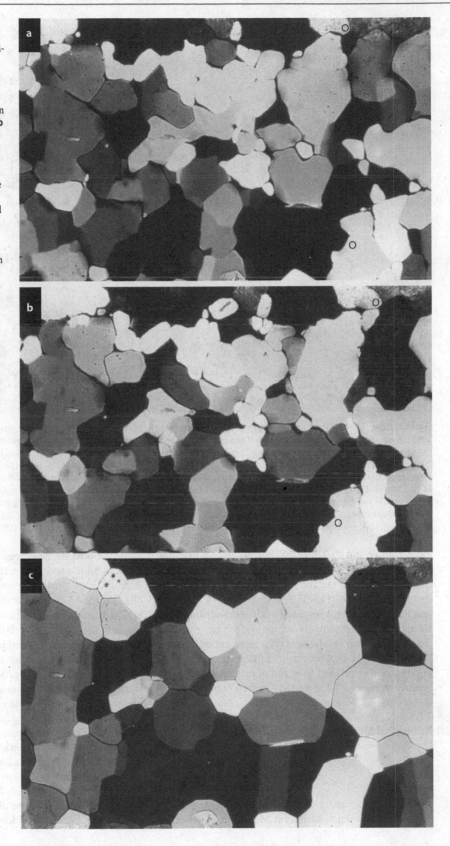

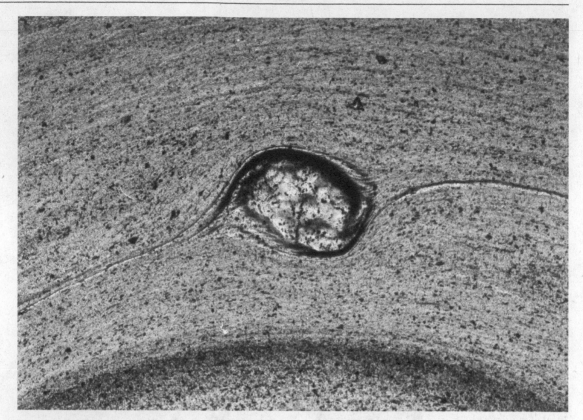

Fig. 11.7. Example of a δ-type mantled porphyroclast developed in a rock analogue experiment in a torsion rig (ten Brink and Passchier 1995). A square polycrystalline aggregate of camphor was embedded in a matrix of octachloropropane (OCP), and both were deformed together at 60 °C and a shear strain rate of 2.6×10^{-3} s^{-1}. Camphor is stronger than OCP at these conditions, and the camphor object rotated in response to non-coaxial flow in the OCP matrix. In the rim of the camphor object, fine-grained material is produced, probably by recrystallisation. This caused development of a fine-grained camphor mantle that deformed into δ-type wings. The entire evolution of this mantled porphyroclast could be studied under the microscope. *Black dots* in the matrix are grains of silicon carbide that act as marker particles. Figure 11.8 shows some results of the analysis of displacement of these marker particles. The curved shape of the wings is due to the circular shape of the specimen chamber (Fig. 11.4). The *dark rim* at the *bottom* of the photograph is the inner grip of the rig (Fig. 11.4). Shear strain in the matrix $\gamma = 94$. Width of view 5 mm. PPL. See also ⊘Video 11.7a. (Photograph courtesy C. ten Brink)

1994; Hafner and Passchier 2000), deformation twinning (Blumenfeld and Wilson 1991), the development of mantled porphyroclasts and the effect of a second phase on deformation patterns (Figs. 11.7, 11.8; Bons 1993; ten Brink and Passchier 1995). A review of analogue modelling and a list of the most commonly used analogue experimental materials is given by Means (1989). Some examples of results of analogue experiments in a linear rig are shown in Fig. 11.6 and ⊘Video 11.6a–e and of experiments in a torsion rig in Figs. 11.7, 11.8 and ⊘Video 11. 7a,b.

11.5
Numerical Modelling

11.5.1
Introduction

Numerical experiments are realised by using computer programs to model deformation. Many techniques of nu-

merical modelling have been used to study deformation in rocks on the scale of the thin section (see Jessell and Bons 2002 for a review). Here, we describe some methods that model the development of geometries of geological microstructures, since this closely approaches the theme of the book. Many relatively simple but useful programs have been developed to study specific geometric processes in rocks, without giving the "virtual materials" true rheological properties. This kind of modelling starts with simple homogeneous deformation that can be modelled in any drawing program (stretching, tilting, rotating), through programs that model simple geological geometries. Examples are flow around rigid objects using analytical equations (Passchier 1987a; Bons et al. 1997; Ramsay and Lisle 2000) and programs for fringe growth (Etchecopar and Malavieille 1987; Kanagawa 1996; Bons 2001; Köhn et al. 2000, 2001a,b; Hilgers et al. 2001; Nollet et al. 2005). Examples of this type of modelling are given in Fig. 6.23 and ⊘Videos 6.23, 6.27.

Fig. 11.8.
Example of the use of marker particles to analyse the deformation pattern in an experiment in the torsion rig. The experiment was on development of δ-type mantled porphyroclasts using a core object of camphor in a matrix of octachloropropane, as shown in Fig. 11.7. The drawings are maps of the four components of finite deformation, determined from displacement of marker particles. The contour of a developed camphor δ-type mantled porphyroclast is shown in each diagram. (After ten Brink and Passchier 1995)

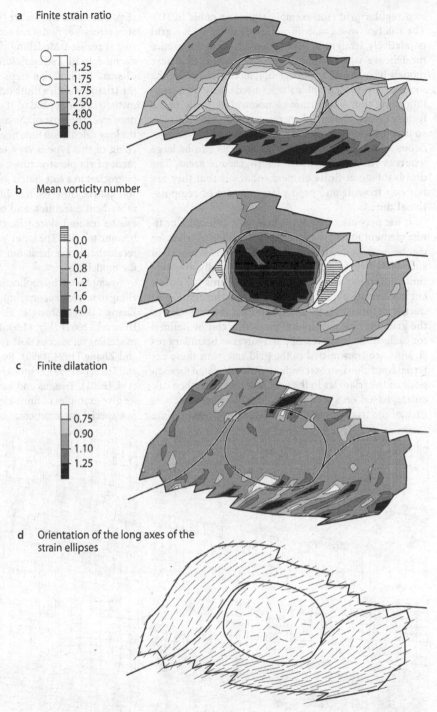

a Finite strain ratio

1.25
1.75
1.75
2.50
4.00
6.00

b Mean vorticity number

0.0
0.4
0.8
1.2
1.6
4.0

c Finite dilatation

0.75
0.90
1.10
1.25

d Orientation of the long axes of the strain ellipses

11.5.2
Finite Element and Finite Difference Modelling

In *finite element* and *finite difference* modelling (Press et al. 1992, 2002; Gersenfeld 1999; Ramsay and Lisle 2000), deformation in a volume of material is approximated by subdividing space into a number of elements with specific rheological properties. The internal deformation is then calculated for imposed external stresses or external strain rates. In both cases, a *grid* of points or *nodes* is defined linked by lines, which build elements, usually triangles. Within an element the deformation is set to be homogeneous. In the original finite difference method all elements have the same size and are arranged in a regular grid (for example square or cubic in 3D). The solution of e.g. a diffusion problem on such a grid is relatively straightforward. Therefore finite difference models are widely used in the Earth Sciences. These models have the disadvantage that all elements have the same size so that complex shapes need a very high resolution and thus an enormous amount of elements. In the finite element method elements can have different sizes so that it is easier to model complex shapes. In homogeneous areas of a model material elements can be large whereas they are small in heterogeneous areas. The disadvantage of finite element models is that they are not easy to write and need a large amount of computational time.

If the network of elements in a finite difference or finite element model is dense enough, groups of triangles can be made to coincide with structures to be modelled such as layers, folds, clasts etc. Commercially available finite difference and finite element programs use different mathematical paths to simulate both the elastic and ductile deformation by deformation of the triangles in the grid. Specific rheological properties can be defined for each of the triangles. Applied forces or boundary restraints are transmitted to the grid and from these external conditions, stress conditions are calculated for each node of the triangles in the grid. The programs then calculate, based on specified rheological properties, how each of the triangles is going to deform in a certain time

step. Once all triangles are "deformed" during one time step, stresses at nodes are recalculated and the next time step is realised. Modelling therefore consists of loading the model with a stress field, and running a large number of small deformation steps which change the shape of the triangles until a finite deformation can be seen. This method can be applied in 3D, but modelling is commonly done in 2D because of the high price of 3D software and the long calculation time needed for 3D modelling. Modelling of this type is very useful to study the development of simple structures, e.g. rotation of a hard porphyroclast in a soft matrix, or folding of a hard layer in a soft matrix. However, it is difficult to work with faults or other brittle features, and only relatively small strains can be realised since triangles of the mesh should not become too flat. The latter problem can be overcome by recalculating the mesh, but this can introduce errors in the modelling.

Examples of the application of finite difference modelling to microstructural studies are given in Wilson and Zhang (1994), Zhang et al. (1996) and Passchier and Druguet (2002) (Fig. 11.9). Examples of finite element modelling on viscous flow around objects are Bjørnerud and Zhang (1994, 1995), Bons et al. (1997), Kenkmann and Dresen (1998), Pennacchioni et al. (2000), Biermeier et al. (2001), Treagus and Lan (2000, 2003, 2004). Below we give examples of finite element modelling embedded in a special environment.

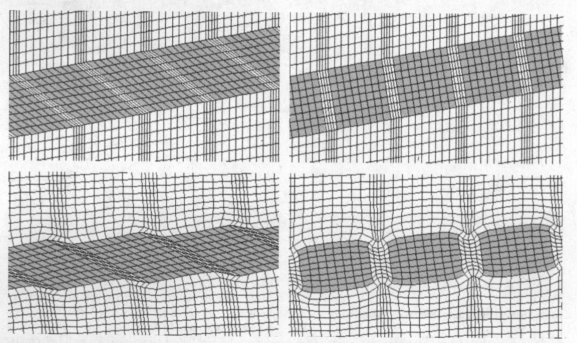

Fig. 11.9. Example of a deformation experiment using a finite difference code to model boudinage. At the *top*, two initial geometries are shown of a grid with two rheologies; a low viscosity (*light grey*) and high viscosity (*dark grey*). The bottom row shows the same grid after a deformation of $R = 2$ by pure shear vertical compression. The experiment shows how the initial orientation of a gap between boudins influences the final shape. (After Passchier and Druguet 2002)

Fig. 11.10.
Example of the initial and final stages of an experiment carried out with ELLE. The grain aggregate at left is allowed to grow without deformation by a decrease in grain boundary energy but grain growth is dominated by high surface energy anisotropy, which prevents most boundaries from moving. Grains with boundaries that can still move eventually dominate the microstructure. Even though the starting microstructure is a foam texture, many of the triple junctions that develop deviate from 120°, which would be the equilibrium angle if there was no surface energy anisotropy. See also ⊘Video 11.10e. (Courtesy Mark Jessell)

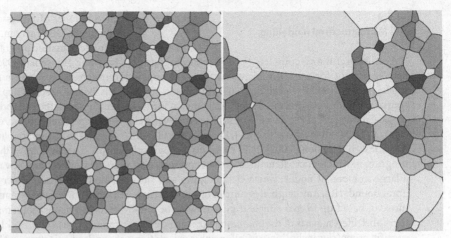

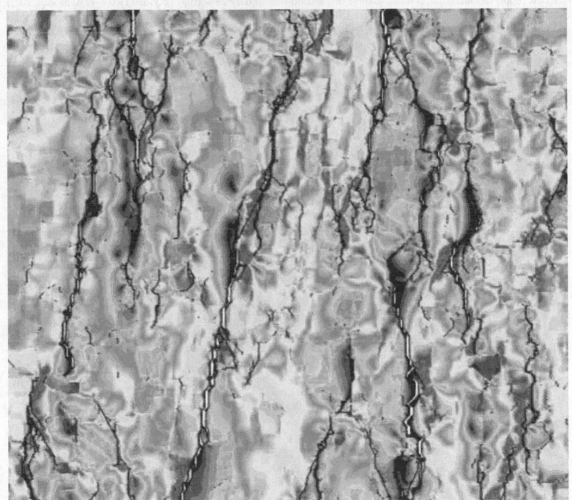

Fig. 11.11. Result of an experiment carried out with ELLE, modelling the development of fractures. The model used is an elastic spring model where grains consist of a number of circular elements that are connected with each other via linear elastic springs. Springs can break if they are stretched beyond a limit value, which leads to the development of larger scale fractures or faults in the model. The material is first loaded uniaxially and then vertically shortened and horizontally extended in pure shear deformation. Different *grey tones* indicate different differential stress (see ⊘Video 11.11b for colour). The developing fractures are hybrid shear fractures at a small angle to the compression direction. During later stages the shear fractures open so that in principal material could precipitate and veins could form. (Courtesy Daniel Köhn)

11.5.3
Full Microstructural Modelling

The most useful and complete type of modelling available at present for work on microstructures is a supermodel known as ELLE (Jessell et al. 2001). ELLE is an open-source computer environment specifically meant for numerical modelling of the development of microstructures (Figs. 11.10, 11.11; ⊘Videos 11.10a–f, 11.11a–c; Jessell et al. 2001; Jessell and Bons 2002). It is based on a graphical program in which microstructures can be represented by a large number of nodes, which delimit grain and sub-grain boundaries. Any original geometry of a microstructure can be set up from a simple digital drawing or photograph. Different parts of the microstructure can be given specific properties including crystallographic orientation and composition. This file is than passed through a number of modules, each of which represents a process such as grain growth, deformation, diffusion etc. A de-

sired number of processes can be switched on for a certain study. Deformation is realised through a finite element deformation program called BASIL (⊘Video 11.10a; Barr and Houseman 1996). The structure is then passed through each process in turn with very short time intervals, before a second step repeats the sequence. Time steps are chosen sufficiently small that the end result is almost similar to simultaneous occurrence. ELLE presently only operates in 2D, and can be set up on most computer platforms. The modules of the ELLE environment are part of an open system and new modules can be added in coordination with the developers. As a result, ELLE is expanded to a more powerful environment all the time and promises to become a major tool in microstructural analysis. The environment has been used to study grain growth and grain boundary anisotropy (⊘Video 11.10a–e; Bons et al. 2001; Piazolo et al. 2002), grain growth in the presence of a fluid (⊘Video 11.10f; Becker et al. 2003) and can even be used to model brittle deformation (Fig. 11.11; ⊘Video 11.10a–c).

From Sample to Section

This final short chapter describes the background to microstructural studies, mainly the way samples should be collected and oriented, and where to find the most suitable sites to take samples for specific studies. Also treated are types of thin sections and problems caused by the fact that thin sections, and most other techniques to study microstructures, are two dimensional sections which have to be used to reconstruct three dimensional features.

12.1
Introduction

A sound analysis of microstructures relies on correct sampling and on the right choice of the direction in which thin sections are cut from samples. This chapter discusses the steps of sample collection, choice of sectioning plane, and problems involved in the interpretation of three-dimensional structures from two-dimensional sections.

12.2
Sampling

The choice of samples for thin sections depends on the topic of interest and methods that are going to be used. In any case, it is important that the main structure in the area is understood in broad lines before samples are taken; thin sections usually contain a lot of information, which is useless if no good field record exists. For example, if a foliation is found in thin section that has not been recognised in the field, it cannot be fitted in a tectonic model and is useless unless a new field trip can be made to identify it. Thin section studies usually give best results if they are undertaken to solve a specific problem that has been defined before the sections were cut (cf. Passchier et al. 1990b). Thin sections cut at random to 'see what it looks like' will be less useful than those prepared with a specific aim.

12.3
Orientation of Hand Specimens

Hand specimens for structural studies should be oriented in the field. This is best done by marking dip and strike of a planar surface of the specimen on that surface (Fig. 12.1). Notice, however, that this still leaves two possible orientations, since the mark could be made on the top surface or on the lower side of a sample. An extra mark is therefore needed; for example, a cross on the top surface or an arrow indicating top (Fig. 12.1b).

Mistakes can easily be made in orienting samples and for the most important samples (e.g. those needed to determine shear sense) it is therefore useful to make a photograph or simple sketch of the sample, its orientation in outcrop and of the marking. Also, samples should be wrapped in paper or plastic bags to avoid breakage and erosion of markings and numbers.

12.4
Where to Sample in Outcrop

If a P-T-t path is to be reconstructed for a specific area (Sect. 1.3, Box 7.5), the most informative lithologies to sample are pelites and metabasites. Pelitic rocks generally develop foliations, which are not easily destroyed by subsequent deformation phases. Metabasites have a weaker memory for a se-

quence of deformation phases but, as for pelites, their mineral content is sensitive to changes in P-T conditions. Psammites usually develop relatively coarse structures, which are difficult to study in thin section, and limestones have a weak memory for a sequence of deformation phases because calcite recrystallises readily, even at low temperature.

Samples should preferably be taken in association with major structures which are understood and of which the relative age is known. If three phases of deformation have been recognised, it is useful to have samples from foliations belonging to each of the phases. If porphyroblasts are visible, these should also be sampled in order to determine their relative age with respect to the foliation. Samples in which overprinting relations are visible are particularly important. Small fold closures or shear zones of each of the phases, if possible from several lithotypes, should also be sampled. The mineral assemblage in such structures may allow an estimation of metamorphic conditions during the deformation phase. If a sample is taken in a major fold, it is useful to know from which limb or part of the hinge it was taken. Shear zones should be sampled in the high strain core, in the low strain rim and in the wall rock; in many cases, it is useful to compare mineral assemblages in the undeformed wall rock with those in a shear zone, in or-

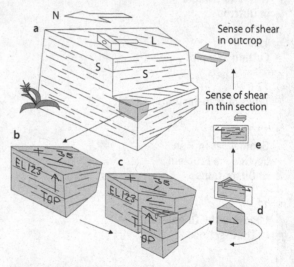

Fig. 12.1. Method to obtain an oriented sample from an outcrop and an oriented thin section from a sample. A sample for structural studies must be oriented, for example as shown in **a**. **b** A *strike-dip symbol* (here 160–35) on the planar top surface of the sample and a *cross* (marking the *top*) fix the orientation of the sample in space. When a thin section is to be cut parallel to the lineation (here 200/15), a chip is cut from the sample with a diamond saw. In order to orient the chip, an *arrow* with a single barb pointing in the direction of the top surface can be used and is drawn on both the sample and the chip. **c** Care should be taken that this *arrow* is copied correctly onto the thin section. **d** Alternatively, a small saw-cut scar can be made in the top surface of the chip, which can be found back in the thin section. **e** Shear sense determined in the thin section, e.g. by shear bands can now be directly related to the sample, and through the sample to the original outcrop. Here, the outcome is thrusting to the NE

der to determine which metamorphic conditions accompanied deformation in the zone, and which ones are older.

If shear sense in a shear zone is to be determined, care should be taken that samples are correctly oriented and that suitable small-scale structures such as porphyroclasts are present in the sample. In many ductile shear zones, fine-grained mylonites and samples representing highly strained parts of the shear zone give better results than samples of coarse-grained or low strain domains.

Different shear sense markers are preferentially found in different lithologies; C'-shear bands in strongly foliated mica-rich mylonite; C/S fabric mainly in granite mylonite; mantled porphyroclasts in deformed granites or pegmatites, and mica fish in quartzite-mylonites. Deformed quartzite should be sampled where possible to determine LPO of quartz; in many shear zones, this is the only suitable shear sense indicator.

If gradients in the style of a structure exist over an area it is useful to take a series of samples across this gradient in a single lithology. In all cases, it is useful to write down in the field for what purpose a sample was taken. Photographs of important sampling sites are also essential and should be taken before the sample is removed. GPS coordinates of sample sites should be obtained in all cases.

Structures like large folds and boudins are spectacular in the field and are consequently often sampled. However, they may be difficult to interpret on the scale of a thin section where only a small fragment of the structure is seen. In many cases it is more advantageous to sample also directly adjacent to such structures where the rock deformed more homogeneously. Care should also be taken to sample the more fine-grained parts of a structure; in coarse-grained rocks only a small (and possibly non-representative) part of the structure can be studied in thin section. For example, if a sample is taken of a mylonite with large feldspar porphyroclasts in a shear zone, and a thin section is made of the matrix between two large feldspar grains, this matrix may give a shear sense opposite to that of the main zone (cf. Bell and Johnson 1992).

One should also not forget to take samples of the 'ordinary matrix rock'; some geologists realise afterwards that they took samples only of the curious and exceptional. If a metamorphic gradient is present or suspected, it is useful to take samples at a regular distance across this gradient for possible later microprobe work.

Finally a remark about geological heritage. It is our opinion that sampling of beautiful prominent structures such as folds, boudins and shear zones destroys field evidence and potential teaching material, and should therefore be avoided; in most cases, it is possible to take similar samples from other less prominent parts of the same outcrop. The same applies to drill cores, which should preferably be taken in parts of the outcrop where they do not destroy the geometry of beautiful structures.

12.5
Cutting Samples

When oriented samples are cut to make thin sections, it is crucial to avoid loss of orientation data. Although several methods can be used, we suggest the following procedure. After a saw cut is made, the orientation of the sample should be copied on both sides of the saw cut in mirror image (Fig. 12.1); we propose a system in which an arrow with a single barb is drawn parallel to the lineation, with the barb indicating the top of the sample. The arrow should be copied from the rock chip that is used to make the thin section onto the glass as shown in Fig. 12.1. Be careful when preparing the rock chips for thin sections that the orientation is not lost or confused. Alternatively, a small saw-cut scar can be made in the top surface of the rock chip used to make the thin section. This scar can be found back in the thin section and reduces the chances of confusion in its orientation.

12.6
Types of Thin Sections

Standard thin sections consist of a basal glass plate of 0.8 to 1.9 mm thick, a slice of sample material, and a cover glass with a thickness of 0.2 mm. The parts of the thin section are assembled with glues with a fixed refractive index, usually between 1.5 and 1.6. The standard thickness of the sample material is 30 µm. The presence of a cover glass is important because the objective lenses of most microscopes have been constructed to give best results when a cover glass is present. Sections should therefore be studied with the cover glass towards the objective lens. Besides standard thin sections, a number of other sections are used for various purposes:

- *Large thin sections*, up to 20 cm long can be made for unusually coarse-grained rocks and large structures; they can be made in the same way as standard sections.
- *Polished sections* and thin sections without cover glass are used for microprobe work and for observation of opaque minerals in reflected light. They usually have a standard thickness of 30 µm and can also be used for normal microtectonic observations. If small scale, delicate structures are to be studied, it can be advantageous to use a polished rather than a standard thin section. Even better results are obtained with sections polished at both sides. Polished thin sections can also be used in other analytical equipment such as the microprobe and SEM, which has the advantage that the same feature can be observed by optical and other techniques (Chap. 10).
- *TEM sections* are prepared with a dissolvable resin such as lakeside, and without a cover glass. After a reconnaissance study of the structure by optical microscope, small rings are glued on the surface with araldite, and subsequently the thin section is emerged

in ethanol, which dissolves the lakeside and leaves the rock chip floating in the fluid; the metal rings remain attached to the chip since araldite does not dissolve in ethanol. The chip can now be broken into fragments to free the metal sample rings with the attached fragment of the rock chip. This fragment can be mounted with the metal ring in an ion-thinning device and subsequently be studied in the TEM (Sect. 10.2.5).

- *Ultrathin sections* with a thickness of 5 μm or even less are used to study minerals with a high birefringence such as calcite. Polishing of the surface is important in this case. Production of ultrathin sections is relatively difficult and requires some experience.

- *Ultrathick sections* are 100 μm thick and are polished on both sides. They are not mounted on a glass slide since they are thick enough to be handled without breaking. Because of their thickness, these sections show high order interference colours and are difficult to use for normal tectonic studies. However, they are suitable for microthermometry of fluid inclusions, since these remain intact in thick sections, but commonly fracture and leak in normal thin sections. Ultrathick sections are also useful to determine the orientation of planar and linear microstructures such as fluid inclusion walls, solid inclusions, grain and subgrain boundaries and axes of microfolds.

- *Thin sections of unconsolidated material* can be made in some cases. Samples can be taken using a small container that is pressed into the outcrop face, peeled of and closed with tape. The samples can even be oriented by measuring the orientation of the container. The sample should be impregnated with a raisin such as araldite before sectioning. Penetration of the impregnation fluid can be improved by applying a vacuum-technique (Stoffel 1976). If water-swelling clay minerals are present, samples can be cut with a diamond saw using isopropyl alcohol and should be impregnated with epoxy between cuts. Dehydrated kerosene can be used as a lubricant during polishing (Chester and Logan 1987).

12.7
Geometries in Thin Section – a Problem of Dimensions

Thin sections are cross-sections through complex three-dimensional structures and are therefore not always easily translated into their three-dimensional equivalents. There is a general tendency to simplify things in the sense that angles, sizes and shapes in thin section are directly assumed to be 'maximum values' in the three-dimensional shape, often with erroneous results. We therefore list some three-dimensional features and the limits within which the geometry of their two-dimensional sections will lie. Figure 12.2 demonstrates the importance of the preparation of several thin sections of different orientation in the case of complex or irregular geometries.

Grain diameter. The presence of grains of variable diameter in thin section does not mean that grains have variable diameter in three dimensions; even if small grains in an aggregate are equidimensional and have equal radius, the grain diameter in thin section will show both small and large grains, and all grain diameters are smaller than or equal to the true diameter (Fig. 12.2a). If the three-dimensional diameter is variable, the distribution of grain size in thin section will be even more irregular. If three-dimensional grain size has to be determined, e.g. the size of recrystallised grains in order to determine the differential stress (Sect. 9.6.2), a correction factor has to be used (Heilbronner and Bruhn 1998; Higgins 2000).

Grain shape. If grains are elliptical in thin section, the ellipticity will lie between the maximum and minimum elliptical ratios of the ellipsoidal shape in three dimensions. Even if all ellipses in the thin section have equal size and orientation, this does not necessarily mean that they represent the maximum ellipticity. An equidimensional shape of grains in thin section may occur if elongate grains are cut normal to their maximum elongation axis. The presence of a shape fabric in thin section may result from a planar or linear fabric in three dimensions. Planar structures are seen as lines in thin section and are therefore indistinguishable from linear structures. An interesting phenomenon that results from this relation may be observed in inclusions of porphyroblasts (Fig. 12.2b). If inclusions have a prolate ellipsoidal shape, as seems to be common in nature, some sections may show elongate aligned inclusions, and others equidimensional ones. The former porphyroblasts may be classified as inter- or syntectonic ones, and the latter erroneously as pretectonic (Fig. 7.9). This illustrates the need to investigate thin sections in at least two orthogonal sections if complex inclusion patterns are studied.

Ellipticity of tubes. Sections through elliptical tubes such as burrows or vermicular intergrowth in symplectites will have an ellipticity that is equal to or exceeding the ellipticity in a section normal to the tube (Fig. 12.2c). In sections parallel to the tube, the structure may resemble layering. Sheath folds (Sect. 5.3.2) can be considered as a special category of tubular structures (Fig. 12.2d); in sections parallel to the tube they resemble tight to isoclinal folds, while in other sections they appear as elliptical shapes.

Thickness of layering. Thickness of layering in thin section corresponds only to true thickness if the section is cut perpendicular to the layering (Fig. 12.2e); in other sections it exceeds true thickness.

Angle between planes and tightness of cylindrical folds. The angle between two planar elements in thin section can be greater or smaller than the true angle (Fig. 12.2f). This principle applies to the angle between a quartz grain

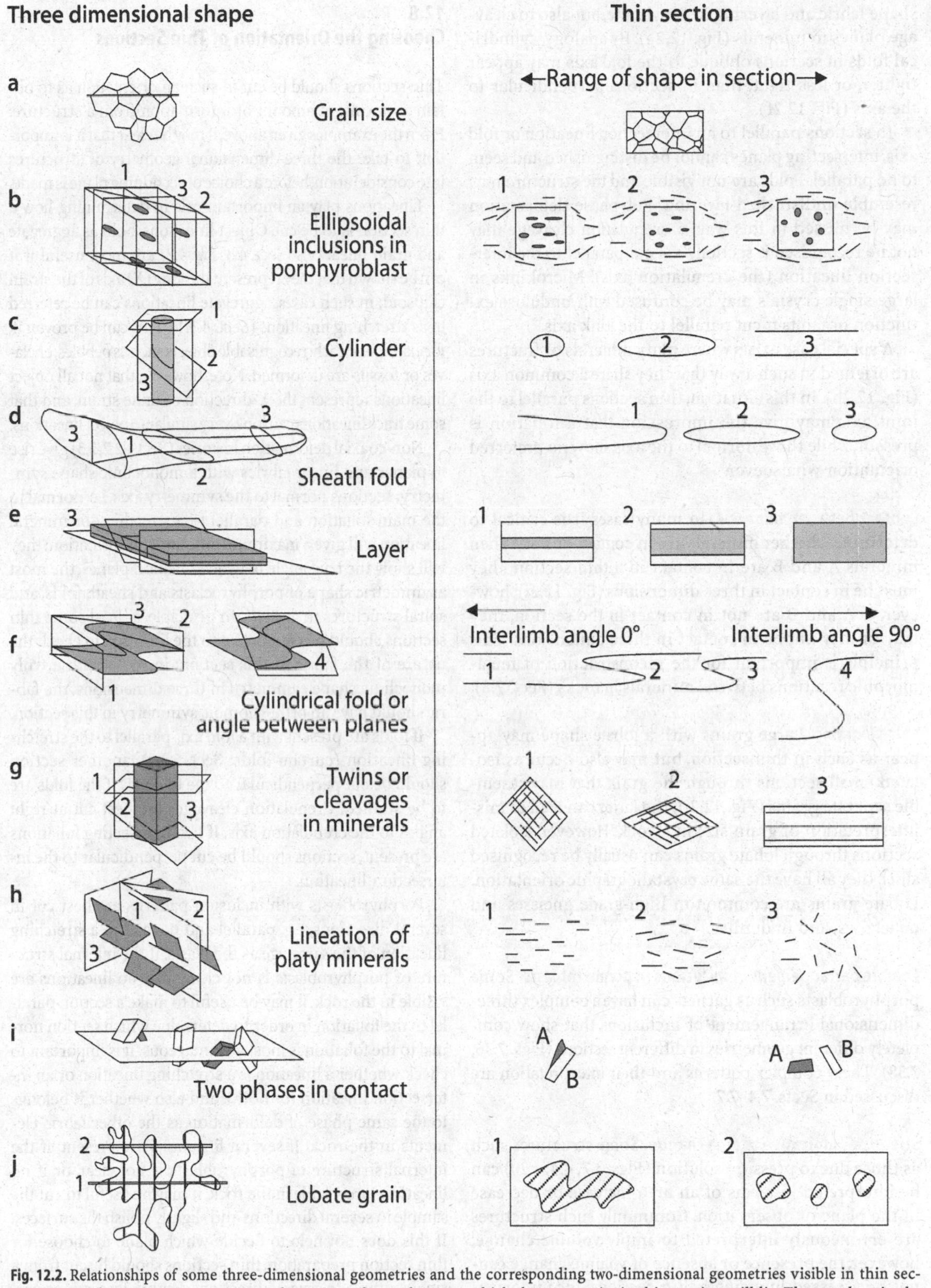

Three dimensional shape

Thin section

←Range of shape in section→

a   Grain size

b   Ellipsoidal inclusions in porphyroblast

c   Cylinder

d   Sheath fold

e   Layer

f   Cylindrical fold or angle between planes

Interlimb angle 0°   Interlimb angle 90°

g   Twins or cleavages in minerals

h   Lineation of platy minerals

i   Two phases in contact

j   Lobate grain

Fig. 12.2. Relationships of some three-dimensional geometries and the corresponding two-dimensional geometries visible in thin section. Two end-member geometries are shown in each case between which the geometries in thin section will lie. The *numbers* in the drawings refer to section planes. Further explanation in text

12.8

shape fabric and layering in a mylonite, but also to cleavage planes in minerals (Fig. 12.2g). By analogy, cylindrical folds in sections oblique to the fold axis may appear tighter, or less tight, than in sections perpendicular to the axis (Fig. 12.2f).

In sections parallel to an intersection lineation or fold axis, intersecting planes cannot be distinguished and seem to be parallel. Folds are not visible and the structure may resemble undisturbed layering. Polyphase deformation may be missed in this way; a crenulation cleavage may not be recognised if sections are cut parallel to the intersection lineation (the crenulation axis). Microkinks in large single crystals may be confused with undulose extinction or faults if cut parallel to the kink axis.

A special case exists where platy minerals or fractures are oriented in such a way that they share a common axis (Fig. 12.2h). In this situation, thin sections parallel to the joint axis may give the impression that a foliation is present, while those normal to the axis show no preferred orientation whatsoever.

Contact between minerals. In many cases it is critical to determine whether minerals are in contact or not. When minerals A and B are in contact in a thin section, they must be in contact in three dimensions (Fig. 12.2i); however, if A and B are not in contact in the section, they may or may not be in contact in three dimensions. This principle is important for the reconstruction of metamorphic reactions between minerals in rocks (Sect. 7.8).

Lobate grains. Large grains with a lobate shape may appear as such in thin section, but may also occur as isolated small sections through the grain that may resemble separate grains (Fig. 12.2j); the latter can lead to misinterpretation of grain size in a rock. However, isolated sections through lobate grains can usually be recognised since they all have the same crystallographic orientation. Lobate grains are common in high-grade gneisses and quartzites, and in dunites.

Complex arrangement of inclusions in porphyroblasts. Some porphyroblasts such as garnets can have a complex three-dimensional arrangement of inclusions that show completely different geometries in different sections (Figs. 7.36, 7.38). These complex patterns and their interpretation are discussed in Sects. 7.4–7.7.

Structures indicative of area change. Some structures such as those due to pressure solution (Figs. 3.7, 4.4, 4.20) can be interpreted in terms of an area increase or decrease in the plane of observation. Commonly such structures are erroneously interpreted to imply volume change; however, the presence or absence of volume change cannot be assessed in a single thin section, since information on the third dimension is lacking (Box 4.8).

12.8
Choosing the Orientation of Thin Sections

Thin sections should be cut in such an orientation as to obtain a maximum amount of information from a structure. From the examples given above, it may be clear that it is important to take the three-dimensional geometry of structures into consideration, before a choice of sectioning plane is made.

Lineations play an important role in determining how a thin section is to be cut. Object lineations, both as aggregate and grain lineations (Sect. 4.3, Fig. 4.2) are most useful if it can be shown that they represent the long (X) axis of the strain ellipsoid; in such cases, aggregate lineations can be referred to as stretching lineations (Sect. 4.3.1). This can be proven if, for example, clearly recognisable objects such as pebbles, enclaves or fossils are deformed. Note, however, that not all object lineations represent the X-direction of finite strain, and that some trace lineations can look very similar to object lineations.

Non-coaxial deformation histories (Sects. 2.7, 5.3) give rise to planar and linear fabrics with a monoclinic shape symmetry; sections normal to the symmetry axis, i.e. normal to the main foliation and parallel to a stretching or mineral lineation will give a maximum amount of information; they will show the true angle between foliation planes, the most asymmetric shape of porphyroclasts and sheath folds, and spiral structures in garnets. In principle, at least some thin sections should be cut normal to the lineation to check the nature of the fabric in this section; in rocks with a truly monoclinic shape symmetry in three dimensions, the fabric should have an orthorhombic symmetry in this section.

If folds are present with a fold axis parallel to the stretching lineation (curtain folds; Sect. 5.3.2), another section should be cut perpendicular to the fold axis if the folds are to be studied. Crenulation cleavages are best cut at right angles to the crenulation axis. If two intersecting foliations are present, sections should be cut perpendicular to the intersection lineation.

Porphyroblasts with inclusion patterns are best cut in several directions, e.g. parallel and normal to a stretching lineation of the same age as the blasts. If the internal structure of porphyroblasts is not clear, or if no lineations are visible in the rock, it may be useful to make a section parallel to the foliation in order to determine which section normal to the foliation is most advantageous. It is important to check whether a lineation is a stretching lineation or an intersection lineation (or both), and also whether it belongs to the same phase of deformation as the other fabric elements in the rock. If several lineations are present, if the internal structure of porphyroblasts is not clear, or if no lineations are visible in the rock, it may be useful to cut the sample in several directions and slightly polish the surfaces. If this does not help to decide which plane to choose for thin section preparation, thin sections should be cut from a single sample in several directions. Especially sections parallel to the foliation may help in such cases.

Glossary

This glossary provides definitions of the main terms used in this book. The definitions do not always cover the significance of all aspects of the term completely, but are meant as a short reference. Additional information on the complete meaning and application of each term can be found in the main text using the index.

δ-type object – *mantled porphyroclast* with thin asymmetric *wings* of mantle material fixed to opposite sides of the porphyroclast and stretching into the matrix. The shape resembles the Greek letter δ

φ-type object – *mantled porphyroclast* with symmetric *wings* (no *stair stepping* is developed)

Θ-type object – *mantled porphyroclast* without *wings*

σ-type object – *mantled porphyroclast* with a geometry resembling the Greek letter σ

acicular – needle-shaped (from the Latin diminutive of 'acus' – needle)

active foliation – foliation that consists of microshear zones or other 'active' planes. It is normally not parallel to the *XY*-plane of finite strain. Examples are *shear band cleavage* and some constrictional crenulation cleavages

advection – transport of something from one region to another, for example transport of heat, *vorticity* or material in solution

aggregate lineation – *object lineation* defined by parallel orientation of elongate aggregates of small grains, which differ from adjacent aggregates in composition, or orientation

aggregate shape preferred orientation – preferred orientation of domains composed of aggregates of small grains in a deformed rock. Also known as a *shape fabric*, either a *planar shape fabric* or *linear shape fabric*, or both. See also *grain shape preferred orientation*

A_k – kinematic dilatancy number

amoeboid fabric – fabric composed of crystals with strongly curved and lobate, interlocking grain boundaries, like an amoeba

anastomosing – term used to describe the shape of some foliations: braided, dividing the rock into lenses (Fig. 4.7)

anhedral crystal – crystal with irregular shape and lack of planar crystal faces

anisotropic – term indicating that a material property is not of equal magnitude in all directions; this may refer to birefringence of light, to growth rate, grain boundary energy, elastic strength, etc. (opposite of isotropic)

annealing – term from metallurgy used to indicate processes of *recovery* and *static recrystallisation* induced by passive heating of a previously deformed material. The term is also sometimes used for the interpretation of microstructures in rocks

annular fluid inclusion – fluid inclusions with a ring-shaped geometry (Fig. 10.17e). From Latin *annulus* – ring

anti-perthite – plagioclase with inclusions of K-feldspar, formed by exsolution of the K-feldspar solid solution in plagioclase

antitaxial fringe – *fringe* in which the growth surface lies between the *fringe* and the *core object*

antitaxial vein – vein with *fibres* growing along their contacts with the wall rock, i.e. along both outer surfaces of the vein (see also *syntaxial vein* and *ataxial vein*)

antithetic – term used for minor faults or *shear bands* with the opposite sense of shear as the major *shear zone* in which they develop

API-gravity – an arbitrary scale for the density of crude liquid petroleum devised jointly by the American Petroleum Institute (API) and the National Bureau of Standards. The scale is in degrees API. A high API gravity means lighter oils

ASPO – aggregate shape preferred orientation

ataxial vein – vein containing *fibres* that lack a localised growth surface

atoll structure – ring-shaped crystal in thin section surrounding a mineral aggregate similar to the matrix. Atoll structures of garnet are common in some micaschists and probably form as *corona* structures, followed by breakdown of the crystal around which they originally formed

atomic number contrast image – SEM-image generated using *backscattered electrons* where grey tones represent different mean atomic weights of elements in the mineral phases

augen – (German for eye) – lens-shaped crystal or crystal aggregate. The term is usually applied to feldspar augen (augengneiss). Most feldspar augen are probably *porphyroclasts* that developed from feldspar megacrysts in intrusive or high-grade metamorphic rocks by partial recrystallisation along the rim in response to deformation

AVA diagram – diagram showing the distribution of *lattice preferred orientation* over a volume of rock (from German 'Achsenverteilungsanalyse')

axial planar foliation – foliation that is approximately parallel to the axial plane of folds. This term is used as an alternative term for *secondary foliation*

backscatter electron mode – observation mode with a sample tilted at a high angle to the incident electron beam, so that electrons are scattered backwards towards a detector at a small angle to the incident beam

bedding vein – *striped shear vein* parallel to bedding in a sedimentary rock

blast – crystal in a metamorphic rock that has grown during metamorphism (see also *porphyroblast*)

blastomylonite – *mylonite* that underwent *static recrystallisation* of part of the fabric after deformation ceased. The term is also used for high-grade mylonite with a relatively coarse recrystallised matrix. Because of the genetic character of the term, its use in not recommended

BLG – recrystallisation – *bulging recrystallisation*

blocky grains – equidimensional crystals grown in dilatation sites

bridge – strip of wall rock connecting two sides of a vein, formed by fracture branching or impingement. Crevasses in glaciers may have bridges that can be used as such

BSE mode – backscattered electron mode

bulging recrystallisation (BLG recrystallisation) – process of local migration of a grain boundary into a neighbouring grain with a higher *dislocation density*, eventually producing new crystals. BLG recrystallisation occurs along the edge of older grains

Burgers vector – vector indicating the displacement of a crystal lattice associated with a *dislocation*

C/S fabric – fabric consisting of C-type *shear band cleavage* and S-planes. S-planes define a shape or mineral foliation that is cut by C-type shear band cleavage or C-planes. See also *Type I S-C mylonite* and *Type II S-C mylonite*

cataclasite – a rock composed of angular rock- and mineral fragments, thought to have formed principally by brittle fracturing without melting

cataclastic lineation – *lineation* in *cataclasite*, usually an *object lineation*

cathodoluminescence – luminescence shown by minerals when excited by a beam of electrons

central object – rigid grain or aggregate alongside of which *fringes* or *strain shadows* formed, commonly a pyrite or magnetite grain

chessboard subgrains – arrangement of approximately square *subgrains*, resembling a chessboard

chiastolite – andalusite with cross-shaped passive inclusion pattern of graphite

chlorite stack – aggregate of chlorite, commonly lens shaped, occurring in *microlithons* of many slates. Chlorite may be alternating with white mica. (001) planes of chlorite are commonly oblique to slaty cleavage. Chlorite stacks are probably pre- or early syntectonic *porphyroblasts*

CIP – *computer-integrated polarisation microscopy*

clast – (1) abbreviation for *porphyroclast*; (2) detrital grain in a sediment

cleavage – *secondary foliation* defined by a preferred orientation of inequant fabric elements. The name is usually restricted to fine-grained rocks such as slates and phyllites, but is also used by some authors as a general name for any *secondary foliation* except coarse layering

cleavage bundle – local area of concentrated cleavage development, probably related with local high strain

cleavage domain – layer or lens with a relatively high content of elongate grains (such as micas or amphiboles) and low content of equidimensional grains (such as quartz, feldspar or carbonate). Together with microlithons they make up a spaced foliation. Micas in cleavage domains commonly have a preferred orientation parallel or at a small angle to the domain

cleavage dome – dome-shaped structure, usually a graphite aggregate, attached to the crystal face of a *porphyroblast*. The structure is probably formed by porphyroblast growth

cleavage front – boundary between rocks without cleavage and cleavage bearing rocks of the same lithology in an orogenic belt

cleavage lamella – synonym for *cleavage domain* (plural: lamellae)

coaxial – term used for flow or progressive deformation with *principal finite strain axes* remaining parallel to *ISA*

Coble creep – solid state diffusion along grain boundaries

compaction band – type of *deformation band* in which fracturing led to volume decrease of the *gouge* or *cataclasite* with respect to the original rock

competent – descriptive but imprecise, interpretative term for highly viscous or relatively strong

complex object – *mantled porphyroclast* with more than one set of *wings*

composite vein – vein with growth both from the walls and in the centre, with two growth surfaces

compositional layering – non-genetic term for an alternation of layers with different lithology. This may be bedding, igneous layering or a layering induced by secondary differentiation processes. See also *differentiated layering*

computer-integrated polarisation microscopy – method to determine *LPO* using an adapted optical microscope and special software

constrained communition – the breaking of grains into ever finer fragments in a fault zone

constrictional strain – type of three-dimensional *strain* for which stretch < 1 along the intermediate strain axis. Constrictional strain produces a *prolate* (flattened cigar-like) strain ellipsoid in the absence of volume change

core object – rigid object in the centre of a strain shadow or strain fringe. Commonly a pyrite or magnetite crystal. See *central object*

core-and-mantle structure – deformed crystalline core, usually a single crystal, surrounded by a mantle of fine-grained material of the same mineral. The structure is thought to develop by preferential *dynamic recrystallisation* in the outer shell of a deforming large single crystal in response to *intracrystalline deformation*. Feldspar core-and-mantle structures are common in rocks subject to low to medium-grade deformation

corona – shell of a single or several minerals around a crystal of another composition. Usually formed by metamorphic reaction

CPL – abbreviation used in this book for crossed polarised light

crack-seal growth – growth of minerals along a fracture, thought to result from periodic fracturing and sealing by growth from a fluid

crenulation – other word for folds with a wavelength of 1 cm or less

crenulation cleavage – secondary spaced foliation with microhinges of an older crenulated foliation in the *microlithons* recognisable with the optical microscope

crinoid type – synonym for *syntaxial fringe*; these commonly form around crinoid stem fragments

critical melt fraction – percentage of melt at which the viscosity of a rock+melt system decreases abruptly by several orders of magnitude (usually reported to be around 30%)

critical resolved shear stress (CRSS) – property of a *slip system* in a crystal. The CRSS defines at which *shear stress*, resolved on the *slip plane*, a dislocation will start to move

crystalloblastic – synonym for *granoblastic*

crystallographic preferred orientation – other term for *lattice preferred orientation*

crystalplastic flow – permanent deformation by *intracrystalline deformation* mechanisms

C'-type shear band cleavage – specific type of *shear band cleavage* commonly developed in strongly foliated and mica-rich mylonites. C'-type shear bands are curved and anastomosing and inclined to the shear zone boundary; a synonym is extensional crenulation cleavage

C-type shear band cleavage – specific type of *shear band cleavage*, particularly common in mylonitised granitoid rocks. C-type shear bands or C-planes are relatively straight and lie parallel to the shear zone boundary. (C from French 'cisaillement' – shear)

curtain fold – type of approximately cylindrical fold common in mylonites with its fold axis subparallel to the *stretching lineation*. Curtain folds lack the closed tubular shape of *sheath folds*. An older name for this structure is oblique fold

decussate fabric – an arrangement of randomly oriented *elongate grains* (such as mica) in a metamorphic rock

deflection fold – structure around a *porphyroblast* in which S_e is deflected through isoclinal folding at both sides of the porphyroblast (Fig. 7.22b)

deflection plane – surface inside a *porphyroblast* where the inclusion pattern abruptly changes orientation without a break in continuity

deformation – change in shape and orientation of objects or volumes of rock from an initial to a final state. We use the term deformation in this general sense, while the word *strain* has a more restricted significance, that is, the change in shape of an object or part of the rock

deformation band – (1) planar narrow fault zone filled with *cataclasite* or *gouge* typically formed in porous sediments close to the surface; (2) dark band in a deformed grain visible under crossed polars, formed by a locally bent crystal lattice

deformation lamella (plural lamellae) – intracrystalline lamella of slightly different optical relief than the host grain, consisting of damaged crystal lattice or arrays of sub-microscopic inclusions

deformation mechanism map – a diagram showing the conditions of *stress* and *homologous temperature* for which specific deformation mechanisms are dominantly, although not exclusively, active; each map is only valid for a specific mineral and grain size

deformation partitioning – local subdivision of the deformation pattern into domains with different deformation parameters such as strain and volume change, or with different dominant active deformation mechanisms

deformation path – path of a volume of rock from the undeformed to the deformed state, as seen in an external reference frame. Also, sum of the displacement paths for a volume of rock

deformation phase – period of deformation during which a group of structures has formed, separated from other structures by overprinting criteria. Successive deformation phases may merge into each other or may be separated by time intervals with little or no deformation, during which metamorphic conditions and the stress field may have changed

deformation tensor – tensor describing *finite deformation*, including *strain* and rotation

deformation twin – twin formed by deformation, common in deformed carbonates and plagioclase feldspar

deviatoric stress – non-hydrostatic or lithostatic component of *stress* defined as $(\sigma_n - \sigma_{mean})$, where σ_n is *normal stress* on a surface and σ_{mean} is the mean stress

diagenetic foliation – foliation, usually parallel to bedding, thought to have formed by diagenetic compaction and dewatering

diagenetic strain – strain due to diagenetic compaction, usually resulting in a shortening normal to bedding and volume loss

differential stress – non-hydrostatic or lithostatic component of *stress*, usually defined as $(\sigma_1 - \sigma_3)$

differentiated layering – secondary, spaced foliation in which *cleavage domains* and *microlithons* are recognisable with the naked eye. Differentiated layering is inferred to have formed by chemical differentiation in an originally more weakly layered or even homogeneous material. In fabric descriptions the non-genetic term *compositional layering* should be used if the origin is not known

diffraction pattern – pattern of crossing lines obtained by diffraction of electrons on lattice planes as observed on a fluorescent screen

diffusion creep – deformation due to migration of *vacancies* through the crystal lattice

dilatancy – here: area increase during deformation of a plane (two dimensions). Sometimes (erroneously) used as a synonym for volume increase

dilatation band – type of *deformation band* in which fracturing led to volume increase of the *gouge* or *cataclasite* with respect to the original rock

disjunctive foliation – secondary spaced foliation without fold hinges in the *microlithons*. If microlithons are wide and continuous, the term *compositional layering* is used as a synonym

dislocation – line defect in a crystal

dislocation climb – movement of a dislocation out of its *slip plane*, normally by migration of *vacancies* to the dislocation site

dislocation creep – movement of *dislocations* in a crystal lattice by *dislocation glide* and climb

dislocation density – total length of all *dislocations* in a volume of material

dislocation glide – movement of *dislocations* in a crystal lattice without *dislocation climb*

displacement path – path traced by a particle in a deforming rock from the undeformed to the deformed state, as seen in an external reference frame

domain shape preferred orientation – fabric of elongate lenses, each of which has a strong *crystallographic preferred orientation* different from that in neighbouring lenses

domainal slaty cleavage – secondary *disjunctive foliation* with lens-shaped *microlithons* in fine-grained rocks

domino boudin – type of asymmetric boudinage with short stubby boudins

dragging microstructure – strong curvature of a boundary between two grains of mineral A where it joins the contact with a mineral B (Fig. 3.34). This structure is interpreted to have formed when the grain boundary migrated and was pinned to some extend at the contact with mineral B

DSPO – *domain shape preferred orientation*

dynamic recrystallisation – recrystallisation during *intracrystalline deformation*. There are three types: *BLG, SGR* or *GBM* recrystallisation

EBSD – *electron backscatter diffraction*

EBSD-pattern – orientation contrast image obtained by *forescattered* or *backscattered* electrons

ecc – abbreviation used for *extensional crenulation cleavage*

EDAX – energy dispersive X-ray analysis. Facility to measure the chemical composition of minerals in an electron microscope

elastic recoil detection analysis – method for quantitative chemical analysis of light elements using particles recoiled out of a sample when struck by an ion-beam

electron channelling OC-image – orientation contrast image obtained by *backscattered electron* mode

electron microprobe – microprobe for quantitative chemical analysis using an electron beam

elongate grains – elongate crystals with non-parallel boundaries grown in dilatation sites

en-echelon – French term meaning a parallel but obliquely dislocated disposition of planar elements, like tiles on a roof (from French 'échelon' – rung of a ladder)

epitaxy – preferential nucleation of a mineral on the lattice of another, with a fixed orientation relationship between the two crystal lattices

equant – said of a crystal having the same or nearly the same diameter in all directions

equigranular fabric – fabric in which all grains have roughly equal size

ERDA – *Elastic Recoil Detection Analysis*

euhedral crystal – crystal with well-developed crystal shape and crystal faces

extensional crenulation cleavage – synonym for *C'-type shear band cleavage*

external foliation – foliation outside and immediately adjacent to a *porphyroblast*. It is commonly referred to by the abbreviation S_e (e for external), and used in comparisons with the internal foliation (S_i)

fabric – the complete spatial and geometrical configuration of all those components that are contained in a rock (Hobbs et al. 1976) and that are penetratively and repeatedly developed throughout the volume of rock under consideration. This includes features such as *foliation, lineation, lattice-preferred orientation* and grain size. The term can also be defined as: the relative orientation of parts of a rock mass

fabric attractor (FA) – line or plane in space towards which material lines and some *fabric elements* rotate

fabric element – part of a fabric such as a *foliation*, *lineation* etc

fabric gradient – a gradual change in the geometry of a fabric in the field or in thin section

fabric skeleton – crest lines on a contoured plot of crystallographic axes in a stereogram, used to describe *lattice-preferred orientation* patterns

Fairbairn lamellae – other term for *deformation lamellae* (Groshong 1988)

false shear band – type of *flanking structure* resembling a *shear band*, but with *slip* and *lift* of opposite sense

feather vein – *fringed* vein segments consisting of a through-going vein and many oblique tapering side-veins

feather-edge structure – pattern of crystallographically determined inclusions resembling a feather edge

fibre – strongly elongate crystal with parallel boundaries grown in dilatation sites

fibre trajectory analysis – reconstruction of the *deformation path* in a volume of rock from the shape of *fibres* in veins or *fringes*

finite deformation – deformation accumulated over a finite period of time

finite displacement vector – vector connecting particles in the undeformed and deformed state for finite deformation

finite strain – strain accumulated over a finite period of time

flame-perthite – *perthite* with flame-shaped lamellae

flanking fold – type of *flanking structure* – folds restricted in occurrence to a domain alongside a fault or vein

flanking structure – domain alongside a fault or vein where foliation is of a different orientation to that further away from the fault or vein. Flanking structures include *flanking folds* and *shear bands*

flattening strain – type of three-dimensional strain for which stretch > 1 along the intermediate strain axis. Flattening strain produces an *oblate* (pancake-like) strain ellipsoid in the absence of volume change

flow – instantaneous movement of material particles making up a deforming volume of rock. If flow is *homogeneous*, the pattern of particle displacements can be described by a simple *velocity field*, or *flow pattern*. A velocity field describes the motion of a population of material particles at a given instant in time. A sequence of velocity fields, describing the accumulation of finite displacements of material particles with time, is a *deformation path* or *progressive deformation history*. Here we must clearly distinguish between the notions of ongoing progressive deformation and the end product of the progressive deformation (*finite deformation*, *finite strain*)

flow law – equation describing the dependence of *strain rate* on parameters such as *stress*, temperature and grain size

flow partitioning – local subdivision of a bulk *flow pattern* into domains with different flow parameters such as *vorticity* and volume change rate

flow pattern – pattern of velocity vectors defining distribution of flow over a volume of rock

flow tensor – tensor describing the velocity of particles at specific points in space

foam-structure – fabric of grain boundaries resembling a foam; it forms by *GBAR*

foliation – planar fabric element that occurs penetratively on a mesoscopic scale in a rock. *Primary foliation* includes bedding and igneous layering; *secondary foliations* are formed by deformation induced processes. Joints are not normally considered as foliations since they are not penetrative on a mesoscopic scale

foliation intersection/inflection axes (FIA) – imaginary axes in a *porphyroblast* formed by curvature or intersection of S_i-planes

foliation track – track of a foliation on an outcrop surface

forescatter electron (FSE) OC image – *orientation contrast* image obtained by *forescattered electron mode*

forescattered electron mode – observation mode with a sample tilted at a small angle to the incident electron beam, so that electrons are scattered forwards towards a detector at a large angle to the incident beam

fracture cleavage – alternative and outdated term for *disjunctive cleavage*. The use of this term is discouraged because of its (commonly erroneous) genetic implication

fringe – body of fibrous crystalline material formed adjacent to a relatively rigid *core object* during *progressive deformation*

fringe structure – structure composed of a *core object* and adjacent *strain fringes*

FSE mode – *forescattered electron mode*

garben – (German for 'stack') – sheafs of elongate minerals such as hornblende, usually arranged along the foliation plane

gauge – (or gage) instrument capable of providing quantitative data. In this book the term 'natural microgauge' refers to a microstructure that can be used to obtain quantitative data from a rock. Don't confuse with *gouge*

GBAR – *grain boundary area reduction*

GBM recrystallisation – *high-temperature grain boundary migration recrystallisation*

generation surface – other term for *main fault vein* – approximately planar vein filled with *pseudotachylyte* and thought to be the site of a fault where pseudotachylyte was generated by frictional heating

geometric softening – softening of a mineral aggregate by development of a *lattice-preferred orientation*

ghost structure – outline of an overgrown mineral in a porphyroblast visible as a local relative scarceness of inclusions

gneissic layering – *compositional layering* in a gneiss

gneissosity – general term for foliation in a gneiss. Use of this term is discouraged because of its vague connotation; several types of foliation (layering, *schistosity*) may occur in the same gneiss

gouge – non-consolidated fractured rock, commonly very fine-grained, formed by brittle deformation at a shallow crustal level along a fault. Don't confuse with *gauge*

grain – equivalent of crystal in metamorphic rocks. In sedimentary rocks a detrital grain may contain several crystals

grain boundary – planar domain of distorted crystalline material between two crystal lattices of different orientation and/or nature

grain boundary area reduction – *grain boundary migration* leading to reduction in the total surface area of *grain boundaries* in an aggregate. The process operates spontaneously in response to the decrease in internal free energy that a grain aggregate gains by decreasing the area of (high-energy) grain boundaries; it leads to straight grain boundaries and large grains

grain boundary migration – the migration of a *grain boundary* through a solid crystalline material

grain boundary misorientation map – map of grain boundaries and the amount of misorientation across them

grain boundary sliding – deformation mechanism where grains slide along grain boundaries. This mechanism is limited to fine-grained rocks and/or relatively high temperature

grain lineation – *object lineation* defined by parallel arrangement of elongate single crystals

grain shape fabric – see *grain shape preferred orientation*

grain shape preferred orientation – preferred orientation of inequidimensional grains in a deformed rock. Also known as a *grain shape fabric*

granoblastic fabric – fabric dominantly formed by equidimensional crystals. An example is a *foam structure*, formed by GBAR in which platy or *acicular* mineral shapes are absent, and most grains have equant shape

growth competition – competition of growing crystals to fill open space beyond the growth surface

growth inclusion – small mineral grain included in a *porphyroblast*, usually parallel or orthogonal to crystal faces. Such inclusions are thought to have formed during growth of the porphyroblast and do not represent part of the matrix that was overgrown by the porphyroblast, as for passive inclusions

growth twin – twin formed during growth of the crystal in which it is contained; this in contrast to a *deformation twin*, which forms during deformation of the crystal

GSPO – *grain shape preferred orientation*

habit – crystal shape, specifically referring to the relative development of individual crystal faces (e.g. prismatic habit; tabular habit)

hardening – increasing resistance to deformation, expressed as increasing *differential stress* at constant *strain rate*, or decreasing strain rate at constant differential stress

helicitic fold – pattern of inclusions in a *porphyroblast* resembling a fold and thought to have formed by overgrowth of a pre-existing fold in the matrix

Hertzian fracture – cone shaped fracture, such as radiating from a stress concentration point at an indentation site

high-temperature grain boundary migration (GBM) recrystallisation – recrystallisation by migrating grain boundaries throughout old grains in response to differences in dislocation density between two grains

homologous temperature – fraction of the melting temperature for a mineral phase in Kelvin. A homologous temperature of 0.7 indicates $7/10^{th}$ of the melting temperature. This is an example of a dimensionless number

hydrolithic weakening – softening of a material due to the presence or introduction of water

hydrostatic pressure – pressure due to the weight of a column of water acting equally in all directions

hyperbolic distribution method – method to determine the *kinematic vorticity number* from the geometry and orientation of mantled porphyroclasts in a mylonite

hypidiomorphic – synonym for *subhedral*, mainly used in igneous rocks

ICPMS – *Inductively Coupled Plasma Mass Spectrometry*

idiomorphic – synonym for *euhedral*, mainly used in igneous rocks

igneous foliation – *primary foliation* defined by a preferred orientation of phenocrysts or other fabric elements in a tectonically undeformed igneous rock

impingement microcrack – small fracture in a grain formed due to local stress concentration where the grain is in contact with another grain or object

inclusion – small crystal included in a larger one of another composition. Inclusions can be subdivided into passive, growth and exsolution inclusions

inclusion band – irregularly shaped surface in a vein composed of fluid and solid inclusions, oriented parallel to the vein-wall rock contact and reflecting the shape of that contact

inclusion trail – surface in a vein composed of fluid and solid inclusions, oriented oblique to the vein-wall rock contact and usually connecting *jogs* in *inclusion bands*

incremental deformation – imaginary, infinitely small deformation

incremental displacement vector – vector connecting particles in the undeformed and deformed state for incremental deformation

inductively coupled plasma mass spectrometry – method of mass spectrometry using ions that are produced by dissolving material and then ionising it with the aid of a high temperature plasma

inequigranular fabric – a fabric showing inhomogeneous distribution of grain size, e.g. bimodal with large grains of approximately equal size in a fine-grained equigranular matrix

injection vein – wedge-shaped or branching vein filled with *pseudotachylyte* and usually connected to one or more *main fault veins* (*generation surfaces*)

instantaneous stretching axes (ISA) – three imaginary orthogonal axes in space, along two of which *stretching rates* are minimal and maximal (known as the shortening and extensional ISA respectively). The third axis is called the intermediate instantaneous stretching axis. ISA exist for any homogeneous flow

intergranular – affecting more than one grain

interlobate fabric – fabric composed of crystals with irregular, lobate grain boundaries

internal foliation – foliation defined by the preferred orientation of passive inclusions in a *porphyroblast*, thought to mimic the orientation and geometry of a foliation that was overgrown by the porphyroblast. The abbreviation S_i (i for internal) is commonly used for internal foliation. S_i is often used in comparison with the *external foliation* S_e

interphase boundary – grain boundary between two different phases, of different chemistry or crystal structure

interstitial – point defect in a crystal lattice; an additional lattice element in between regular lattice units

intracrystalline deformation – deformation by movement of *vacancies* or *dislocations* in the crystal lattice

intragranular – inside a single grain

ISA – instantaneous stretching axes

isochore – a line on a PT diagram showing the variation of the pressure of a fluid with temperature, when the volume of the fluid is kept constant

isotropic – term indicating that a material property is of equal magnitude in all directions; this may refer to birefringence of light, to growth rate, grain boundary energy, elastic strength, etc. (opposite of anisotropic)

jog – step in a planar structure such as a fault

kelyphitic structure – a type of *symplectitic* reaction rim bordering olivine; also used for symplectitic amphibole-plagioclase *coronas* (from Greek 'κελυφος' – pod)

kinematic dilatancy number – (A_k) measure of the *dilation* rate of flow, normalised against strain rate; dimensionless number

kinematic vorticity number – (W_k) measure of the *vorticity* of flow, normalised against *strain rate*; dimensionless number

LA-ICPMS – *Laser Ablation Inductively Coupled Plasma Mass Spectrometry*

laser ablation inductively coupled plasma mass spectrometry (LA-ICPMS) – use of a laser to ablate material from a sample to be analysed by plasma mass spectrometry

lattice-preferred orientation (LPO) – statistical preferred orientation of the crystal lattices of a population of crystals in a rock. In older texts also referred to as *texture*, crystallographic fabric or crystallographic preferred orientation

left-over grain – one of a group of small grains with identical orientation that lie within or at the edges of another large grain (Fig. 3.34). They are interpreted as relics of a large single old grain that is incompletely overgrown by a neighbour

lepidoblastic fabric – planar fabric defined by the preferred orientation of tabular or platy crystals (from Greek 'λεπις' – scale)

lift – measure of deformation in a *flanking structure*. Elevation of a marker plane above its cut-off point against a vein or fracture, measured normal to the straight far-field part of the marker plane (Fig. 5.46a)

line defect – dislocation

linear shape fabric – fabric defined by the elongate shape of grains or grain aggregates. A type of *shape preferred orientation*

lineation – linear fabric element that occurs penetratively on a mesoscopic scale in a rock. *Striations* and *fibres* are not normally considered to be lineations since they do not occur penetratively on a mesoscopic scale

lineation track – track of a *lineation* on an outcrop surface, visible when a lineation is cut obliquely by the surface

lithostatic pressure – pressure at a point in the Earth due to the weight of the overlying column of rock. Lithostatic pressure at a point is uniform in all directions by definition

loop bedding – soft sediment deformation structure resembling an isoclinal fold

LPO – *lattice-preferred orientation*

LS-tectonite – deformed rock containing both planar and linear shape fabric elements

L-tectonite – deformed rock with a *linear shape fabric*; no *foliation* is present

magmatic flow – flow by displacement of melt, with consequent rigid body rotation of crystals but without *crystalplastic* deformation

main fault vein – approximately planar vein filled with *pseudotachylyte* and thought to be the site of a fault where pseudotachylyte was generated by frictional heating

mantled porphyroclast – *porphyroclast* with an elongated mantle of the same mineral composition, stretched in the direction of the foliation. It is inferred to have formed at the expense of the porphyroclast by recrystallisation in its rim. Mantled porphyroclasts are common in *mylonites* and are used to determine sense of shear. See also *δ-type object*, *σ-type object* and *strain shadow*

material line – line that consists of material particles: as opposed to (imaginary) space lines such as coordinate axes

matrix – 1. method of notation for components of a tensor; 2. fine-grained ground mass in a rock

M-domain – alternative term for *cleavage lamella* in a micaschist (M stands for mica-rich, in contrast to *Q(uartz)-domain*

mean stress – hydrostatic or lithostatic component of *stress*, of equal magnitude in any direction

median line/surface – line (surface) dividing a vein into parts that grew in opposite directions

metamorphic cycle – cycle of changing pressure and temperature experienced by a volume of rock

metamorphic event – episode of metamorphism, characterised by changes in mineral assemblage in a volume of rock

mica-fish or mica fish – lensoid mica grain common in mylonites. A type of *mineral fish*. Their asymmetry can be used to determine sense of shear

microfault induced microcracks – other term for *microscopic feather fractures*

microgauge – see *gauge*

microlithon – layer or lens with a relatively small degree of preferred orientation as compared to *cleavage domains*. A crenulated older foliation may be present in microlithons. Together with cleavage domains, microlithons make up a *spaced foliation*

microscopic feather fractures (mff) – small wedge-shaped tensional veins in fault zones, branching from and oblique to Y-shears or fault planes (see *Riedel shear*)

millipede structure – inclusion pattern in a *porphyroblast* resembling an Australian millipede (*Bellaria* sp.)

mimetic growth – growth of minerals controlled by pre-existing grain arrangements (Vernon 1976)

mineral fish – elongate lozenge- or lens-shaped single crystal (*porphyroclast*), usually oblique to the *mylonitic foli-ation*, embedded in a finer grained matrix in a *mylonite*

mineral lineation – *grain lineation* composed of crystals of the same mineral with an *acicular* crystal habit such as amphiboles, pyroxenes, sillimanite and tourmaline

moat – monomineralic *corona*

monocrystalline ribbon – ribbon in a metamorphic rock composed of a single crystal, usually formed by deformation followed by growth

mortar structure – *porphyroclast* surrounded by an fine-grained aggregate of the same mineral. Mortar structure is similar to *core-and-mantle* structure but the new grains are smaller. A mortar structure is not necessarily formed by recrystallisation, but possibly by cataclasis or a combination of both. The use of this term is discouraged because of its (commonly erroneous) genetic implication of "mechanically crushed rock"

mylonite – strongly deformed rock from a ductile shear zone, commonly with a planar foliation and usually with a *stretching lineation*. Evidence for high strain such as *quartz ribbons* and *porphyroclasts* in a more fine grained matrix are common. Mylonite was originally (Lapworth 1885) defined as a brittle fault rock, but mylonites are now thought to have formed predominantly by crystal-plastic flow of the matrix, although some minerals suspended in the matrix may show brittle fracturing. The word derives from the Greek 'μυλων', a mill

mylonitic foliation – type of *secondary foliation* developed in *mylonites*

myrmekite – symplectite of quartz and plagioclase, commonly adjacent to K-feldspar, indicating replacement

Nabarro-Herring creep – solid state diffusion through the crystal lattice

nematoblastic fabric – linear fabric defined by preferred orientation of *acicular* or prismatic crystals (from Greek 'νημα' – thread)

new grain – recrystallised grain or grain of a new mineral formed by metamorphic reaction

Newtonian flow – flow in which strain rate depends linearly on differential stress

non-coaxial – term used for *flow* or *progressive deformation* in which *material lines* that were initially parallel to *ISA* rotate to another orientation

non-Newtonian flow – *flow* in which *strain rate* depends on *differential stress* through a non-linear relation (e.g. stress to the power of a constant n)

normal stress – component of *stress* acting on a plane (a vector), normal to that plane

NRA – *Nuclear Reaction Analysis*

nuclear microprobe – microprobe for quantitative chemical analysis using a proton beam

nuclear reaction analysis – method for accurate quantitative chemical analysis using proton-induced secondary reaction products

object lineation – lineation formed by preferred orientation of elongate monocrystalline or polycrystalline objects

object-centre path – inferred path in space that a *central object* must have followed to produce the shape of an adjacent *fringe*. The centre path is defined for a single fringe and is determined in a reference frame fixed to this fringe

oblique fabric or oblique foliation – shape- preferred orientation oblique to mylonitic- or other foliation. Oblique fabrics are a type of *grain shape preferred orientation* that commonly develops in response to *non-coaxial* flow

oblique fold – old name for *curtain fold*

oblique-S$_i$ porphyroblast – *porphyroblast* with a straight *inclusion pattern* at an angle with the foliation in the matrix

ODF – orientation distribution function. Function describing the three-dimensional orientation of crystals in a rock

opening trajectory – the inferred trajectory in space that an imaginary point on the wall of a vein has traced during the process of vein opening, in a reference frame fixed to one of the vein walls or the *median line*

ophitic microstructure – common igneous fabric that consists of elongate plagioclase grains which are completely surrounded by pyroxene grains. This microstructure is found in basaltic and gabbroic rocks

oppositely concave microfolds (OCMs) – structure at both ends of a *'millipede'* structure, where the *external foliation* bends inwards towards a *porphyroblast* (Fig. 7.22). See also *deflection fold*

orientation contrast (OC) – *SEM*-image generated using *backscattered electrons* where grey tones represent different orientation of material in the sample

oscillatory zoning – growth zoning in undeformed crystals, common in *phenocrysts* in igneous rocks

Ostwald ripening – growth of few grains to a larger size at the cost of their neighbours. The process is similar to *GBAR* but more commonly used to describe a porous material with a pore fluid or melt between grains

overprinting – the partial erasure of a feature in the rock fabric by a younger feature

palaeopiezometer – a *microgauge* used to measure *differential stress*

palaeostress – *stress* value reached in a rock at some time in geological history

particle size distribution (PSD) – presentation of the number of grains in groups of specific grain diameter in an aggregate

partitioning – see *flow partitioning* and *deformation partitioning*

passive foliation – foliation that is rotating passively towards the *fabric attractor*. Passive foliations are subparallel to the *XY*-plane of finite strain. Examples are most types of continuous foliations. See also *active foliation*

passive inclusion – mineral grain included in a *porphyroblast* or larger grain in the rock matrix. Such inclusions are thought to represent a fragment of the matrix which was overgrown by the porphyroblast and passively included in its crystal lattice without significant displacement or rotation

P-domain – Alternative term for cleavage- or *M(mica)-domain* in rocks with spaced cleavage. P stands for phyllosilicates

pencil cleavage – term used for a structure where two intersecting foliations are of equal grade of development, causing the rock to fracture into pencil-shaped fragments upon weathering. The term is mostly used for diagenetic cleavage intersected at a high angle by tectonic cleavage (Fig. B.4.5)

penetrative fabric element – a fabric element that occurs penetratively throughout a rock. A *foliation* is an example of a penetrative fabric element. We use the word penetrative to mean: at the scale of observation. In thin section this means: down to the scale of individual mineral grains

perthite – K-feldspar with inclusions of plagioclase, formed by exsolution of the albite solid solution in K-feldspar

phenocryst – large, commonly *euhedral* single crystal in an igneous rock or pseudotachylyte, thought to have formed by growth from a melt

phonon – quantized mode of vibration in a crystal lattice

phyllonite – micaceous *mylonite*, sometimes (erroneously) used for *ultramylonite* (from Greek 'φυλλον' – leaf)

PIGE – *Proton Induced Gamma-ray Emission*

pinning microstructure – strongly indented boundary between two grains of a mineral A where it is attached to a small grain of mineral B (Fig. 3.34). The structure is thought to be caused by pinning of a migrating grain boundary on the grain of B.

pinning of dislocations – immobilisation of parts of *dislocations* because they are fixed to or unable to pass certain structures in a material

PIXE – *Proton Induced X-ray Emission*

pixel – smallest element of an image on a computer screen

planar shape fabric – planar fabric defined by the shape of deformed flattened grains or grain aggregates (see also *linear shape fabric*). A type of shape preferred orientation

plane strain – type of three-dimensional strain for which $Y/Z = X/Y$. In the absence of volume change this means that $Y = 1$

platelet lineation – *lineation* defined by planar minerals such as micas which share a common axis

platy quartz – monocrystalline *quartz ribbon* with few or no intracrystalline deformation structures, common in high-grade gneisses

poikiloblast – *poikiloblastic crystal*

poikiloblastic crystal – crystal containing a large volume fraction of passive inclusions; such crystals have a 'spongy' aspect (from Greek 'ποικιλος' – various)

point defect – defect in a crystal lattice where a lattice element is missing (*vacancy*) or additional (*interstitial*)

polygonal arc – arc-shaped distribution of non-deformed mica grains, probably formed by *static recrystallisation* of a folded mica foliation

polygonal fabric – fabric composed of crystals with straight grain boundaries, commonly with *anhedral* or *subhedral* crystals

pore fluid factor – ratio of the pore fluid pressure to the vertical or lithostatic stress

porphyroblast – (with b, sometimes abbreviated as 'blast') single crystal of a diameter exceeding the surrounding matrix and inferred to have grown in a solidified rock in response to changes in metamorphic conditions. The word derives from Greek: Porphyro – refers to the large crystals of feldspar in a porphyry, from Greek 'πορφυριτες λιθος' – purple stone; blast comes from Greek 'βλαστος' – bud

porphyroclast – (with c, sometimes abbreviated as 'clast') single crystal of a size exceeding the mean grain size in the surrounding matrix and inferred to represent a remnant of an originally coarse grained rock. Porphyroclasts are common in *mylonites*. The word derives from Greek porphyry (see *porphyroblast*) and 'κλαστος', shattered, from 'κλαν' – to break

porphyroclast aspect ratio method – method to determine the *kinematic vorticity number* from the geometry of mantled porphyroclasts in a *mylonite*

power-law flow – special type of *non-Newtonian flow*

PPL – abbreviation used in this book for plain polarised light

ppm – parts per million

pressure fringe – alternative, but genetic term for *strain fringe*

pressure shadow – alternative, but genetic term for *strain shadow*

pressure solution – localised dissolution of material induced by enhanced solubility of solids in response to the presence of a *differential stress* field

primary foliation – foliation that was present in a sedimentary or igneous rock before deformation. Primary foliation includes bedding and igneous layering

principal strain axes – orthogonal axes along which *shear strain* is zero. Along two of these axes, *stretch* values are maximal and minimal. The third axis is called the intermediate principal strain axis

principal strain values – values of *stretch* along the principal strain axes

principal stress axes – orthogonal axes normal to planes on which *shear stress* is zero. Along two of these axes normal stress is maximal and minimal. The third axis is called the intermediate principal stress axis

principal stress values – values of *normal stress* along principal stress axes

principal stretches – values of *stretch* along principal strain axes. Also known as principal strain values

process zone – domain in front of a propagating fault where microcracks form in grains, then link to a network, then separate to form brittle fault rock

progressive deformation – process of the accumulation of deformation with time

protomylonite – weakly to moderately deformed rock in a shear zone, transitional between the undeformed wall rock and a *mylonite*. Also a mylonite with 10–50% matrix

proton induced gamma-ray emission – method for accurate quantitative chemical analysis using proton induced gamma-rays

proton induced X-ray emission – method for accurate quantitative chemical analysis using proton induced X-rays

pseudomorph – shape of a crystal aggregate or crystal, inherited from an older crystal or object which has been replaced

pseudotachylyte – (or pseudotachylite) dark brittle fault rock occurring in veins and fractures in host rocks with low porosity. Pseudotachylyte is thought to form by local melting of a host rock along a fault in response to seismic activity on the fault and associated local generation of frictional heat. The name was derived from 'tachylyte' (basalt glass) occurring in settings that could not be explained by igneous activity

pure shear flow – a special type of coaxial flow, usually defined as having no *stretch* along the intermediate principal ISA

pyrite-type fringe – synonym for *antitaxial fringe*. Such fringes commonly develop around pyrite grains

Q-domain – alternative term for quartz-rich *microlithon*. See also *M-domain*

quarter structure – asymmetric microstructure around a *porphyroclast* with similar geometry in opposite quadrants. These structures can be used as a *shear sense indicator*

quartz ribbon – highly elongated disc- or lens-shaped crystal or aggregate of quartz, common in *mylonites* and high-grade rocks. In thin section the crystals are ribbon-shaped, hence the name. Quartz ribbons form by flattening of originally equidimensional quartz grains, or possibly by migration of grain boundaries to form single large grains from more fine-grained parent aggregates. They may exhibit undulose extinction, or be recrystallised into polycrystalline ribbons

reaction softening – softening induced by the growth of new minerals which are more easily deformable than minerals of the host rock

recoil time-of-flight spectrometry – as *ERDA*, measuring the time it takes particles to reach the detector

recovery – general name for the processes in a crystal or crystal aggregate that lead to a decrease in the combined length of included *dislocations*, and rearrangement of dislocations into *subgrain boundaries*. The process of recovery is driven by a decrease in the internal free energy of a crystal or crystal aggregate. Processes of *dynamic recrystallisation* and *GBAR* are not included in recovery as described in this text. Recovery is also active during the process of *static recrystallisation*

recrystallisation – rearrangement of crystalline matter to a modified set of crystals by migration and modification of grain boundaries. Recrystallisation does not necessarily involve chemical changes. It usually involves a decrease or increase in the crystal size

re-entrant zone – zone of passive inclusions along the ribs of a euhedral *porphyroblast*. An example are the cross-shaped re-entrant zones in chiastolite

reservoir zone – volume of rock containing *pseudotachylyte* injection veins rheology – study of the deformation and flow of matter, more specifically the mechanics of flow

ribbon – highly elongated lens-shaped single crystal or monomineralic aggregate

ribbon mylonite – strongly foliated *mylonite* mainly composed of parallel monophase lenses. Common in high grade shear zones. Transitional to striped gneiss

Riedel shears – sets of subsidiary faults in brittle fault zones, subdivided into groups with different shape, shear sense and orientation such as R, Y and P shears. Named after clay deformation experiments by W. Riedel

rock analogues – crystalline materials used in deformation experiments as analogues for rocks

S-C fabric – other way to write *C/S fabric*, which is the original nomenclature

scanning electron microscope – electron microscope where an image is produced by emission or scattering of electrons from the surface of a sample

schistosity – *secondary foliation* defined by preferred orientation of inequant fabric elements in a medium to coarse-grained rock. Individual foliation-defining elements (e.g. micas) are visible with the naked eye. Sometimes used as a general term for any secondary foliation

screw dislocation – *dislocation* with a *Burgers vector* parallel to the dislocation line

secondary electron – electron that has been detached from atoms in a sample in an electron microscope or similar device

secondary electron mode – mode of observation in an *SEM* using secondary electrons that have been generated in the sample

secondary foliation – *foliation* developed in response to deformation and/or metamorphic processes in a rock in the solid state

secondary ionization mass spectrometry – method of mass spectrometry using ions that are produced by bombardment of a sample with a primary ion beam

sector zoning – preferential incorporation of passive inclusions in specific zones of a crystal

SEM – *scanning electron microscope*

separatrix – imaginary surface in inhomogeneous *flow*, separating different *flow patterns*

seriate fabric – a fabric showing complete grain size gradation of fine to coarse

SGR – *recrystallisation* – *subgrain rotation recrystallisation*

shape preferred orientation (SPO) – preferred orientation of elongate or planar aggregates or grains in a volume of deformed rock. There are *grain-*, *aggregate-* and *domain shape preferred orientations*

shear band – minor shear zone with *lift* and *slip* of the same sense

shear band boudinage – type of asymmetric boudinage with elongate boudins separated by *shear bands*

shear band cleavage – structure with shear bands, with a similar aspect as *crenulation cleavage*. The older foliation is apparently extended and not shortened as in crenulation cleavage. Shear band cleavages are common in *mylonites* and can be divided into two types; *C-type* and *C'-type* (the latter is also known as *extensional crenulation cleavage*)

shear sense indicator – structure with a monoclinic symmetry that can be used to find the sense of shear in a *mylonitic* or *protomylonitic* rock

shear strain (γ) – (1) tangent of the angle by which a line A, originally orthogonal to a line B, rotated with respect to B in finite deformation; (2) - tangent of the angle by which a line initial normal to the boundary of a shear zone rotated with respect to that boundary in finite deformation. Strain in shear zones is commonly expressed as shear strain

shear stress – component of *stress* acting on a plane (a vector), parallel to that plane

shear vein – vein formed by opening with a major shear component

shear zone – planar zone of relatively intense deformation in which progressive deformation is *non-coaxial*

sheath fold – strongly non-cylindrical fold in the form of a sheath, usually oriented parallel to a stretching lineation. Sheath folds are common in shear zones and especially in *mylonites*

SHRIMP – (1) type of seafood; (2) sensitive high resolution ion microprobe – *SIMS* type instrument that can analyse small parts of individual grains. Used to date parts of zoned or inhomogeneous mineral grains

sigmoid – lozenge-shaped polycrystalline lens in strongly deformed rock with a shape similar to that of a *σ-type* mantled porphyroclast. A sigmoid is an aggregate of small grains without a central clast

simple shear flow – a type of *non-coaxial* plane strain *flow* in which a plane exists (the flow plane) that does not deform. This plane lies at 45° to the shortening and extensional *ISA*, and contains the intermediate ISA

SIMS – *Secondary Ionization Mass Spectrometry*

slaty cleavage – *secondary foliation*, either continuous or spaced in fine-grained rocks (slates and phyllites). Spacing up to 0.01 mm is included in the definition of slaty cleavage

slickenfibre vein – vein composed of *fibres* subparallel to the vein wall, deposited on *slickensides*

slickenfibres – fibrous grains along a fault surface or *slickenside*, parallel to the fault and usually parallel to the direction of latest movement along the fault

slickenlines – scratches or linear markings on a *slickenside* that indicate slip direction along the slickenside. Also known as *striae* (singular *stria*)

slickenside – smoothed or polished fault surface

slickolite – surface in a rock, commonly rich in insoluble material and of irregular shape, which differs from a *stylolite* in that teeth are oblique to the surface. Slickolites have a shear component and grade into *slickensides* and *stylolites*

slip – displacement along a fault or crystallographic plane

slip plane – crystallographic plane along which a *dislocation* can move

slip system – combination of crystallographic plane and direction in which a *dislocation* can move, defined by the dislocation and its *Burgers vector*

snowball garnet – as *spiral Si-garnet* but with apparent rotation angle $> 180°$

softening – decreasing resistance to deformation, expressed as decreasing *differential stress* at constant *strain rate*, or increasing strain rate at constant differential stress

solution transfer – displacement of matter through an aqueous solution in a rock. This process is usually associated with *pressure solution* and precipitation

spaced foliation – *secondary foliation* containing *microlithons* and *cleavage lamellae*

spin – rotational component of *flow*, more specifically the angular velocity of instantaneous stretching axes in an external reference frame

spiral S$_i$-garnet – garnet with a spiral-shaped inclusion pattern (S$_i$) in thin section. The three-dimensional shape may be more complex. The angle of apparent relative rotation between S$_i$ in the centre of the garnet and S$_e$ is less than 180°

SPO – *shape preferred orientation*

stacking fault – strip of misfitted crystal lattice between two partial *dislocations* in a crystal

stair-stepping – two planes show stair-stepping if they are parallel to each other but offset across a *porphyroclast*. Stair-stepping is common in recrystallised wings around porphyroclasts and can be used to determine *sense of shear*

static recrystallisation – general term for *recovery* and *grain boundary migration* processes, driven by remaining *dislocations* and a large surface of grain boundaries, mainly after deformation. It involves *GBAR* and minor *SGR*- and *GBM* recrystallisation and recovery, and leads to removal of undulose extinction, straightening of grain boundaries and grain growth

steady-state fabric – fabric in a deforming rock which does not change essentially with further deformation. *Oblique foliation* of elongate dynamically recrystallised grains is an example of a steady state fabric

S-tectonite – deformed rock characterised by a planar fabric (no lineation is present)

strain – tensorial quantity describing change in shape; a strained situation is commonly represented as an ellipsoid, comparing with an unstrained situation represented by a sphere. Three principal stretches along the axes of the strain ellipsoid define the magnitude of three-dimensional strain. Strain is a more restricted term than deformation, which also includes rotational and translational components

strain cap – strongly foliated domain adjacent to a rigid object, usually enriched in mica or insoluble minerals. Strain caps generally occur together with *strain shadows*; the former lie in the shortening direction and the latter in the extension direction around the object

strain ellipse – representation of strain in a plane. A circle with radius 1 deforms into a strain ellipse

strain ellipsoid – representation of strain in three dimensions. A sphere with radius 1 deforms into a strain ellipsoid. The symmetry axes of the ellipsoid are the principal strain axes

strain fringe – type of *strain shadow* containing fibrous material precipitated adjacent to a stiff or rigid object. The fringe is usually composed of another mineral than the rigid object

strain hardening – hardening of a rock with increasing strain; see *hardening*

strain rate – *strain* per time unit

strain shadow – cone-shaped domain adjacent to a *porphyroclast* or *porphyroblast* in the direction of the foliation and usually composed of another mineral than the porphyroclast or -blast. It forms by rearrangement of material in response to inhomogeneous deformation of the matrix adjacent to the porphyroclast or -blast. Strain shadows are usually massive or contain equidimensional crystals; if they contain *fibres*, the term strain fringe is used instead. A mantled porphyroclast differs from a strain shadow in that the mantle has the same mineral composition as the porphyroclast and is inferred to have formed at the expense of the porphyroclast by recrystallisation

strain softening – softening of a rock with increasing strain; see *softening*

strain-free – descriptive term for an optically undeformed-looking crystal lattice, i.e. without *undulose extinction* and *subgrains*. A strain free grain may be undeformed, or have lost intracrystalline deformation features by *recovery* or *recrystallisation*

stress – tensorial quantity with six independent variables describing the orientation and magnitude of force vectors acting on planes of any orientation at a specific point in a volume of rock

stress corrosion cracking – propagation of a fracture aided by chemical reactions at the fracture tip

stretch – change in length of a line: new length, divided by original length

stretching lineation – *object lineation* defined by aggregates or deformed single grains inferred to have formed by stretching grains or grain aggregates

stretching rate – *stretch* per time unit

striae (singular stria) – alternative term for *slickenlines*

striation – linear stripes or scratches on a fault plane, formed by movement on the fault

striped gneiss – gneiss with planar compositional layering interpreted as a *mylonite*, formed at high metamorphic grade

striped shear vein – *shear vein* with an internal layering and in many cases a linear fabric subparallel to the vein walls. The linear fabric is usually due to corrugated *inclusion trails*, not *fibres*

structure – geometrically distinct feature in a rock; if penetratively developed, it is known as a fabric element

stylolite – surface in a rock, commonly rich in insoluble material and of irregular shape, which formed by localised removal of material by pressure solution. In 3D the surface can carry many narrow teeth normal or oblique to the surface which gave the structure its name. From Latin *stylus* – pen, and Greek λιθος – stone

subgrain – volume of crystalline material surrounded by *subgrain boundaries*

subgrain boundary (or subgrain wall) – planar array of *dislocations* separating two volumes of crystalline material with the same composition but with slightly misoriented crystal lattices (usually less than 5°)

subgrain rotation recrystallisation – (SGR-recrystallisation) dynamic recrystallisation through rotation of *subgrains* by addition of dislocations. Subgrains rotate with respect to each other and develop into new grains

subgrain wall – *subgrain boundary*

subhedral crystal – crystal with irregular crystal form but with some well developed crystal faces (see also *anhedral*, *euhedral* and *hypidiomorphic*)

submagmatic flow – deformation involving flow of melt and crystals, assisted by *crystalplastic deformation*

subophitic microstructure – common igneous fabric that consists of elongate plagioclase grains which are partly surrounded by pyroxene grains and partly in contact with other plagioclase grains. This microstructure is found in basaltic and gabbroic rocks

superfault – normal fault with exceptionally strong development of fault rocks, thought to have formed by a catastrophic event such as caldera collapse or meteorite impact

superplastic deformation – deformation in which very high strains are reached without development of elongated grains or a lattice-preferred orientation

suture – surface separating two parts of a *strain fringe* with different orientation of *fibres*

symplectite – lamellar or *vermicular* intergrowth of at least two minerals, usually produced by metamorphic replacement

syntaxial fringe – *fringe* in with the growth surface lies between the fringe and the matrix (see also *crinoid type fringe*)

syntaxial vein – vein with *fibre* growth from the walls towards the *median line*

syntaxy – type of *epitaxy* where the crystal lattices of the overgrown phase and the new phase are parallel

synthetic – term used for minor faults or shear bands with the same sense of displacement as the major shear zone in which they develop

Taber growth – growth of *fibres* in a vein due to *advection* of material through a porous wall rock. Vein walls can be pushed apart by the growing fibres

Taber vein – fibrous vein formed by *Taber growth*

tectonic event – period of deformation recognisable over a large area, distinct and separable from earlier and later events. Tectonic events may correspond to one or more *deformation phases*

tectonic strain – finite strain of a volume of rock accumulated after diagenesis

TEM – *transmission electron microscope*

tension gash – vein formed by dilatation

textural sector zoning – pattern of inclusions in a *porphyroblast* controlled by crystallographic directions

texture – synonym for microfabric or microstructure in most geological literature. In the non-geological literature on metals and ceramics the term is used for *lattice-preferred orientation*. Because of this conflicting use, this term has been largely avoided in the text (see Box 1.1)

T-fracture – type of subsidiary tensional fracture in a fault zone. Open T-fractures are known as *microscopic feather fractures*

thermal ionization mass spectrometry – method of mass spectrometry using ions that are produced by heating and ionising sample material

tiling – structure of imbricate large grains, arranged as tiles on a roof; this structure may occur in igneous rocks with phenocrysts and in *mylonites* with *porphyroclasts*, and can be used to determine *sense of shear*

tilt wall – type of *subgrain boundary* that consists of an array of *edge dislocations* with the same *Burgers vector*

TIMS – *Thermal Ionization Mass Spectrometry*

total strain – complete *finite strain* undergone by a volume of rock from its origin as a sediment or igneous rock, including *diagenetic strain*

trace lineation – *lineation* formed by parallel arrangement of linear features without volume, such as fold axes or intersection lines

tracking – the ability of the boundaries of *fibres* or *elongate grains* to follow the opening path of a vein during growth

transgranular fracture – relatively planar fracture crossing several grains

transposition – erasure of older fabric elements by strong deformation and/or metamorphic processes

transmission electron microscope – electron microscope which produces an image from electrons that have passed through a very thin sample

truncation plane – surface inside a *porphyroblast* where an inclusion pattern is interrupted and truncated against an inclusion pattern with another orientation

Tuttle lamellae – other term for healed microfractures (Groshong 1988)

twist wall – type of *subgrain boundary* that consists of two intersecting sets of *screw dislocations* with different *Burgers vectors*

Type I crossed girdle – preferred orientation pattern of crystallographic *axes* in a stereogram defined by two small circles connected by a central girdle (Fig. 4.40b)

Type I S-C mylonite – other term for *C/S fabric* as found in mylonitised granites (Lister and Snoke 1984)

Type II crossed girdle – preferred orientation pattern of crystallographic *axes* in a stereogram resembling the letter X (Fig. 4.40b)

Type II S-C mylonite – term mainly used for *stair stepping* wings around mica fish in quartzite *mylonite*, and interpreted as a type of *C/S fabric* (Lister and Snoke 1984)

ultramylonite – extremely fine-grained *mylonite* or mylonite with 90–100 vol-% matrix and 0–10 vol-% *porphyroclasts*

undulose extinction – irregular extinction of a single crystal under crossed polars due to a distorted crystal lattice with a high concentration of defects. Undulose extinction should not be confused with zoning

unswept core – part of a grain that has not been swept by a grain boundary during grain boundary migration

vacancy – *point defect* in a crystal lattice; a missing lattice element in between regular lattice units

vermicular – worm-shaped

vorticity – rotational component of *flow*, measured as the mean angular velocity of material lines with respect to *ISA*

window structure – type of grain boundary microstructure in which a grain boundary between two grains of a mineral A bulges between two grains of a mineral B (Fig. 3.34). The structure is inferred to form by *grain boundary migration*

wing crack – small curved fracture, usually filled with vein material and with a tapering horn-shape

winged object – 1. bird; 2. plane; 3. mantled porphyroclast in which the mantle has been deformed into wings

wings – term used for the appendages present on both sides of porphyroclasts trending parallel to the foliation. Sometimes referred to as "tails"

W_k – kinematic vorticity number of flow

W_m – mean kinematic vorticity number for a deformation path

X-, Y- and Z-axes – *principal strain axes*. X is the maximum, Y the intermediate and Z the minimum stretch

X-ray tomography – technique to investigate the internal structure of a material using the absorption of X-rays in the sample in different directions

yield stress – *differential stress* value, above which a material starts to deform permanently. Below the yield stress deformation is elastic. Since stress and elasticity are tensors, the yield stress is not a single number in most materials

Z contrast image – *atomic number contrast image*

Zener pinning – pinning of grain boundaries on small solid inclusions, thus inhibiting grain growth

References

Abbassi MR, Mancktelow NS (1992) Single layer buckle folding in non-linear materials I. Experimental study of fold development from an isolated initial perturbation. J Struct Geol 14:85–104

Aerden DGAM (1994) Kinematics of orogenic collapse in the Variscan Pyrenees deduced from microstructures in porphyroblastic rocks from the Lys-Caillaouas massif. Tectonophysics 238:139–160

Aerden DGAM (1995) Porphyroblast non-rotation during crustal extension in the Variscan Lys-Caillaouas Massif, Pyrenees. J Struct Geol 17:709–725

Aerden DGAM (1996) The pyrite-type strain fringes from Lourdes (France): indicators of Alpine thrust kinematics in the Pyrenees. J Struct Geol 18:75–91

Allen AR (1979) Mechanism of frictional fusion in fault zones. J Struct Geol 1:231–243

Allen FM, Smith BK, Busek PR (1987) Direct observation of dissociated dislocations in garnet. Science 238:1695–1697

Allison I, LaTour TE (1977) Brittle deformation of hornblende in a mylonite: a direct geometrical analogue of ductile deformation by translation gliding. Can J Earth Sci 14:1953–1958

Allison I, Barnett RL, Kerrich R (1979) Superplastic flow and changes in crystal chemistry of feldspars. Tectonophysics 53:41–46

Alsop GI, Holdsworth RE (2004) The geometry and topology of natural sheath folds: a new tool for structural analysis. J Struct Geol 26:1561–1589

Altenberger U, Wilhelm S (2000) Ductile deformation of K-feldspar in dry eclogite facies shear zones in the Bergen Arcs, Norway. Tectonophysics 320:107–121

Alvarez W, Engelder T, Lowrie W (1976) Formation of spaced cleavage and folds in brittle limestone by dissolution. Geology 4:698–701

Ando J, Fujino K, Takeshita T (1993) Dislocation microstructures in naturally deformed silicate garnets. Phys Earth Planet In 80:105–116

Andrews LM, Railsback LB (1997) Controls on stylolite development; morphologic, lithologic, and temporal evidence from bedding-parallel and transverse stylolites from the U.S. Appalachians. J Geol 105:59–73

Angelier J (1984) Tectonic analysis of fault slip data sets. J Geophys Res 89:5835–5848

Angelier J (1989) From orientation to magnitudes in palaeostress determinations using fault slip data. J Struct Geol 11:37–50

Antonellini MA, Aydin A, Pollard DA (1994) Microstructure of deformation bands in porous sandstones at Arches National Park, Utah. J Struct Geol 16:941–959

Arbaret L, Diot H, Bouchez JL (1996) Shape fabrics of particles in low concentration suspensions; 2D analogue experiments and application to tiling in magma. J Struct Geol 18:941–950

Arbaret L, Mancktelow NS, Burg JP (2001) Effect of shape and orientation on rigid particle rotation and matrix deformation in simple shear flow. J Struct Geol 23:113–125

Argles TW, Platt JP (1999) Stepped fibres in sillimanite-bearing veins: valid shear-sense indicators in high grade rocks? J Struct Geol 21:153–159

Arzi AA (1978) Critical phenomena in the rheology of partially-molten rocks. Tectonophysics 44:1953–1958

Ashworth JR (1980) Deformation mechanisms in mildly shocked chondritic diopside. Meteoritics 15:105–115

Ashworth JR (1985) Transmission electron microscopy of L-group chondrites, 1. Natural shock effects. Earth Planet Sc Lett 73:17–32

Atkinson BK (1982) Subcritical crack propagation in rocks; theory, experimental results and application. J Struct Geol 4:41–56

Atkinson BK (1984) Subcritical crack growth in geological materials. J Geophys Res 89:4077–4114

Augustithitis SS (1973) Atlas of the textural patterns of granites, gneisses and associated rock types. Elsevier, Amsterdam

Augustsson C, Bahlburg H (2003) Cathodoluminescence spectra of detrital quartz as provenance indicators for Paleozoic metasediments in southern Andean Patagonia. J S Am Earth Sci 16:15–26

Austrheim H, Boundy TM (1994) Pseudotachylytes generated during seismic faulting and eclogitization of the deep crust. Science 265:82–83

Austrheim H, Erambert M, Boundy TM (1996) Garnets recording deep crustal earthquakes. Earth Planet Sc Lett 139:223–238

Avé Lallemant HG (1978) Deformation of diopside and websterite. Tectonophysics 48:1–27

Avé Lallemant HG, Carter NL (1970) Syntectonic recrystallization of olivine and modes of flow in the upper mantle. Bull Geol Soc Am 81:2203–2220

Aydin A (1978) Small faults formed as deformation bands in sandstone. Pure Appl Geophys 116:913–930

Aydin A (2000) Fractures, faults, and hydrocarbon entrapment, migration and flow. Mar Petrol Geol 17:797–814

Aydin A, Johnson AM (1978) Development of fault zones as zones of deformation bands and as slip surfaces in sandstone. Pure Appl Geophys 116:931–942

Aydin A, Johnson AM (1983) Analysis of faulting in porous sandstones. J Struct Geol 5:19–31

Azor A, Fernando Simancas J, Exposito I, Lodeiro FG, Martinez Poyatos DJ (1997) Deformation of garnets in a low-grade shear zone. J Struct Geol 19:1137–1148

Babaie HA, LaTour TE (1994) Semibrittle and cataclastic deformation of hornblende-quartz rocks in a ductile shear zone. Tectonophysics 229:19–30

Babaie HA, LaTour TE (1998) Semi-brittle deformation and cataclastic flow of hornblende-quartz rock in a ductile shear zone. In: Snoke A, Tullis J, Todd VR (eds) Fault related rocks – a photographic atlas. Princeton University Press, New Jersey, pp 180–183

Baily JE, Hirsch PB (1962) The recrystallization process in some polycrystalline metals. Proceedings of the Royal Society of London A 267:11–30

Bak J, Korstgard J, Sorensen K (1975) A major shear zone within the Nagssugtoqidian of West Greenland. Tectonophysics 27:191–209

Barbarand J, Pagel M (2001) Cathodoluminescence study of apatite crystals. Am Miner 86:473–484

Barber DJ, Wenk HR (1991) Dauphiné twinning in deformed quartzites: implications of an in-situ TEM study of the α-β-phase transformation. Phys Chem Miner 17:492–502

Barber DJ, Wenk HR (2001) Slip and dislocation behaviour in dolomite. Eur J Mineral 13:221–243

Barbin V, Schvoerer M (1997) Cathodoluminescence and geosciences. Comptes Rendus de l'Académie des Sciences – Series IIA – Earth Planet Sc Lett 325:157–169

Bard JP (1986) Microtextures of igneous and metamorphic rocks. Reidel, Dordrecht

Barker AJ (1990) Metamorphic textures and microstructures. Blackie, Glasgow

Barker AJ (1994) Interpretation of porphyroblast inclusion trails: limitations imposed by growth kinetics and strain rates. J Metam Geol 12:681–694

Barker AJ (1998) Introduction to metamorphic textures and microstructures. Second edition. Blackie, Glasgow

Barnhoorn A, Bystricky M, Burlini L, Kunze K (2004) The role of recrystallisation on the deformation behaviour of calcite rocks: Large strain torsion experiments on Carrara marble. J Struct Geol 26:885–903

Barr TD, Houseman GA (1996) Deformation fields around a fault embedded in a non-linear ductile medium. Geophys J Int 125:473–490

Bartozzi M, Boyle AP, Prior DJ (2000) Automated grain boundary detection and classification in orientation contrast images. J Struct Geol 22:1569–1579

Bascou J, Barruol G, Vauchez A, Mainprice D, Egydio-Silva M (2001) EBSD-measured lattice preferred orientations and seismic properties of eclogites. Tectonophysics 342:61–80

Bascou J, Tomassi A, Mainprice D (2002) Plastic deformation and development of clinopyroxene lattice preferred orientations in eclogites. J Struct Geol 24:1357–1368

Bauer P, Palm S, Handy MR (2000a) Strain localization and fluid pathways in mylonite: inferences from in situ deformation of a water-bearing quartz analogue (norcamphor). Tectonophysics 320:141–165

Bauer P, Rosenberg C, Handy MR (2000b) See-through deformation experiments on brittle viscous norcamphor at controlled temperature, strain rate and applied confining pressure. J Struct Geol 22:281–289

Beach A (1975) The geometry of en-echelon vein arrays. Tectonophysics 28:245–263

Beach A (1979) Pressure solution as a metamorphic process in deformed terrigenous sedimentary rocks. Lithos 12:51–58

Beam EC, Fisher DM (1999) An estimate of kinematic vorticity from rotated elongate porphyroblasts. J Struct Geol 21:1553–1559

Beaumont-Smith CJ (2001) The role of conjugate crenulation cleavage in the development of 'millipede' microstructures. J Struct Geol 23:973–978

Becker JK, Koehn D, Walte N, Jessell M, Bons PD, Passchier CW, Evans L (2003) Numerical simulation of disequilibrium structures in solid-melt systems during grain growth. J Virtual Explorer 11

Behr HJ (1965) Zur Methodik tektonischer Forschung im kristallinen Grundgebirge. Ber Geol Ges DDR Gesamtgeb Geol Wiss 10:163–179

Behrmann JH (1983) Microstructure and fabric transition in calcite tectonites from the Sierra Alhamilla (Spain). Geol Rdsch 72:605–618

Behrmann JH (1987) A precautionary note on shear bands as kinematic indicators. J Struct Geol 9:659–666

Behrmann JH, Mainprice D (1987) Deformation mechanisms in a high-temperature quartz feldspar mylonite: evidence for superplastic flow in the lower continental crust. Tectonophysics 140:297–305

Behrmann JH, Platt JP (1982) Sense of nappe emplacement from quartz c-axis fabrics. Earth Plan Sci Lett 59:208–215

Bell TH (1981) Foliation development – the contribution, geometry and significance of progressive bulk inhomogeneous shortening. Tectonophysics 75:273–296

Bell TH (1985) Deformation partitioning and porphyroblast rotation in metamorphic rocks: a radical reinterpretation. J Metam Geol 3:109–118

Bell TH (1986) Foliation development and refraction in metamorphic rocks: reactivation of earlier foliations and decrenulation due to shifting patterns of deformation partitioning. J Metam Geol 4:421–444

Bell TH (1998) Recrystallization of biotite by subgrain rotation. In: Snoke A, Tullis J, Todd VR (eds) Fault related rocks – a photographic atlas. Princeton University Press, New Jersey, pp 272–273

Bell TH, Chen A (2002) The development of spiral-shaped inclusion trails during multiple metamorphism and folding. J Metam Geol 20:397–412

Bell TH, Cuff C (1989) Dissolution, solution transfer, diffusion versus fluid flow and volume loss during deformation/metamorphism. J Metam Geol 7:425–447

Bell TH, Etheridge MA (1973) Microstructure of mylonites and their descriptive terminology. Lithos 6:337–348

Bell TH, Etheridge MA (1976) The deformation and recrystallization of quartz in a mylonite zone, Central Australia. Tectonophysics 32:235–267

Bell TH, Forde A (1995) On the significance of foliation patterns preserved around folds by mineral growth. Tectonophysics 246:171–181

Bell TH, Hayward N (1991) Episodic metamorphic reactions during orogenesis: the control of deformation partitioning on reaction sites and duration. J Metam Geol 9:619–640

Bell TH, Hickey KA (1997) Distribution of pre-folding linear indicators of movement direction around the Spring Hill Synform, Vermont: significance for mechanism of folding in this portion of the Appalchians. Tectonophysics 274:275–294

Bell TH, Hickey KA (1999) Complex microstructures preserved in rocks with a simple matrix: significance for deformation and metamorphic processes. J Metam Geol 17:521–536

Bell TH, Johnson SE (1989) Porphyroblast intrusion trails: the key to orogenesis. J Metam Geol 3:109–118

Bell TH, Johnson SE (1990) Rotation of relatively large rigid objects during ductile deformation: well established fact or intuitive prejudice? Austr Journ Earth Sci 37:441–446

Bell TH, Johnson SE (1992) Shear sense: a new approach that resolves conflicts between criteria in metamorphic rocks. J Metam Geol 10:99–124

Bell TH, Mares VM (1999) Correlating deformation and metamorphism around orogenic arcs. Am Mineralogist 84:1727–1740

Bell TH, Rubenach MJ (1980) Crenulation cleavage development – evidence for progressive bulk inhomogeneous shortening from "millipede" microstructures in the Robertson River Metamorphics. Tectonophysics 68:T9–T15

Bell TH, Rubenach MJ (1983) Sequential porphyroblast growth and crenulation cleavage development during progressive deformation. Tectonophysics 92:171–194

Bell TH, Rubenach MJ, Fleming PD (1986a) Porphyroblast nucleation, growth and dissolution in regional metamorphic rocks as a function of deformation partitioning during foliation development. J Metam Geol 4:37–67

Bell IA, Wilson CJL, McLaren AC, Etheridge MA (1986b) Kinks in mica: role of dislocations and (001) cleavage. Tectonophysics 127:49–65

Bell TH, Forde A, Hayward N (1992a) Do smoothly curved, spiral-shaped inclusion trails signify porphyroblast rotation? Geology 20:59–62

Bell TH, Forde A, Hayward N (1992b) Reply on 'Do smoothly curved, spiral-shaped inclusion trails signify porphyroblast rotation?'. Geology 20:1055–1056

Bell TH, Johnson SE, Davis B, Forde A, Hayward N, Wilkins C (1992c) Porphyroblast inclusion-trail orientation data: eppure non son girate. J Metam Geol 10:295–300

Bell TH, Hickey KA, Wang J (1997) Spiral and staircase inclusion trail axes within garnet and staurolite porphyroblasts from schists of the Bolton Syncline, Connecticut: timing of porphyroblast growth and the effect of fold development. J Metam Geol 15:467–478

Bell TH, Hickey KA, Upton GJG (1998) Distinguishing and correlating multiple phases of metamorphism across a multiply deformed region using the axes of spiral, staircase and sigmoidal inclusion trails in garnet. J Metam Geol 16:767–794

Bell TH, Ham AP, Hickey KA (2003) Early formed regional antiforms and synforms that fold younger matrix schistosities: their effect on sites of mineral growth. Tectonophysics 367:253–278

Benn K, Allard B (1989) Preferred mineral orientations related to magmatic flow in ophiolite layered gabbros. J Petrol 30:925–946

Beny C, Guilhaumou N, Touray JC (1982) Native sulphur-bearing fluid inclusions in the CO_2-H_2S-H_2O-S system – microthermometry and Raman microprobe (MOLE) analysis – thermochemical interpretations. Chem Geol 37:113–127

Berger A, Stünitz H (1996) Deformation mechanisms and reaction of hornblende: examples from the Bergell tonalite (Central Alps). Tectonophysics 257:149–174

Berlenbach JW, Roering C (1992) Sheath-fold-like structures in pseudotachylytes. J Struct Geol 14:847–856

Berthé D, Choukroune P, Jegouzo P (1979a) Orthogneiss, mylonite and non-coaxial deformation of granites: the example of the South Armoricain shear zone. J Struct Geol 1:31–42

Berthé D, Choukroune P, Gapais D (1979b) Orientations préférentielles du quartz et orthogneissification progressive en régime cisaillant: l'exemple du cisaillement sud-armoricain. Bull Minéral 102:265–272

Best MG (1982) Igneous and metamorphic petrology. Freeman, New York

Bestmann M, Prior DJ (2003) Intragranular dynamic recrystallisation in naturally deformed calcite marble: a case study by means of misorientation analysis. J Struct Geol 25:1597–1613

Bestmann M, Kunze K, Matthews A (2000) Evolution of a calcite marble shear zone complex on Thassos Island, Greece: microstructural and textural fabrics and their kinematic significance. J Struct Geol 22:1789–1807

Bestmann M, Prior DJ, Veltkamp KTA (2004) Development of single-crystal sigma-shaped quartz porphyroclasts by dissolution-precipitation creep in a calcite marble shear zone. J Struct Geol 26:869–883

Beutner EC (1978) Slaty cleavage and related strain in Martinsburg slate, Delaware Water Gap, New Jersey. Am J Sci 278:1–23

Beutner EC (1980) Slaty cleavage unrelated to tectonic dewatering: the Siamo and Michigamme slates revisited. Bull Geol Soc Am 91:171–178

Beutner EC, Charles EG (1985) Large volume loss during cleavage formation, Habburg sequence, Pennsylvania. Geology 13:803–805

Beutner EC, Diegel FA (1985) Determination of fold kinematics from syntectonic fibers in pressure shadows, Marinsburg Slate, New Jersey. Am J Sci 285:16–50

Beutner EC, Fisher DM, Kirkpatrick JL (1988) Kinematics of deformation at a thrust fault ramp from syntectonic fibers in pressure shadows. In: Mitra G, Wojtal S (eds) Geometries and mechanisms of thrusting, with special reference to the Appalachians. Geol Soc Am Special Pap 222, pp 77–88

Bhagat SS, Marshak S (1990) Microlithon alteration associated with development of solution cleavage in argillaceous limestone: textural, trace-elemental and stable-isotopic observations. J Struct Geol 12:165–176

Biermann C (1979) Investigations into the development of microstructures in amphibole-bearing rocks from the Seve Köli nappe complex. PhD Thesis, Leiden State Univ

Biermann C (1981) (100) deformation twins in naturally deformed amphiboles. Nature 292:821–823

Biermann C, van Roermund HLM (1983) Defect structures in naturally deformed clinoamphiboles – a TEM study. Tectonophysics 95:267–278

Biermeier C, Stüwe K, Barr TD (2001) The rotation rate of cylindrical objects during simple shear. J Struct Geol 23:765–776

Bishop RR (1996) Grain boundary migration in experimentally deformed quartz aggregates: the relationship between dynamically recrystallized grain size and steady state flow stress. B Sc thesis, Brown University, pp 36

Bjørnerud MG, Magloughlin JF (2004) Pressure-related feedback processes in the generation of pseudotachylytes. J Struct Geol 26:2317–2323

Bjørnerud MG, Zhang H (1994) Rotation of porphyroclasts in non-coaxial deformation: Insights from computer simulations. J Metamorphic Geol 12:135–139

Bjørnerud MG, Zhang H (1995) Flow mixing, object-matrix coherence, mantle growth and the development of porphyroclast tails. J Struct Geol 17:1347–1350

Blacic JD, Christie JM (1984) Plasticity and hydrolytic weakening of quartz single crystals. J Geophys Res 89:4223–4239

Blenkinsop J (1991a) Correlation of paleotectonic fracture and microfracture orientations in cores with seismic anisotropy at Cajon Pass drill hole, southern California. J Geophys Res 95:11143–11150

Blenkinsop TG (1991b) Cataclasis and processes of particle size reduction. Pure Appl Geophys 136:59–86

Blenkinsop TG (2000) Deformation microstructures and mechanisms in minerals and rocks. Kluwer Academic Publishers, Dordrecht, 150 pp

Blenkinsop TG, Drury MR (1988) Stress estimates and fault history from quartz microstructures. J Struct Geol 10:673–684

Blenkinsop TG, Rutter EH (1986) Cataclastic deformation of quartzite in the Moine thrust zone. J Struct Geol 8:669–682

Blenkinsop TG, Sibson RH (1991) Aseismic fracturing and cataclasis involving reaction softening within core material from the Cajon Pass drillhole. J Geophys Res 97:5135–5144

Blenkinsop TG, Treloar PJ (1995) Geometry, classification and kinematics of S-C fabrics. J Struct Geol 17:397–408

Blumenfeld P (1983) Le "tuilage des mégacristaux" – un critère d'écoulement rotationnel pour les fluidalités des roches magmatiques – application au granite de Barbey. Bull Soc Geol Fr 25:309–318

Blumenfeld P, Bouchez JL (1988) Shear criteria in granite and migmatite deformed in the magmatic and solid states. J Struct Geol 10:361–372

Blumenfeld PR, Wilson CJL (1991) Boundary migration and kinking in sheared naphthalene. J Struct Geol 13:471–484

Blumenfeld P, Mainprice D, Bouchez JL (1985) Glissement de direction dominant dans le quartz de filons de granite, cisaillés en conditions sub-solidus (Vosges, France). CR Acad Sci, Sér II 301:1303–1308

Blumenfeld P, Mainprice D, Bouchez JL (1986) C-slip in quartz from subsolidus deformed granite. Tectonophysics 127:97–115

Boiron MC, Moissette A, Cathelineau M, Banks D, Monin C, Dubessy J (1999) Detailed determination of paleofluid chemistry: an integrated study of sulphate-volatile rich brines and aqueo-carbonic fluids in quartz veins from Ouro Fino (Brazil). Chem Geol 154:179–192

Boland JN, van Roermund HLM (1983) Mechanisms of exsolution in omphacites from high temperature, type B, eclogites. Phys Chem Miner 9:30–37

Bons PD (1993) Experimental deformation of polyphase rock analogues. Geol Ultraject 110:1–112

Bons PD (2000) The formation of veins and their microstructures. In: Jessell MW, Urai JL (eds) Stress, strain and structure, a volume in honour of W. D. Means. J Virt Explorer 2

Bons PD (2001) The formation of large quartz veins by rapid ascent of fluids in mobile hydrofractures. Tectonophysics 336:1–17

Bons PD, Jessell MW (1997) Experimental simulation of the formation of fibrous veins by localised dissolution-precipitation creep. Mineral Mag 61:53–63

Bons PD, Jessell MW (1999) Micro-shear zones in experimentally deformed octachloropropane. J Struct Geol 21:323–334

Bons PD, Urai JL (1992) Syndeformational grain growth: microstructures and kinetics. J Struct Geol 14:1101–1109

Bons PD, Jessell MW, Passchier CW (1993) The analysis of progressive deformation in rock analogues. J Struct Geol 15:403–411

Bons PD, Barr TD, Ten Brink CE (1997) The development of deltaclasts in non-linear viscous materials: a numerical approach. Tectonophysics 270:29–41

Borg IY, Heard HC (1969) Mechanical twinning and slip in experimentally deformed plagioclase. Contrib Mineral Petrol 23:128–135

Borg IY, Heard HC (1970) Experimental deformation of plagioclase. In: Paulitsch P (ed) Experimental and natural rock deformation. Springer-Verlag, Berlin Heidelberg New York, pp 375–403

Borges FS, White SH (1980) Microstructural and chemical studies of sheared anorthosites, Roneval, South Harris. J Struct Geol 2:273–280

Borradaile GJ (1981) Particulate flow of rock and the formation of cleavage. Tectonophysics 72:305–321

Borradaile GJ, McArthur J (1990) Experimental calcite fabrics in a synthetic weaker aggregate by coaxial and non-coaxial deformation. J Struct Geol 12:351–364

Borradaile GJ, Bayly MB, Powell CMA (1982) Atlas of deformational and metamorphic rock fabrics. Springer-Verlag, Berlin Heidelberg New York

Bos B, Peach CJ, Spiers CJ (2000) Frictional-viscous flow of simulated fault gouge caused by the combined effects of phyllosilicates and pressure solution. Tectonophysics 327:173–194

Bose S, Marques FO (2004) Controls on the geometry of tails around rigid circular inclusions: insights from analogue modelling in simple shear. J Struct Geol 26:2145–2156

Bouchez JL (1977) Plastic deformation of quartzites at low temperatures in an area of natural strain gradient. Tectonophysics 39:25–50

Bouchez JL (1978) Preferred orientations of quartz <a> axes in some tectonites: kinematic inferences. Tectonophysics 49:25–30

Bouchez JL, Lister GS, Nicolas A (1983) Fabric asymmetry and shear sense in movement zones. Geol Rdsch 72:401–419

Bouchez JL, Delas C, Gleizes G, Nedelec A, Cuney M (1992) Submagmatic microfractures in granites. Geology 20:35–38

Boullier AM, Bouchez JL (1978) Le quartz en rubant dans les mylonites. Bull Soc Geol Fr 20:253–262

Boullier AM, Guéguen Y (1975) SP-mylonites: origin of some mylonites by superplastic flow. Contrib Mineral Petrol 50:93–104

Boullier AM, Guéguen Y (1998a) Anorthosite mylonites produced by superplastic flow. In: Snoke A, Tullis J, Todd VR (eds) Fault related rocks – a photographic atlas. Princeton University Press, New Jersey, pp 514–515

Boullier AM, Guéguen Y (1998b) Peridotite mylonite produced by superplastic flow. In: Snoke A, Tullis J, Todd VR (eds) Fault related rocks – a photographic atlas. Princeton University Press, New Jersey, pp 592–593

Boullier AM, France-Lanord C, Dubessy J, Adamy J, Champenois M (1991) Linked fluid and tectonic evolution in the High Himalaya mountains (Nepal). Contrib Min Pet 107:358–372

Boullier AM, Firdaous K, Boudier AM (1997) Fluid circulation related to deformation in the Zabargad gneisses (Red Sea Rift). Tectonophysics 279:281–302

Boundy TM, Austrheim H (1998) Deep crustal eclogite-facies pseudotachylytes. In: Snoke A, Tullis J, Todd VR (eds) Fault related rocks – a photographic atlas. Princeton University Press, New Jersey, pp 126–127

Brandon MT, Cowan DS, Feehan JG (1994) Fault-zone structures and solution-mass-transfer cleavage in Late Cretaceous nappes, San Juan Islands, Washington. In: Swanson DA, Haugerud RA (eds) Geologic field trips in the Pacific Northwest. Geol Soc Am Ann Mtg Seattle, Washington

Braun G (1994) A statistical geometric method for quantitative texture analysis. In: Bunge HJ, Siegesmund S, Skrotzki W, Weber K (eds) Textures of geological materials. DGM Informationsges, Oberursel, pp 29–60

Breese MBH, Jamieson DN, King PJC (1996) Materials analysis using a nuclear microprobe. Wiley & Sons, New York

Brodie KH (1998a) High-temperature mylonites: I. Metabasic mylonites. In: Snoke A, Tullis J, Todd VR (eds) Fault related rocks – a photographic atlas. Princeton University Press, New Jersey, pp 428–431

Brodie KH (1998b) Retrogressive mylonitic metabasic fault rocks. In: Snoke A, Tullis J, Todd VR (eds) Fault related rocks – a photographic atlas. Princeton University Press, New Jersey, pp 402–403

Brodie KH, Rutter EH (1985) On the relationship between deformation and metamorphism with special reference to the behaviour of basic rocks. In: Thompson AB, Rubie DC (eds) Metamorphic reactions: kinetics, textures, and deformation. Advances in Phys Geochem 4:138–179

Brodie KH, Rutter EH (2000) Deformation mechanisms and rheology; why marble is weaker than quartzite. J Geol Soc, London 157:1093–1096

Brokmeier HG (1994) Application of neutron diffraction to measure preferred orientations of geological materials. In: Bunge HJ, Siegesmund S, Skrotzki W, Weber K (eds) Textures of geological materials. DGM Informationsges, Oberursel, pp 327–344

Brown WL, Macaudière J (1984) Microfracturing in relation to atomic structure of plagioclase from a deformed meta-anorthosite. J Struct Geol 6:579–586

Brunel M (1980) Quartz fabrics in shear zone mylonites: evidence for a major imprint due to late strain increments. Tectonophysics 64:T33–T44

Brunel M (1986) Ductile thrusting in the Himalayas: shear sense criteria and stretching lineations. Tectonics 5:247–265

Buatier M, van Roermund HLM, Drury M, Lardeaux JM (1991) Deformation and recrystallization mechanisms in naturally deformed omphacites from the Sesia-Lanzo zone; geophysical consequences. Tectonophysics 195:11–27

Bucher K, Frey M (1994) Petrogenesis of metamorphic rocks. Springer-Verlag, Berlin Heidelberg New York

Buck P (1970) Verformung von Hornblende-Einkristallen bei Drucken bis 21 kb. Contrib Min Pet 28:62–71

Burg JP, Laurent P (1978) Strain analysis of a shear zone in a granodiorite. Tectonophysics 47:15–42

Burg JP, Wilson CJL, Mitchell JC (1986) Dynamic recrystallization and fabric development during the simple shear deformation of ice. J Struct Geol 8:857–870

Burkhard M (1990) Ductile deformation mechanisms in micritic limestones naturally deformed at low temperatures (150–350 °C). In: Knipe RJ, Rutter EH (eds) Deformation mechanisms, rheology and tectonics. Geol Soc Special Pub 54, pp 241–257

Burkhard M (1993) Calcite twins, their geometry, appearance and significance as stress-strain markers and indicators of tectonic regime: a review. J Struct Geol 15:351–368

Burlini L, Kunze K (2000) Fabric and seismic properties of Carrara marble mylonite. Phys Chem Earth 25:133–139

Burlini L, Marquer D, Challandes N, Mazzola S, Zangarini N (1998) Seismic properties of highly strained marbles from the Splügenpass, central Alps. J Struct Geol 20:277–292

Burruss RC (1987) Diagenetic paleotemperatures from aqueous fluid inclusions: re-equilibration of inclusions in carbonates cements by burial heating. Mineral Mag 51:477–481

Burruss RC (1991) Practical aspects of fluorescence microscopy of petroleum fluid inclusions. In: Barker CE, Kopp OC (eds) Luminescence microscopy: quantitative and qualitative aspects. SEPM Short Course 25:1–7

Burruss RC, Cercone KR, Harris PM (1985) Timing of hydrocarbon migration: evidence from fluid inclusions in calcite cements, tectonics and burial history. In: Schneidermann N, Harris PM (eds) Carbonates cements. Soc Econ Paleontol Mineral Special Publ 26, pp 277–289

Burton KV, O'Nions RK (1991) High resolution garnet chronometry and the rates of metamorphic processes. Earth Planet Sci Lett 107:649–671

Busa MD, Gray NH (1992) Rotated staurolith porphyroblasts in the Littleton Schist at Bolton, Connecticut, USA. J Metam Geol 10:627–636

Buseck PR (ed) (1992) Minerals and reactions at the atomic scale: Transmission electron microscopy. Geochim Cosmochim Acta 57:4537–4538

Bystricky M, Mackwell S (2001) Creep of dry clinopyroxene aggregates. J Geoph Res 106:13443–13454

Calvo JP, Rodriguez-Pascua M, Martin-Velazquez S, Jimenez, De Vicete G (1998) Microdeformation of lacustrine laminite sequences from Late Miocene formations of SE Spain: an interpretation of loop bedding. Sedimentology 45:279–292

Camacho A, Vernon RH, Fitz Gerald JD (1995) Large volumes of anhydrous pseudotachylyte in the Woodroffe Thrust, eastern Musgrave Ranges, Australia. J Struct Geol 17:371–383

Carlson WD (1989) The significance of intergranular diffusion to the mechanism and kinetics of porphyroblast crystallization. Contrib Min Petr 103:1–24

Carlson WD, Gordon CL (2004) Effects of matrix grain size on the kinetics of intergranular diffusion. J Metamorph Geol 22:733

Carlson WD, Rowe T, Ketcham RA, Colbert MW (2003) Geological applications of high resolution X-ray computed tomography in petrology, meteoritics and palaeontology. In: Mees F, Swennen R, van Geet M, Jacobs P (eds) Applications of X-ray computed tomography in the geosciences. 215, pp 7–22

Carmichael DM (1969) On the mechanism of prograde metamorphic reactions in quartz-bearing pelitic rocks. Contrib Mineral Petrol 20:244–267

Carstens H (1969) Dislocation structures in pyropes from Norwegian and Czech garnet peridotites. Contrib Min Pet 24:348–353

Carstens H (1971) Plastic stress relaxation around solid inclusions in pyrope. Contrib Min Pet 32:289–294

Carter NL (1971) Static deformation of silica and silicates. J Geophys Res 76:5514–5540

Carter NL, Avé Lallemant HG (1970) High temperature flow of dunite and peridotite. Bull Geol Soc Am 81:2181–2202

Carter NL, Kirby SH (1978) Transient creep and semibrittle behaviour of crystalline rocks. Pure Appl Geophys 116:807–839

Carter NL, Raleigh CB (1969) Principal stress directoins from plastic flow in crystals. Bull Geol Soc Am 80:1213–1264

Carter NL, Tsenn MC (1987) Flow properties of continental lithosphere. Tectonophysics 136:27–63

Casey M (1981) Numerical analysis of X-ray texture data: an implementation in Fortran allowing triclinic or axial specimen symmetry and most crystal symmetries. Tectonophysics 78:51–64

Casey M, Dietrich D, Ramsay JG (1983) Methods for determining deformation history for chocolate tablet boudinage with fibrous crystals. Tectonophysics 92:211–239

Casey M, Kunze K, Olgaard DL (1998) Texture of Solnhofen limestone deformed to high strains in torsion. J Struct Geol 20:255–267

Cashman S, Cashman K (2000) Cataclasis and deformation-band formation in unconsolidated marine terrace sand, Humboldt County, California. Geology 28:111–114

Cashman KV, Ferry JM (1988) Crystal size distribution (CSD) in rocks and the kinetics and dynamics of crystallization: III Metamorphic crystallization. Contrib Mineral Petrol 99:401–415

Ceriani S, Mancktelow NS, Pennacchioni G (2003) Analogue modelling of the influence of shape and particle/matrix interface lubrication on the rotational behaviour of rigid particles in simple shear. J Struct Geol 25:2005–2021

Cesbron F, Ohnenstetter D, Blanc PH, Rauer O, Sichere MC (1993) Incorporation de terres rares dans les zircons de synthèse: étude par cathodoluminescence. Comptes Rendus de l'Académie des Sciences Paris 316:1231–1238

Champness PE (ed) (2001) Electron diffraction in the transmission electron microscope. BIOS Scientific Publishers, Oxford

Chan YC, Crespi JM (1999) Albite porphyroblasts with sigmoidal inclusion trails and their kinematic implications: an example from the Taconic Allochthon, west-central Vermont. J Struct Geol 21:1407–1417

Chatterjee KK (1994) Micro- and meso-scale deformation structures in experimental fault zones. J Struct Geol 16:1463–1476

Chen J, Wang QC, Zhai MG, Ye K (1996) Plastic deformation of garnet in eclogite. Sci China Ser D 39(1):18–25

Chester M, Chester JS (1998) Ultracataclasite structure and friction processes of the San Andreas Fault. Tectonophysics 295:199–221

Chester FM, Logan JM (1987) Composite planar fabric of gouge from the Punchbowl Fault, California. J Struct Geol 9:621–634

Chester FM, Logan JM (1998) Shear band fabric in gouge. In: Snoke A, Tullis J, Todd VR (eds) Fault related rocks – a photographic atlas. Princeton University Press, New Jersey, pp 68–69

Chester FM, Friedman M, Logan JM (1985) Foliated cataclasites. Tectonophysics 111:139–146

Chester FM, Evans JP, Biegel RL (1993) Internal structure and weakening mechanisms of the San Andreas Fault. J Geophys Res 98:771–786

Choukroune P (1971) Contribution à l'étude des mécanismes de la déformation avec schistosité grace aux cristallisations syncinématiques dans les "zones abritées" ("pressure shadows"). Bull Soc Geol Fr 13:257–271

Choukroune P, Lagarde JL (1977) Plans de schistosité et déformation rotationelle: l'exemple du gneiss de Champtoceaux (Massif Armoricain). CR Acad Sci Paris 284:2331–2334

Choukroune P, Seguret M (1968) Exemple de relations entre joints de cisaillement, fentes de tension, plis et schistosité (autochthone de la nappe de Gavarnie – Pyrenées Centrales). Rev Géogr Phys Géol Dyn 10:239–247

Christensen JN, Rosenfeld JL, De Paulo DJ (1989) Rates of tectonometamorphic processes from rubidium and strontium isotopes in garnet. Science 244:1465–1468

Christensen JN, Halliday AN, Lee DC, Hall CM (1995) In situ Sr isotopic analysis by laser ablation. Earth Planet Sci Letters 136:79–85

Christie JM (1958) Dynamic interpretation of the fabric of a dolomite from the Moine thrust-zone in north-west Scotland. Am J Sci 256:159–170

Christie JM, Ardell AJ (1974) Substructures of deformation lamellae in quartz. Geology 2:405–408

Christie JM, Green HW (1964) Several new slip mechanisms in quartz. EOS 45:103

Christie JM, Ord A (1980) Flow stress from microstructures of my-lonites: example and current assessment. J Geophys Res 85 (B11):6253–6262

Cladouhos TT (1999a) Shape preferred orientations of survivor grains in fault gouge. J Struct Geol 21:419–436

Cladouhos TT (1999b) A kinematic model for deformation within brittle shear zones. J Struct Geol 21:437–448

Clark MB, Fisher DM (1995) Strain partitioning and crack-seal growth of chlorite-muscovite aggregates during progressive noncoaxial strain – an example from the slate belt of Taiwan. J Struct Geol 17:461–474

Clark MB, Fisher DM, Chia-Yu L (1993) Kinematic analysis of the Hfuehshan Range: a large-scale pop-up structure. Tectonics 12:205–217

Clarke GL (1990) Pyroxene microlites and contact metamorphism in pseudotachylyte veinlets from MacRobertson Land, East Antarctica. Aust J Earth Sci 37:1–8

Clarke GL, Norman A (1993) Generation of pseudotachylite under granulite facies conditions, and its preservation during cooling. J Metamorph Geol 11:319–335

Clarke GL, Collins WJ, Vernon RH (1990) Successive overprinting granulite facies metamorphic events in the Anmatjira Ranges, central Australia. J Metam Geol 8:65–88

Cobbold P (1976) Mechanical effects of anisotropy during large finite deformations. Bull Soc Geol Fr 18:1497–1510

Cobbold PR, Quinquis H (1980) Development of sheath folds in shear regimes. J Struct Geol 2:119–126

Cobbold PR, Cosgrove JW, Summers JM (1971) Development of internal structures in deformed anisotropic rocks. Tectonophysics 12:23–53

Coe RS, Kirby SH (1975) The orthoenstatite to clinoenstatite transformation by shearing and reversion by annealing: mechanism and potential applications. Contrib Mineral Petrol 52:29–55

Coelho S, Passchier CW, Grasemann B (2005) Geometric description of flanking structures. J Struct Geol 27, in press

Colletta B, Letouzey J, Pinedo R, Ballard JF, Bale P (1991) Computerized X-ray tomography analysis of sandbox models: Examples of thin-skinned thrust systems. Geology 19:1063–1067

Collinson JD, Thompson DB (1982) Sedimentary structures. George Allen and Unwin, London, U.K., 194 p

Conrad RE, Friedman M (1976) Microscopic feather fractures in the faulting process. Tectonophysics 33:187–198

Cosgrove JW (1976) The formation of slaty cleavage. J Geol Soc Lond 262:153–176

Cosgrove JW (1993) The interplay between fluids, folds and thrusts during the deformation of a sedimentary succession. J Struct Geol 15:491–500

Costin LS (1983) A microcrack model for the deformation and failure of brittle rock. J Geophys Res 88:9485–9492

Cox FC (1969) Inclusions in garnet: discussion and suggested mechanisms of growth for syntectonic garnets. Geol Mag 106:57–62

Cox SF (1987) Antitaxial crack-seal vein microstructures and their relationship to displacement paths. J Struct Geol 9:779–787

Cox SF (1995) Faulting processes at high fluid pressures: An example of fault valve behavior from the Wattle Gully Fault, Victoria, Australia. J Geophys Res 100:12841–12859

Cox SF, Etheridge MA (1983) Crack-seal fibre growth mechanisms and their significance in the development of oriented layer silicate microstructures. Tectonophysics 92:147–170

Cox SJ, Wall VJ, Etheridge MA, Potter TF (1991) Deformation and metamorphic processes in the formation of mesothermal, vein-hosted gold deposits: examples from the Lachlan fold belt in central Victoria, Australia. Ore Geol Rev 6:391–423

Craddock JP, Nielson KJ, Malone DH (2000) Calcite twinning strain constraints on the emplacement rate and kinematic pattern of the upper plate of the Heart Mountain Detachment. J Struct Geol 22:983–991

Craig J, Fitches WR, Maltman AJ (1982) Chlorite-mica stacks in low-strain rocks from Central Wales. Geol Mag 119:243–256

Crawford ML, Hollister LS (1985) Metamorphic fluids: the evidence from fluid inclusions. In: Walther JV, Wood BJ (eds) Advances in Physical Geochemistry, Vol. 5, pp 1–35

Cruden AR, Tobisch OT, Launeau P (1999) Magmatic fabric evidence for conduit-fed emplacement of tabular intrusion: Dinkey Creed pluton, central Sierra Nevada batholith, California. J Geoph Res 104:10511–10530

Culshaw N, Fyson W (1984) Quartz ribbons in high grade gneiss: modifications of dynamically formed quartz c-axis preferred orientations by oriented grain growth. J Struct Geol 6:663–668

Cumbest RJ, van Roermund HLM, Drury MR, Simpson C (1989) Burgers vector determination in clinoamphibole by computer simulation. Am Mineral 74:586–592

Curewitz D, Karson JA (1999) Ultracataclasis, sintering, and frictional melting in pseudotachylytes from East Greenland. J Struct Geol 21:1693–1713

Daly JS, Cliff RA, Yardley BWD (1989) Evolution of metamorphic belts. Geol Soc Lond Spec Publ 43:566

Dalziel IW, Bailey SW (1968) Deformed garnets in a mylonitic rock from the Grenville front and their tectonic significance. Am J Sci 266:542–562

Darot M, Reushle ÂT, Gueguen Y (1985) Fracture parameters of Fontainebleau sandstones: experimental study using high temperature controlled atmosphere double torsion apparatus. In: Ashworth E (ed) Research and engineering applications in rocks masses. Balkema, Rotterdam, pp 463–470

Darwin C (1846) Geological observations in South America. Smith-Elder

Davatzes NC, Aydin A, Eichhubl P (2003) Overprinting faulting mechanisms during the development of multiple fault sets in sandstone, Chimney Rock fault array, Utah, USA. Tectonophysics 363:1–18

Davidson SG, Anastasio DJ, Bebout GE, Holl JE, Hedlund CA (1998) Volume loss and metasomatism during cleavage formation in carbonate rocks. J Struct Geol 20:707–726

Davies RM, Griffin WL, O'Reilly SY, Andrew AS (2003) Unusual mineral inclusions and carbon isotopes of alluvial diamonds from Bingara, eastern Australia. Lithos 69:51–66

Davis GH (1984) Structural geology of rocks and regions. Wiley, New York

Davis GH (1999) Structural geology of the Colorado Plateau region of Southern Utah with special emphasis on deformation bands. Geol Soc Am Special Pap 342, pp 1–157

Davis GH, Gardulski AF, Lister GS (1987) Shear zone origin of quartzite mylonite and mylonitic pegmatite in the Coyote Mountains metamorphic core complex, Arizona. J Struct Geol 9:289–297

Davison I (1995) Fault slip evolution determined from crack-seal veins in pull-aparts and their implications for general slip models. J Struct Geol 17:1025–1034

Dayan H (1981) Deformation studies of the folded mylonites of the Moine Thrust, Eriboll District, Northwest Scotland. PhD Thesis, Univ Leeds

de Bresser JHP (1989) Calcite c-axis textures along the Gavarnie thrust zone, central Pyrenees. Geol Mijnb 68:367–376

de Bresser JHP (1991) Intracrystalline deformation of calcite. Geol Ultraject 79:1–191

de Bresser JHP, Spiers CJ (1993) Slip systems in calcite single crystals deformed at 300–800 °C. J Geophys Res 98:6397–6409

de Bresser JHP, Spiers CJ (1997) Strength characteristics of the r, f, and c slip systems in calcite. Tectonophysics 272:1–23 de Bresser JHP, Peach CJ, Reijs JPJ, Spiers CJ (1998) On dynamic recrystallization during solid state flow; effects of stress and temperature. Geophys Res Lett 25:3457–3460

de Bresser JHP, Ter Heege JH, Spiers CJ (2001) Grain size reduction by dynamic recrystallization: can it result in major rheological weakening? Intern J Earth Sci 90:28–45

de Bresser JHP, Evans B, Renner J (2002) On estimating the strength of calcite rocks under natural conditions. In: De Meer S, Drury MR, de Bresser JHP, Pennock GM (eds) Deformation mechanisms, rheology and tectonics: current status and future perspectives. GSL Special Pub 200, pp 309–329

de Hoff RT, Rhines FN (1968) Quantitative microscopy. McGraw Hill, New York

de Paor DC (1983) Orthographic analysis of geological structures – 1 Deformation theory. J Struct Geol 5:255–277

de Roo JA, Weber K (1992) Laminated veins and hydrothermal breccia as markers of low-angle faulting, Rhenish Massif, Germany. Tectonophysics 208:413–430

de Vivo B, Frezzotti ML (eds) (1994) Fluid inclusions in minerals: methods and applications. Short Course of the IMA Working Group "Inclusions in Minerals". Pontignano - Siena, 1–4

Debat P, Soula JC, Kubin L, Vidal JL (1978) Optical studies of natural deformation microstructures in feldspars (gneiss and pegmatites from Occitania, Southern France. Lithos 11:133–145

Dell'Angelo LN, Tullis J (1988) Experimental deformation of partially melted granitic aggregates. J Metam Geol 6:495–516

Dell'Angelo LN, Tullis J (1989) Fabric development in experimentally sheared quartzites. Tectonophysics 169:1–21

den Brok B (1992) An experimental investigation into the effect of water on the flow of quartzite. Geol Ultraject 95

den Brok SWJ (1998) Effect of microcracking on pressure-solution strain rate: the Gratz grain-boundary model. Geology 26:915–918

den Brok B, Kruhl J (1996) Ductility of garnet as an indicator of extremely high temperature deformation: discussion. J Struct Geol 18:1369–1373

den Brok SWJ, Zahid M, Passchier CW (1998a) Cataclastic solution creep of very soluble brittle salt as a rock analogue. Earth Planet Sc Lett 163:83–95

den Brok SWJ, Zahid M, Passchier CW (1998b) Stress-induced grain-boundary migration in very soluble brittle salt. J Struct Geol 21:147–151

den Brok SWJ, Morel J, Zahid M (2002) In situ experimental study of roughness development at a stressed solid/fluid interface. In: De Meer S, Drury MR, de Bresser JHP, Penock GM (eds) Deformation mechanisms, rheology and tectonics: current status and future perspectives. GSL Special Publ 200, pp 73–83

Dennis AJ, Secor DT (1987) A model for the development of crenulations in shear zones with applications from the Southern Appalachian Piedmont. J Struct Geol 9:809–817

Dewers T, Ortoleva P (1990) Differentiated structures arising from mechano-chemical feedback in stressed rocks. Earth-Sci Rev 29:283–298

Dewers T, Ortoleva P (1991) Influences of clay minerals on sandstone cementation and pressure solution. Geology 19:1045–1048

Di Toro G, Pennacchioni G (2004) Superheated friction-induced melts in zoned pseudotachylytes within the Adamello tonalites (Italian Southern Alps). J Struct Geol 26:1783–1801

Dietrich D, Grant PR (1985) Cathodoluminescence petrography of syntectonic quartz fibres. J Struct Geol 7:541–554

Dietrich D, Song H (1984) Calcite fabrics in a natural shear environment, the Helvetic nappes of western Switzerland. J Struct Geol 6:19–32

Dingley DJ, Field DP (1997) Electron backscatter diffraction and orientation imaging microscopy. Mater Sci Technol 13:69–78

Doherty R (1980) Dendritic growth. In: Hargraves RB (ed) Physics of magmatic processes. Princeton University Press, New Jersey, pp 576–600

Dollinger G, Blacic JD (1975) Deformation mechanisms in experimentally and naturally deformed amphiboles. Earth Planet Sci Lett 26:409–416

Dornbusch HJ, Weber K, Skrotzki W (1994) Development of microstructure and texture in high-temperature mylonites from the Ivrea Zone. In: Bung HJ, Siegesmund S, Skrotzki W, Weber K (eds) Textures of geological materials. DGM Informationsgesellschaft, Oberursel, pp 187–201

Doukhan N, Sautter V, Doukhan JC (1994) Ultradeep, ultramafic mantle xenoliths – transmission electron microscopy preliminary results. Phys Earth Planet In 82:195–207

Drury MR (1993) Deformation lamellae in metals and minerals. In: Boland JN, Fitz Gerald JD (eds) Defects and processes in the solid state: geoscience applications – The McLaren Volume. Elsevier Science Publishers BV, pp 195–212

Drury MR, Humphreys FJ (1987) Deformation lamellae as indicators of stress level. Abstr Trans Am Geophys Un (EOS) 68:1471

Drury MR, Humphreys FJ (1988) Microstructural shear criteria associated with grain boundary sliding during ductile deformation. J Struct Geol 10:83–90

Drury MR, Urai JL (1990) Deformation-related recrystallisation processes. Tectonophysics 172:235–253

Drury MR, Humphreys FJ, White SH (1985) Large strain deformation studies using polycrystalline magnesium as a rock analogue. Part II: dynamic recrystallization mechanisms at high temperatures. Phys Earth Planet In 40:208–222

Du Bernard X, Eichhubl P, Aydin A (2002) Dilatation bands: a new form of localized failure in granular media. Geophys Res Lett 29/24, doi: 10.1029/2002GL015966

Dubessy J, Poty B, Ramboz C (1989) Advances in C-O-H-N-S fluid geochemistry based on micro-Raman spectrometric analysis of fluid inclusions. Eur J Mineral 1:517–534

Dubessy J, Buschaert S, Lamb W, Pironon J, Thiéry R (2001) Methane-bearing aqueous fluid inclusions: Raman analysis, thermodynamic modelling and application to petroleum basins. Chem Geol 173:193–205

Duebendorfer M, Christensen CH (1998) Plastic-to-brittle deformation of microcline during deformation and cooling of a granitic pluton. In: Snoke A, Tullis J, Todd VR (eds) Fault related rocks – a photographic atlas. Princeton University Press, New Jersey, pp 176–179

Dunlap WJ, Hirth G, Teyssier C (1997) Thermomechanical evolution of a ductile duplex. Tectonics 16:983–1000

Dunn DE, La Fountain LJ, Jackson RE (1973) Porosity dependence and mechanism of brittle fracture in sandstones. J Geophys Res 78:2403–2417

Dunne WM, Hancock PL (1994) Palaeostress analysis of small-scale brittle structures. In: Hancock P (ed) Continental deformation. Pergamon Press, Oxford, pp 101–120

Durney DW (1972) Solution transfer, an important geological deformation mechanism. Nature 235:315–317

Durney DW, Kisch HJ (1994) A field classification and intensity scale for first generation cleavages. AGSO J Aust Geol Geop 15:257–295

Durney DW, Ramsay JG (1973) Incremental strains measured by syntectonic crystal growths. In: DeJong KA, Scholten R (eds) Gravity and tectonics. Wiley, New York, pp 67–95

Eggleton RA, Buseck PR (1980) The orthoclase-microcline inversion: a high-resolution transmission electron microscope study and strain analysis. Contrib Miner Petrol 74:123–133

Egydio-Silva M, Mainprice D (1999) Determination of stress directions from plagioclase fabrics in high grade deformed rocks (Além Paraíba shear zone, Ribeira fold belt, southeastern Brazil). J Struct Geol 21:1751–1771

Eisbacher GH (1970) Deformation mechanics of mylonitic rocks and fractured granites in Cobequid Mountains, Nova Scotia. Can Bull Geol Soc Am 81:2009–2020

El Ali A, Barbin V, Calas G, Cervelle B, Ramseyer K, Bouroulec J (1993) Mn^{2+}-activated luminescence in dolomite, calcite and magnesite: quantitative determination of manganese and site distribution by EPR and CL spectroscopy. Chem Geol 104:189–202

Elliott D (1973) Diffusion flow laws in metamorphic rocks. Bull Geol Soc Am 84:2645–2664

Ellis MA (1986) The determination of progressive deformation histories from antitaxial syntectonic crystal fibres. J Struct Geol 8:701–709

Engelder T (1984) The role of pore water circulation during the deformation of foreland fold and thrust belts. J Geophys Res 89:4319–4325

Engelder T, Marshak S (1985) Disjunctive cleavage formed at shallow depths in sedimentary rocks. J Struct Geol 7:327–344

England PC, Richardson SW (1977) The influence of erosion upon the mineral facies of rocks from different metamorphic environments. J Geol Soc Lond 134:201–213

England PC, Thompson AB (1984) Pressure-temperature-time paths of regional metamorphism 1. Heat transfer during the evolution of regions of thickened continental crust. J Petrol 25:894–928

Ermanovics IF, Helmstaedt H, Plant AG (1972) An occurrence of Archean pseudotachylite for southeastern Manitoba. Can J Earth Sci 9:257–265

Erskine BG, Heidelbach F, Wenk HR (1993) Lattice preferred orientations and microstructures of deformed Cordilleran marbles: correlation of shear indicators and determination of strain paths. J Struct Geol 15:1189–1206

Erslev EA (1988) Normalized center-to-center strain analysis of packed aggregates. J Struct Geol 10:201–210

Erslev E, Mann C (1984) Pressure solution shortening in the Martinsburg Formation, New Jersey Prov Pennsylvania. Acad Sci 58:84–88

Essene EJ (1989) The current status of thermobarometry in metamorphic rocks. In: Daly JS, Cliff RA, Yardley, BWD (eds) Evolution of metamorphic belts. Geol Soc Spec Publ 43:1–44, Blackwell, Oxford

Etchecopar A (1977) A plane kinematic model of progressive deformation in a ploycrystalline aggregate. Tectonophysics 39:121–139

Etchecopar A (1984) Etude des états de contrainte en tectonique cassante et simulation de déformations plastiques (approche mathématique). Thesis, Uni Sci Technique du Languedoc, Montpellier

Etchecopar A, Malavieille J (1987) Computer models of pressure shadows: a method for strain measurement and shear sense determination. J Struct Geol 9:667–677

Etchecopar A, Vasseur G (1987) A 3-D kinematic model of fabric development in polycrystalline aggregates: comparisons with experimental and natural examples. J Struct Geol 9:705–718

Etheridge MA (1975) Deformation and recrystallization of orthopyroxene from the Giles complex, Central Australia. Tectonophysics 25:87–114

Etheridge MA (1983) Differential stress magnitudes during regional deformation and metamorphism: upper bound imposed by tensile fracturing. Geology 11:231–234

Etheridge MA, Oertel G (1979) Strain measurements from phyllosilicate preferred orientation – a precautionary note. Tectonophysics 60:107–120

Etheridge MA, Wilkie JC (1979) Grainsize reduction, grain boundary sliding and the flow strength of mylonites. Tectonophysics 58:159–178

Etheridge MA, Wilkie JC (1981) An assessment of dynamically recrystallized grainsize as a paleopiezometer in quartz-bearing mylonite zones. Tectonophysics 78:475–508

Etheridge MA, Paterson MS, Hobbs BE (1974) Experimentally produced preferred orientation in synthetic mica aggregates. Contrib Mineral Petrol 44:275–294

Etheridge MA, Wall VJ, Cox SF, Vernon RH (1983) The role of the fluid phase during regional metamorphism and deformation. J Metam Geol 1:205–226

Evans JP (1988) Deformation mechanisms in granitic rocks at shallow crustal levels. J Struct Geol 10:437–443

Evans JP (1990) Textures, deformation mechnisms and the role of fluids in the cataclastic deformation of granite rocks. Spec Publ Geol Soc Lond 54:29–39

Evans JP (1998) Deformation of granitic rocks at shallow crustal levels. In: Snoke A, Tullis J, Todd VR (eds) Fault related rocks – a photographic atlas. Princeton University Press, New Jersey, pp 38–39

Evans P, Chester FM (1995) Fluid-rock interaction and weakening of fault of the San Andreas system: Inferences from San Gabriel fault rock geochemistry and microstructures. J Geophys Res 100:13007–13020

Evans MA, Dunne WM (1991) Strain factorization and partitioning in the North Mountain thrust sheet, central Appalachians, USA. J Struct Geol 13:21–36

Evans MA, Groshong Jr RH (1994) A computer program for the calcite strain-gauge technique. J Struct Geol 16:277–281

Evans B, Renner J, Hirth G (2001) A few remarks on the kinetics of static grain growth in rocks. Int J Earth Sciences 90:88–103

Exner U, Mancktelow NS, Grasemann B (2004) Progressive development of s-type flanking folds in simple shear. J Struct Geol 26:2191–2201

Fabbri O, Lin A, Tokushige H (2000) Coeval formation of cataclasite and pseudotachylyte in a Miocene forearc granodiorite, southern Kyushu, Japan. J Struct Geol 22:1015–1025

Fairbairn HW (1941) Deformation lamellae in quartz from the Ajibik Formation, Michigan. Geol Soc Am Bull 52:1265–1278

Farver JR, Yund RA (1991a) Oxygen diffusion in quartz: dependence on temperature and water fugacity. Chem Geol 90:55–70

Farver JR, Yund RA (1991b) Measurement of oxygen grain boundary diffusion in natural, fine-grained quartz aggregates. Geochim Cosmochim Acta 55:1597–1607

Ferguson CC (1980) Displacement of inert mineral grains by growing porphyroblasts: a volume balance constraint. Contrib Mineral Petrol 91:541–544

Ferguson CC (1981) A strain reversal method for estimating extension from fragmented rigid inclusions. Tectonophysics 79:T43–T52

Ferguson CC (1987) Fracture and deformation histories of stretched belemnites and other rigid-brittle inclusions in tectonites. Tectonophysics 139:255–273

Ferguson CC, Lloyd GE (1980) On the mechanical interaction between a growing porphyroblast and its surrounding matrix. Contrib Mineral Petrol 75:339–352

Fernandez A (1987) Preferred orientation developed by rigid markers in two-dimensional simple shear strain: a theoretical and experimental study. Tectonophysics 136:151–158

Fernandez A (1988) Strain analysis from shape preferred orientation in flow. Bulletin of the Geological Institut of the University of Uppsala 14:61–67

Fernandez A, Feybesse JR, Mezure JF (1883) Theoretical and experimental study of fabric development by different shaped markers in two-dimensional simple shear. Bull Soc Geol Fr 25:319–326

Ferrill DA (1991) Calcite twin widths and intensities as metamorphic indicators in natural low-temperature deformation of limestone. J Struct Geol 13:667–676

Ferrill DA (1998) Calcite twin width and intensities as metamorphic indicators in natural low-temperature deformation of limestone. J Struct Geol 13:667–675

Ferrill DA, Groshong Jr RH (1993) Determination conditions in the northern Subalpine Chain, France, estimated from deformations modes in coarse-grained limestones. J Struct Geol 15:995–1006

Ferrill DA, Morris AP, Evans MA, Burkhard M, Groshong Jr RH, Onasch CM, Trepmann CA, Stöckhert B (2001) Mechanical twinning of jadeite – an indication of synseismic loading beneath the brittle-ductile transition. Intern J Earth Sci 90:4–13

Ferrill DA, Morris PA, Evans MA, Burkhard M, Groshong Jr RH, Onasch CM (2004) Calcite twin morphology: a low-temperature deformation geothermometer. J Struct Geol 26:1521–1529

Finch AA, Klein J (1999) The causes and petrological significance of cathodoluminescence emissions from alkali feldspars. Contrib Min Pet 135:234–243

Fisher DM (1990) Orientation history and rheology in slates, Kodiak and Afognak Islands, Alaska. J Struct Geol 12:483–498

Fisher DM, Anastasio DJ (1994) Kinematic analysis of a large-scale leading edge fold, Lost River Range, Idaho. J Struct Geol 16:337–354

Fisher DM, Brantley SL (1992) Models of quartz overgrowth and vein formation; deformation and episodic fluid flow in an ancient subduction zone. J Geophys Res 97:20043–20061

Fisher D, Bryne T (1990) The character and distribution of mineralized fractures in the Kodiak Formation, Alaska: implications for fluid flow in an underthrust sequence. J Geophys Res 95 (B6):9069–9080

Fisher QJ, Knipe RJ (2001) The permeability of faults within siliciclastic petroleum reservoirs of the North Sea and Norwegian Continental Shelf. Mar Petrol Geol 18:1063–1081

Fisher DM, Brantley SL, Everett M, Dzvonik J (1995) Cyclic fluid flow through a regionally extensive fracture network within the Kodiak accretionary prism. J Geophys Res 100:12881–12894

Fitches WR, Cave R, Craig J, Maltman AJ (1986) Early veins as evidence of detachment in the lower Palaeozoic rocks of the Welsh Basin. J Struct Geol 8:607–620

Fitz Gerald JD, Stünitz H (1993) Deformation of granitoids at low metamorphic grade I. Reaction and grain size reduction. Tectonophysics 221:269–297

Fitz Gerald JD, Etheridge MA, Vernon RH (1983) Dynamic recrystallization in a naturally deformed albite. Text Microstruct 5:219–237

Fleming PD, Offler R (1968) Pre-tectonic metamorphic crystallization in the Mt. Lofty Ranges, South Australia. Geol Mag 105:356–359

Fletscher RC, Pollard DD (1981) Anticrack model for pressure solution surfaces. Geology 9:419–424

Fliervoet TF, White SH (1995) Quartz deformation in a very fine grained quartzo-feldspathic mylonite: a lack of evidence for dominant grain boundary sliding deformation. J Struct Geol 17:1095–1109

Fliervoet TF, White SH, Drury MR (1997) Evidence for dominant grain-boundary sliding deformation in greenschist- and amphibolite-grade polymineralic ultramylonites from the Redbank Deformed Zone, Central Australia. J Struct Geol 19:1495–1520

Flood RH, Vernon RH (1988) Microstructural evidence of orders of crystallization in granitoid rocks. Lithos 21:237–245

Fowler TJ (1996) Flexural-slip generated bedding-parallel veins from central Victoria, Australia. J Struct Geol 18:1399–1415

Fowler TJ, Winsor CN (1997) Characteristics and occurence of bedding-parallel slip surface and laminated veins in chevron folds from the Bendigo-Castlemaine goldfields: implications for flexural slip folding. J Struct Geol 19:799–815

Fredrich J, Evans B (1992) Strength recovery along simulated faults by solution transfer processes. In: Tillerson JR, Wawersik WR (eds) Rock mechanics. Balkema, Rotterdam, pp 121–130

Freeman B (1985) The motion of rigid ellipsoidal particles in slow flows. Tectonophysics 113:163–183

Freeman B, Lisle RJ (1987) The relationship between tectonic strain and the three-dimensional shape of pebbles in deformed conglomerates. J Geol Soc London 144:635–639

Frejvald M (1970) The problem of platy quartz in rocks of crystalline basement. Acta Univ Carolina 2:95–103

Friedman M, Logan JM (1970) Microscopic feather fractures. Bull Geol Soc Am 81:3417–3420

Frondel C (1934) Selective incrustation of crystal forms. Am Mineral 19:316–329

Fry N (1979) Density distribution techniques and strained length methods for determination of finite strains. J Struct Geol 1:221–230

Fry N (2001) Stress space: striated faults, deformation twins, and their constraints on paleostress. J Struct Geol 23:1–9

Fueten F (1997) A computer controlled rotating polarizer stage for the petrographic microscope. Comput Geosci 23:203–208

Fueten F, Robin PYF (1992) Finite element modelling of the propagation of a pressure solution cleavage seam. J Struct Geol 14:953–962

Fueten F, Robin PYF, Schweinberger M (2002) Finite element modelling of the evolution of pressure solution cleavage. J Struct Geol 24:1055–1064

Furusho M, Kanagawa K (1999) Transformation-induced strain localization in a lherzolite mylonite from the Hidaka metamorphic belt of central Hokkaido, Japan. Tectonophysics 313:411–432

Gallagher JJ, Friedman M, Handin J, Sowers GM (1974) Experimental studies relating to microfractures in sandstone. Tectonophysics 21:203–247

Gamond JF (1983) Displacement features associated with fault zones:a comparison between observed examples and experimental models. J Struct Geol 5:33–45

Gandais M, Willaime C (1984) Mechanical properties of feldspars. In: Brown WL (ed) Feldspars and feldspathoids. NATO Asi, Reidel, Dordrecht Ser C137:207–246

Gapais D (1989) Shear structures within deformed granites: mechanical and thermal indications. Geology 17:1144–1147

Gapais D, Brun JP (1981) A comparison of mineral grain fabrics and finite strain in amphibolites from eastern Finland. Can J Earth Sci 18:995–1003

Gapais D, White SH (1982) Ductile shear bands in a naturally deformed quartzite. Text Microstruct 5:1–17

Garcia Celma A (1982) Domainal and fabric heterogeneities in the Cap de Creus quartz mylonites. J Struct Geol 44:443–456

Garcia Celma A (1983) C-axis and shape-fabrics in quartz-mylonites of Cap de Creus (Spain); their properties and development. PhD Thesis, Utrecht Univ

Gates AE, Glover L (1989) Alleghanian tectono-thermal evolution of the dextral transcurrent hylas zone, Virginia Piedmont, USA. J Struct Geol 11:407–419

Gaviglio P (1986) Crack-seal mechanismin a limestone: A factor of deformation in strike-slip faulting. Tectonophysics 131:241–255

Gay NC (1968) Pure shear and simple shear deformation of inhomogeneous viscous fluids 1 Theory. Tectonophysics 5:211–234

George SC, Krieger FW, Eadington PJ, Quezada RA, Greenwood PF, Eisenberg LI, Hamilton PJ, Wilson MA (1997) Geochemical comparison of oil-bearing fluid inclusions and produced oil from the Toro Sandstone, Papua New Guinea. Org Geochem 26:155–173

George CS, Ruble TE, Dutkiewicz and A, Eadington PJ (2001) Assessing the maturity of oil trapped in fluid inclusions using molecular geochemistry data and visually-determined fluorescence colours. Appl Geochem 16:451–473

Gershenfeld N (1999) Mathematical modelling. Cambridge University Press. 994 pp

Ghosh SK (1987) Measure of non-coaxiality. J Struct Geol 9:111–114

Ghosh SK, Ramberg H (1976) Reorientation of inclusions by combination of pure shear and simple shear. Tectonophysics 34:1–70

Gifkins RC (1976) Grain boundary sliding and its accommodation during creep and superplasticity. Metall Trans A 7A:1225–1232

Gillhaus A, Habermann D, Meijer J, Richter DK (2000) Cathodoluminescence spectroscopy and micro-PIXE: combined high-resolution Mn analyses in dolomites: First results. Nucl Instr Meth Phys Res B 161–163:842–845

Giuliani G, Dubessy J, Banks DA, Hoang Quang Vinh, Lhomme T, Pironon J, Garnier V, Phan Trong Trinh, Pham van Long, Ohnenstetter D, Schwarz D (2003) CO_2-H_2S COS-S_8-AlO(OH)-bearing fluid inclusions in ruby from marble-hosted deposits in Luc Yen area, North Vietnam. Chem Geol 194:167–185

Gleason GC, Tullis J (1995) A flow law for dislocation creep of quartz aggregates determined with the molten salt cell. Tectonophysics 247:1–23

Gleason GC, Tullis J, Heidelbach F (1993) The role of dynamic recrystallization in the development of lattice preferred orientations in experimentally deformed quartz aggregates. J Struct Geol 15:1145–1168

Godard G, van Roermund HLM (1995) Deformation-induced clinopyroxene fabrics from eclogites. J Struct Geol 17:1425–1443

Goldstein AG (1988) Factors affecting the kinematic interpretation of asymmetric boudinage in shear zones. J Struct Geol 10: 707–715

Goldstein RH, Reynolds TJ (1994) Systematics of fluid inclusions in diagenetic minerals. SEPM Short Course 31

Goldstein JI, Newbury DE, Echlin P, Joy DC, Romig Jr AD, Lyman CE, Fiori C, Lifshin E (1992) Scanning electron microscopy and X-ray microanalysis, 2nd ed. Plenum Press, New York

Goldstein A, Pickens J, Klepeis K, Linn F (1995) Finite strain heterogeneity and volume loss in slates of the Taconic Allochthon, Vermont, U.S.A. J Struct Geol 17:1207–1216

Goldstein A, Knight J, Kimball K (1998) Deformed graptolites, finite strain and volume loss during cleavage formation in rocks of the taconic slate belt, New York and Vermont, U.S.A. J Struct Geol 20:1769–1782

González-Casado JM, García-Cuevas C (1999) Calcite twins from microveins as indicators of deformation history. J Struct Geol 21:875–889

González-Casado JM, García-Cuevas C (2002) Strain analysis from calcite e-twins in the Cameros basin, NW Iberian Chain, Spain. J Struct Geol 24:1777–1788

González-Casado JM, Jiménez-Berrocoso A, García-Cuevas C, Elorza J (2003) Strain determinations using inoceramid shells as strain markers: a comparison of the calcite strain gauge technique and the Fry method. J Struct Geol 25:1773–1778

Goodchild JS, Fueten F (1998) Edge detection in petrographic images using the rotating polarizer stage. Comput Geosci 24: 745–751

Goodwin LB, Tikoff B (2002) Competency contrast, kinematics, and the development of foliations and lineations in the crust. J Struct Geol 24:1065–1085

Goodwin LB, Wenk HR (1990) Intracrystalline folding and cataclasis in biotite of the Santa Rosa mylonite zone – HVEM and TEM observations. Tectonophysics 172:201–214

Gorton NT, Walker G, Burley SD (1996) Experimental analysis of the composite blue CL emission in quartz. J Luminesc 72–74:669–671

Goscombe B, Passchier CW (2003) Asymmetric boudins as shear sense indicators – an assessment from field data. J Struct Geol 25:575–589

Gottstein G, Mecking H (1985) Recrystallization. In: Wenk HR (ed) Preferred orientation in deformed metals and rocks – an introduction to modern texture analysis. Academic Press, New York, pp 183–218

Götze J (2000) Cathodoluminescence microscopy and spectroscopy in applied mineralogy. Freiberger Forschungshefte, C 485, Freiberg TU Bergakademie

Götze J, Plötze M, Habermann D (2001) Origin, spectral characteristics and practical applications of the cathodoluminescence (CL) of quartz: a review. Mineral Petrol 71:225–250

Götze J, Plötze M, Götte T, Neuser RD, Richter DK (2002) Cathodoluminescence (CL) and electron paramagnetic resonance (EPR) studies of clay minerals. Mineral Petrol 76:195–212

Gower RJW, Simpson C (1992) Phase boundary mobility in naturally deformed, high-grade quartzofeldspathic rocks: evidence for diffusional creep. J Struct Geol 14:301–314

Graham RH (1978) Quantitative deformation studies in the Permian rocks of Alpes-Maritimes. Goguel Symp, BRGM, pp 220–238

Grant PR, White SH (1978) Cathodoluminescence and microstructure of quartz overgrowths on quartz. Scanning Electron Microsc 1:789–794

Grasemann B, Stüwe K (2001) The development of flanking folds during simple shear and their use as kinematic indicators. J Struct Geol 23:715–724

Grasemann B, Fritz H, Vannay JC (1999) Quantitative kinematic flow analysis from the Main Central Thrust Zone (NW-Himalaya, India): implications for a decelerating strain path and the extrusion of an orogenic wedge. J Struct Geol 21:837–853

Grasemann B, Stüwe K, Vannay JC (2003) Sense and non-sense of shear in flanking structures. J Struct Geol 25:19–34

Gratier JP, Renard F, Labaume F (1999) How pressure solution creep and fracturing processes interact in the upper crust to make it behave in both a brittle and viscous manner. J Struct Geol 21: 1189–1197

Gratier JP, Muquet L, Hassani R, Renard F (2005) Experimental microstylolites in quartz and modelled application to natural stylolitic structures J Struct Geol 27:89–100

Gratz AJ (1991) Solution-transfer compaction of quartzites: progress towards a rate law. Geology 19:901–904

Gray DR (1977) Morphologic classification of crenulation cleavages. J Geol 85:229–235

Gray DR (1978) Cleavages in deformed psammitic rocks from Southeastern Australia: their nature and origin. Bull Geol Soc Am 89:577–590

Gray DR (1979) Microstructure of crenulation cleavages: an indicator of cleavage origin. Am J Sci 279:97–128

Gray DR (1981) Compound tectonic fabrics in singly folded rocks from SW Virginia, USA. Tectonophysics 78:229–248

Gray NH, Busa MD (1994) The three-dimensional geometry of simulated porphyroblast inclusion trails: inert-marker, viscous-flow models. J Metam Geol 12:575–587

Gray DR, Durney DW (1979a) Crenulation cleavage differentiation: implication of solution-deposition processes. J Struct Geol 1: 73–80

Gray DR, Durney DW (1979b) Investigation on the mechanical significance of crenulation cleavage. Tectonophysics 58:35–79

Gray DR, Willman CE (1991) Thrust-related strain gradients and thrusting mechanisms in a chevron-folded sequence, southeastern Australia. J Struct Geol 13:691–710

Green HW, Griggs DT, Christie JM (1970) Syntectonic and annealing recrystallization of fine-grained quartz aggregates. In: Paulitsch P (ed) Experimental and natural rock deformation. Springer-Verlag, Berlin Heidelberg New York, pp 272–335

Gregg WJ (1985) Microscopic deformation mechanisms associated with mica film formation in cleaved psammitic rocks. J Struct Geol 7:45–56

Gregg WJ (1986) Deformation of chlorite-mica aggregates in cleaved psammitic and pelitic rocks from Islesboro, Maine, USA. J Struct Geol 8:59–68

Griggs DT, Turner FJ, Heard HC (1960) Deformation of rocks at 500–800 °C. In: Griggs DT, Handin J (eds) Rock deformation. Mem Geol Soc Am 79:39–104

Grocott J (1977) The relationship between Precambrian shear belts and modern fault systems. J Geol Soc Lond 133:257–262

Grocott J (1981) Fracture geometry of pseudotachylite generation zones – study of shear fractures formed during seismic events. J Struct Geol 3:169–179

Groshong Jr RH (1972) Strain calculated from twinning in calcite. Geol Soc Am Bull 83:2025–2048

Groshong Jr RH (1974) Experimental test of the least-squares strain gauge calculation using twinned calcite. Geol Soc Am 85, pp 1855–1864

Groshong Jr RH (1988) Low temperature deformation mechanisms and their interpretation. Bull Geol Soc Am 100:1329–1360

Groshong Jr RH, Pfiffner OA, Pringle LR (1984a) Strain partitioning in the Helvetic thrust belt of eastern Switzerland from the leading edge to the internal zone. J Struct Geol 6:5–18

Groshong Jr RH, Teufel LW, Gasteiger C (1984b) Precision and accuracy of the calcite strain-gauge technique. Geol Soc Am Bull 95:357–363

Gueguen Y, Boullier AM (1975) Evidence of superplasticity in mantle peridotites. NATO Petrophys Proc, Wiley & Academic Press, New York, pp 19–33

Guermani A, Pennacchioni G (1998) Brittle precursors of plastic deformation in a granite: an example from the Mont Blanc Massif (Helvetic, western Alps). J Struct Geol 20:135–148

Guidotti CV, Johnson SE (2002) Pseudomorphs and associated microstructures of western Maine, USA. J Struct Geol 24:1139–1156

Guillier B, Baby P, Colletta B, Mendez E, Limachi R, Letouzey J, Specht M (1995) Analyse géometrique et cinematique d'un "duplex" issu d'un modèle analogique visualise en 3D par tomographie aux rayons X. Comptes Rendus de l'Academie des Sciences, Serie II. Cr Acad Sci II A 321:901–908

Guillopé M, Poirier JP (1979) Dynamic recrystallization during creep of single-crystalline halite: an experimental study. J Geophys Res 84:5557–5567

Guzetta G (1984) Kinematics of stylolite formation and physics of the pressure-solution process. Tectonophysics 101:383–394

Habermann D, Neuser RD, Richter DK (1996) REE-activated cathodoluminescence of calcite and dolomite: high-resolution spectrometric analysis of CL emission (HRS-CL). Sediment Geol 101:1–7

Habermann D, Neuser R, Richter DK (1998) Low limit of Mn^{2+}-activated cathodoluminescence of calcite: state of the art. Sed Geol 116:13–24

Hacker BR, Christie JM (1990) Brittle/ductile and plastic/cataclastic transition in experimentally deformed and metamorphosed amphibolite. In: AG Duba, WB Durham, JW Handin, HF Wang (eds) The brittle-ductile transition in rocks. AGU Geophys Monogr Ser 56:127–148

Hacker B, Yin A, Christie JM, Snoke AW (1990) Differential stress, strain rate, and temperatures of mylonitization in the Ruby Mountains, Nevada: implications for the rate and duration of uplift. J Geophys Res 95:8569–8580

Hacker B, Yin A, Christie JM, Davis GA (1992) Stress magnitude, strain rate, and rheology of extended middle continental crust inferred from quartz grain sizes in the Whipple Mountains, California. Tectonics 11:36–46

Hadizadeh J, Tullis J (1992) Cataclastic flow and semi-brittle deformation of anorthosite. J Struct Geol 14:57–63

Hafner M, Passchier CW (2000) Development of S-C' type cleavage in paraffin waxusing a circular shear rig. J Virt Explorer 2:16–17

Hailemariam H, Mulugeta G (1998) Temperature-dependent rheology of bouncing putties used as rock analogs. Tectonophysics 294:131–141

Hall MG, Lloyd GE (1981) The SEM examination of geological samples with a semiconductor back-scattered electron detector. Am Mineral 66:362–368

Hallbauer DK, Wagner H, Cook NGW (1973) Some observations concerning the microscopic and mechanical behavior of quartzite specimens in stiff, triaxial compression tests. Int J Rock Mech Min Sci 10:713–726

Hanchar JM, Miller CF (1993) Zircon zonation patterns as revealed by cathodoluminescence and backscattered electron images: Implications for interpretation of complex crustal histories. Chem Geol 110:1–13

Hanchar JM, Rudnick RL (1995) Revealing hidden structures: the application of cathodoluminescence and back-scattered electron imaging to dating zircons from lower crustal xenoliths. Lithos 36:289–303

Handy MR (1989) Deformation regimes and the rheological evolution of fault zones in the lithosphere – the effects of pressure, temperature, grainsize and time. Tectonophysics 163:119–152

Handy MR (1992) Correction and addition to "The solid-state flow of polymineralic rocks". J Geophys Res 97:1897–1899

Handy MR, Wissing SB, Streit LE (1999) Frictional-viscous flow in mylonite with varied bimineralic composition and its effect on lithospheric strength. Tectonophysics 303:175–191

Hanmer S (1982) Microstructure and geochemistry of plagioclase and microcline in naturally deformed granite. J Struct Geol 4:197–213

Hanmer S (1984a) Strain-insensitive foliations in polymineralic rocks. Can J Earth Sci 21:1410–1414

Hanmer S (1984b) The potential use of planar and elliptical structures as indicators of strain regime and kinematics of tectonic flow. Geol Surv Can Pap 84:133–142

Hanmer S (1988) Great Slave Lake Shear Zone, Canadian Shield: reconstructed vertical profile of a crustal-scale fault zone. Tectonophysics 149:245–264

Hanmer S (1990) Natural rotated inclusions in non-ideal shear. Tectonophysics 176:245–255

Hanmer S (2000) Matrix mosaics, brittle deformation, and elongate porphyroclasts: granulite facies microstructures in the Striding-Athabasca mylonite zone, western Canada. J Struct Geol 22:947–967

Hanmer S, Passchier CW (1991) Shear sense indicators: a review. Geol Surv Can Pap 90:1–71

Hanmer S, Williams M, Kopf C (1995) Modest movements, spectacular fabrics in an intracontinental deep-crustal strike-slip fault: Striding-Athabasca mylonite zone, NW Canadian Shield. J Struct Geol 17:493–507

Harley SL (1989) The origins of granulites: a metamorphic perspective. Geol Mag 126:215–247

Harper TR, Szymanski JS (1991) The nature and determination of stress in the accessible lithosphere. In: Whitmarsh RB, Bott MHP, Fairhead JD, Kusznir JD (eds) Tectonic stress in the lithosphere. Philos Trans R Soc London A 337:5–24

Harris LB, Cobbold PR (1985) Development of conjugate shear bands during bulk simple shearing. J Struct Geol 7:37–44

Harris JH, van der Pluijm BA (1998) Relative timing of calcite twinning strain and fold thrust belt development; Hudson Valley fold-thrust belt, New York, USA. J Struct Geol 20:21–33

Hartwig G (1925) Praktisch-geologische Beschreibung des Kalisalzbergwerkes "Rössing-Barnten" bei Hildesheim. Jber Niedersächs Geol Ver 17:1–74

Harvey PK, Ferguson CC (1973) Spherically arranged inclusions in post-tectonic garnet porphyroblasts. Min Mag 39:85–88

Hay RS, Evans B (1987) Chemically induced grain boundary migration in calcite: temperature dependence, phenomenology, and possible applications to geologic systems. Contrib Min Pet 97:127–141

Hayward N (1990) Determination of early fold axis orientations within multiply deformed rocks using porphyroblasts. Tectonophysics 179:353–369

Hayward N (1992) Microstructural analysis of the classical spiral garnet porphyroblasts of south-east Vermont: evidence for non-rotation. J Metam Geol 10:567–587

Hedlund CA, Anastasio DJ, Fisher DM (1994) Kinematics of fault-related folding in a duplex, Lost River Range, Idaho, USA. J Struct Geol 16:571–584

Heidelbach F, Kunze K, Wenk HR (2000) Texture analysis of a re-crystallised quartzite using electron diffraction in the scanning electron microscope. J Struct Geol 22:91–104

Heilbronner R (2000) Optical orientation imaging. In: Jessell MW, Urai JL (eds) Stress, strain and structure, a volume in honour of W. D. Means. J Virt Explorer 2

Heilbronner R, Bruhn D (1998) The influence of three-dimensional grain size distributions on the rheology of polyphase rocks. J Struct Geol 20:695–705

Heilbronner R, Herwegh M (1997) Time slicing, an image process-ing technique to visualize the temporal development of fabrics. J Struct Geol 19:861–874

Heilbronner R, Tullis J (2002) The effect of static annealing on micro-structures and crystallographic preferred orientations of quartz-ites experimentally deformed in axial compression and shear. In: De Meer S, Drury MD, de Bresser JHP Pennock GM (eds) Defor-mation mechanisms, rheology and tectonics: current status and future perspectives. Special Publication 200, GSL, pp 191–218

Heilbronner-Panozzo R, Pauli C (1993) Integrated spatial and ori-entation analysis of quartz c-axes by computer-aided microscopy. J Struct Geol 15:369–382

Heilbronner-Panozzo R, Pauli C (1994) Orientation and misorien-tation imaging: integration of microstructural and textural analy-sis. In: Bunge HJ, Siegesmund S, Skrotzki W, Weber K (eds) Tex-tures of geological materials. DGM Informationsges, Oberursel, pp 147–164

Helmstead H, Anderson OL, Gavasci AT (1972) Petrofabric studies of eclogite, spinel-websterite, and spinel-lherzolite xenoliths from kimberlite-bearing breccia pipes in southeastern Utah and north-eastern Arizona. J Geophys Res 77:4350–4365

Henderson JR, Henderson MN, Wright TO (1990) Water-sill hypoth-esis for the origin of certain quartz veins in the Meguma Group, Nova Scotia, Canada. Geology 18:654–657

Herwegh M, Handy MR (1996) The evolution of high-temperature mylonitic microfabrics: evidence from simple shearing of a quartz analogue (norcamphor). J Struct Geol 18:689–710

Herwegh M, Handy MR (1998) The origin of shape preferred orien-tations in mylonite: inferences from in-situ experiments on poly-crystalline norcamphor. J Struct Geol 20:681–694

Herwegh M, Handy MR, Heilbronner R (1997) Temperature- and strain-rate-dependent microfabric evolution in monomineralic mylonite: evidence from in situ deformation of norcamphor. Tectonophysics 280:83–106

Herwegh M, Handy MR, Heilbronner R (2000) Evolution of mylonitic microfabrics. In: Jessell MW, Urai JL (eds) Stress, strain and struc-ture, a volume in honour of W.D. Means. J Virt Explorer 2

Herwegh M, Xiao X, Evans B (2003) The effect of dissolved magnesi-um on diffusion creep in calcite. Earth Planet Sci Lett 212:457–470

Hesthammer J, Johansen TES, Watts L (2000) Spatial relationships with-in fault damage zones in sandstone. Mar Petrol Geol 17:873–893

Heydari F (2000) Porosity loss, fluid flow, and mass transfer in lime-stone reservoirs. Application to the upper Jurassic Smackover formation, Mississippi. AAPG Bull 84:100–118

Hibbard MJ (1987) Deformation of incompletely crystallised magma systems: granitic gneisses and their tectonic implications. J Geol 951:543–561

Hickey KA, Bell TH (1999) Behaviour of rigid ductile objects during deformation and metamorphism: a test using schists from the Bolton Syncline, Connecticut, USA. J Metam Geol 17:211–228

Hickman SH, Evans B (1995) Kinetics of pressure solution at halite-silica interfaces and intergranular clay films. J Geophys Res 100:13113–13132

Higgins MD (2000) Measurements of crystal size distributions. Am Mineral 85:1105–1116

Hilgers C, Urai JL (2002) Microstructural observations on natural syntectonic fibrous veins: implications for the growth process. Tectonophysics 352:257–274

Hilgers C, Urai JL, Post AD, Bons PD (1997) Fibrous vein microstruc-ture; experimental and numerical simulation. Aardkundige Mededelingen 8:107–109

Hilgers C, Köhn D, Bons PD, Urai JL (2001) Development of crystal morphology during unitaxial growth ina progessively widening vein: II. Numerical simulations of the evolution of antitaxial fi-brous veins. J Struct Geol 23:873–885

Hippertt JFM (1993) 'V'-pull-apart microstructures: a new shear sense indicator. J Struct Geol 15:1393–1404

Hippertt JFM (1994) Grain boundary microstructures in micaceous quartzite: Significance for fluid movement and deformation processes in low metamorphic grade shear zones. Geology 102: 331–348

Hippertt JF, Hongn FD (1998) Deformation mechanisms in the my-lonite/ultramylonite transition. J Struct Geol 20:1435–1448

Hippertt J, Rocha A, Lana C, Egydio-Silva M, Takeshita T (2001) Quartz plastic segregation and ribbon development in high-grade striped gneisses. J Struct Geol 23:67–80

Hirth G, Tullis J (1992) Dislocation creep regimes in quartz aggre-gates. J Struct Geol 14:145–159

Hirth G, Teyssier C, Dunlap WJ (2001) An evaluation of quartzite flow laws based on comparisons between experimentally and naturally deformed rocks. Int J Earth Sciences 90:77–87

Hisada E (2004) Clast-size analysis of impact-generated pseudotachy-lite from Vredefort Dome, South Africa. J Struct Geol 26:1419–1424

Ho N, Peacor DR, van der Pluijm BA (1995) Reorientation mecha-nisms of phyllosilicates in the mudstone-to-slate transition at Lehigh Gap, Pennsylvania. J Struct Geol 17:345–356

Ho N, Peacor DR, van der Pluijm BA (1996) Contrasting roles of de-trital and authigenic phyllosilicates during slaty cleavage devel-opment. J Struct Geol 18:615–623

Ho NC, van der Pluijm BA, Peacor DR (2001) Static recrystallization and preferred orientation of phyllosilicates: Michigamme For-mation, northern Michigan, USA. J Struct Geol 23:887–893

Hobbs BE (1985) The geological significance of microfabric. In: Wenk HR (ed) Preferred orientation in deformed metals and rocks. Academic Press, New York

Hobbs BE, Means WD, Williams PF (1976) An outline of structural geology. Wiley, New York

Hobbs BE, Means WD, Williams PF (1982) The relationship between foliation and strain: an experimental investigation. J Struct Geol 4:411–428

Holcombe RJ, Little TA (2001) A sensitive vorticity gauge using ro-tated prophyroblasts, and its application to rocks adjacent to the Alpine fault, New Zealand. J Struct Geol 23:979–989

Holl JE, Anastasio DJ (1995) Cleavage development within a fore-land fold and thrust belt, southern Pyrenees, Spain. J Struct Geol 17:357–369

Hongn FD, Hippertt JFM (2001) Quartz crystallographic and mor-phologic fabrics during folding/transposition in mylonites. J Struct Geol 23:81–92

Hooper RJ, Hatcher RD (1988) Mylonites from the Towaliga fault zone, central Georgia: products of heterogeneous non-coaxial deformation. Tectonophysics 152:1–17

Horii H, Nemat-Nasser SJ (1985) Compression-induced microcracks growth in brittle solids: axial splitting and shear failure. J Geophys Res 90:3105–3125

Hoshikuma A (1996) Grain growth and superplasticity; their implication to earth science. J Geol Soc Jpn 102:232–239

Houseknecht DW (1988) Intergranular pressure solution in four quartzose sandstones. J Sed Petrol 58:228–246

Hull D (1975) Introduction to dislocations. Pergamon Press, Oxford

Humphreys FJ (1997) A unified theory of recovery, recrystallization and grain growth, based on the stability and growth of cellular microstructures – I. The basic model. Acta Materialia 45:4231–4240

Humphreys JF, Hatherley M (1995) Recrystallization and related annealing phenomena. Elsevier Science, New York, 498 pp

Humphreys FJ, Huang Y, Brough I, Harris C (1999) Electron backscatter diffraction of grain and subgrain. J Microsc 195(3): 212–216

Hunter RH (1987) Textural equilibrium in layered igneous rocks. In: Parsons I (ed) Origins of igneous layering. Reidel, Dordrecht, pp 473–503

Hutton DHW (1982) A tectonic model for the emplacement of the Main Donegal granite, NW Ireland. J Geol Soc Lond 139:615–631

Ildefonse B, Mancktelow NS (1993) Deformation around rigid particles: The influence of slip at the particle/matrix interface. Tectonophysics 221:345–359

Ildefonse B, Lardeux JM, Caron JM (1990) The behaviour of shape-preferred orientations in metamorphic rocks: amphiboles and jadeites from the Monte Mucrone area, Sesia-Lanzo zone, Italian western Alps. J Struct Geol 12:1005–1012

Ildefonse B, Launeau B, Fernandez A, Bouchez JL (1992a) Effect of mechanical interactions on development of shape preferred orientations: a two-dimensional experimental approach. J Struct Geol 14:73–83

Ildefonse B, Sokoutis D, Mancktelow NS (1992b) Mechanical interactions between rigid particles in a deforming ductile matrix. Analogue experiments in simple shear flow. J Struct Geol 14: 1253–1266

Ilg BR, Karlstrom KE (2000) Porphyroblast inclusion trail geometries in the Grand Canyon: evidence for non-rotation and rotation? J Struct Geol 22:231–243

Imon R, Okudaira T, Fujimoto A (2002) Dissolution and precipitation processes in the deformed amphibolites: an example from the ductile shear zone of the Ryoke metamorphic belt, SW Japan. J Metam Geol 20:297–308

Imon R, Okudaira T, Kanagawa K (2004) Development of shape- and lattice-preferred orientations of amphibole grains during initial cataclastic deformation and subsequent deformation by dissolution, precipitation creep in amphibolites from the Ryoke metamorphic belt, SW Japan. J Struct Geol 26:793–805

Ingerson E, Tuttle OF (1945) Relations of lamellae and crystallography of quartz and fabric directions in some deformed rocks. Trans Am Geophys Union 26:95–105

Ingrin J, Doukhan N, Doukhan JC (1991) High-temperature deformation of diopside single crystals 2, TEM observations in the induced defect microstructures. J Geoph Res 77:4350–4365

Ishii K (1988) Grain-growth and re-orientation of phyllo-silicate minerals during the development of slaty cleavage in the South Kitakami Mountains, NE Japan. J Struct Geol 10:145–154

Ishii K, Sawaguchi T (2002) Lattice- and shape-preferred orientation of orthopyroxene porphyroclasts in peridotites: an application of two-dimensional numerical modeling. J Struct Geol 24:517–530

Jaeger JC (1963) Extensional failures in rocks subject to fluid pressure. J Geophys Res 68:6066–6067

Jamison WR, Spang JH (1976) Use of calcite twin lamellae to infer differential stress. Geol Soc Am 87:868–872

Jeffery GB (1922) The motion of ellipsoidal particles immersed in a viscous fluid. Proceedings of the Royal Society of London A102:161–179

Jensen LN, Starkey J (1985) Plagioclase microfabrics in a ductile shear zone from the Jotun Nappe, Norway. J Struct Geol 7:527–541

Jessell MW (1986) Grain boundary migration and fabric development in experimentally deformed octachloropropane. J Struct Geol 8:527–542

Jessell MW (1987) Grain-boundary migration microstructures in a naturally deformed quartzite. J Struct Geol 9:1007–1014

Jessell MW (1988a) Simulation of fabric development in recrystallizing aggregates: 1 Description of the model. J Struct Geol 10: 771–778

Jessell MW (1988b) Simulation of fabric development in recrystallizing aggregates: 2 Example model runs. J Struct Geol 10: 779–793

Jessell MW, Bons PD (2002) The numerical simulation of microstructure. Geol Soc London Spec Publ 200:1137–14

Jessell MW, Willman CE, Gray DR (1994) Bedding parallel veins and their relationship to folding. J Struct Geol 16:753–767

Jessell M, Bons P, Evans L, Barr T, Stüwe K (2001) Elle: the numerical simulation of metamorphic and deformation microstructures. Comput Geosci 27:17–30

Jessell MW, Kostenko O, Jamtveit B (2003) The preservation potential of microstructures during static grain growth. J Metam Geol 21:481–491

Jezek J, Melky R, Schulmann K, Venera Z (1994) The behaviour of rigid triaxial ellipsoidal particles in viscous flows-modeling of fabric evolution in a multiparticle system. Tectonophysics 229:165–180

Ji S (1998a) Deformation microstructure of natural plagioclase. In: Snoke A, Tullis J, Todd VR (eds) Fault related rocks – a photographic atlas. Princeton University Press, New Jersey, pp 276–277

Ji S (1998b) Kink bands and recrystallization in plagioclase. In: Snoke A, Tullis J, Todd VR (eds) Fault related rocks – a photographic atlas. Princeton University Press, New Jersey, pp 278–279

Ji S, Mainprice D (1987) Experimental deformation of sintered albite above and below the order- disorder transition. Geodynam Acta 1:113–124

Ji S, Mainprice D (1988) Naturally deformed fabrics of plagioclase: implications for slip systems and seismic anisotropy. Tectonophysics 147:145–163

Ji S, Mainprice D (1990) Recrystallization and fabric development in plagioclase. Geology 98:65–79

Ji S, Martignole J (1994) Ductility of garnet as a indicator of extremely high temperature deformation. J Struct Geol 16:985–996

Ji S, Martignole J (1996) Ductility of garnet as an indicator of extremely high temperature deformation: Reply. J Struct Geol 18:1375–1379

Ji S, Zhao P (1993) Location of tensile fracture within rigid-brittle inclusions in a ductile flowing matrix. Tectonophysics 220:23–31

Ji S, Zhao P (1994) Strength of two-phase rocks: a model based on fiber-loading theory. J Struct Geol 16:253–262

Ji S, Mainprice D, Boudier F (1988) Sense of shear in high temperature movement zones from the fabric asymmetry of plagioclase feldspar. J Struct Geol 10:73–81

Ji S, Saruwatari K, Mainprice D, Wirth R, Xu Z, Xia B (2003) Microstructures, petrofabrics and seismic properties of ultra high-pressure eclogites from Sulu region, China: implications for rheology of subducted continental crust and origin of mantle reflections. Tectonophysics 370:49–76

Ji S, Jiang Z, Rybacki E, Wirth R, Prior D, Xia B (2004) Strain softening and microstructural evolution of anorthite aggregates and quartz-anorthite layered composites deformed in torsion. Earth Planet Sc Lett 222:377–390

Jiang D (2001) Reading history of folding from porphyroblasts. J Struct Geol 23:1327–1335

Jiang D, Williams PF (2004) Reference frame, angular momentum, and porphyroblast rotation. J Struct Geol 26:2211–2224

Joanny V, Villeurbanne R, van Roermund H, Lardeaux JM (1991) The clinopyroxene/plagioclase symplectite in retrograde eclogites: a potential geothermobarometer. Geol Rdsch 80:303–320

Joesten R, Fisher G (1988) Kinetics of diffusion-controlled mineral growth in the Christmas Mountains (Texas) contact aureole. Bull Geol Soc Am 100:714–732

Johnson SE (1990) Lack of porphyroblast rotation in the otago schists, New Zealand: implications for crenulation cleavage development, folding and deformation partitioning. J Metam Geol 8:13–30

Johnson TE (1991) Nomenclature and geometric classification of cleavage transected folds. J Struct Geol 13:261–274

Johnson SE (1993a) Unravelling the spirals: a serial thin-section study and three-dimensional computer-aided reconstruction of spiral-shaped inclusion trails in garnet porphyroblasts. J Metam Geol 11:621–634

Johnson SE (1993b) Testing models for the development of spiral-shaped inclusion trails in garnet porphyroblasts: to rotate or not to rotate, that is the question. J Metam Geol 11:635–659

Johnson SE (1999a) Near-orthogonal foliation development in orogens: meaningless complexity, or reflection of fundamental dynamic processes? J Struct Geol 21:1183–1187

Johnson SE (1999b) Porphyroblast microstructures; a review of current and future trends. Amer Mineral 84:1711–1726

Johnson SE, Bell TH (1996) How usefull are 'millipede' and other similar porphyroblast microstructures for determining syn-metamorphic deformation histories? J Metam Geol 14:15–28

Johnson SE, Moore RR (1996) De-bugging the 'millipede' porphyroblast microstructure: a serial thin-section study and 3-D computer animation. J Metam Geol 14:3–14

Johnson SE, Vernon RH (1995a) Inferring the timing of porphyroblast growth in the absence of continuity between inclusion trails and matrix foliations: can it be reliably done? J Struct Geol 17:1203–1206

Johnson SE, Vernon RH (1995b) Stepping stones and pitfalls in the determination of an anticlockwise P-T-t-deformation path: the low-P, high-T Cooma Complex, Australia. J Metam Geol 13:165–183

Johnson SE, Williams ML (1998) Determining finite longitudinal strains from opposedly-concave microfolds in and around porphyroblasts: a new quantitative method. J Struct Geol 20:1521–1530

Jordan PG (1987) The deformational behaviour of bimineralic limestone – halite aggregates. Tectonophysics 135:185–197

Jordan PG (1988) The rheology of polymineralic rocks – an approach. Geol Rdsch 77:285–294

Joy DC, Newbury DE, Davidson DL (1982) Electron channelling patterns in the scanning electron microscope. J Appl Phys 53: R82–R122

Jung H, Karato SI (2001a) Effects of water on dynamically recrystallised grain-size of olivine. J Struct Geol 23:1337–1344

Jung H, Karato SI (2001b) Water induced fabric transitions in olivine. Science 293:1460–1463

Jung WS, Ree JH, Park Y (1999) Non-rotation of garnet porphyroblasts and 3-D inclusion trail data: an example from the Imjingang belt, South Korea. Tectonophysics 307:381–395

Kaibysheva OA, Pshenichniuka AI, Astanin VV (1998) Superplasticity resulting from cooperative grain boundary sliding. Acta Materialia 46:4911–4916

Kamb WB (1959) Theory of preferred orientation developped by crystallisation under stress. J Geol 67:153–170

Kanagawa K (1991) Change in dominant mechanisms for phyllosilicate preferred orientation during cleavage development in the Kitakami Slates of NE Japan. J Struct Geol 13:927–945

Kanagawa K (1996) Simulated Pressure Fringes, Vorticity, and Progressive Deformation. In: de Paor DG (ed) Structural geology and personal computers. Elsevier Science, Oxford, pp 259–283

Kanaori Y (1986) A SEM cathodoluminescence study of quartz in mildly deformed granite from the region of the Atotsugawa Fault, central Japan. Tectonophysics 131:133–146

Kanaori Y, Kawakami S, Yairi K (1991) Microstructure of deformed biotite defining foliation in cataclasite zones in granite, central Japan. J Struct Geol 13:777–786

Kano K, Sato H (1988) Foliated Fault gouges: examples from the shear zones of the Sakai-Toge and Narai Faults, central Japan. J Geol Soc Jpn 94:453–456

Karato SI (1984) Grain-size distribution and rheology of the upper mantle. Tectonophysics 104:155–176

Karato SI (1987) Scanning electron microscope observations of dislocations in olivine. Phys Chem Min 14:245–248

Karato SI, Toriumi M, Fujii T (1980) Dynamic recrystallization of olivine single crystals during high temperature creep. Geophys Res Lett 7:649–652

Karato SI, Paterson MS, Fitz Gerald JD (1986) Rheology of synthetic olivine aggregates: influence of grain size and water. J Geophys Res 91:8151–8176

Karcz Z, Scholz CH (2003) The fractal geometry of some stylolites from the Calcare Massiccio Formation, Italy. J Struct Geol 25: 1301–1316

Karlstrom KE, Miller CF, Kingsbury JA, Wooden JL (1993) Pluton emplacement along a ductile thrust zone, Pine mountains, south-eastern California: Interaction between deformational and solidification processes. Bull Geol Soc Am 105:213–230

Kelley SP, Arnaud NO, Turner SP (1994) High spatial resolution $^{40}Ar/^{39}Ar$ investigations using an ultra-violet laser probe extraction technique. Geochim Cosmochim Acta 58:3519–3525

Kempe U, Götze J (2002) Cathodoluminescence (CL) behaviour and crystal chemistry of apatite from rare-metal deposits. Mineral Mag 66:135–156

Kenis I, Urai JL, van der Zee W, Sintubin M (2004) Mullions in the High-Ardenne Slate Belt (Belgium): numerical model and parameter sensitivity analysis. J Struct Geol 26:1677–1169

Kenkmann T (2000) Processes controlling the shrinkage of porphyroclasts in gabbroic shear zones. J Struct Geol 22:471–487

Kenkmann T, Dresen G (1998) Stress gradients around porphyroclasts: palaeopiezometric estimates and numerical modelling. J Struct Geol 20:163–173

Kennedy LA, Logan JM (1997) The role of veining and dissolution in the evolution of fine grained mylonites; the McConnell Thrust, Alberta. J Struct Geol 19:785–797

Kennedy LA, Logan JM (1998) Microstructures of cataclasites in a limestone-on-shale thrust fault: implications for low-temperature recrystallization of calcite. Tectonophysics 295:167–186

Kennedy LA, White JC (2001) Low-temperature recrystallization in calcite: mechanisms and consequences. Geology 29:1027–1030

Kerrich R (1986) Fluid infiltration into fault zones: chemical, isotopic, and mechanical effects. Pure Appl Geophys 124:225–268

Khazanehdari J, Rutter EH, Casey M, Burlini L (1998) The role of crystallographic fabric in the generation of seismic anisotropy and reflectivity of high strain zones in calcite rocks. J Struct Geol 20:293–299

Kidan TW, Cosgrove JW (1996) The deformation of multilayers by layer-normal compression; an experimental investigation. J Struct Geol 18:461–474

Killick AM (1990) Pseudotachylite generated as a result of a drilling "burn-in". Tectonophysics 171:221–227

Killick AM, Thwaites AM, Germs GJB, Schoch AE (1988) Pseudotachylite associated with a bedding-parallel fault zone between the Witwatersrand and Ventersdorp Supergroups, South Africa. Geol Rundsch 77:329–344

Kilsdonk B, Wiltschko DV (1988) Deformation mechanisms in the southeastern ramp region of the Pine Mountain Block, Tennessee. Geol Soc Am Bulletin 100:653–664

Kirby SH, Stern LA (1993) Experimental dynamic metamorphism of mineral single crystals. J Struct Geol 15:1223–1240

Kirkwood D, Malo M, St-Julien P, Therrien P (1995) Vertical and fold-axis parallel extension within a slate belt in a transpressive setting, nothern Appalachians. J Struct Geol 17:329–343

Kisch HJ (1991) Development of slaty cleavage and degree of very-low-grade metamorphism: a review. J Metam Geol 9:735–750

Kisch HJ (1998) Criteria for incipient slaty and crenulation cleavage development in Tertiary flysch of the Helvetic zone of the Swiss Alps. J Struct Geol 20:601–615

Kleinschodt R, Duyster JP (2002) Deformation of garnet: an EBSD study on granulites from Sri Lanka, India and the Ivrea Zone. J Struct Geol 24:1829–1844

Kleinschodt R, McGrew AJ (2000) Garnet plasticity in the lower continental crust: implications for deformation mechanisms based on microstructures and SEM electron channeling pattern analysis. J Struct Geol 22:795–809

Knipe RJ (1979) Chemical changes during slaty cleavage development. Bull Minéral 102:206–210

Knipe RJ (1981) The interaction of deformation and metamorphism in slates. Tectonophysics 78:249–272

Knipe RJ (1989) Deformation mechanisms – recognition from natural tectonites. J Struct Geol 11:127–146

Knipe RJ, Law RD (1987) The influence of crystallographic orientation and grain boundary migration on microstructural and textural development in a S-C mylonite. Tectonophysics 135:155–169

Kobberger G, Zulauf G (1995) Experimental folding and boudinage under pure constrictional conditions. J Struct Geol 17:1055–1063

Koch PS (1983) Rheology and microstructures of experimentally deformed quartz aggregates. PhD Thesis, Univ of California, Los Angeles

Koch PS, Christie JM (1981) Spacing of deformation lamellae as a palaeopiezometer. Abstr Trans Am Geophys Un (EOS) 62:1030

Kohlstedt DL, Goetze C, Durham WB, Vandersande JB (1976) A new technique for decorating dislocations in olivine. Science 19:1045–1091

Kohlstedt DL, Evans B, Mackwell SJ (1995) Strength of the lithosphere: constraints imposed by laboratory experiments. J Geophys Res 100:17587–17602

Köhn D, Passchier CW (2000) Shear sense indicators in striped bedding-veins. J Struct Geol 22:1141–1151

Köhn D, Hilgers C, Bons PD, Passchier CW (2000) Numerical simulation of fibre growth in antitaxial strain fringes. J Struct Geol 22:1311–1324

Köhn D, Aerden DGAM, Bons PD, Passchier CW (2001a) Computer experiments to investigate complex fibre patterns in natural antitaxial strain fringes. J Metam Geol 19:217–231

Köhn D, Bons PD, Hilgers C, Passchier CW (2001b) Animations of progressive fibrous vein and fringe formation. J Virt Explorer 3, Animations in Geology

Köhn D, Bons PD, Hilgers C, Passchier CW (2003) Development of antitaxial strain frings during non-coaxial deformation: an experimental study. J Struct Geol 25:263–275

Kollé JJ, Blacic JD (1982) Deformation of single-crystal clinopyroxenes, 1. Mechanical twinning in diopside and hedenbergite. J Geophys Res 87:4019–4034

Krabbendam M, Urai JL, van Vliet LJ (2003) Grain size stabilisation by dispersed graphite in a high-grade quartz mylonite: an example from Naxos (Greece). J Struct Geol 25:855–866

Kranz RL (1983) Microcracks in rocks: a review. Tectonophysics 100:449–480

Kranz RL, Scholz CH (1977) Critical dilatant volume of rocks at the onset of tertiary creep. J Geophys Res 82:4893–4897

Kraus J, Williams PF (1998) Relationship between foliation development, porphyroblast growth and large-scale folding in a metaturbidite suite, Snow Lake, Canada. J Struct Geol 20:61–67

Kraus J, Williams PF (2001) A new spin on 'non-rotating' porphyroblasts: implications of cleavage refractions and reference frames. J Struct Geol 23:963–971

Kretz R (1969) On the spatial distribution of crystals in rocks. Lithos 2:39–65

Kreutzberger ME, Peacor DR (1988) Behaviour of illite and chlorite during pressure solution of shaly limestone of the Kalkberg Formation, Catskill, New York. J Struct Geol 10:803–812

Krieger-Lassen NC (1996) The relative precision of crystal orientations measured from electron backscattering patterns. J Microsc 181:72–81

Kriegsman LM (1993) Geodynamic evolution of the Pan-African lower crust in Sri Lanka. Geol Ultraject 114:208

Krohe A (1990) Local variations in quartz (c)-axis orientations in non-coaxial regimes and their significance for the mechanics of S-C fabrics. J Struct Geol 12:995–1004

Kronenberg AK (1994) Hydrogen speciation and chemical weakening of quartz. In: Heaney PJ, Prewitt CT, Gibbs GV (eds) Silica: physical behavior, geochemistry, and materials applications. Mineral Soc Am Rev Mineral 29:123–176

Kronenberg AK, Shelton GL (1980) Deformation microstructures in experimentally deformed Maryland diabase. J Struct Geol 2:341–354

Kronenberg AK, Kirby SH, Pikston JC (1990) Basal slip and mechanical anisotropy of biotite. J Geophys Res 95:19257–19278

Kroustrup JP, Gundersen HJG, Væth M (1988) Stereological analysis of three-dimensional structure organization of surfaces in multiphase specimens: statistical methods and model-inferences. J Microsc 149:135–152

Kruhl J (1987a) Preferred lattice orientations of plagioclase from amphibolite and greenschist facies rocks near the Insubric Line (Western Alps). Tectonophysics 135:233–242

Kruhl JH (1987b) Preferred orientations of plagioclase from amphibolite and greenstone facies rocks near the Insubric Line, Western Alps. Tectonophysics 135:233–242

Kruhl JH (1996) Prism- and basal-plane parallel subgrain boundaries in quartz; a microstructural geothermobarometer. J Metam Geol 14:581–589

Kruhl JH (1998) Reply: prism- and basal-plane parallel subgrain boundaries in quartz: a microstructural geothermobarometer. J Metamorph Petrol 16:142–146

Kruhl JH (2001) Crystallographic control on the development of foam textures in quartz, plagioclase and analogue material. In: Dresen G, Handy M (eds) Deformation mechanisms, rheology and microstructures. Int J Earth Sci 90:104–117

Kruhl JH, Peternell M (2002) The equilibration of high-angle grain boundaries in dynamically recrystallized quartz: the effect of crystallography and temperature. J Struct Geol 24:1125–1137

Kruse R, Stünitz H (1999) Deformation mechanisms and phase distribution in mafic high-temperature mylonites from the Jotun Nappe, Southern Norway. Tectonophysics 303:223–249

Kruse R, Stünitz H, Kunze K (2001) Dynamic recrystallisation processes in plagioclase porphyroclasts. J Struct Geol 23:1781–1802

Kurz W, Neubauer F, Unzog W, Genser J, Wang X (2000) Microstructural and textural development of calcite marbles during polyphase deformation of Penninic units within the Tauern Window (Eastern Alps). Tectonophysics 316:327–342

Küster M, Stöckhert B (1999) High differential stress and sublithostatic pore fluid pressure in the ductile regime – microstructural evidence for short-term post-seismic creep in the Sesia Zone, Western Alps. Tectonophysics 303:263–277

Labaume P, Berty C, Laurent PH (1991) Syn-diagenetic evolution of shear structures in superficial nappes: an example from the Northern Apennines (NW Italy). J Struct Geol 13:385–398

Lacassin R, Mattauer M (1985) Kilometre scale sheath fold at Mattmark and implications for transport direction in the Alps. Nature 315:739–742

Lacombe O, Laurent P (1996) Determination of deviatoric stress tensors based on inversion of calcite twin data from experimentally deformed monophase samples. Preliminary results. Tectonophysics 255:189–202

Lafrance B, Vernon RH (1993) Mass transfer and microfracturing in gabbroic mylonites of the Guadalupe igneous complex, California. In: Boland JN, Fitz Gerald JD (eds) Defects and processes in the solid state: geoscience applications. Elsevier, Amsterdam, pp 151–167

Lafrance B, Vernon RH (1998) Coupled mass transfer and microfracturing in gabbroic mylonites. In: Snoke A, Tullis J, Todd VR (eds) Fault related rocks – a photographic atlas. Princeton University Press, New Jersey, pp 204–207

Lafrance B, White JC, Williams PF (1994) Natural calcite c-axis fabrics: an alternate interpretation. Tectonophysics 229:1–18

Lafrance B, John BE, Scoates JS (1996) Syn-emplacement recrystallisation and deformation microstructures in the Poe Mountain anorthosite, Wyoming. Contrib Min Pet 122:431–440

Lafrance B, John BE, Frost BR (1998) Ultra high-temperature and subsolidus shear zones: examples from the Poe Mountain anorthosite, Wyoming. J Struct Geol 20:945–955

Lagarde JL, Dallain C, Ledru P, Courrioux G (1994) Strain patterns within the late Variscan granitic dome of Velay, French Massif Central. J Struct Geol 16:839–852

Langdon T (1995) The characteristics of superplastic-like flow in ceramics. In: Bradt RC (ed) Plastic deformation of ceramics. Plenum Press, New York, pp 251–268

Lanzirotti A, Hanson GN (1997) An assessment of the utility of staurolite in U-Pb dating of metamorphism. Contr Mineral Petrol 129:352–365

Lapworth C (1885) The highland controversy in British geology: its causes, course and consequences. Nature 32:558–559

Laubach SE, Reed RM, Olson JE, Lander RH, Bonnell LM (2004) Coevolution of crack-seal texture and fracture porosity in sedimentary rocks: cathodoluminescence observations of regional fractures. J Struct Geol 26:967–982

Laurent P (1987) Shear-sense determination on striated faults from e twin lamellae in calcite. J Struct Geol 9:591–596

Laurent P, Bernard P, Vasseur G, Etchecopar A (1981) Stress tensor determination from the study of e twins in calcite: a linear programming method. Tectonophysics 78:651–660

Laurent P, Tourneret C, Laborde O (1990) Determining deviatoric stress tensors from calcite twins: applications to monophased synthetic and natural polycrystals. Tectonophysics 9:379–389

Laurent P, Kern H, Lacombe O (2000) Determination of deviatoric stress tensors based on inversion of calcite twin data from experimentally deformed monophase samples. Part II. Axial and triaxial stress experiments.Tectonophysics 327:131–148

Law RD (1987) Heterogeneous deformation and quartz crystallographic fabric transitions: natural examples from the Moine Thrust zone at the Stack of Glencoul, northern Assynt. J Struct Geol 9:819–834

Law RD (1990) Crystallographic fabrics: a selective review of their applications to research in structural geology. In: Knipe RJ, Rutter EH (eds) Deformation mechanisms, rheology and tectonics. Geol Soc Spec Publ 54, pp 335–352

Law RD (1998) Oblique grain-shape fabrics in a mylonitic quartz vein. In: Snoke A, Tullis J, Todd VR (eds) Fault related rocks – a photographic atlas. Princeton University Press, New Jersey, pp 264–265

Law RD, Knipe RJ, Dayan H (1984) Strain path partitioning within thrust sheets: microstructural and petrofabric evidence from the Moine Thrust zone at Loch Eriboll, northwest Scotland. J Struct Geol 6:477–497

Law RD, Casey M, Knipe RJ (1986) Kinematic and tectonic significance of microstructures and crystallographic fabrics within quartz mylonites from the Assynt and Eriboll regions of the Moine thrust zone, NW Scottland. Trans R Soc Edinb Earth Sci 77:99–125

Law RD, Schmid SM, Wheeler J (1990) Simple shear deformation and quartz crystallographic fabrics: a possible natural example from the Torridon area of NW Scotland. J Struct Geol 12:29–45

Law RD, Morgan SS, Casey M, Sylvester AG, Nyman M (1992) The Papoose Flat Pluton, California: a reassessment of its emplacement history in the light of new microstructural and crystallographic observations. T Roy Soc Edin-Earth 83:361–375

Law RD, Searle MP, Simpson RL (2004) Strain, deformation temperatures and vorticity of flow at the top of the Greater Himalayan Slab, Everest Massif, Tibet. J Geol Soc Lond 161:305–320

Lawrence RD (1970) Stress analysis based on albite twinning of plagioclase feldspars. Bull Geol Soc Am 81:2507–2512

Lee JH, Peacor DR, Lewis DD, Wintsch RP (1984) Chlorite – illite/ muscovite interlayered and interstratified crystals: a TEM/AEM study. Contrib Mineral Petrol 88:372–385

Lee JH, Peacor DR, Lewis DD, Wintsch RP (1986) Evidence for syntectonic crystallization for the mudstone to slate transition at Lehigh Gap, Pennsylvania, USA. J Struct Geol 8:767–780

Legros F, Cantagrel JM, Devouard B (2000) Pseudotachylyte at the base of the Arequipa volcanic landslide deposit: implications for emplacement mechanisms. Geology 108:601–611

Lehner FK (1995) A model of intergranular pressure solution in open systems. Tectonophysics 245:153–170

Lehnert R (2000) Beyond imagination – image formation based on Raman spectroscopy. GIT Imaging Microsc 2/2000:6–9

Leiss B, Molli G (2003) 'High-temperature' texture in naturally deformed Carrara marble from the Alpi Apuane, Italy. J Struct Geol 25:649–658

Leiss B, Ullemeyer K, Weber K, Brokmeier HG, Bunge HJ, Drury M, Faul U, Fueten F, Frischbutter A, Klein H et al. (2000) Recent developments and goals in texture research of geological materials. J Struct Geol 22:1531–1540

Li G, Peacor DR, Merriman RJ, Roberts B, van der Pluijm BA (1994) TEM and AEM constraints on the origin and significance of chlorite-mica stacks in slates: an example from Central Wales, U.K. J Struct Geol 16:1339–1357

Lifshitz IM, Slyozov VV (1961) The kinetics of precipitation from supersaturated solid solutions. J Phys Chem Solids 19:35–50

Lin A (1994) Glassy pseudotachylyte veins from the Fuyun fault zone, northwest China. J Struct Geol 16:71–84

Lin A (1996) Injection veins of crushing-originated pseudotachylyte and fault gouge formed during seismic faulting. Engineering Geology 43:213–224

Lin A (1997) Ductile deformation of biotite in foliated cataclasites, Iida-Matsukawa Fault, central Japan. J Asian Earth Sci 15:407–411

Lin A (1998) Glassy and microlitic pseudotachylytes In: Snoke A, Tullis J, Todd VR (eds) Fault related rocks – a photographic atlas. Princeton University Press, New Jersey, pp 112–121

Lin A (1999) S-C cataclasite in granitic rock. Tectonophysics 304: 257–273

Lin A (2001) S-C fabrics developed in cataclastic rocks from the Nojima fault zone, Japan and their implications for tectonic history. J Struct Geol 23:1167–1178

Lin A, Shimamoto T (1998) Selective melting processes as inferred from experimentally generated pseudotachylytes. J Asian Earth Sci 16:533–545

Lin A, Miyata T, Wan T (1998) Tectonic characteristics of the central segment of the Tancheng-Lujiang fault zone, Shandong Peninsula, eastern China. Tectonophysics 293:85–104

Lister GS (1977) Discussion: crossed-girdle c-axis fabrics in quartzites plastically deformed by plane strain and progressive simple shear. Tectonophysics 39:51–54

Lister GS (1993) Do smoothly curved, spiral shaped inclusion trails signify porphyroblast rotation? Comment. Geology 21:479–480

Lister GS, Dornsiepen UF (1982) Fabric transitions in the Saxony granulite terrain. J Struct Geol 41:81–92

Lister GS, Hobbs BE (1980) The simulation of fabric development during plastic deformation and its application to quartzite: the influence of deformation history. J Struct Geol 2:355–371

Lister GS, Paterson MS (1979) The simulation of fabric development during plastic deformation and its application to quartzite: fabric transitions. J Struct Geol 1:99–115

Lister GS, Price GP (1978) Fabric development in a quartz-feldspar mylonite. Tectonophysics 49:37–78

Lister GS, Snoke AW (1984) S-C Mylonites. J Struct Geol 6:617–638

Lister GS, Williams PF (1979) Fabric development in shear zones: theoretical controls and observed phenomena. J Struct Geol 1:283–297

Lister GS, Williams PF (1983) The partitioning of deformation in flowing rock masses. Tectonophysics 92:1–33

Lister GS, Paterson MS, Hobbs BE (1978) The simulation of fabric development in plastic deformation and its application to quartzite: the model. Tectonophysics 45:107–158

Lister GS, Boland JN, Zwart HJ (1986) Step-wise growth of biotite porphyroblasts in pelitic schists of the western Lys-Caillaouas massif (Pyrenees). J Struct Geol 8:543–562

Lloyd GE (1987) Atomic number and crystallographic contrast images with the SEM: a review of backscattered electron techniques. Mineral Mag 51:3–19

Lloyd G (1994) An appreciation of the SEM electron channeling technique for microstructural analysis of geological materials. In: Bunge HJ, Siegesmunde S, Skrotzki W, Weber K (eds) Textures of geological materials. DGM Informationgesellschaft Verlag, Oberursel, pp 109–126

Lloyd GE (2000) Grain boundary contact effects during faulting of quartzite: an SEM/EBSD analysis. J Struct Geol 22:1675–1693

Lloyd GE (2004) Microstructural evolution in a mylonitic quartz simple shear zone: the significant roles of dauphine twinning and misorientation. In: Alsop GI, Holdsworth RE, McCaffrey K, Hand M (eds) Transports and flow processes in shear zones. GSL Special Pub 224:39–61

Lloyd GE, Ferguson CC (1981) Boudinage structures: some new interpretations based on elastic-plastic finite element simulations. J Struct Geol 3:117–128

Lloyd GE, Freeman B (1991) SEM electron channelling analysis of dynamic recrystallization in a quartz grain. J Struct Geol 13:945–954

Lloyd GE, Freeman B (1994) Dynamic recrystallisation of quartz and quartzites. J Struct Geol 16:867–881

Lloyd GE, Hall MG (1981) Application of scanning electron microscopy to the study of deformed rocks. Tectonophysics 78:687–698

Lloyd GE, Knipe RJ (1992) Deformation mechanisms accommodating faulting of quartzite under upper crustal conditions. J Struct Geol 14:127–143

Lloyd GE, Hall MG, Cockayne B, Jones DW (1981) Selected-area electron-channelling patterns from geological materials: specimen preparation, indexing and representation of patterns and applications. Can Mineral 19:505–518

Lloyd GE, Ferguson CC, Reading K (1982) A stress-transfer model for the development of extension fracture boudinage. J Struct Geol 43:355–372

Lloyd GE, Law RD, Mainprice D, Wheeler J (1992) Microstructural and crystal fabric evolution during shear zone formation. J Struct Geol 14:1079–1100

Lloyd GE, Farmer AB, Mainprice D (1997) Misorientation analysis and the formation and orientation of subgrain and grain boundaries. Tectonophysics 279:55–78

Lofgren GE (1971a) Spherulitic structures in glassy and crystalline rocks. J Geophys Res 76:5635–5648

Lofgren GE (1971b) Experimentally produced devitrification textures in natural rhyolite glass. Geol Soc Am Bull 82:111–124

Lofgren GF (1974) An experimental study of plagioclase crystal morphology: isothermal crystallization. Am J Sci 247:243–273

Logan JM, Friedman M, Higgs NG, Dengo C, Shimamoto T (1979) Experimental studies of simulated gouge and their application to studies of natural fault zones. US Geol Surv Open-file Rep 79–1239:305–343

Logan JM, Dengo CA, Higgs NG, Wang ZZ (1992) Fabrics of experimental fault zones: their development and relationship to mechanical behavior. In: Evans B, Wong TF (eds) Fault mechanics and transport properties of rocks. New York Academic Press, pp 33–68

Lotze F (1957) Steinsalz und Kalisalze I. Gebrüder Bornträger, Berlin, pp 465

Loudon R (2001) The Raman effect in crystals. Adv Phys 50:813–864

Luan FC, Paterson MS (1992) Preparation and deformation of synthetic aggregates of quartz. J Geophys Res 97:301–320

Lüders V (1996) Contribution of infrared microscopy to fluid inclusion studies in some opaque minerals (wolframite, stibnite, bournonite); metallogenic implications. Econ Geol 91:1462–1468

Lüders V, Ziemann M (1999) Possibilities and limits of infrared light microthermometry applied to studies of pyrite-hosted fluid inclusions. Chem Geol 154:169–178

Luneau P, Cruden AR, (1998) Magmatic fabric acquisition mechanisms in a syenite: results of a combined anisotropy of magnetic susceptibility and image analysis study. J Geophys Res 103:5067–5089

Lüneburg CM, Lebit HDW (1998) The development of a single cleavage in an area of repeated folding. J Struct Geol 20:1531–1548

Macaudiere J, Brown WL, Ohnenstetter D (1985) Microcrystalline textures resulting from rapid crystallization in a pseudotachylite melt in a meta-anorthosite. Contrib Mineral Petrol 89:39–51

Machel HG (1985) Fibrous gypsum and fibrous anhydrite in veins. Sedimentology 32:443–454

Mackinnon P, Fueten F, Robin PY (1997) A fracture model for quartz ribbons in straight gneisses. J Struct Geol 19:1–14

Maddock RH (1986) Partial melting of lithic porphyroclasts in fault-generated pseudotachylites. Neues Jahrb Miner Abh 155:1–14

Maddock RH (1992) Effects of lithology, cataclasis and melting on the composition of fault-generated pseudotachylytes in Lewisian gneiss, Scotland. Tectonophysics 204:261–278

Maddock RH (1998) Pseudotachylyte formed by frictional fusion. In: Snoke A, Tullis J, Todd VR (eds) Fault related rocks – a photographic atlas. Princeton University Press, New Jersey, pp 80–87

Maddock RH, Grocott J, van Nes M (1987) Vesicles, amygdales and similar structures in fault-generated pseudotachylites. Lithos 20:419–432

Magloughlin JF (1989) The nature and significance of pseudotachylite from the Nason terrane, North Cascade Mountains, Washington. J Struct Geol 11:907–917

Magloughlin JF (1992) Microstructural and chemical changes associated with cataclasis and friction melting at shallow crustal levels: The cataclasite-pseudotachylyte connection. In: Magloughlin JF, Spray JG (eds) Frictional melting processes and products in geological materials. Tectonophysics 204:243–260

Magloughlin JF (1998a) Amygdules within pseudotachylyte. In: Snoke A, Tullis J, Todd VR (eds) Fault related rocks – a photographic atlas. Princeton University Press, New Jersey, pp 92–93

Magloughlin JF (1998b) Amygdules and microbreccia collapse structure in a pseudotachylyte. In: Snoke A, Tullis J, Todd VR (eds) Fault related rocks – a photographic atlas. Princeton University Press, New Jersey, pp 94–95

Magloughlin JF, Spray JG (1992) Frictional melting processes and products in geological materials: introduction and discussion. Tectonophysics 204:197–204

Main I, Mair K, Kwon O, Elphick S, Ngwenya B (2001) Experimental constraints on the mechanical and hydraulic properties of deformation bands in porous sandstones; a review. In: Holdsworth RE, Strachan RA, Magloughlin JF, Knipe RJ (eds) The nature and tectonic significance of fault zone weakening. Geol Soc Special Pub 186, pp 43–63

Mainprice D, Nicolas A (1989) Development of shape and lattice preferred orientations: application to the seismic anisotropy of the lower crust. J Struct Geol 11:175–190

Mainprice D, Bouchez JL, Blumenfeld P, Tubia JM (1986) Dominant c-slip in naturally deformed quartz: implications for dramatic plastic softening at high temperature. Geology 14:819–822

Mainprice D, Lloyd GE, Casey M (1993) Individual orientation measurements in quartz polycrystals: advantages and limitations for texture and petrophysical property determinations. J Struct Geol 15:1169–1188

Mainprice D, Popp T, Gueguen Y, Huenges E, Rutter EH, Wenk HR, Burlini L (2003) Physical properties of rocks and other geomaterials, a special volume to honour professor H. Kern. Tectonophysics 370:1–311

Mainprice D, Bascou J, Cordier P, Tommasi A (2004) Crystal preferred orientations of garnet: comparison between numerical simulations and electron back-scattered diffraction (EBSD) measurements in naturally deformed eclogites. J Struct Ga 26:2089–2102

Mair K, Main I, Elphick S (2000) Sequential growth of deformation bands in the laboratory. J Struct Geol 22:25–42

Mair K, Elphick S, Main I (2002) Influence of confining pressure on the mechanical and structural evolution of laboratory deformation bands. Geophys Res Lett 29 doi: 10.1029/2001GL013964

Malavieille J, Cobb F (1986) Cinématique des déformations ductiles dans trois massifs métamorphiques de l'Ouest des Etats-Unis: Albion (Idaho), Raft River et Grouse Creek (Utah). Bull Soc Geol France 2:885–898

Malavieille J, Etchecopar A, Burg JP (1982) Analyse de la géométrie des zones abritées: simulation et application à des examples naturels. CR Acad Sci Paris 294:279–284

Mancktelow NS (1988a) An automated machine for pure shear deformation of analogue materials in plane strain. J Struct Geol 10:101–108

Mancktelow NS (1988b) The rheology of paraffin wax and its usefulness as an analogue for rocks. Bull Geol Instn Univ Uppsala 14:181–193

Mancktelow NS (1994) On volume change and mass transport during the development of crenulation cleavage. J Struct Geol 16:1217–1232

Mancktelow NS, Abbassi MR (1992) Single layer buckle folding in non-linear materials II. Comparison between theory and experiment. J Struct Geol 14:105–120

Mancktelow NS, Arbaret L, Pennacchioni G (2002) Experimental observations on the effect of interface slip on rotation and stabilisation of rigid particles in simple shear and a comparison with natural mylonites. J Struct Geol 24:567–585

March A (1932) Mathematische Theorie der Regelung nach der Korngestalt bei affiner Deformation. Z Krist 81:285–297

Mares VM, Kronenberg AK (1993) Experimental deformation of muscovite. J Struct Geol 15:1061–1076

Marfunin AS (ed) (1995) Methods and instrumentations. Advanced mineralogy, vol 2. Springer-Verlag, Berlin

Mariano AN, Roeder PL (1989) Wöhlerite: chemical composition, cathodoluminescence and environment of crystallization. Can Mineral 27:709–720

Marjoribanks RW (1976) The relation between microfabric and strain in a progressively deformed quartzite sequence from central Australia. Tectonophysics 32:269–293

Markwort L, Kip B, DaSilva E, Rousell B (1995) Raman imaging of heterogeneous polymers: a comparison of global versus point illumination. Appl Spectrosc 49:1411–1430

Marques FO, Coelho S (2001) Rotation of rigid elliptical cylinders in viscous simple shear flow: Analogue experiments. J Struct Geol 23:609–617

Marques FO, Coelho S (2003) 2-D shape preferred orientations of rigid particles in transtensional viscous flow. J Struct Geol 25:841–854

Marrett R, Peacock DCP (1999) Strain and stress. J Struct Geol 21:1057–1063

Marshak S, Mitra G (1988) Basic methods of structural geology. Prentice Hall, Englewood Cliffs, New Jersey

Marshall DJ (1988) Cathodoluminescence of geological materials. Unwin Hyman, Boston

Marshall DB, McLaren AC (1977a) Deformation mechanisms in experimentally deformed plagioclase feldspars. Phys Chem Mineral 1:351–370

Marshall DB, McLaren AC (1977b) The direct observation and analysis of dislocations in experimentally deformed plagioclase feldspars. J Mater Sci 12:893–903

Martelat JE, Schulmann K, Lardeaux JM, Nicollet C, Cardon H (1999) Granulite microfabrics and deformation mechanisms in southern Madagascar. J Struct Geol 21:671–688

Martini JEJ (1992) The metamorphic history of the Vredefort dome at approximately 2 Ga as revealed by coesite-stishovite-bearing pseudotachylites. J Metamorph Geol 10:517–527

Masch L, Wenk JT, Preuss E (1985) Electron microscopy study of hyalomylonites – evidence for frictional melting in landslides. Tectonophysics 115:131–160

Massonne HJ (2001) First find of coesite in the ultrahigh-pressure metamorphic region of the Central Erzgebirge, Germany, and their metamorphic evolution. Eur J Mineral 13:565–570

Masuda T, Ando S (1988) Viscous flow around a rigid spherical body: a hydrodynamical approach. Tectonophysics 148:337–346

Masuda T, Fujimura A (1981) Microstructural development of fine-grained quartz aggregates by syntectonic recrystallisation. Tectonophysics 72:105–128

Masuda T, Kimura N (2004) Can a Newtonian viscous-matrix model be applied to microboudinage of columnar grains in quartzose tectonites? J Struct Geol 26:1749–1754

Masuda T, Kuriyama M (1988) Successive "mid-point" fracturing during microboudinage: an estimate of the stress-strain relation during a natural deformation. Tectonophysics 147:171–177

Masuda T, Mochizuki S (1989) Development of snowball structure: numerical simulation of inclusion trails during synkinematic porphyroblast growth in metamorphic rocks. Tectonophysics 170:141–150

Masuda T, Shibutani T, Igarashi T, Kuriyama M (1989) Microboudin structure of piedmontite in quartz schists: a proposal for a new indicator of relative palaeo-differential stress. Tectonophysics 163:169–180

Masuda T, Shibutani T, Kuriyama M, Igarashi T (1990) Development of microboudinage:an estimate of changing differential stress with increasing strain. Tectonophysics 178:379–387

Masuda T, Koike T, Yuko T, Morikawa T (1991) Discontinuous grain growth of quartz in metacherts: the influence of mica on a microstructural transition. J Metam Geol 9:389–402

Masuda T, Michibayashi, Ohta H (1995a) Shape preferred orientation of rigid particles in a viscous matrix; re-evaluation to determine kinematic parameters of ductile deformation. J Struct Geol 17:115–129

Masuda T, Mizuno N, Kobayashi M, Nam TN, Otoh S (1995b) Stress and strain estimates for Newtonian and non-Newtonian material in a rotational shear zone. J Struct Geol 17:451–454

Masuda T, Kimura N, Hara Y (2003) Progress in microboudin method for palaeo-stress analysis of metamorphic tectonites; application of mathematically refined expression. Tectonophysics 364:2–8

Masuda T, Kimura N, Fu B, Li X (2004) Validity of the microboudin method for palaeo-stress analysis: application to extraordinarily long sodic amphibole grains in a metachert from Aksu, China. J Struct Geol 26:203–206

Mattauer M (1973) Les déformations des matériaux de l'écorse terrestre. Hermann, Paris

Matter A, Ramseyer K (1985) Cathodoluminescence microscopy as a tool for provenance studies of sandstones. In: Zuffa GG (ed) Provenance of arenites. Reidel, Dordrecht, pp 191–211

Mauler A, Burlini L, Kunze K, Burg JP, Philippot P (2000a) P-wave anisotropy in eclogites and relationship to omphacite crystallographic fabric. Phys Chem Earth 25:119–126

Mauler A, Bystricky M, Kunze K, Mackwell S (2000b) Microstructure and lattice preferred orientations in experimentally deformed clinopyroxene aggregates. J Struct Geol 22:1633–1648

Mawer CK (1987) Mechanics of formation of gold-bearing quartz veins, Nova Scotia, Canada. Tectonophysics 135:99–119

Mawer CK, Williams PF (1991) Progressive folding and foliation development in a sheared coticule-bearing phyllite. J Struct Geol 13:539–557

Maxwell JC (1962) Origin of slaty and fracture cleavage in the Delaware Water Gap area, New Jersey and Pennsylvania. In: Engel AEJ, James HL, Leonard BF (eds) Petrological studies: a volume in honour of A. F. Buddington. Geol Soc Ame, Boulder, Colorado, pp 281–311

McCaig AM (1987) Deformation and fluid-rock interaction in metasomatic dilatant shear bands. Tectonophysics 135:121–132

McCaig A (1998) Fluid flow mechanisms in mylonites: evidence from compositional zoning patterns In: Snoke A, Tullis J, Todd VR (eds) Fault related rocks – a photographic atlas. Princeton University Press, New Jersey, pp 208–209

McClay KR (1976) The rheology of plasticine. Tectonophysics 33:T7–T15

McClay KR (1977) Pressure solution and coble creep in rocks and minerals: a review. J Geol Soc Lond 134:57–70

McEwen TJ (1981) Brittle deformation in pitted pebble conglomerates. J Struct Geol 3:25–37

McLaren AC (1991) Transmission electron microscopy of minerals and rocks. Cambridge University Press

McLaren AC, Etheridge MA (1976) A transmission electron microscope study of naturally deformed orthopyroxene. I Slip mechanisms. Contrib Mineral Petrol 57:163–177

McLaren AC, Fitz Gerald, JD (2000) Microstructural changes and deformation during the phase transformations in solid ammonium nitrate. In: Jessell MW, Urai JL (eds) Stress, strain and structure, a volume in honour of W. D. Means. J Virt Explorer 2

McLaren AC, Pryer LL (2001) Microstructural investigation of the interaction and interdependence of cataclastic and plastic mechanisms in feldspar crystals deformed in the semi-brittle field. Tectonophysics 335:1–15

McLelland J (1984) The origin of ribbon lineation within the southern Adirondacks, USA. J Struct Geol 6:147–157

McMillan PF, Hofmeister AM (1988) Infrared and Raman spectroscopy. In: Hawthorne FC (ed) Spectroscopic methods in mineralogy and geology. Rev Mineral 18, Mineral Soc Am, Washington DC, pp 99–159

McNaught MA (2002) Estimating uncertainty in normalized Fry plots using a bootstrap approach. J Struct Geol 24:311–322

McNulty BA, Tobisch OT, Cruden AR, Gilder S (2000) Multi-stage emplacement of the Mt Givens pluton, central Sierra Nevada batholith, California. Bull Geol Soc Am 112:103–119

Means WD (1976) Stress and strain. Springer-Verlag, Heidelberg

Means WD (1977) Experimental contributions to the study of foliations in rocks: a review of research since 1960. Tectonophysics 39:329–354

Means WD (1979) Stress and strain. Springer-Verlag, Berlin Heidelberg New York

Means WD (1980) High-temperature simple shearing fabrics: a new experimental approach. J Struct Geol 2:197–202

Means WD (1981) The concept of steady – state foliation. Tectonophysics 78:179–199

Means WD (1982) An unfamiliar Mohr-circle construction for finite strain. Tectonophysics 89:T1–T6

Means WD (1983) Application of the Mohr-circle construction to problems of inhomogeneous deformation. J Struct Geol 5:279–286

Means WD (1987) A newly recognized type of slickenside striation. J Struct Geol 9:585–590

Means WD (1989) Synkinematic microscopy of transparent polycrystals. J Struct Geol 11:163–174

Means WD (1994) Rotational quantities in homogeneous flow and the development of small-scale structure. J Struct Geol 16:437–445

Means WD, Jessell MW (1986) Accommodation migration of grain boundaries. Tectonophysics 127:67–86

Means WD, Li T (2001) A laboratory simulation of fibrous veins: some first observations. J Struct Geol 23:857–863

Means WD, Ree JH (1988) Seven types of sub-grain boundaries in octachloropropane. J Struct Geol 10:765–770

Means WD, Xia ZG (1981) Deformation of crystalline materials in thin section. Geology 9:538–543

Means WD, Hobbs BE, Lister GS, Williams PF (1980) Vorticity and non-coaxiality in progressive deformations. J Struct Geol 2:371–378

Means WD, Williams PF, Hobbs BE (1984) Incremental deformation and fabric development in a KCl–mica mixture. J Struct Geol 6:391–398

Meere PA, Mulchrone KF (2003) The effect of sample size on geological strain estimation from passively deformed clastic sedimentary rocks. J Struct Geol 25:1587–1595

Menéndez B, Zhu W, Wong TF (1996) Micromechanics of brittle faulting and cataclastic flow in Berea sandstone. J Struct Geol 18:1–16

Mercier JCC (1985) Olivine and pyroxenes. In: Wenk HR (ed) Preferred orientation in deformed metals and rocks: an introduction to modern texture analysis. Academic Press, Orlando, pp 407–430

Mercier JC, Anderson DA, Carter NL (1977) Stress in the lithosphere: inferences from steady-state flow of rocks. Pure Appl Geophys 115:119–226

Michibayashi K (1993) Syntectonic development of a strain-independent steady-state grain size during mylonitization. Tectonophysics 222:151–164

Mineeva RV, Speranskii AV, Bao YN, Bershov LV, Ryabchikov ID, Chukichev MV (2000) Diamond crystals from Peoples Republic of China: An electron spin resonance and cathodoluminescence experimental study. Geochem Internat 38:323–330

Misch P (1969) Paracrystalline microboudinage of zoned grains and other criteria for synkinematic growth of metamorphic minerals. Am J Sci 267:43–63

Misch P (1970) Paracrystalline microboudinage in a metamorphic reaction sequence. Bull Geol Soc Am 81:2483–2486

Misch P (1971) Porphyroblasts and 'crystallisation force': some textural criteria. Bull Geol Soc Am 83:1203–1204

Mitra G (1978) Ductile deformation zones and mylonites: the mechanical processes involved in the deformation of crystalline basement rocks. Am J Sci 278:1057–1084

Mitra S (1987) Regional variations in deformation mechanisms and structural styles in the central Appalachian orogenic belt. Geol Soc Am 98, pp 569–590

Mitra G (1998) Progressive development of foliation in a brittle deformation zone. In: Snoke A, Tullis J, Todd VR (eds) Fault related rocks – a photographic atlas. Princeton University Press, New Jersey, pp 52–53

Mitra G, Yonkee WA (1985) Relationship of spaced cleavage to folds and thrusts in the Idaho-Utah-Wyoming thrust belt. J Struct Geol 7:361–374

Miyake A (1993) Rotation of biotite porphyroblasts in pelitic schist from the Nukata area, central Japan. J Struct Geol 15:1303–1313

Miyashiro A (1973) Metamorphism and metamorphic belts. Allen and Unwin, London

Miyazaki K (1991) Ostwald ripening of garnet in high P/T metamorphic rocks. Contrib Min Pet 108:118–128

Mollema PN, Antonellini MA (1996) Compaction bands: a structural analog for anti-mode I cracks in aeolian sandstone. Tectonophysics 267:209–228

Möller A, Mezger K, Schenk V (1998) Crustal age domains and the evolution of the continental crust in the Mozambique Belt of Tanzania: combined Sm-Nd, Rb-Sr, and Pb-Pb isotopic evidence. J Petrol 39:749–783

Molli G, Heilbronner R (1999) Microstructures associated with static and dynamic recrystallization of Carrara marble (Alpi Apuane, NW Tuscany, Italy). Geologie en Mijnbouw 78:119–126

Molnar P, England P (1990) Temperatures, heat flux, and frictional stress near major thrust faults. J Geophys Res 95:4833–4856

Montardi Y, Mainprice D (1987) A transmission electron microscopic study of the natural plastic deformation of plagioclases (An68–70). Bull Mineral 110:1–14

Montel JM, Veschambre M, Nicollet C (1994) Datation de la monazite a la microsonde electronique. CR Acad des Sciences, Serie II 318:1489–1495

Montel JM, Foret S, Veschambre M, Nicollet C, Provost A (1996) Electron microprobe dating of monazite. Chem Geol 131:37–53

Moore AC (1970) Descriptive terminology for the textures of rocks in granulite facies terrains. Lithos 3:123–127

Moore DE, Lockner DA (1995) The role of microcracking in shear-fracture propagation in granite. J Struct Geol 17:95–111

Mora CI, Ramseyer (1992) Cathodoluminescence of coexisting plagioclases, Boehls Butte anorthosite: CL activators and fluid flow paths. Am Min 77:258–1265

Morgan SS, Law RD, Nyman MW (1998) Laccolith-like emplacement model for the Papoose Flat pluton based on porphyroblast-matrix analysis. Geol Soc Am Bull 110:96–110

Morrison-Smith DJ (1976) Transmission electron microscopy of experimentally deformed hornblende. Am Mineralogist 61: 272–280

Mosar J (1989) Deformation interne dans les Prealpes medianes (Suisse). Eclogae Geol Helv 82:765–793

Mügge O (1928) Ueber die Entstehung faseriger Minerale und ihrer Aggregationsformen. Neues Jahrbuch für Mineralogie, Geologie und Paläontologie 58A:303–438

Mukherjee AK (1971) The rate controlling mechanism in superplasticity. Mater Sci Eng 8:83–89

Mulchrone KF, O'Sullivan F, Meere PA (2003) Finite strain estimation using the mean radial length of elliptical objects with bootstrap confidence intervals. J Struct Geol 25:529–539

Mulchrone KF, Grogan S, Prithwijit De (2005) The relationship between magmatic tiling, fluid flow and crystal fraction. J Struct Geol 27:179–197

Müller W, Aerden D, Halliday AN (2000) Isotopic dating of strain fringe increments: duration and rates of deformation in shear zones. Science 288:2195–2198

Nakashima S, Matayoshi H, Yuko T, Michibayashi K, Masuda T, Kuroki N, Yamagishi H, Ito Y, Nakamura A (1995) Infrared microspectroscopy analysis of water distribution in deformed and metamorphosed rocks. Tectonophysics 245:263–276

Nam TN, Otoh S, Masuda T (1999) In-situ annealing experiments of octachloropropane as a rock analogue: kinetics and energetics of grain growth. Tectonophysics 304:57–70

Narahara DK, Wiltschko DV (1986) Deformation in the hinge region of a chevron fold, Valley and Ridge Province, central Pennsylvania. J Struct Geol 8:157–168

Nasdala L (2002) Raman mapping – a tool for revealing internal structures of minerals. Acta Universitatis Carolinae – Geologica 46:61–62

Nasdala L, Banerjee A, Häger T, Hofmeister W (2001) Laser-Raman micro-spectroscopy in mineralogical research. Microsc Anal, Eur ed 70:7–9

Nasdala L, Zhang M, Kempe U, Panczer G, Gaft M, Andrut M, Plötze M (2003) Spectroscopic methods applied to zircon. In: Hanchar JM, Hoskin PWO (eds) Zircon. Rev Mineral Geochem 53, Mineral Soc Am, Washington DC, pp 427–467

Nasdala L, Götze J, Hanchar JM, Gaft M, Krbetschek MR (2004a) Luminescence techniques in the earth sciences. In: Beran A, Libowitzky E (eds) Spectroscopic methods in mineralogy. EMU Notes in Mineralogy 6, European Mineralogical Union

Nasdala L, Smith DC, Kaindl R, Ziemann M (2004b) Raman spectroscopy: analytical perspectives in mineralogical research. In: Beran A, Libowitzky E (eds) Spectroscopic methods in mineralogy. EMU Notes in Mineralogy 6, European Mineralogical Union

Nazé L, Doukhan N, Doukhan JC, Latrons K (1987) A TEM study of lattice defects in naturally and experimentally deformed orthopyroxenes. Bull Minéral 110:497–512

Nemcok M, Kovác D, Lisle RJ (1999) A stress inversion procedure for polyphase calcite twin and fault/slip data sets. J Struct Geol 21: 597–611

Nes E, Ryum N, Hunderi O (1985). On Zener drag. Acta Metall 33:11

Neurath C, Smith RB (1982) The effect of material properties on growth rates of folding and boudinage: experiments with wax models. J Struct Geol 4:215–229

Newman J (1994) The influence of grain size distribution on methods for estimating paleostress from twinning in carbonates. J Struct Geol 16:1589–1601

Newman J, Lamb WM, Drury MR, Vissers RLM (1999) Deformation processes in a peridotite shear zone: reaction-softening by an H_2O-deficient, continuous net transfer reaction. Tectonophysics 303:193–222

Ngoc Nam T, Otoh S, Masuda T (1999) In-situ annealing experiments of octachloropropane as a rock analogue: kinetics and energetics of grain growth. Tectonophysics 304:57–70

Nicholson R (1991) Vein morphology, host rock deformation and the origin of the fabrics of echelon mineral veins. J Struct Geol 13:635–641

Nicholson R, Ejiofor IB (1987) The three-dimensional morphology of arrays of echelon and sigmoidal mineral-filled fractures: data from North Cornwall. GSL 144:79–83

Nicholson R, Pollard DD (1985) Dilation and linkage of echelon cracks. J Struct Geol 7:583–590

Nicolas A, Christensen NI (1987) Formation of anisotropy in upper mantle peridotites: a review. In: Fuchs K, Froidevaux C (eds) Composition, structure and dynamics of the lithosphere – asthenosphere system. Am Geophys Un Geodyn Ser 16:111–123

Nicolas A, Poirier JP (1976) Crystalline plasticity and solid state flow in metamorphic rocks. Wiley, New York

Nicolas A, Reuber I, Benn K (1988) A new magma chamber model based on structural studies in the Oman ophiolite. Tectonophysics 151:87–105

Nishikawa O, Takeshita T (1999) Dynamic analysis and two types of kink bands in quartz veins deformed under subgreenschist conditions. Tectonophysics 301:21–34

Nishikawa O, Takeshita T (2000) Progressive lattice misorientation and microstructural development in quartz veins deformed under subgreenschist conditions. J Struct Geol 22:259–276

Nishikawa O, Saiki K, Wenk HR (2004) Intra-granular strains and grain boundary morphologies of dynamically recrystallized quartz aggregates in a mylonite. J Struct Geol 26:127–141

Nollet S, Urai JL, Bons PD, Hilgers C (2005) Numerical simulations of polycrystal growth in veins. J Struct Geol 27:217–230

Norrell GT, Teixell A, Harper GD (1989) Microstructure of serpentinite mylonites from the Josephine ophiolite and serpentinitization in retrogressive shear zones, California. Bull Geol Soc Am 101:673–682

Nyman MW, Law RD, Smelik EA (1992) Cataclastic deformation for the development of core mantle structures in amphibole. Geology 20:455–458

O'Brien DK, Wenk HR, Ratschbacher L, You Z (1987) Preferred orientation of phyllosilicates in phyllonites and ultramylonites. J Struct Geol 9:719–730

O'Hara K (1992) Major- and trace-element constraints on the petrogenesis of a fault-related pseudotachylyte, western Blue Ridge province, North Carolina. In: Magloughlin JF, Spray JG (eds) Frictional melting processes and products in geological materials. Tectonophysics 204:279–288

O'Hara KD (2001) A pseudotachylyte geothermometer. J Struct Geol 23:1345–1357

Odonne F (1994) Kinematic behaviour of an interface and competence contrast: analogue models with different degrees of bonding between deformable inclusions and their matrix. J Struct Geol 16:997–1006

Oertel G (1970) Deformation of a slaty, lapillar tuff in the Lake district, England. Bull Geol Soc Am 81:1173–1188

Oertel G (1983) Construction of crossed girdles by superposing four subfabrics, each with a single maximum. Geol Rdsch 72:451–467

Oertel G (1985) Phyllosilicate textures in slates. In: Wenk HR (ed) Preferred orientation in deformed metals and rocks: an introduction to modern texture analysis. Academic Press, Orlando, pp 431–440

Ogilvie SR, Glover PWJ (2001) The petrophysical properties of deformation bands in relation to their microstructure. Earth and Planetary Scientific Letters 193:129–142

Ohlmacher GC, Aydin A (1997) Mechanics of vein, fault and solution surface formation in the Appalachian Valley and Ridge, northeastern Tennessee, U.S.A.: implications for fault friction, state of stress and fluid pressure. J Struct Geol 19:927–944

Ohnenstetter D, Cesbron F, Remond G, Caruba R, Claude JM (1991) Émissions de cathodoluminescence de deux populations de zircons naturels: téntative d'interpretation. CR Acad Sci Paris 313:641–647

Okudaira T, Takeshita T, Hara I, Ando J (1995) A new estimate of the conditions for transition from basal <a> to prism [c] slip in naturally deformed quartz. Tectonophysics 250:31–46

Okudaira T, Takeshita T, Toriumi M, Kruhl JH (1998) Prism- and basal-plane parallel subgrain boundaries in quartz; a microstructural geothermobarometer; discussion and reply. In: Wallis S, Banno S, Komatsu M (eds) Mid to deep crustal processes in the island arc setting; the metamorphic belts of Japan. J Metam Geol 16:141–146

Olesen NØ (1982) Heterogeneous strain of a phyllite as revealed by porphyroblast-matrix relationship. J Struct Geol 4:481–490

Olesen NØ (1998) Plagioclase porphyroclasts in a high-temperature shear-zone. In: Snoke A, Tullis J, Todd VR (eds) Fault related rocks – a photographic atlas. Princeton University Press, New Jersey, pp 506–507

Olgaard DL, Evans B (1988) Grain growth in synthetic marbles with added mica and water. Contrib Mineral and Petrol 100:246–260

Oliver NHS, Bons PD (2001) Mechanisms of fluid flow and fluid-rock interaction in fossil metamorphic hydrothermal systems inferred fromvein wallrock patterns, geometry and microstructure. Geofluids 1:137–162

Oliver DH, Goodge JW (1996) Leucoxene fish as a micro-kinematic indicator. J Struct Geol 18:1493–1495

Olsen TS, Kohlstedt DL (1984) Analysis of dislocations in some naturally deformed plagioclase feldspars. Phys Chem Miner 11:153–160

Olsen TS, Kohlstedt DL (1985) Natural deformation and recrystallization of some intermediate plagioclase feldspars. Tectonophysics 111:107–131

Olson JE, Pollard DD (1991) The initiation and growth of en echelon veins. J Struct Geol 13:595–608

Onasch CM, Davis TL (1988) Strain determination using cathodoluminescence of calcite overgrowths. J Struct Geol 10:301–304

Orange DL, Geddes DS, Moore JC (1993) Structural and fluid evolution of a young accretionary complex: the Hoh rock assemblage of the western Olympic Peninsula, Washington. Geol Soc Am 105:1053–1075

Ord A, Christie JM (1984) Flow stresses from microstructures in mylonitic quartzites of the Moine thrust zone, Assynt area, Scotland. J Struct Geol 6:639–655

Orzol J, Trepmann CA, Stöckhert B, Shi G (2003) Critical shear stress for mechanical twinning of jadeite – an experimental study. Tectonophysics 372:135–145

Ozawa K (1989) Stress-induced Al-Cr zoning of spinel in deformed peridotites. Nature 338:141–144

Pabst A (1931) 'Pressure shadows' and the measurement of the orientations of minerals in rocks. Am Miner 16:55–61

Padmanabhan KA, Davies GJ (1980) Superplasticity. Materials Research and Engineering, Springer-Verlag, Berlin

Pagel M, Barbin V, Blanc P, Ohnenstetter D (eds) (2000) Cathodoluminescence in geosciences. Springer-Verlag, Berlin

Panozzo R (1983) Two-dimensional analysis of shape fabric using projections of lines in a plane. Tectonophysics 95:279–294

Panozzo R (1984) Two-dimensional strain from the orientation of lines in a plane. J Struct Geol 6:215–222

Panozzo R, Hürlimann H (1983) A simple method for the quantitative discrimination of convex and convex-concave lines. Microscopica Acta 87:169–176

Paola S, Spalla MI (2000) Contrasting tectonic records in the pre-Alpine metabasites of the Southern Alps (lake Como, Italy). J Geodynamics 30:167–189

Park Y, Hanson B (1999) Experimental investigation of Ostwald-ripening rates of forsterite in the haplobasaltic system. J Volcanology & Geothermal Research 90:103–113

Park Y, Means WD (1996) Direct observation of deformation processes in crystal mushes. J Struct Geol 18:847–858

Park Y, Ree JH, Kim S (2001) Lattice preferred orientation in deformed - the annealed material: observations from experimental and natural polycrystalline aggregates. Int J Earth Sciences 90:127–135

Parnell J, Honghan C, Middleton D, Haggan T, Carey P (2000) Significance of fibrous mineral veins in hydrocarbon migration: fluid inclusion studies. J Geochem Explor 69–70:623–627

Passchier CW (1982a) Mylonitic deformation in the Saint-Barthélemy Massif, French Pyrenees, with emphasis on the genetic relationship between ultramylonite and pseudotachylyte. GUA Pap Geol Ser 1 16:1–173

Passchier CW (1982b) Pseudotachylyte and the development of ultramylonite bands in the Saint-Barthélemy Massif, French Pyrenees. J Struct Geol 4:69–79

Passchier CW (1984) The generation of ductile and brittle shear bands in a low-angle mylonite zone. J Struct Geol 6:273–281

Passchier CW (1985) Water deficient mylonite zones – an example from the Pyrenees. Lithos 18:115–127

Passchier CW (1986a) Mylonites in the continental crust and their role as seismic reflectors. Geol Mijnb 65:167–176

Passchier CW (1986b) Flow in natural shear zones: the consequences of spinning flow regimes. Earth Plan et Sci Lett 77:70–80

Passchier CW (1987a) Efficient use of the velocity gradients tensor in flow modelling. Tectonophysics 136:159–163

Passchier CW (1987b) Stable positions of rigid objects in non-coaxial flow: a study in vorticity analysis. J Struct Geol 9:679–690

Passchier CW (1988a) Analysis of deformation paths in shear zones. Geol Rdsch 77:309–318

Passchier CW (1988b) The use of Mohr circles to describe non-coaxial progressive deformation. Tectonophysics 149:323–338

Passchier CW (1990a) Reconstruction of deformation and flow parameters from deformed vein sets. Tectonophysics 180:185–199

Passchier CW (1990b) A Mohr circle construction to plot the stretch history of material lines. J Struct Geol 12:513–515

Passchier CW (1991a) The classification of dilatant flow types. J Struct Geol 13:101–104

Passchier CW (1991b) Geometric constraints on the development of shear bands in rocks. Geol Mijnb 70:203–211

Passchier CW (1994) Mixing in flow perturbations: a model for development of mantled porphyroclasts in mylonites. J Struct Geol 16:733–736

Passchier CW (1998) Monoclinic model shear zones. J Struct Geol 20:1121–1137

Passchier CW (2001) Flanking structures. J Struct Geol 23:951–962

Passchier CW, Druguet E (2002) Numerical modelling of asymmetric boudinage. J Struct Geol 24:1789–1803

Passchier CW, Simpson C (1986) Porphyroclast systems as kinematic indicators. J Struct Geol 8:831–844

Passchier CW, Sokoutis D (1993) Experimental modelling of mantled porphyroclasts. J Struct Geol 15:895–910

Passchier CW, Speck PJHR (1994) The kinematic interpretation of obliquely-transected porphyroblasts: an example from the Trois Seigneurs Massif, France. J Struct Geol 16:971–984

Passchier CW, Urai JL (1988) Vorticity and strain analysis using Mohr diagrams. J Struct Geol 10:755–763

Passchier CW, Hoek JD, Bekendam RF, de Boorder H (1990a) Ductile reactivation of Proterozoic brittle fault rocks: an example from the Vestfold Hills. East Antarctica Prec Res 47:3–16

Passchier CW, Myers JS, Kröner A (1990b) Field geology of high-grade gneiss terrains. Springer-Verlag, Berlin Heidelberg New York

Passchier CW, Trouw RAJ, Zwart HJ, Vissers RLM (1992) Porphyroblast rotation: eppur si muove? J Metam Geol 10:283–294

Passchier CW, ten Brink CE, Bons PD, Sokoutis D (1993) Delta-objects as a gauge for stress sensitivity of strain rate in mylonites. Earth Plan et Sci Lett 120:239–245

Passchier CW, Trouw RAJ, Ribeiro A, Paciulo F (2002) Tectonic evolution of the southern Kaoko Belt, Namibia. J Afr Earth Sci 35: 61–75

Passier ML (1991) Simple shear experimenten met het gesteente analoog paraffine was. Rep, Utrecht Univ (unpubl.)

Paterson MS (1995) A theory for granular flow accommodated by material transfer via an intergranular fluid. Tectonophysics 245:135–151

Paterson SR, Tobisch OT (1992) Rates of processes in magmatic arcs: implications for the timing and nature of pluton emplacement and wall rock deformation. J Struct Geol 14:291–300

Paterson SR, Vernon RH (1995) Bursting the bubble of balloonig plutons: A return to nested diapirs emplaced by multiple processes. Bull Geol Soc Am 107:1356–1380

Paterson SR, Vernon RH, Tobisch OT (1989) A review of criteria for the identification of magmatic and tectonic foliations in granitoids. J Struct Geol 11:349–364

Paterson SR, Fowler T, Schmidt KL, Yoshinobu AS, Yuan ES, Miller RB (1998) Interpreting magmatic fabric patterns in plutons. Lithos 44:53–82

Pauli C, Schmid SM, Heilbronner RP (1996) Fabric domains in quartz mylonites: localized three dimensional analysis of microstructure and texture. J Struct Geol 18:1183–1203

Peacock DCP, Sanderson DJ (1995) Pull-aparts, shear fractures and pressure solution. Tectonophysics 241:1–13

Pennacchioni G, Fasolo L, Cecchi MM, Salasnich L (2000) Finite-element modelling of simple shear flow in Newtonian and non-Newtonian fluids around a circular rigid particle. J Struct Geol 22:683–692

Pennacchioni G, Di Toro G, Mancktelow NS (2001) Strain-insensitive preferred orientation of porphyroclasts in Mont Mary mylonites. J Struct Geol 23:1281–1298

Petit JP (1987) Criteria for the sense of movement on fault surfaces in brittle rocks. J Struct Geol 9:597–608

Petit JP, Matthauer M (1995) Paleostress superimposition deduced from mesoscale structures in limestone: the Matelles exposure, Languedoc, France. J Struct Geol 17:245–256

Pfiffner OA, Burkhard M (1987) Determination of paleo-stress axes orientations from fault, twin and earthquake data. Ann Tecton 1: 48–57

Pfiffner OA, Ramsay JG (1982) Constraints on geological strain rates – arguments from finite strain states of naturally deformed rocks. J Geophys Res 87 B1:311–321

Philip H, Etchecopar A (1978) Exemple de variations de direction de cristallisation fibreuse dans un champ de contraintes unique. Bull Soc Géol Fr 20:263–268

Phillipot P, van Roermund HLM (1992) Deformation processes in eclogitic rocks: evidence for the rheological delamination of the oceanic crust in deeper levels of subduction zones. J Struct Geol 14:1059–1078

Phillips ER (1974) Myrmekite – one hundred years later. Lithos 7: 181–194

Phillips GN (1980) Water activity changes across an amphibolite-granulite facies transition, Broken Hill, Australia. Contrib Mineral Petrol 75:377–386

Philpotts AR (1964) Origin of pseudotachylytes. Am J Sci 262: 1008–1035

Piazolo S, Passchier CW (2002a) Controls on lineation development in low to medium-grade shear zones: a study from the Cap de Creus peninsula, NE Spain. J Struct Geol 24:25–44

Piazolo S, Passchier CW (2002b) Experimental modelling of viscous inclusions in a circular high-strain rig: implications for the interpretation of shape fabrics and deformed enclaves. J Geophys Res 107, B10, 2242, ETG 11:1–15

Piazolo S, Ten Grotenhuis SM, Passchier CW (2001) New apparatus for controlled general flow modelling of analogue materials. Geol Soc Am Mem 193:235–244

Piazolo S, Bons PD, Passchier CW (2002) The influence of matrix rheology and vorticity on fabric development of populations of rigid objects during plane strain deformation. Tectonophysics 351:315–329

Pieri M, Kunze K, Burlini L, Stretton I, Olgaard DL, Burg JP, Wenk HR (2001) Texture development of calcite by deformation and dynamic recrystallization at 1 000 K during torsion experiments of marble to large strains. Tectonophysics 330:119–140

Platt JP (1984) Secondary cleavages in ductile shear zones. J Struct Geol 6:439–442

Platt JP, Behrmann JH (1986) Structures and fabrics in a crustal-scale shear zone, Betic Cordillera, SE Spain. J Struct Geol 8:15–33

Platt JP, Vissers RLM (1980) Extensional structures in anisotropic rocks. J Struct Geol 2:397–410

Plinius Secundus G (0079) Naturalis historia. Rome

Poirier JP (1980) Shear localization and shear instability in materials in the ductile field. J Struct Geol 2:135–142

Poirier JP (1985) Creep of crystals: high-temperature deformation processes in metals, ceramics and minerals. Cambridge Univ Press, Cambridge

Poirier JP, Guillope M (1979) Deformation induced recrystallisation of minerals. B Mineral 102:67–74

Post RL (1977) High-temperature creep of Mt Burnet dunite. Tectonophysics 42:75–110

Post A, Tullis J (1998) The rate of water penetration in experimentally deformed quartzite: Implications for hydrolytic weakening. Tectonophysics 295:117–137

Post A, Tullis J (1999) A recrystallized grain size paleopiezometer for experimentally deformed feldspar aggregates. Tectonophysics 303:159–173

Post AD, Tullis J, Yund RA (1996) Effects of chemical environment on dislocation creep of quartzite. J Geophys Res 101:22143–22155

Powell CMA (1979) A morphological classification of rock cleavage. Tectonophysics 58:21–34

Powell D, Treagus JE (1969) On the geometry of S-shaped inclusion trails in garnet porphyroblasts. Mineral Mag 36:453–456

Powell D, Treagus JE (1970) Rotational fabrics in metamorphic minerals. Mineral Mag 37:801–814

Powell CMA, Vernon RH (1979) Growth and rotation history of garnet porphyroblasts with inclusion spirals in a Karakoram schist. Tectonophysics 54:25–43

Press WH, Teukolsky SA, Vetterling WT, Flannery BP (1992) Numerical recipes in C. Cambridge University Press. 1020 pp

Press WH, Teukolsky SA, Vetterling WT, Flannery BP (2002) Numerical recipes in C++. The art of scientific computing. Cambridge University Press, 1032 pp

Price NJ, Cosgrove JW (1990) Analysis of geological structures. Cambridge Univ Press, Cambridge

Prior DJ (1987) Syntectonic porphyroblast growth in phyllites: textures and processes. J Metam Geol 5:27–39

Prior DJ (1993) Sub-critical fracture and associated retrogression of garnet during mylonitic deformation. Contrib Min Pet 113: 545–556

Prior DJ (1999) Problems in determining the orientation of crystal misorientation axes for small angular misorientations, using electron backscatter diffraction in the SEM. J Microsc 195:217–225

Prior D, Trimby PW, Weber UD, Dingley DJ (1996) Orientation contrast imaging of microstructures in rocks using forescatter detectors in the scanning electron microscope. Mineral Mag 60:859–869

Prior DJ, Boyle AP, Brenker F, Cheadle MC, Day A, Lopez G, Potts GJ, Reddy S, Spiess R, Timms N, Trimby P, Wheeler J, Zetterstrom L (1999) The application of electron backscatter diffraction and orientation contrast imaging in the SEM to textural problems in rocks. Am Miner 84:1741–1759

Prior DJ, Wheeler J, Brenker FE, Harte B, Matthews M (2000) Crystal plasticity of natural garnet: new microstructural evidence. Geology 28:1003–1006

Prior DJ, Wheeler J, Peruzzo L, Spiess R, Storey C (2002) Some garnet microstructures: an illustration of the potential of orientation maps and misorientation analysis in microstructural studies. J Struct Geol 24:999–1001

Pryer LL (1993) Microstructures in feldspars from a major crustal thrust zone: the Grenville Front, Ontario, Canada. J Struct Geol 15: 21–36

Pryer LL, Robin PYF (1995) Retrograde metamorphic reactions in deforming granites and the origin of flame perthite. J Metam Geol 14:645–658

Pryer LL, Robin PYF (1996) Differential stress control on the growth and orientation of flame perthite: a palaeostress-direction indicator. J Struct Geol 18:1151–1166

Pryer LL, Robin PYF, Lloyd GE (1995) An SEM electron channelling study of flame perthite from the Killarney Granite, Southwestern Grenville Front, Ontario. Can Miner 33:333–347

Putnis A, McConnell JDC (1980) Principles of mineral behaviour. Blackwell, Elsevier, Oxford

Railsback LB (1998) Evaluation of spacing of stylolites and its implication for self-organizations of pressure dissolution. J Sediment Petrol 68:2–7

Railsback LB, Andrews LM (1995) Tectonic stylolites in the 'undeformed' Cumberland Plateau of southern Tennessee. J Struct Geol 17:911–922

Rajlich P (1991) Kinematics of crenulation cleavage development: example from the Upper Devonian rocks from northeast of the Bohemian Massif, Czechoslovakia. Tectonophysics 190: 193–208

Raleigh CB (1965) Glide mechanisms in experimentally deformed minerals. Science 150:739–741

Raleigh CB, Talbot JL (1967) Mechanical twinning in naturally and experimentally deformed diopside. Am J Sci 265:151–165

Ramberg H (1955) Natural and experimental boudinage and pinch-and-swell structures. J Geol 63:512–526

Ramsay JG (1962) The geometry and mechanics of formation of "similar" type folds. J Geol 70:309–327

Ramsay JG (1967) Folding and fracturing of rocks. McGraw Hill, New York

Ramsay JG (1980a) Shear zone geometry: a review. J Struct Geol 2: 83–101

Ramsay JG (1980b) The crack-seal mechanism of rock deformation. Nature 284:135–139

Ramsay JG (1981) Tectonics of the Helvetic nappes. In: McClay KR, Price NJ (eds) Thrust and nappe tectonics. Spec Publ Geol Soc Lond 9:293–309

Ramsay JG, Graham RH (1970) Strain variation in shear belts. Can J Earth Sci 7:786–813

Ramsay JG, Huber MI (1983) The techniques of modern structural geology, 1: Strain analysis. Academic Press, London

Ramsay JG, Huber MI (1987) The techniques of modern structural geology, 2: Folds and fractures. Academic Press, London

Ramsay JG, Lisle RJ (2000) The techniques of modern structural geology, 3: Applications of continuum mechanics. Academic Press, London

Ramsay JG, Wood DS (1973) The geometric effects of volume change during deformation processes. Tectonophysics 16:263–277

Ranalli G (1984) Grain size distribution and flow stress in tectonites. J Struct Geol 6:443–448

Randle V (1992) Microtexture determination and its applications. The Institute of Materials, London

Ratschbacher L, Wenk HR, Sintubin M (1991) Calcite textures: examples from nappes with strain-path partitioning. J Struct Geol 13:369–384

Ratteron P, Doukhan N, Jaoul O, Dukhan JC (1994) High-temperature deformation of diopside IV: predominance of {110} glide above 1 000 °C. Phys Earth Planet Interiors 82:209–222

Rawling GC, Goodwin LB (2003) Cataclasis and particulate flow in faulted, poorly lithified sediments. J Struct Geol 25:317–331

Ray R (1982) Creep in polycrystalline aggregates by matter transport through a liquid phase. J Geophys Res 87:4731–4739

Ray SK (1999) Transformation of cataclastically deformed rocks to pseudotachylyte by pervasion of frictional melt: inferences from clast-size analysis. Tectonophysics 301:283–304

Ray SK (2004) Melt-clast interaction and power-law size distribution of clasts in pseudotachylytes. J Struct Geol 26:1831–1843

Ree JH (1990) High temperature deformation of octachloropropane – dynamic grain growth and lattice reorientation. Geol Soc Spec Publ 54:363–368

Ree JH (1991) An experimental steady-state foliation. J Struct Geol 13:1001–1011

Ree JH (1994) Grain boundary sliding and development of grain boundary openings in experimentally deformed octachloropropane. J Struct Geol 16:403–418

Ree JH (2000) Grain boundary sliding grain boundary sliding in experimental deformation of octachloropropane. In: Jessell MW, Urai JL (eds) Stress, strain and structure, a volume in honour of W. D. Means. J Virt Explorer 2

Reed SJB (1996) Electron microprobe analysis and scanning electron microscopy in geology. Cambridge University Press

Reimold WU (1995) Impact cratering – a review, with special reference to the economic importance of impact structures and the southern African impact crater record. Earth Moon Planets 70:21–45

Reks IJ, Gray DR (1982) Pencil structure and strain in weakly deformed mudstone and siltstone. J Struct Geol 42:161–176

Reks IJ, Gray DR (1983) Strain patterns and shortening in a folded thrust sheet: an example from the southern Appalachians. Tectonophysics 93:99–128

Renard T, Schmittbuhl J, Gratier JP, Meakin P, Merino E (2004) Three-dimensional roughness of stylolites in limestones. J Geophys Res 109, B03209

Reynard B, Gillet P, Willaime C (1989) Deformation mechanisms in naturally deformed glaucophanes: a TEM and HREM study. Eur J Miner 1:611–624

Rhede D, Wendt I, Förster HJ (1996) A three-dimensional method for calculating independent chemical U/Pb- and Th/Pb-ages of accessory minerals. Chem Geol 27:247–253

Rice AHN, Mitchell JI (1991) Porphyroblast textural sector-zoning and matrix-displacement. Mineral Mag 55:379–396

Richter DK, Zinkernagel U (1981) Zur Anwendung der Kathodolumineszenz in der Karbonatpetrographie. Geol Rdsch 70:1276–1302

Rickard MJ (1961) A note on crenulated rocks. Geol Mag 98:324–332

Ridley J (1986) Modelling of the relations between reaction enthalpy and the buffering of reaction progress in metamorphism. Mineralog Mag 50:375–384

Riedel W (1929) Zur Mechanik geologischer Brucherscheinungen. Zentralblatt für Mineralogie, Geologie und Paläontologie 1929B:354–368

Ring U, Brandon MT (1999) Ductile strain, coaxial deformation and mass loss in the Franciscan complex: implications for exhumation processes in subduction zones. In: Ring U, Brandon MT, Lister GS, Willett S (eds) Exhumation processes: normal faulting, ductile flow and erosion. Geol Soc Spec Publ 154:55–86

Ring U, Brandon MT, Ramthun A (2001) Solution-mass-transfer deformation adjacent to the Glarus Thrust, with implications for the tectonic evolution of the Alpine wedge in eastern Switzerland. J Struct Geol 23:1491–1505

Roberts D, Strömgård KE (1972) A comparison of natural and experimental strain patterns around fold hinge zones. Tectonophysics 14:105–120

Robin PY (1978) Pressure solution at grain-to-grain contacts. Geochim Cosmochim Acta, 42:1383–1389

Robin PY (1979) Theory of metamorphic segregation and related processes. Geochim Cosmochim Acta 43:1587–1600

Robin PF, Torrance JG (1987) Statistical analysis of the effect of sample size on palaeostrain calculation. I. Single face measurements. Tectonophysics 138:311–317

Rodriguez-Pascua MA, Calvo JP, De Vicente G, Gómez-Gras D (2000) Soft-sediment deformation structures interpreted as seismites in lacustrine sediments of the Prebetic Zone, SE Spain, and their potential use as indicators of earthquake magnitudes during the Late Miocene. Sed Geol 135:117–135

Roedder E (1984) Fluid inclusions. Rev Mineral Mineral Soc Am 12:108

Rooney TP, Rieker RE, Gavasci AT (1975) Hornblende deformation features. Geology 4:364–366

Roper PJ (1972) Structural significance of "button" or "fish scale" texture in the phyllonitic schist of the Brevard zone. Geol Soc Am Bull 83:853–860

Rosenberg CL, Handy MR (2001) Mechanisms and orientation of melt segregation paths during pure shearing of a partially molten rock analog (norcamphor-benzamide). J Struct Geol 23:1917–1932

Rosenberg CL, Stünitz H (2003) Deformation and recrystallization of plagioclase along a temperature gradient: an example from the Bergell tonalite. J Struct Geol 25:389–408

Rosenfeld JL (1968) Garnet rotations due to major Paleozoic deformations in Southeast Vermont. In: Zen EA (ed) Studies of Appalachian geology. Wiley, New York, pp 185–202

Rosenfeld JL (1970) Rotated garnets in metamorphic rocks. Geol Soc Am Spec Pap 129:102

Rosenfeld JL (1985) Schistosity. In: Wenk HR (ed) Preferred orientation in deformed metals and rocks: an introduction to modern texture analysis. Academic Press, Orlando, pp 441–461

Rosin SM, Johnson EL, Mancktelow N (2004) An optical method for the determination of <a> axis orientations in deformed aggregates of quartz. J Struct Geol 26:2059–2064

Ross JV (1973) Mylonitic rocks and flattened garnets in the southern Okanagan of British Columbia. Can J Earth Sci 10:1–17

Ross JV, Avé Lallemant HG, Carter NL (1980) Stress-dependence of recrystallized-grain and subgrain size in olivine. Tectonophysics 70:39–61

Rossetti F, Ranalli G, Faccenna C (1999) Rheological properties of paraffin as an analogue material for viscous crustal deformation. J Struct Geol 21:413–417

Rousell DH (1981) Fabric and origin of gneissic layers in anorthositic rocks of the St. Charles sill, Ontario. Can J Earth Sci 18:1681–1693

Rowe KJ, Rutter EH (1990) Paleostress estimation using calcite twinning: experimental calibration and application to nature. J Struct Geol 12:1–18

Roy AB (1978) Evolution of slaty cleavage in relation to diagenesis and metamorphism: a study from the Hunsrückschiefer. Bull Geol Soc Am 89:1775–1785

Rubie DC (1983) Reaction-enhanced ductility: the role of solid-solid univariant reactions in deformation of the crust and mantle. Tectonophysics 96:331–352

Rubie DC (1990) Mechanisms of reaction enhanced deformability in minerals and rocks. In: Barber DJ, Meredith PG (eds) Deformation processes in minerals, ceramics and rocks. Hyman, London, pp 262–295

Rutter EH (1976) The kinetics of rock deformation by pressure solution. Phil Trans R Soc Lond A283:203–219

Rutter EH (1983) Pressure solution in nature, theory and experiment. J Geol Soc Lond 140:725–740

Rutter EH (1986) On the nomenclature of mode of failure transitions in rocks. Tectonophysics 122:381–387

Rutter EH (1995) Experimental study of the influence of stress, temperature, and strain on the dynamic recrystallization of Carrara marble. J Geophys Res 100:24651–24663

Rutter EH, Hadizadeh J (1991) On the influence of porosity on the low-temperature brittle-ductile transition in siliciclastic rocks. J Struct Geol 13:609–614

Rutter EH, Neumann DHK (1995) Experimental deformation of partially molten Westerly granite under fluid-absent conditions, with implications for the extraction of granitic magmas. J Geophys Res 100:15697–15715

Rutter EH, Maddock RH, Hall SW, White SH (1986) Comparative microstructures of natural and experimentally produced clay-bearing fault gouges. Pure Appl Geophys 124:3–30

Rutter EH, Casey M, Burlini L (1994) Preferred crystallographic orientation development during the plastic and superplastic flow of calcite rocks. J Struct Geol 16:1431–1446

Ryan CG (1995) The nuclear microprobe as a probe of earth structure and geological processes. Nucl Instr Meth 104:377–394

Ryan CG, Griffin WL (1993) The nuclear microprobe as a tool in geology and mineral exploration. Nucl Instr Meth B77:381–398

Ryan CG, Jamieson DN, Griffin WL, Cripps G, Szymanski R (2001) The new CSIRO GEMOC nuclear microprobe: first results, performance and recent applications. Nucl Instr Meth B181:12–19

Rybacki E, Dresen G (2000) Dislocation and diffusion creep of synthetic anorthite aggregates. J Geophys Res 105:17–36

Rybacki E, Dresen G (2004) Deformation mechanism maps for feldspar rocks. Tectonophysics 382:173–187

Saha D (1998) Local volume change vs overall volume constancy during crenulation cleavage development in low grade rocks. J Struct Geol 20:587–599

Saltzer SD, Hodges KV (1988) The Middle Mountain shear zone, southern Idaho: kinematic analysis of an early Tertiary high-temperature detachment. Bull Geol Soc Am 100:96–103

Samanta SK, Mandal N, Chakraborty C, Majumder K (2002) Simulation of inclusion trail patterns within rotating synkinematic porphyroblasts. Comput Geosci 28:297–308

Sammis C, King G, Biegel R (1987) The kinematics of gouge deformation. Pure Appl Geophys 125:777–812

Sander B (1950) Einführung in die Gefügekunde der geologischen Körper, Band II: Die Korngefüge. Springer-Verlag, Wien Berlin Heidelberg

Sawaguchi T, Ishii K (2003) Three-dimensional numerical modeling of lattice- and shape-preferred orientation of orthopyroxene porphyroclasts in peridotites. J Struct Geol 25:1425–1444

Sawyer EW (2001) Melt segregation in the continental crust: distribution and movement of melt in anatectic rocks. J Metam Geol 19:291–309

Scherrer NC, Engi M, Gnos E, Jacob V, Liechti A (2000) Monazite analysis; from sample preparation to microprobe age dating and REE qualification. Schweiz Min Petr Mitt 80:93–105

Schmid SM (1982) Microfabric studies as indicators of deformation mechanisms and flow laws operative in mountain building. In: Hsu KJ (ed) Mountain building processes. Academic Press, London, pp 95–110

Schmid SM (1994) Textures of geological materials: computer model predictions versus empirical interpretations based on rock deformation experiments and field studies. In: Bunge HJ, Siegesmund S, Skrotzki W, Weber K (eds) Textures of geological materials. DGM Informationsges, Oberursel, pp 279–301

Schmid SM, Casey M (1986) Complete fabric analysis of some commonly observed quartz C-axis patterns. Geophys Monogr 36: 263–286

Schmid SM, Handy MR (1991) Towards a genetic classification of fault rocks: geological usage and tectonophysical implications. In: Müller DW, McKenzie JA, Weissert H (eds) Controversies in modern geology, evolution of geological theories in sedimentology, earth history and tectonics. Academic Press, London, pp 339–361

Schmid DW, Podladchikov YY (2004) Are isolated stable rigid clasts in shear zones equivalent to voids? Tectonophysics 384:233–242

Schmid SM, Boland JN, Paterson MS (1977) Superplastic flow in fine grained limestone. Tectonophysics 43:257–291

Schmid SM, Paterson MS, Boland JN (1980) High temperature flow and dynamic recrystallization in Carrara Marble. Tectonophysics 65:245–280

Schmid SM, Casey M, Starkey J (1981) The microfabric of calcite tectonites from the Helvetic nappes (Swiss Alps). In: McClay KR, Price NJ (eds) Thrust and nappe tectonics. Spec Publ Geol Soc Lond 9:151–158

Schmid SM, Zingg A, Handy M (1987) The kinematics of movements along the Insubric Line and the emplacement of the Ivrea Zone. Tectonophysics 135:47–66

Schofield AN, Wroth CP (1968) Critical state soil mechanics. McGraw-Hill, London

Scholz CH (1988) The brittle-plastic transition and the depth of seismic faulting. Geol Rdsch 77:319–328

Scholz CH (2002) The mechanics of earthquakes and faulting. Cambridge University Press, 471 pp

Schoneveld C (1977) A study of some typical inclusion patterns in strongly paracrystalline rotated garnets. Tectonophysics 39: 453–471

Schoneveld C (1979) The geometry and the significance of inclusion patterns in syntectonic porphyroblasts. PhD Thesis, Leiden State Univ

Schweitzer J, Simpson C (1986) Cleavage development in dolomite of the Elbrook formation, southwest Virginia. Bull Geol Soc Am 97:778–786

Secor DT (1965) Role of fluidpressure in jointing. Am J Sci 263: 633–646

Sedgwick A (1835) Remarks on the structure of large mineral masses, and especially on the chemical changes produced in the aggregation of stratified rocks during different periods after their deposition. Trans Geol Soc Lond 2nd Ser 3:416–486

Seifert KE (1964) The genesis of plagioclase twinning in the Nonewang granite. Am Mineral 49:297–320

Sellars CM (1978) Recrystallization of metals during hot deformation. Phil Trans R Soc Lond A288:147–158

Selverstone J (1993) Micro- to macro-scale interactions between deformation and metamorphism, Tauern Window, Eastern Alps. Schweiz Mineral Petrogr Mitt 73:229–239

Seydoux-Guillaume AM, Wirth R, Nasdala L, Gottschalk M, Montel JM, Heinrich W (2002) An XRD, TEM and Raman study of experimentally annealed natural monazite. Phys Chem Miner 29:240–253

Shand SJ (1916) The pseudotachylyte of Parijs (Orange Free State) and its relation to "trap-shotten gneiss" and "flinty crush-rock". Quarterly J Geol Soc Lond 72:198–221

Shaosheng J, Mainprice D (1988) Natural deformation fabrics of plagioclase: implications for slip systems and seismic anisotropy. Tectonophysics 147:145–163

Shea WT, Kronenberg AK (1992) Rheology and deformation mechanisms of an isotropic micaschist. J Geophys Res 97: 15201–15237

Shelley D (1979) Plagioclase preferred orientation, foreshore group metasediments, Bluff, New Zealand. Tectonophysics 58:279–290

Shelley D (1985) Determining paleo-flow directions from groundmass fabrics in the Lyttleton radial dykes, New Zealand. J Volc Geoth Res 25:69–79

Shelley D (1989) Plagioclase and quartz preferred orientations in a low-grade schist: the roles of primary growth and plastic deformation. J Struct Geol 11:1029–1038

Shelley D (1992) Calcite twinning and determination of paleostress orientations: three methods compared. Tectonophysics 206: 193–201

Shelley D (1993) Igneous and metamorphic rocks under the microscope. Chapman and Hall, London

Shelley D (1994) Spider texture and amphibolite preferred orientations. J Struct Geol 16:709–717

Shelley D (1995) Asymetric shape preferred orientations as shear-sense indicators. J Struct Geol 17:509:517

Shepherd TJ, Rankin AH, Alderton DHM (1985) A practical guide to fluid inclusion studies. Blackie, Glasgow

Shewmon PG (1969) Transformations in metals. McGraw-Hill, New York

Shigematsu N (1999) Dynamic recrystallization in deformed plagioclase during progressive shear deformation. Tectonophysics 305:437–452

Shimamoto T (1989) The origin of S-C mylonites and a new fault-zone model. J Struct Geol 11:51–64

Shimamoto T, Nagahama H (1992) An argument against the crush origin of pseudotachylytes based on the analysis of clast-size distribution. J Struct Geol 14:999–1006

Shimamoto T, Kanaori Y, Asai K (1991) Cathodoluminiscence observations on low-temperature mylonites: potential for detection of solution-precipitation microstructures. J Struct Geol 13:967–973

Shimizu I (1995) Kinetics of pressure solution creep in quartz: theoretical considerations. Tectonophysics 245:121–134

Sibson RH (1975) Generation of pseudotachylyte by ancient seismic faulting. Geophys J R Astr Soc 43:775–794

Sibson RH (1977a) Kinetic shear resistance, fluid pressures and radiation efficiency during seismic faulting. Pure Appl Geophys 115: 387–399

Sibson RH (1977b) Fault rocks and fault mechanisms. J Geol Soc Lond 133:191–213

Sibson RH (1980) Transient discontinuities in ductile shear zones. J Struct Geol 2:165–171

Sibson RH (1983) Continental fault structure and the shallow earthquake source. J Geol Soc Lond 140:741–769

Sibson RH (1990) Conditions for fault-valve behaviour. In: Knipe RJ, Rutter EH (eds) Deformation mechanisms, rheology and tectonics. Geol Soc Special Pub 54, pp 15–28

Siddans AWB (1972) Slaty cleavage – a review of research since 1815. Earth Sci Rev 8:205–232

Siddans AWB (1977) The development of slaty cleavage in part of the French Alps. Tectonophysics 39:533–557

Sie SH, Suter GF, Chekhmir A, Green TH (1995) Microbeam recoil detection for the study of hydration of minerals. Nucl Instr Meth 104:261–266

Siegesmund S, Helming K, Kruse R (1994) Complete texture analysis of a deformed amphibolite: comparison between neutron diffraction and U-stage data. J Struct Geol 16:131–142

Simpson C (1985) Deformation of granitic rocks across the brittle-ductile transition. J Struct Geol 7:503–511

Simpson C, de Paor DG (1993) Strain and kinematic analysis in general shear zones. J Struct Geol 15:1–20

Simpson C, Schmid SM (1983) An evaluation of criteria to determine the sense of movement in sheared rocks. Bull Geol Soc Am 94:1281–1288

Simpson C, Wintsch RP (1989) Evidence for deformation-induced K-feldspar replacement by myrmekite. J Metam Geol 7:261–275

Sintubin M (1994a) Clay fabrics in relation to the burial history of shales. Sedimentology 41:1161–1169

Sintubin M (1994b) Phyllosilicate preferred orientation in relation to strain path determination in the lower Paleozoic Stavelot-Venn Massif (Ardennes, Belgium). Tectonophysics 237:215–231

Sintubin M (1997) Quantitative approach of the cleavage evolution in the Dinant Allochthon (Ardennes, France-Belgium). Ann Soc Geol Bel 119:145–158

Sintubin M (1998) Mica (110) pole figures in support of Marchian behaviour of phyllosilicates during the development of phylloslilicate preferred orientation. Mater Sci Forum 273–275: 601–606

Sintubin M, Wenk HR, Phillips DS (1995) Texture development in platy minerals: comparison of Bi2223 aggregates with phyllosilicate fabrics. Mater Sci Eng A202:157–171

Skrotzki W (1992) Defect structure and deformation mechanisms in naturally deformed hornblende. Phys Stat Sol 131:605–624

Skrotzki W (1994) Mechanisms of texture development in rocks. In: Bunge HJ, Siegesmund S, Skrotzki W, Weber K (eds) Textures of geological materials. DGM Informationsges, Oberursel, pp 167–186

Slack JF, Palmer MR, Stevens BPJ, Barnes RG (1993) Origin and significance of tourmaline rich rocks in the Broken Hill district, Australia. Econ Geol 88:505–541

Smirnova EP, Zezin RB, Saparin GV, Obyden SK (1999) Color cathodoluminescence of natural diamond with a curvilinear zonality and orbicular core. Dokl Earth Sci 366:522–525

Smith JV (1974) Feldspar minerals, 2 vols. Springer-Verlag, Berlin Heidelberg New York

Smith JV (2002) Structural analysis of f low-related textures in lavas. Earth-Sci Rev 57:279–297

Smith JV (2005) Textures recording transient porosity in synkinematic quartz veins, South Coast, New South Wales, Australia. J Struct Geol 27:357–370

Smith JV, Brown WL (1988) Feldspar minerals. Crystal structures, physical, chemical, and microtextural properties. Springer-Verlag, Berlin

Smith CS, Guttman L (1953) Measurement of internal boundaries in three-dimensional structures by random sectioning. Trans Am Inst Min Eng J Metals 197:81–87

Smith DC, Carabatos-Nédelec C (2001) Raman spectroscopy applied to crystals: phenomena and principles, concepts and conventions. In: Lewis IR, Edwards HGM (eds) Handbook of Raman spectroscopy. From the research laboratory to the process line. Practical spectroscopy series 28, Marcel Dekker Inc, New York, pp 349–422

Snoke A, Tullis J, Todd VR (1998) Fault related rocks – a photographic atlas. Princeton University Press, New Jersey, 617 pp

Sokoutis D (1990) Experimental mullions at single and double interfaces. J Struct Geol 12:365–373

Solar GS, Brown M (1999) The classic high-T-low-P metamorphism of west-central Maine: is it post-tectonic or syntectonic? Evidence from porphyroblast-matrix relations. Can Mineral 37:311–333

Solomon SF (1989) The early diagenetic origin of lower Carboniferous mottled limestones (pseudobreccias). Sedimentology 36:399–418

Sorby HC (1853) On the origin of slaty cleavage. Edinb New Philos J 55:137–148

Sosman RB (1927) The properties of silica. Am Chem Soc Monogr Ser, Chemical Catalog Company, New York

Southwick DL (1987) Bundled slaty cleavage in laminated argillite, North-Central Minnesota. J Struct Geol 9:985–993

Spalla MI, Carminati E, Ceriani S, Oliva A, Battaglia D (1999) Influence of deformation partitioning and metamorphic re-equilibration on P-T path reconstruction in the pre-Alpine basement of central Southern Alps (Northern Italy). J Metamorphic Geol 17:319–336

Spalla MI, Siletto GB, di Paola S, Gosso G (2000) The role of structural and metamorphic memory in the distinction of tectono-metamorphic units: the basement of the Como lake in the Southern Alps. J Geodynamics 30:191–204

Spang JH, Oldershaw AE, Groshong Jr RH (1974) The nature of thin twin lamellae in calcite. EOS 55:419

Spear FS, Daniel CG (1998) Three-dimensional imaging of garnet porphyroblast sizes and chemical zoning: nucleation and growth history in the garnet zone. Geol Mater Res 1:1–44

Spear FS, Selverstone J (1983) Quantitative P-T paths from zoned minerals: theory and tectonic applications. Contrib Mineral Petrol 83:348–357

Spear FS, Kohn MJ, Florence FP (1990) A model for garnet and plagioclase growth in pelitic schists: implications for thermobarometry and P-T path determinations. J Metam Geol 8:683–696

Spencer S (1991) The use of syntectonic fibres to determine strain estimates and deformation paths: an appraisal. Tectonophysics 194:13–34

Spiers CJ, Brzesowsky RH (1993) Densification behaviour of wet granular salt: theory versus experiment. In: Kakihana H, Hardy HR Jr, Hoshi T, Toyokura K (eds) Seventh Symposium on Salt 1:83–92

Spiers CJ, Schutjens PMTM, Brzesowsky RH, Peach CJ, Liezenberg JL, Zwart HJ (1990) Experimental determination of constitutive parameters governing creep of rocksalt by pressure solution. Geol Soc Spec Publ 54:215–227

Spiess R, Peruzzo L, Prior DJ, Wheeler J (2001) Development of garnet porphyroblasts by multiple nucleation, coalescence and boundary misorientation-driven rotations. J Metam Geol 19:269–290

Spötl C (1991) Cathodoluminescence of magnesite: examples from the Eastern Alps. Geology 19:52–55

Spray JG (1987) Artificial generation of pseudotachylite using friction welding apparatus: simulation of melting on a fault plane. J Struct Geol 9:49–60

Spray JG (1988) Generation and crystallization of an amphibolite shear melt: an investigation using radial friction welding apparatus. Contrib Mineral Petrol 99:464–475

Spray JG (1992) A physical basis for the frictional melting of some rock-forming minerals. Tectonophysics 204:205–221

Spray JG (1995) Pseudotachylyte controversy: fact or fiction? Geology 23:1119–1122

Spray JG (1997) Superfaults. Geology 25:579–582

Spray JG, Thompson LM (1995) Friction melt distribution in a terrestrial multi-ring impact basin. Nature 373:1119–1122

Spray JG, Kelley SP, Reimold WU (1995) Laser probe $^{40}Ar/^{39}Ar$ dating of coesite- and stishovite-bearing pseudotachylytes and the age of the Vredefort impact event. Meteoritics 30:335–343

Spry A (1969) Metamorphic textures. Pergamon Press, Oxford

Stallard A (1998) Episodic porphyroblast growth in the Fleur de Lys Supergroup, Newfoundland: timing relative to the sequential development of multiple crenulation cleavages. J Metam Geol 16:711–728

Stallard A, Hickey K (2001a) Shear zone vs folding origin for spiral inclusion trails in the Canton Schist. J Struct Geol 23:1845–1864

Stallard A, Hickey K (2001b) Fold mechanisms in the Canton Schists: constraints on the contribution of flexural flow. J Struct Geol 23:1865–1881

Stallard A, Hickey K (2002) A comparison of microstructural and chemical patterns in garnet from the Fleur de Lys Supergroup, Newfoundland. J Struct Geol 24:1109–1123

Stanley RS (1990) The evolution of mesoscopic imbricate thrust faults-an example from the Vermont Foreland, USA. J Struct Geol 12:227–241

Steinhardt CK (1989) Lack of porphyroblast rotation in noncoaxially deformed schists from Petrel Cove, South Australia, and its implications. Tectonophysics 158:127–140

Stel H (1981) Crystal growth in cataclasites: diagnostic microstructures and implications. Tectonophysics 78:585–600

Stesky RM (1978) Mechanisms of high temperature frictional sliding in Westerly granite. Can J Earth Sci 15:361–375

Stesky RM, Brace WF, Riley DK, Robin PYF (1974) Friction in faulted rock at high temperature and pressure. Tectonophysics 23:177–203

Stewart LK (1997) Crenulation cleavage development by partitioning of deformation into zones of progressive shearing (combined shearing, shortening and volume loss) and progressive shortening (no volume loss): quantification of solution shortening and intermicrolithon-movement. Tectonophysics 281:125–140

Stipp M, Tullis J (2003) The recrystallized grain size piezometer for quartz. Geophys Res Lett 30:2088 doi:10.1029/2003GL018444

Stipp M, Stünitz H, Heilbronner R, Schmid SM (2002) The eastern Tonale fault zone: a "natural laboratory" for crystal plastic deformation of quartz over a temperature range from 250 to 700 °C. J Struct Geol 24:1861–1884

Stockdale PB (1922) Stylolites: their nature and origin. Indiana University Studies 9:1–97

Stoffel G (1976) Het vervaardigen van slijpplaatjes van ongeconsolideerde sedimenten. Museologia 6:12–16

Strating EHH, Vissers RLM (1994) Structures in natural serpentinite gouges. J Struct Geol 16:1205–1215

Strehlau J (1986) A discussion of the depth extent of rupture in large continental earthquakes. Earthquake Source Mechanics. Geophys Monogr 37:131–146

Stünitz H (1993) Transition from fracturing to viscous flow in a naturally deformed metagabbro. In: Boland JN, Fitz Gerald JF (eds) Defects and processes in the solid state: geoscience applications. Elsevier, Amsterdam, pp 121–150

Stünitz H (1998) Syndeformational recrystallization: dynamic or compositionally induced? Contrib Mineral Petrol 131:219–236

Stünitz H, Fitz Gerald JD (1993) Deformation of granitoids at low metamorphic grade, II. Granular flow in albite-rich mylonites. Tectonophysics 221:299–324

Stünitz H, Tullis J (2001) Weakening and strain localization produced by syn-deformational reaction of plagioclase. Int J Earth Sciences 90:136–148

Stünitz H, Fitz Gerald JD, Tullis J (2003) Dislocation generation, slip systems, and dynamic recrystallization in experimentally deformed plagioclase single crystals. Tectonophysics 372:215–233

Suhr G (1993) Evaluation of upper mantle microstructures in the Table Mountain massif (Bay of Islands ophiolite). J Struct Geol 15:1273–1292

Suzuki K, Adachi M, Tanaka T (1991) Middle precambrian provenance of Jurassic sandstone in the Mino Terrane, central Japan: Th-U-total Pb evidence from an electron microprobe monazite study. Sediment Geol 75:141–147

Swanson MT (1992) Fault structure, wear mechanisms and rupture processes in pseudotachylyte generation. Tectonophysics 204:223–242

Swanson MT (1998) Pseudotachylyte-bearing strike-slip faults in mylonitic rocks In: Snoke A, Tullis J, Todd VR (eds) Fault related rocks – a photographic atlas. Princeton University Press, New Jersey, pp 104–107

Taber S (1916) The origin of veins of the asbestiform minerals. P Natl Acad Sci Usa 2:659–664

Taber S (1918) The origin of veinlets in the Silurian and Devonian strata of central New York. J Geology 26:56–73

Takagi H (1998) Foliated fault gouges. In: Snoke A, Tullis J, Todd VR (eds) Fault related rocks – a photographic atlas. Princeton University Press, New Jersey, pp 58–61

Takagi H, Goto K, Shigematsu N (2000) Ultramylonite bands derived from cataclasite and pseudotachylyte in granites, northeast Japan. J Struct Geol 22:1325–1339

Takahashi M, Nagahama H, Masuda T, Fujimura A (1998) Fractal analysis of experimentally, dynamically recrystallized quartz grains and its possible application as a strain ratemeter. J Struct Geol 20:269–275

Takeshita T, Tome C, Wenk HR, Kocks UF (1987) Single-crystal yield surface for trigonal lattices: application to texture transitions in calcite polycrystals. J Geophys Res 92:12917–12930

Takeshita T, Wenk HR, Lebensohn R (1999) Development of preferred orientation and microstructure in sheared quartzite; comparison of natural data and simulated results. Tectonophysics 312:133–155

Talbot GL (1965) Crenulation cleavage in the Hunsrückschiefer of the Middle Moselle region. Geol Rdsch 54:1026–1043

Talbot CJ (1970) The minimum strain ellipsoid using deformed quartz veins. Tectonophysics 9:47–76

Tan BK, Gray DR, Steward I (1995) Volume change accompanying cleavage development in graptolitic shales from Gisborne, Victoria, Australia. J Struct Geol 17:1387–1394

Tanaka H (1992) Cataclastic lineations. J Struct Geol 14:1239–1252

Tanner PWG (1989) The flexural slip mechanism. J Struct Geol 11:635–655

Tanner PWG (1992a) Morphology and geometry of duplexes formed during flexural flow folding. J Struct Geol 14:1173–1192

Tanner PWG (1992b) Vein morphology, host rock deformation and the origin of the fabrics of echelon mineral veins; discussion. J Struct Geol 14:373–375

Tapp B, Wickham J (1987) Relationships of rock cleavage fabrics to incremental and acccumulated strain in the Conococheague Formation, USA. J Struct Geol 9:457–472

Tapponier P, Brace WF (1976) Development of stress-induced microcracks in Westerly granite. Int J Rock Mech Min 13:103–112

ten Brink CE (1996) Development of porphyroclast geometries during non-coaxial flow. Geologica Ultraiectina 142:163

ten Brink CE, Passchier CW (1995) Modelling of mantled porphyroclasts using non-Newtonian rock analogue materials. J Struct Geol 17:131–146

ten Grotenhuis SM, Passchier CW, Bons PD (2002) The influence of strain localisation on the rotation behaviour of rigid objects in experimental shear zones. J Struct Geol 24:485–501

ten Grotenhuis SM, Trouw RAJ, Passchier CW (2003) Evolution of mica fish in mylonitic rocks. Tectonophysics 372:1–21

ten Have T, Heynen WMM (1985) Cathodoluminiscence activation and zonation in carbonate rocks: an experimental approach. Geol Mijnb 64:297–310

ter Heege JH, de Bresser,JHP, Spiers CJ (2002) The influence of dynamic recrystallization on the grain size distribution and rheological behaviour of Carrara marble deformed in axial compression. In: De Meer S, Drury MR, de Bresser JHP, Pennock GM (eds) Deformation mechanisms, rheology and tectonics: current status and future perspectives. Geological Society, London, Special Publications 200:119–136

Terry MP, Robinson P, Hamilton MA, Jercinovici MJ (2000) Monazite geochronology of UHP and HP metamorphism, deformation, and exhumation, Nordøyane, Western Gneiss Region, Norway. Am Miner 85:1651–1664

Teufel LW (1981) Localization of deformation along faults; implications to fault zone permeability. Geol Soc Am 13:565

Thompson AB, England PC (1984) Pressure-temperature-time paths of regional metamorphism, II. Their inference and interpretation using mineral assemblages in metamorphic rocks. J Petrol 25:928–955

Thompson LM, Spray JG (1994) Pseudotachylytic rock distribution and genesis within the Sudbury impact structure. In: Dressler BO, Grieve RAF, Sharpton VL (eds) Large meteorite impacts and planetary evolution. Boulder, Colorado, Geological Society of America, Special Paper 293:275–287

Tikoff B, Teyssier C (1994) Strain and fabric based on porphyroclast interaction. J Struct Geol 16:477–491

Toyoshima T (1990) Pseudotachylyte from the Main Zone of the Hidaka metamorphic belt: Hokkaido, northern Japan. J Metamorph Geol 8:507–523

Toyoshima T (1998) Gabbro mylonite developed along a crustal-scale decollement. In: Snoke A, Tullis J, Todd VR (eds) Fault related rocks – a photographic atlas. Princeton University Press, New Jersey, pp 426–427

Treagus SH (1983) A theory of finite strain variation through contrasting layers and its bearing on cleavage refraction. J Struct Geol 5:351–368

Treagus SH (1985) The relationship between foliation and strain: an experimental investigation: discussion. J Struct Geol 7:119–122

Treagus SH (1999) Are viscosity ratios of rocks measurable from cleavage refraction? J Struct Geol 21:895–901

Treagus SH, Lan L (2000) Pure shear deformation of square objects, and applications to geological strain analysis. J Struct Geol 22:105–122

Treagus SH, Lan L (2003) Simple shear of deformable square objects. J Struct Geol 25:1993–2003

Treagus SH, Lan L (2004) Deformation of square objects and boudins. J Struct Geol 26:1361–1376

Treagus SH, Sokoutis D (1992) Laboratory modelling of strain variation across rheological boundaries. J Struct Geol 14:405–424

Trepied L, Doukhan JC, Paquet J (1980) Subgrain boundaries in quartz: theoretical analysis and microscopic observations. Phys Chem Miner 5:201–218

Trepmann CA, Stöckhert B (2001) Mechanical twinning of jadeite – an indication of synseismic loading beneath the brittle-ductile transition. Int J Earth Sci 90:4–13

Trepmann CA, Stöckhert B (2003) Quartz microstructures developed during non-steady state plastic flow at rapidly decaying stress and strain rate. J Struct Geol 25:2035–2053

Trimby PW, Prior DJ (1999) Microstructural imaging techniques: a comparison between light and scanning electron microscopy. Tectonophysics 303:71–81

Trimby PW, Prior DJ, Wheeler J (1998) Grain boundary hierarchy development in a quartz mylonite. J Struct Geol 20:917–935

Trimby PW, Drury MR, Spiers CJ (2000) Misorientations across etched boundaries in deformed rock salt: a study using electron backscatter diffraction. J Struct Geol 22:81–89

Trouw RAJ (1973) Structural geology of the Marsfjällen area, Caledonides of Västerbotten, Sweden. Sver Geol Unders Ser C 689:1–115

Tse ST, Rice JR (1986) Crustal earthquake instability in relation to the depth variation of frictional slip properties. J Geophys Res 91:9452–9472

Tsurumi J, Hosonuma H, Kanagawa K (2003) Strain localization due to a positive feedback of deformation and myrmekite-forming reaction in granite and aplite mylonites along the Hatagawa Shear Zone of NE Japan. J Struct Geol 25:557–574

Tucillo ME, Essene EJ, van der Pluijm BA (1990) Growth and retrograde zoning in garnets from high grade metapelites: implications for pressure-temperature paths. Geology 18:839–842

Tullis J (1970) Quartz: preferred orientation in rocks produced by Dauphiné twinning. Science 186:1342–1344

Tullis TE (1976) Experiments on the origin of slaty cleavage and schistosity. Bull Geol Soc Am 87:745–753

Tullis J (1977) Preferred orientation of quartz produced by slip during plane strain. Tectonophysics 39:87–102

Tullis TE (1980) The use of mechanical twinning in minerals as a measure of shear stress magnitudes. J Geophys Res 85:6263–6268

Tullis J (1983) Deformation of feldspars. In: Ribbe PH (ed) Feldspar mineralogy. Mineral Soc Am Rev Mineral 2:297–323

Tullis TE, Wood DS (1975) Correlation of finite strain from both reduction bodies and preferred orientation of mica in slate from Wales. Bull Geol Soc Am 86:632–638

Tullis J, Yund RA (1977) Experimental deformation of dry Westerly Granite. J Geophys Res 82:5705–5718

Tullis J, Yund RA (1980) Hydrolitic weakening of experimentally deformed Westerly granite and Hale albite rock. J Struct Geol 2:439–451

Tullis J, Yund RA (1985) Dynamic recrystallisation of feldspar: a mechanism for ductile shear zone formation. Geology 13:238–241

Tullis J, Yund RA (1987) Transition from cataclastic flow to dislocation creep of feldspar: mechanisms and microstructures. Geology 15:606–609

Tullis J, Yund RA (1991) Diffusion creep in feldspar aggregates: experimental evidence. J Struct Geol 13:987–1000

Tullis J, Yund RA (1992) The brittle-ductile transition in feldspar aggregates: an experimental study. In: Evans B, Wong TF (eds) Fault mechanics and transport properties of rocks: a festschrift in honour of W. F. Brace, pp 89–117

Tullis J, Yund RA (1998) Cataclastic flow of feldspar aggregates. In: Snoke A, Tullis J, Todd VR (eds) Fault related rocks – a photographic atlas. Princeton University Press, New Jersey, pp 200–201

Tullis J, Christie JM, Griggs DT (1973) Microstructures and preferred orientations of experimentally deformed quartzites. Bull Geol Soc Am 84:297–314

Tullis JT, Snoke AW, Todd VR (1982) Significance of petrogenesis of mylonitic rocks. Geology 10:227–230

Tullis J, Dell'Angelo L, Yund RA (1990) Ductile shear zones from brittle precursors in feldspathic rocks: the role of dynamic recrystallization. In: Hobbs BE, Heard HC (eds) Mineral and rock deformation: laboratory studies. AGU, Geophys Monogr 56:67–81

Tullis J, Stünitz H, Teyssier C, Heilbronner R (2000) Deformation microstructures in quartzo-feldspathic rocks. In: Jessell MW, Urai JL (eds) Stress, strain and structure; a volume in honour of W.D. Means Vol 2. J Virt Explorer

Tungatt PD, Humphreys FJ (1981a) An in situ optical investigation of the deformation behaviour of sodium nitrate – an analogue of calcite. Tectonophysics 78:661–676

Tungatt PD, Humphreys FJ (1981b) Transparent analogue materials as an aid to understanding high-temperature polycrystalline plasticity. In: Horsewell A, Leffers T, Lilholt H (eds) Deformation of polycrystals: mechanisms and microstructures. Riso Nat Lab, Roskilde, pp 393–398

Tungatt PD, Humphreys FJ (1984) The plastic deformation and dynamic recrystallisation of polycrystalline sodium nitrate. Acta Metall 32:1625–1635

Turner FJ (1953) Nature and dynamic interpretation of deformation lamellae in calcite of three marbles. Am J Sci 251:276–298

Turner FJ (1968) Metamorphic petrology, mineralogical and field aspects. McGraw Hill, New York, 403 pp

Turner FJ, Weiss LE (1963) Structurals analysis of metamorphic tectonites. McGraw Hill, New York

Turner FJ, Griggs DT, Heard HC (1954) Experimental deformation of calcite crystals. Bull Geol Soc Am 65:883–934

Twiss RJ (1977) Theory and applicability of a recrystallized grain size paleopiezometer. Pageoph 115:227–244

Twiss RJ (1986) Variable sensitivity piezometric equations for dislocation density and subgrain diameter and their relevance to olivine and quartz. In: Heard HC, Hobbs BE (eds) Mineral and rock deformation: laboratory studies, the Paterson volume. Geophys Monogr 36:247–261, Am Geophys Union, Washington DC

Twiss RJ, Moores EM (1992) Structural geology. Freeman, New York

Ullemeyer K, Helming K, Siegesmund S (1994) Quantitative texture analysis of plagioclase. In: Bunge HJ, Siegesmund S, Skrotzki W, Weber K (eds) Textures of geological materials. DGM Informationsges, Oberursel, pp 83–92

Ullemeyer K, Braun G, Dahms M, Kruhl JH, Olesen NO, Siegesmund S (2000) Texture analysis of a muscovite-bearing quartzite: a comparison of some currently used techniques. J Struct Geol 22:1541–1557

Ulrich S, Schulmann K, Casey M (2002) Microstructural evolution and rheological behaviour of marbles deformed at different crustal levels. J Struct Geol 24:979–995

Underhill JR, Woodcock NH (1987) Faulting mechanisms in high-porosity sandstones; New Red Sandstone, Arran, Scotland. In: Jones ME, Preston MF (eds) Deformation of sediments and sedimentary rocks. Geological Society Special Publications 29:91–105

Urai JL (1983) Water assisted dynamic recrystallization and weakening in polycrystalline bischofite. Tectonophysics 96:125–157

Urai JL, Humphreys FJ (1981) The development of shear zones in polycrystalline camphor. Tectonophysics 78:677–685

Urai JL, Humphreys FJ, Burrows SE (1980) In-situ studies of the deformation and dynamic recrystallization of rhombohedral camphor. J Materials Sci 15:1231–1240

Urai J, Means WD, Lister GS (1986) Dynamic recrystallization of minerals. In: Heard HC, Hobbs BE (eds) Mineral and rock deformation: laboratory studies, the Paterson volume. Geophys Monogr 36:161–200, Am Geophys Union, Washington DC

Urai JL, Williams PF, van Roermund HLM (1991) Kinematics of crystal growth in syntectonic fibrous veins. J Struct Geol 13:823–836

van Daalen M, Heilbronner R, Kunze K (1999) Orientation analysis of localized shear deformation in quartz fibres at the brittle-ductile transition. Tectonophysics 303:83–107

vandenDriessche J, Brun JP (1987) Rolling structures at large shear strain. J Struct Geol 9:691–704

van den Kerkhof AM, Olsen SN (1990) A natural example of superdense CO_2 inclusions: microthermometry and Raman analysis. Geochim Cosmochim Acta 54:895–901

van der Molen I, Paterson MS (1979) Experimental deformation of partially melted granite. Contrib Miner Petrol 70:299–318

van der Pluijm BA (1984) An unusual 'crack-seal' vein geometry. J Struct Geol 6:593–597

van der Pluijm BA (1991) Marble mylonites in the Bancroft shear zone, Ontario, Canada: microstructures and deformation mechanisms. J Struct Geol 13:1125–1136

van der Pluijm BA, Kaars-Sijpesteijn CH (1984) Chlorite-mica aggregates: morphology, orientation, development and bearing on cleavage formation in very-low-grade rocks. J Struct Geol 6:399–408

van der Pluijm BA, Ho NC, Peacor DR (1994) High-resolution X-ray texture goniometry. J Struct Geol 16:1029–1032

van der Pluijm BA, Ho NC, Peacor DR, Merriman RJ (1998) Contradictions of slate formation resolved? Nature 392:348

van der Wal D, Vissers RMD, Drury MR (1992) Oblique fabrics in porphyroclastic Alpine peridotites: a shear sense indicator for upper mantle flow. J Struct Geol 14:839–846

van der Wal D, Chopra P, Drury M, Fitz Gerald J (1993) Relationships between dynamically recrystallized grain size and deformation conditions in experimentally deformed olivine rocks. Geophys Res Lett 20:1479–1482

van Roermund HLN (1983) Petrofabrics and microstructures of omphacites in a high temperature eclogite from the Swedish Caledonides. Bull Minéral 106:709–713

van Roermund HLM (1992) Thermal and deformation induced omphacite microstructures from eclogites; implications for the formation and uplift of HP metamorphic terrains. Trends Miner 1:117–151

van Roermund HLM, Boland JN (1981) The dislocation substructures of naturally deformed omphacites. Tectonophysics 78:403–418

van Roermund HLM, Boland JN (1983) Retrograde P-T trajectories of high-temperature eclogites deduced from omphacite exsolution microstructures. Bull Minéral 106:723–726

Vannucchi P (2001) Monitoring paleo-fluid pressure through vein microstructures. J Geodyn 32:567–581

Vernon RH (1965) Plagioclase twins in some mafic gneisses from Broken Hill, Australia. Miner Mag 35:488–507

Vernon RH (1975) Deformation and recrystallization of a plagioclase grain. Am Mineralogist 60:884–888

Vernon RH (1976) Metamorphic processes. Allen and Unwin, London

Vernon RH (1978) Porphyroblast-matrix microstructural relationships in deformed metamorphic rocks. Geol Rundschau 67:288–305

Vernon RH (1981) Optical microstructure of partly recrystallized calcite in some naturally deformed marbles. Tectonophysics 78:601–612

Vernon RH (1988) Microstructural evidence of rotation and non-rotation of mica porphyroblasts. J Metam Geol 6:595–601

Vernon RH (1989) Porphyroblast-matrix microstructural relationships – recent approaches and problems. In: Daly JS, Cliff RA, Yardley BWD (eds) Evolution of metamorphic belts. Geol Soc London, Spec Publ 43:83–102

Vernon RH (2000) Review of microstrucrtural evidence of magmatic and solid-state flow. Electron Geosci 5:2

Vernon RH, Flood RH (1987) Contrasting deformation and metamorphism of S and I type granitoids in the Lachlan Fold Belt, Eastern Australia. Tectonophysics 147:127–143

Vernon RH, Williams VA, D'Arcy WF (1983) Grain-size reduction and foliation development in a deformed granitoid batholith. Tectonophysics 92:123–145

Vernon RH, Etheridge MA, Wall VJ (1988) Shape and microstructure of microgranitoid enclaves: Indicators of magma mingling and flow. Lithos 22:1–11

Vernon RH, Collins WJ, Paterson SR (1993a) Pre-foliation metamorphism in low-pressure/high-temperature terrains. Tectonophysics 219:241–256

Vernon RH, Paterson SR, Foster D (1993b) Growth and deformation of porphyroblasts in the Foothills terrane, central Sierra Nevada, California: negotiating a microstructural minefield. J Metam Geol 11:203–222

Vidal JL, Kubin L, Debat P, Soula JL (1980) Deformation and dynamic recrystallisation of K-feldspar augen in orthogneiss from Montagne Noir, Occitania. Lithos 13:247–257

Visser P, Mancktelow NS (1992) The rotation of garnet porphyroblasts around a single fold, Lukmanier Pass, Central Alps. J Struct Geol 14:1193–1202

Vissers RLM (1989) Asymmetric quartz c-axis fabrics and flow vorticity: a study using rotated garnets. J Struct Geol 11:231–244

Voegelé V, Ando JI, Cordier P, Liebermann RC (1998a) Plastic deformation of silicate garnet. I. High pressure experiments. Phys Earth Planet In 108:305–318

Voegelé V, Cordier P, Sautter V, Sharp TG, Lardeaux JM, Marques FO (1998b) Plastic deformation of silicate garnets II. Deformation microstructures in natural samples. Phys Earth Planet In 108:319–338

Vollbrecht A, Rust S, Weber K (1991) Development of microcracks in granites during cooling and uplift: examples from the Variscan basement in NE Bavaria. J Struct Geol 13:787–799

Vrolijk P (1987) Tectonically driven fluid flow in the Kodiak accretionary complex, Alaska. Geology 15:466–469

Wagner F, Wenk HR, Kern V, van Houtte P, Esling C (1982) Development of preferred orientaion in plane strain deformed limestone, experiment and theory. Contrib Mineral Petrol 80:132–139

Waldron HM, Sandiford M (1988) Deformation volume and cleavage development in metasedimentary rocks from the Ballarat slate belt. J Struct Geol 10:53–62

Walker AN, Rutter EH, Brodie KH (1990) Experimental study of grain-size sensitive flow of synthetic, hot-pressed calcite rocks. In: Knipe RJ, Rutter EH (eds) Deformation mechanisms, rheology and tectonics. Geol Soc Special Pub 54, pp 259–284

Wallace R-C (1976) Partial fusion along the Alpine Fault Zone, New Zealand. Geol Soc Am Bull 87:1225–122

Wallis SR (1992a) Vorticity analysis in a metachert from the Sanbagawa belt, SW Japan. J Struct Geol 14:271–280

Wallis SR (1992b) Comment on 'Do smoothly curved, spiral-shaped inclusion trails signify porphyroblast rotation?' Geology 20:1054–1056

Wallis SR (1995) Vorticity analysis and recognition of ductile extension in the Sambagawa belt, SW Japan. J Struct Geol 17:1077–1093

Wallis SR, Platt JP, Knott SD (1993) Recognition of syn-convergence extension in accretionary wedges with examples from the Calabrian Arc and the Eastern Alps. Am J Sci 293:463–494

Walte NP, Bons PD, Passchier CW, Köhn D (2003) Disequilibrium melt distribution during static recrystallization. Geology 31:1009–1012

Wang Z, Ji S (1999) Deformation of silicate garnets: Brittle ductile transition and its geological implications. Can Miner 37:525–541

Wang X, Liou JG (1991) Regional ultrahigh-pressure coesite-bearing eclogite-terrane in central China: Evidence from country rocks, gneiss, marble and metapelite. Geology 19:933–936

Wang W, Scholz CH (1995) Micromechanics of rock friction: 3. Quantitative modelling of base friction. J Geophys Res 100:4243–4247

Wang X, Liou JG, Mao HK (1989) Coesite-bearing eclogite from the Dabie Mountains in central China. Geology 17:1085–1088

Weber K (1976) Gefügeuntersuchungen an transversalgeschieferten Gesteinen aus dem östlichen Rheinischen Schiefergebirge. Geol Jahrb Reihe D 15:1–99

Weber K (1981) Kinematic and metamorphic aspects of cleavage formation in very low grade metamorphic slates. Tectonophysics 78:291–306

Weber JC, Ferrill DA, Roden-Tice MK (2001) Calcite and quartz microstructural geothermometry of low-grade metasedimentary rocks, Northern Range, Trinidad. J Struct Geol 23:93–112

Weijermars R (1986) Flow behaviour and physical chemistry of bouncing putties and related polymers in view of tectonic laboratory applications. Tectonophysics 124:325–358

Wenk HR (1985) Carbonates. In: Wenk HR (ed) Preferred orientation in deformed metals and rocks: an introduction to modern texture analysis. Academic Press, Orlando, pp 11–47

Wenk HR, Takeshita T, van Houtte P, Wagner F (1986a) Plastic anisotropy and texture development in calcite polycrystals. J Geophys Res 91:3861–3869

Wenk HR, Bunge HJ, Jansen E, Pannetier J (1986b) Preferred orientation of plagioclase-neutron diffraction and U-stage data. Tectonophysics 126:271–284

Wenk HR, Takeshita T, Bechler E, Erskine BG, Matthies S (1987) Pure shear and simple shear calcite textures. Comparison of experimental, theoretical and natural data. J Struct Geol 9:731–746

Wheeler J (1987a) The significance of grain-scale stresses in the kinetics of metamorphism. Contrib Miner Petrol 17:397–404

Wheeler J (1987b) The determination of true shear senses from the deflection of passive markers in shear zones. J Geol Soc Lond 144:73–77

Wheeler J (1992) Importance of pressure solution and Coble creep in the deformation of polymineralic rocks. J Geophys Res 97:4579–4586

Wheeler J, Prior DJ, Jiang Z, Speiss R, Trimby PW (2001) The petrological significance of misorientations between grains. Contrib Min Pet 141:109–124

White S (1975) Tectonic deformation and recrystallization of oligoclase. Contrib Min Pet 50:287–304

White SH (1976) The role of dislocation processes during tectonic deformation with special reference to quartz. In: Strens RJ (ed) The physics and chemistry of minerals and rocks. Wiley, London, pp 75–91

White SH (1977) Geological significance of recovery and recrystallization processes in quartz. Tectonophysics 39:143–170

White SH (1979a) Grain and sub-grain size variations across a mylonite zone. Contrib Mineral Petrol 70:193–202

White SH (1979b) Large strain deformation: report on a tectonic studies group discussion meeting held at Imperial College, London; introduction. J Struct Geol 4:333–339

White JC, Barnett RL (1990) Microstructural signatures and glide twins in microcline, Hemlo, Ontario. Can Miner 28:757–769

White SH, Johnston DC (1981) A microstructural and microchemical study of cleavage lamellae in a slate. J Struct Geol 3:279–290

White SH, Knipe RJ (1978) Microstructure and cleavage development in selected slates. Contrib Mineral Petrol 66:165–174

White JC, Mawer CK (1988) Dynamic recrystallisation and associated exsolution in perthites: evidence of deep crystal thrusting. J Geophys Res 93:325–337

White JC, White SH (1981) On the structure of grain boundaries in tectonites. Tectonophysics 78:613–628

White SH, Burrows SE, Carreras J, Shaw ND, Humphreys FJ (1980) On mylonites in ductile shear zones. J Struct Geol 2:175–187

Whitmeyer SJ, Simpson C (2003) High strain-rate deformation fabrics characterize a kilometers-thick Paleozoic fault zone in the Eastern Sierras Pampeanas, central Argentina. J Struct Geol 25: 909–922

Wickham JS (1973) An estimate of strain increments in a naturally deformed carbonate rock. Am J Sci 237:23–47

Wickham J, Anthony M (1977) Strain paths and folding of carbonate rocks near Blue Ridge, central Appalachians. Bull Geol Soc Am 88:920–924

Wiesmayr G, Grasemann B (2005) Sense and non-sense of shear in flanking structures with layer-parallel shortening: implications for fault-related folds. J Struct Geol 27:249–264

Wilcox RE, Harding TP, Seely DR (1973) Basic wrench tectonics. Am Assoc Petrol Geol Bull 57:74–96

Williams PF (1972a) Development of metamorphic layering and cleavage in low grade metamorphic rocks at Bermagui, Australia. Am J Sci 272:1–47

Williams PF (1972b) 'Pressure shadow' structures in foliated rocks from Bermagui, New South Wales. J Geol Soc Aust 18:371–377

Williams PF (1985) Multiply deformed terrains – problems of correlation. J Struct Geol 7:269–280

Williams PF (1990) Differentiated layering in metamorphic rocks. Earth-Sci Rev 29:267–281

Williams ML (1994) Sigmoidal inclusion trails, punctuated fabric development, and interactions between metamorphism and deformation. J Metam Geol 12:1–21

Williams ML, Burr JL (1994) Preservation and evolution of quartz phenocrysts in deformed rhyolites from the Proterozoic of southwestern North America. J Struct Geol 16:203–222

Williams ML, Jercinovic MJ (2002) Microprobe monazite geochronology: putting absolute time into microstructural analysis. J Struct Geol 24:1013–1028

Williams PF, Jiang D (1999) Rotating garnets. J Metam Geol 17: 367–378

Williams PF, Jiang D (2001) The role of initial perturbations in the development of folds in a rock-analogue. J Struct Geol 23: 845–856

Williams PF, Price GP (1990) Origin of kink bands and shear-band cleavage in shear zones: an experimental study. J Struct Geol 12: 145–164

Williams PF, Urai JL (1989) Curved vein fibres – an alternative explanation. Tectonophysics 158:311–333

Williame C, Christie JM, Kovacs MP (1979) Experimental deformation of K-feldspar single crystals. Bull Mineral 102:168–177

Williams H, Turner FJ, Gilbert CM (1954) Petrography – an introduction to the study of rocks in thin sections, 1st edn. Freeman, IMC, San Francisco

Williams H, Turner FJ, Gilbert CM (1982) Petrography – an introduction to the study of rocks in thin sections, 2nd edn. Freeman, IMC, San Francisco

Williams ML, Jercinovic MJ, Terry MP (1999) Age mapping and dating of monazite on the electron microprobe: deconvoluting multistage tectonic histories. Geology 27:367–378

Williams ML, Scheltema KE, Jercinovic MJ (2001) High-resolution compositional mapping of matrix phases: implications for mass transfer during crenulation cleavage development in the Moretown Formation, western Massachusetts. J Struct Geol 23:923–939

Willis DG (1977) A kinematic model of preferred orientation. Bull Geol Soc Am 88:883–894

Wilson RW (1971) On syntectonic porphyroblast growth. Tectonophysics 11:239–260

Wilson CJL (1975) Preferred orientation in quartz ribbon mylonites. Bull Geol Soc Am 86:968–974

Wilson CJL (1980) Shear zones in a pegmatite: a study of albite-mica-quartz deformation. J Struct Geol 2:203–209

Wilson CJL (1982) Texture and grain growth during the annealing of ice. Text Microstruct 5:19–31

Wilson CJL (1984) Shear bands, crenulations and differentiated layering in ice-mica models. J Struct Geol 6:303–320

Wilson CJL (1994) Crystal growth during a single-stage opening event and its implications for syntectonic veins. J Struct Geol 16:1283–1296

Wilson C, Marmo B (2000) Flow in polycrystalline ice. In: Jessell MW, Urai JL (eds) Stress, strain and structure, a volume in honour of W. D. Means. J Virt Explorer 2

Wilson CJL, Sim HM (2002) The localization of strain and c-axis evolution in anisotropic ice. J Glaciology 48:601–610

Wilson CJR, Zhang Y (1994) Comparison between experiment and computer modelling of plane-strain simple-shear ice deformation. J Glaciol 40:48–55

Wilson CJL, Burg JP, Mitchell JC (1986) The origin of kinks in polycrystalline ice. Tectonophysics 127:27–48

Wilson CJL, Russell-Head DS, Sim HM (2003) The application of an automated fabric analyser system to the textural evolution of folded ice layers in shear zones. Annal Glaciology 37:7–17

Wiltschko DV, Morse JW (2001) Crystallization pressure versus 'crack seal' as the mechanism for banded veins. Geology 29:79–82

Wiltschko DV, Medwedeff DA, Millson HE (1985) Distribution and mechanisms of strain within rocks on the Northwest ramp of Pine Mountain block, southern Appalachian foreland: a field test of theory. Bull Geol Soc Am 96:426–435

Wintsch RP (1986) The possible effects of deformation on chemical processes in metamorphic fault zones. In: Thompson AB, Rubie DC (eds) Metamorphic reactions, kinetics textures and deformation. Springer-Verlag, Berlin Heidelberg New York, pp 251–268

Wintsch RP (1998) Strengthening of fault breccia by K-feldspar cementation. In: Snoke A, Tullis J, Todd VR (eds) Fault related rocks – a photographic atlas. Princeton University Press, New Jersey, pp 42–43

Wintsch RP, Yi K (2002) Dissolution and replacement creep: a significant deformation mechanism in mid-crustal rocks. J Struct Geol 24:1179–1193

Wintsch RP, Kvale CM, Kisch HD (1991) Open system constant volume development of slaty cleavage, and strain induced replacement reactions in the Martinsburg Formation, Lehigh Gap, Pennsylvania. Bull Geol Soc Am 103:916–927

Wintsch RP, Christoffersen R, Kronenberg AK (1995) Fluid-rock reaction weakening of fault zones. J Geophys Res 100:13021–13032

Wojtal SF, Mitra G (1986) Strain hardening and strain softening in fault zones from foreland thrusts. Geol Soc Am 97:674–687

Wong TF, Biegel R (1985) Effects of pressure on the micromechanics of faulting in San Marcos gabbro. J Struct Geol 7:737–749

Wood DS (1974) Current views of the development of slaty cleavage. Annu Rev Earth Sci 2:1–35

Wood DS, Oertel G (1980) Deformation in the Cambrian slate belt of Wales. J Geol 88:309–326

Wood DS, Oertel G, Singh J, Bennett HF (1976) Strain and anisotropy in deformed rocks. Phil Trans R Soc Lond A283:27–42

Woodland BG (1982) Gradational development of domainal slaty cleavage, its origin and relation to chlorite porphyroblasts in the Martinsburg Formation, eastern Pennsylvania. Tectonophysics 82:89–124

Woodland BG (1985) Relationship of concretions and chlorite-muscovite porphyroblasts to the development of domainal cleavage in low-grade metamorphic deformed rocks from north-central Wales, Great Britain. J Struct Geol 7:205–216

Worley B, Powell R, Wilson CJL (1997) Crenulation cleavage formation:Evolving diffusion, deformation and equilibration mechanisms with increasing metamorphic grade. J Struct Geol 19:1121–1135

Wright TO, Henderson JR (1992) Volume loss during cleavage formation in the Meguma Group, Nova Scotia, Canada. J Struct Geol 14:281–290

Wright TO, Platt LB (1982) Pressure dissolution and cleavage in the Martinsburg shale. Am J Sci 282:122–135

Wu S, Groshong Jr RH (1991a) Low temperature deformation of sandstone, southern Appalachian fold-thrust belt. Geol Soc Am Bull 103:861–875

Wu S, Groshong Jr RH (1991b) Strain analysis using quartz deformation bands. Tectonophysics 190:269–282

Xypolias P, Koukouvelas IK (2001) Kinematic vorticity and strain rate patterns associated with ductile extrusion in the Chelmos shear zone (external Hellenides, Greece). Tectonophysics 338:59–77

Yacobi C, Holt DB (1990) Cathodoluminescence microscopy of inorganic solids. Plenum, New York

Yang C, Homman NPO, Malmaqvist KG, Johansson L, Halden NM, Barbin V (1995) Ionoluminescence. A new tool for the nuclear microprobe in geology. Scan Microsc 9:43–62

Yardley BWD (1989) An introduction to metamorphic petrology. Longman Earth Sci Series, Wiley, New York

Yardley BWD, MacKenzie WS, Guilford C (1990) Atlas of metamorphic rocks and their textures. Longman, New York

Yardley BWD, Condliffe E, Lloyd GE, Harris DHM (1996) A polyphase garnet from Western Ireland: evidence of immiscibility in the grossular-almandine series. Eur J Mineral 8:383–392

Yuan ES, Paterson SR (1993) Evaluating flow from structures in plutons. GSA Abstr with Progr 25:305

Yund RA, Tullis J (1991) Compositional changes of minerals associated with dynamic recrystallisation. Contrib Mineral Petrol 108:346–355

Zelin MG, Krasilnikov NA, Valiev RZ, Grabski MW, Yang HS, Mukherjee AK (1994) On the microstructural aspects of the nonhomogeneity of superplastic deformation at the level of grain groups. Acta Mater 42:119–126

Zeuch DH (1982) Ductile faulting, dynamic recrystallization, and grain size sensitive flow of olivine. Tectonophysics 83:293–308

Zhang S, Karato S (1995) Lattice preferred orientation of olivine aggregates deformed in simple shear. Nature 375:774–777

Zhang J, Wong TF, Yanagidani T, Davis DM (1990) Pressure-induced microcracking and grain crushing in porous rocks. J Geophys Res 95:341–352

Zhang J, Dirks PHGM, Passchier CW (1994) Extensional collapse and uplift in a polymetamorphic granulite terrain in the Archaean and Palaeoproterozoic of north China. Precambrian Res 67:37–57

Zhang Y, Hobbs BE, Ord A, Mühlhaus HB (1996) Computer simulation of single-layer buckling. J Struct Geol 18:643–655

Zhu XK, O'Nions RK, Belshaw NS, Gibb AJ (1997) Lewisian crustal history from in situ SIMS mineral chronometry and related metamorphic textures. Chem Geol 136:205–218

Zulauf J, Zulauf G (2004) Rheology of plasticine used as rock analogue: the impact of temperature, composition and strain. J Struct Geol 26:725–737

Zulauf G, Zulauf J, Hastreiter P, Tomandl B (2003) A deformation apparatus for three-dimensional coaxial deformation and its application to rheologically stratified analogue material. J Struct Geol 25:469–480

Zwart HJ (1960) The chronological succession of folding and metamorphism in the central Pyrenees. Geol Rdsch 50:203–218

Zwart HJ (1962) On the determination of polymetamorphic mineral associations, and its application to the Bosost area (central Pyrenees). Geol Rdsch 52:38–65

Zwart HJ, Calon T (1977) Chloritoid crystals from Curaglia: growth during flattening or pushing aside. Tectonophysics 39:477–486

Zwart HJ, Oele JA (1966) Rotated magnetite crystals from the Rocroi Massif (Ardennes). Geol Mijnb 45:70–74

Index